Cyclic Oxidation of High Temperature Materials

Mechanisms, testing methods, characterisation and life time estimation

Proceedings of an EFC Workshop, Frankfurt/Main, 1999

EARLIER VOLUMES IN THIS SERIES

Number 1: Corrosion in the Nuclear Industry
Prepared by the Working Party on Nuclear Corrosion

Number 2: Practical Corrosion Principles: a manual of corrosion experiments
Prepared by the Working Party on Corrosion Education

Number 3: General Guidelines for Corrosion Testing of Materials for Marine Applications
Prepared by the Working Party on Marine Corrosion

Number 4: Guidelines on Electrochemical Corrosion Measurements
Prepared by the Working Party on Physico-Chemical Methods of Corrosion Testing

Number 5: Illustrated Case Histories of Marine Corrosion
Prepared by the Working Party on Marine Corrosion

Number 6: Corrosion Education Manual
Prepared by the Working Party on Corrosion Education

Number 7: Corrosion Problems Related to Nuclear Waste Disposal
Prepared by the Working Party on Nuclear Corrosion

Number 8: Microbial Corrosion
Prepared by the Working Party on Microbial Corrosion

Number 9: Microbiological Degradation of Materials — and Methods of Protection
Prepared by the Working Party on Microbial Corrosion

Number 10: Marine Corrosion of Stainless Steels: Chlorination and Microbial Effects
Prepared by the Working Party on Marine Corrosion

Number 11: Corrosion Inhibitors
Prepared by the Working Party on Inhibitors

Number 12: Modifications of Passive Films
Prepared by the Working Party on Surface Science and Mechanisms of Corrosion and Protection

Number 13: Predicting CO_2 Corrosion in the Oil and Gas Industry
Prepared by the Working Party on Corrosion in Oil and Gas Production

Number 14: Guidelines for Methods of Testing and Research in High Temperature Corrosion
Prepared by the Working Party on Corrosion by Hot Gases and Combustion Products

Number 15: Microbial Corrosion (Proceedings of the 3rd International EFC Workshop)
Prepared by the Working Party on Microbial Corrosion

Number 16: Guidelines on Materials Requirements for Carbon and Low Alloy Steels for H_2S-Containing Environments in Oil and Gas Production
Prepared by the Working Party on Corrosion in Oil and Gas Production

Number 17: Corrosion Resistant Alloys for Oil and Gas Production: Guidance on General Requirements and Test Methods for H_2S Service
Prepared by the Working Party on Corrosion in Oil and Gas Production

Number 18: Stainless Steel in Concrete: State of the Art Report
Prepared by the Working Party on Corrosion of Reinforcement in Concrete

Number 19: Sea Water Corrosion of Stainless Steels — Mechanisms and Experiences
Prepared by the Working Parties on Marine Corrosion and Microbial Corrosion

Number 20: Organic and Inorganic Coatings for Corrosion Prevention — Research and Experiences
Papers from EUROCORR '96

Number 21: Corrosion-Deformation Interactions
CDI '96 in conjunction with EUROCORR '96

Number 22: Aspects of Microbially Induced Corrosion
Papers from EUROCORR '96 and the EFC Working Party on Microbial Corrosion

Number 23: CO_2 Corrosion Control in Oil and Gas Production — Design Considerations
Prepared by the Working Party on Corrosion in Oil and Gas Production

Number 24: Electrochemical Rehabilitation Methods for Reinforced Concrete Structures – A State of the Art Report
Prepared by the Working Party on Corrosion of Reinforcement in Concrete

Number 25: Corrosion of Reinforcement in Concrete — Monitoring, Prevention and Rehabilitation
Papers from EUROCORR '97

Number 26: Advances in Corrosion Control and Materials in Oil and Gas Production
Papers from EUROCORR '97 and EUROCORR '98

European Federation of Corrosion Publications

NUMBER 27

Cyclic Oxidation of High Temperature Materials

Mechanisms, testing methods, characterisation and life time estimation

Proceedings of an EFC Workshop, Frankfurt/Main, 1999

Edited by

M. Schütze & W. J. Quadakkers

Published for the European Federation of Corrosion by IOM Communications

Book Number 726
Published in 1999 by IOM Communications Ltd
1 Carlton House Terrace, London SW1Y 5DB

IOM Communications Ltd
is a wholly-owned subsidiary of
The Institute of Materials

ISBN 1-86125-100-9

Design and production by
SPIRES Design Partnership

Made and printed in Great Britain

Contents

Part 2 – Test Procedures and Parameters Influencing Cyclic Oxidation Behaviour 109

Part 3 – Long Term Behaviour and Near-Service Conditions 271

Part 4 – The Behaviour of Coatings Under Cycling Conditions 337

European Federation of Corrosion Publications

Series Introduction

The EFC, incorporated in Belgium, was founded in 1955 with the purpose of promoting European co-operation in the fields of research into corrosion and corrosion prevention.

Membership is based upon participation by corrosion societies and committees in technical Working Parties. Member societies appoint delegates to Working Parties, whose membership is expanded by personal corresponding membership.

The activities of the Working Parties cover corrosion topics associated with inhibition, education, reinforcement in concrete, microbial effects, hot gases and combustion products, environment sensitive fracture, marine environments, surface science, physico–chemical methods of measurement, the nuclear industry, computer based information systems, the oil and gas industry, the petrochemical industry and coatings. Working Parties on other topics are established as required.

The Working Parties function in various ways, e.g. by preparing reports, organising symposia, conducting intensive courses and producing instructional material, including films. The activities of the Working Parties are co-ordinated, through a Science and Technology Advisory Committee, by the Scientific Secretary.

The administration of the EFC is handled by three Secretariats: DECHEMA e. V. in Germany, the Société de Chimie Industrielle in France, and The Institute of Materials in the United Kingdom. These three Secretariats meet at the Board of Administrators of the EFC. There is an annual General Assembly at which delegates from all member societies meet to determine and approve EFC policy. News of EFC activities, forthcoming conferences, courses etc. is published in a range of accredited corrosion and certain other journals throughout Europe. More detailed descriptions of activities are given in a Newsletter prepared by the Scientific Secretary.

The output of the EFC takes various forms. Papers on particular topics, for example, reviews or results of experimental work, may be published in scientific and technical journals in one or more countries in Europe. Conference proceedings are often published by the organisation responsible for the conference.

In 1987 the, then, Institute of Metals was appointed as the official EFC publisher. Although the arrangement is non-exclusive and other routes for publication are still available, it is expected that the Working Parties of the EFC will use The Institute of Materials for publication of reports, proceedings etc. wherever possible.

The name of The Institute of Metals was changed to The Institute of Materials with effect from 1 January 1992.

The EFC Series is now published by the wholly-owned subsidiary of The Institute of Materials, IOM Communications Ltd.

A. D. Mercer
EFC Series Editor,
The Institute of Materials, London, UK

EFC Secretariats are located at:

Dr B A Rickinson
European Federation of Corrosion, The Institute of Materials, 1 Carlton House Terrace, London, SW1Y 5DB, UK

Mr P Berge
Fédération Européene de la Corrosion, Société de Chimie Industrielle, 28 rue Saint-Dominique, F-75007 Paris, FRANCE

Professor Dr G Kreysa
Europäische Föderation Korrosion, DECHEMA e. V., Theodor-Heuss-Allee 25, D-60486, Frankfurt, GERMANY

Preface

During actual service high temperature materials are frequently subjected to environmental attack and mechanical stresses. The high temperature resistance of these materials relies to a large extent on the formation of a protective oxide scale on their surfaces and on their ability to maintain the protective effect during long term operation. Failure of these protective scales followed by enhanced corrosive attack mostly occurs as a result of externally applied stresses in combination with stresses induced by temperature changes during plant start-up and shut-down. In most laboratory investigations, materials behaviour with respect to protective scale formation is often tested only under isothermal conditions. However, for the assessment of the suitability of high temperature materials in practical applications, the high temperature oxidation behaviour at constant temperature is only of limited importance. The oxidation behaviour under temperature cycling gives a much better indication of what has to be expected when the materials are exposed to service conditions.

Therefore, cyclic oxidation has become one of the most frequently used testing procedures because it is a simple laboratory test that simulates the conditions frequently encountered during actual service. The method has widely been applied in many industrial and scientific laboratories, but up to now neither standards nor general recommendations exist which can be used by all investigators. This makes ranking of materials and comparison of results obtained in different laboratories extremely difficult. In recent years refinements of the cyclic oxidation testing method, e.g. by incorporating new techniques, such as acoustic emission or thin layer activation measurements, have taken place. Furthermore, well developed models concerning the failure of protective oxide scales are now available.

Taking all this into account, a workshop was organised by the working party 'Corrosion by Hot Gases and Combustion Products' of the EFC which took place at Karl-Winnacker-Institut der DECHEMA e.V. on 25 and 26 February 1999, following the tradition of workshops in various areas of high temperature corrosion held by the EFC working party since the beginning of the 1980s. One of these workshops (whose results are published in EFC Publication No. 14) included a first attempt towards standardisation of various high temperature corrosion testing methods, however cyclic oxidation testing was only one of the various aspects in this publication. The 1999 Workshop was devoted solely to this topic and the aims were to summarise the present state-of-the-art in respect to this type of materials testing, and to discuss possible further steps towards standardisation of the cyclic oxidation testing method as a reliable tool for high temperature materials characterisation. The present volume contains the papers of this Cyclic Oxidation Workshop and is structured according to the various presentation and poster sessions:

- Oxidation and Failure Mechanisms under Cyclic Conditions
- Test Procedures and Parameters Influencing the Cyclic Oxidation Behaviour

- Long-Term Behaviour and Near-Service Conditions
- The Behaviour of Coatings under Cycling Conditions
- Conclusions on the Present State of Knowledge

In their papers the authors report work which has been performed very recently in the field of materials and coating behaviour under cyclic conditions, and thus provide a comprehensive survey of the present knowledge and understanding of this clearly focused, technologically important topic. The contribution of each author is gratefully acknowledged and thanks are also due to the referees of the papers.

M. Schütze
Karl-Winnacker-Institut der DECHEMA e.V., Frankfurt, Germany
Chairman of EFC Working Party "Corrosion by Hot Gases and Combustion Products"

W. J. Quadakkers
Forschungzentrum Jülich, Institute for Materials and Processes in Energy Systems, Jülich, Germany

Foreword

Most of the contributions compiled in this volume were presented at the workshop 'Cyclic Oxidation Testing as a Tool for High Temperature Materials Characterisation'. This workshop had very timely objectives, viz. to summarise the knowledge about effects of thermal cycling in oxidation, to gather information about present studies on cyclic oxidation, and to draw consequences concerning methods of testing and research on high temperature corrosion of materials. For decades, most researchers had carried out oxidation studies with more or less fortuitous interruptions and investigated the corroded specimens after quenching or cooling without considering effects of thermal cycling. But such effects are common in industrial practice; shutdowns, for example, can strongly affect the lifetime of high temperature materials. Only materials for heating elements, resistance wires and turbine blades have been tested systematically under cycling conditions by producers and users, although generally without a base of fundamental and systematic conceptions. It was now certainly the time to bring together engineers concerned with problems due to thermal cycling and researchers doing oxidation studies.

To summarise the problem shortly and simply: high temperature alloys are to be protected against the atmosphere, which may be oxidising, sulfidising, chloridising and/or carburising, by a protective oxide layer. This oxide is a brittle phase, with generally a lower thermal expansion coefficient and this leads to stresses on cooling or heating. The stresses may cause creep of the metal substrate and/or cracking and spallation of the oxide scale. The cracks may heal by continued oxide growth. These processes were described very expertly and discussed in depth in a book from the chairman of the workshop, Michael Schütze, entitled 'Protective Oxide Scales and their Breakdown' (Wiley, 1997).

In a first fundamental contribution to the present volume H. E. Evans (Univ. of Birmingham) puts forward finite element analyses on the oxide spallation due to crack formation at the oxide/metal interface and emphasizes the effect of metal creep which impairs the cracking. V. Guttmann (JRC Petten) considers the effects of surface pretreatment and demonstrates that a rough irregular surface leads to cracking and spallation, due to stresses which are more or less vertical to the interface. Another most interesting effect is described by F. S. Pettit (Univ. of Pittsburg) for alumina-forming alloys, i.e. that the presence of water vapour leads to enhanced cracking and spalling. Certainly, internal corrosion processes, such as internal oxidation and nitriding are also enhanced by thermal cycling, as shown for Ni-base alloys by H. J. Christ *et al.* (Univ. of Siegen), when the ingress of oxygen and nitrogen into the materials occurs especially at specimen edges where geometrically induced growth stresses reach high values. Cyclic oxidation is a main problem for Fe–Cr–Al foils used as catalyst supports in automotive exhaust systems; breakaway oxidation starts when the Al is consumed by oxide growth and spallation. A joint project of TU Clausthal, University of Liverpool, Universidad Complutense Madrid, and Krupp VDM Werdohl was conducted on the influence of cycle time and temperature on the start of breakaway. W. J. Quadakkers (FZ Jülich) shows that the effect of cycling depends on alloy creep strength and dimensions — for samples of 1–2 mm thickness the lifetime was affected by cycling, but not for thin foils where creep is possible!

Cyclic oxidation tests have been conducted by H. P. Martinz *et al.* (Plansee A. G., Reutte) on the ODS alloy PM 1000 with various surface states but in contrast to the results of V. Guttmann it was concluded that the smoother surfaces were oxidised more strongly with subsequent severe spallation than the rougher surfaces.

From Oak Ridge National Laboratory very valuable experiences are reported on the effects of cycle frequency and on the failure of alumina scales on Fe–Cr–Al alloys. Three experiments are discussed: (i) isothermal, (ii) discontinuous exposures with cycles of 20 h to > 300 h, and (iii) rapid cycling. Differences were observed only after a 'critical mass gain' which is different for the various alloys. Experience from 30 years was compiled in NASA Lewis Research Center, where a model has been developed, describing the stepwise process of scale growth, diffusion and spallation which allows life predictions. The extensive testing has led to a statistical model of compositional effects and a comprehensive database of commercial alloys. The authors consider cooling temperature and humidity level as important testing parameters and state that heating and cooling rates are not likely to affect scale spallation. Even more complex conditions arise in testing of hot corrosion behaviour under cyclic conditions as reported from Oak Ridge National Laboratory the corrosion attack, as expected, depends not only on the overall contaminant flux, but also on the initial amount of salt deposited during each recoating cycle and deposit removal before salt recoating.

So far I have commented on the papers in the sequence of their presentation and in print but now I would like to summarise certain aspects. Very welcome are some contributions from alloy and steel producers, who generally have been using cyclic oxidation tests for a long time. The materials tested are steels for automotive exhaust systems at UGINE, heat resistant cast alloy steels at AVESTA and HAYNES, heat resistant cast alloys at POSE-MARRE and thin strip or wires of Fe–Cr–Al alloys at KANTHAL AB. The testing conditions certainly were different in each company and were adjusted to simulate a variety of industrially relevant conditions — this clearly shows that it will not be possible to demand from all laboratories that they perform cyclic oxidation tests under identical circumstances.

The latter statement stems from V. A. C. Haanappel (JRC Ispra) and is derived from the evaluation of a questionnaire which was distributed several years ago and returned from some laboratories. He presented a checklist with the most relevant points for performing cyclic oxidation tests, which is divided into three major sections: specimen characteristics, test characteristics and specimen evaluation methods. This checklist is intended to help in setting up a Code of Practice for cyclic oxidation testing and for developing documents in which a detailed description of all aspects of cyclic oxidation testing is given.

A further aspect for summarising in this review is the testing of special materials, coatings and composites. Most important in this group are the tests of the system: alloy/bond coat/thermal barrier coating, used for protection of components in aircraft engines and gas turbines. The failure is generally by oxidation of the bond coat (MCrAlY) and rupture at the interface of the thermally grown oxide on the bond coat and the thermal barrier coating ($ZrO_2 \cdot Y_2O_3$); the great difficulty in testing is to define a sensible test specimen. Cyclic tests have been conducted at Forschungszentrum

Jülich, DLR Köln, and Silesian Technical University Katowice. Further coatings tested are (i) TiAl and $TiSi_2$ at DECHEMA, which proved to be protective in a sulfidising atmosphere, (ii) NiAl at Cambridge University, (iii) Cr_2O_3 and Nd_2O_3 films at CRNS Dijon, and (iv) SiC–C/C composites at TU Clausthal. One may say that in these tests wide variations of coating materials, coating techniques, substrates, corrosion conditions etc. were present but the aspect of cyclic oxidation testing was not so important.

To come back to the parameters of cyclic oxidation testing, we have seen that many are relevant: surface pretreatment and roughness, presence of water vapour, internal corrosion, specimen shape i.e. presence of edges and corners, cycle time and frequency, cooling temperature, rates of heating and cooling, and accordingly standardisation of testing is difficult. It had been stated before that it will not be possible to demand from all laboratories that cyclic oxidation tests must follow identical rules. However, it should be possible to develop guidelines, to help researchers in conducting the tests in a reasonable way and to consider and document all parameters. A step in this direction was taken in the last contribution by J. Nicholls and M. J. Bennett (Univ. of Cranfield), giving 'Guidelines for test standardisation, aimed at the assessment of service behaviour'. The authors distinguish three stages: (1) stable oxide growth, (2) spalling and healing of the scale, and (3) breakaway oxidation. The effect of cycling parameters is most pronounced in the second stage. In this stage, the scale reaches a critical thickness and starts to spall, then the scale-forming element is rapidly consumed and the near-surface region is depleted of that element, self-repair is no longer possible and internal oxidation and corrosion starts, leading to breakaway corrosion. Again, the significance of the creep strength of the metallic substrate for oxide adherence is pointed out, relaxation of oxidation induced stresses can occur via creep of the metal substrate and consequently, the effect of cooling on scale spalling differs from material to material, depending on the creep strength. The significance of careful corrosion product characterisation is pointed out, thus it is not sufficient to use only gravimetric data to obtain reliable results. Concerning testing methods for the testing of materials for aircraft engines and gas turbines and implications of various test methods are discussed, i.e. method of heating, cooling and heating rates, maximum test temperature, duty cycle at temperature and test environment.

So the workshop and this book present a lot of fine papers, with many results and advanced understanding, valuable as well for industrial practice as for progress in high temperature corrosion science. The many influences and parameters that affect cyclic oxidation have been clarified and some guidelines for testing and evaluation have been proposed. As put forward in the final discussion, further steps towards a standardisation of testing would be most desirable and the high temperature corrosion community would be most grateful to those experts that have arranged and shaped this workshop for further activities in this aspect, e.g. the founding of a project aiming at the definition of guidelines and/or standards for thermal cycling testing.

This volume will be an excellent base for further work and such developments, we have to be most grateful to M. Schütze and W. J. Quadakkers for arranging the workshop and editing its proceedings!

H. J. Grabke
Past Chairman of the EFC Working Party: Corrosion by Hot Gases and Combustion Products, 1986–1998

Part 1

Oxidation and Failure Mechanisms under Cyclic Conditions

1
Interfacial Crack Growth During Temperature Changes

H. E. EVANS
School of Metallurgy and Materials, The University of Birmingham, Birmingham, B15 2TT, UK

ABSTRACT
Finite-element analyses have been performed of the oxide spallation process in which decohesion occurs by the growth of a wedge crack along the oxide/metal interface. For conditions of direct cooling, the growth of the interfacial crack is inhibited by stress relaxation at its tip associated with alloy creep but the final stage of crack growth occurs rapidly at intermediate temperatures where creep is limited. Results are provided for the alumina-forming Haynes 214 alloy cooled from 1100°C from which it is shown that high cooling rates impair spallation resistance. These general cracking kinetics where an extended quiescent period is followed by rapid propagation gives confidence in the validity of the Critical Strain Energy Criterion for predicting spallation. Initial results are also provided on a chromia-forming austenitic steel of the effect of thermal cycling on stress development within the oxide layer and across the oxide/metal interface. The importance of creep relaxation is again emphasised since this allows in-plane tensile stresses to develop in the oxide layer on the return to temperature. The significance of these on oxide cracking and oxidation resistance is discussed.

1. Introduction

High-temperature alloys rely on the early formation of a dense surface layer, usually of chromia or alumina, for oxidation protection. Whereas the intrinsic growth rates of such oxides are acceptably low, spallation of the protective layer exposes the depleted substrate and can lead to high metal wastage rates. Spallation tends to occur when the oxide layer experiences in-plane compression which, for simple thermal loading conditions, occurs during cooling for the usual case that the thermal expansion coefficient of the oxide, α_{ox}, is less than that, α_m, of the alloy.

For oxidation-resistant alloys with sensibly flat surfaces, i.e. those for which the substrate radius of curvature is far greater than the oxide thickness, spallation occurs by the growth of a wedge crack along the oxide/metal interface and the critical temperature drop, ΔT_C, for oxide spallation decreases with increasing oxide thickness [1–3]. This dependence is consistent with the critical strain-energy criterion (CSEC) [4] for interfacial fracture for which a simple energy balance between the energy required for interface fracture and that stored within the oxide layer gives [1]:

$$\Delta T_c = \left[\frac{\gamma_F}{\xi E_{ox} (\Delta\alpha)^2 (1 - \upsilon_{ox})} \right]^{1/2} \tag{1}$$

Here, γ_F is the energy required to produce interface fracture, ξ is the thickness of the oxide layer, $\Delta\alpha$ (= $\alpha_m - \alpha_{ox}$) is the difference in thermal expansion coefficient between metal and oxide, E_{ox} and υ_{ox} are, respectively, the Young's modulus and Poisson's ratio of the oxide.

Useful progress has been made in recent years in modelling the growth of this interfacial crack and in estimating the critical temperature drop, ΔT_c, to initiate spallation through the use of finite element methods [5–7]. A common result from this body of work is that stress relaxation by substrate creep can retard the growth of the interfacial crack and delay the onset of spallation. Some of this work on the Ni-based, alumina-forming Haynes 214 alloy will be reviewed in this paper together with previously unpublished studies of the distribution of stress and displacement along the oxide/metal interface. It has been argued previously [7] and will be reiterated here that the success of the CSEC in predicting spallation is due to this retardation of crack growth by creep.

In many practical cases, of course, a single downward temperature transient will not initiate oxide spallation but repeated complete cycles, i.e. cool and re-heat to temperature, will be damaging [8–12]. Understanding of the spallation process under such conditions is limited although various, largely empirical, predictive routes have been proposed [13,14]. Mechanistic modelling would be a preferred approach, however, since, in principle, this offers the ability to extrapolate beyond a data base and also insights into the material and test parameters of importance. An essential requirement of any such method is that it should have the ability to model crack kinetics since, ultimately, it is the growth of cracks which causes oxide spallation. An interesting approach to the incorporation of cracks into a thermal cycling model has recently been presented by Chan [15] who considered the development of wing cracks from pre-existing shear cracks within the oxide layer, i.e. essentially the same geometry as proposed for wedge cracking [1]. Chan's analysis assumes elastic fracture behaviour and is reasonable for fracture within a refractory oxide layer, i.e. delamination. There are difficulties with its application to oxide/metal interface cracks, however, simply because of the influence of substrate creep on stress relaxation and crack blunting. As outlined above, the concepts of elastic fracture mechanics are unlikely to be valid under such conditions and numerical methods will be needed to model the situation. The present paper reports the first scoping results of such a study on a chromia-forming austenitic steel where particular emphasis is placed on the development, under thermal cycling conditions, of oxide and interface stresses in a finite-element model within which cracking is permitted.

2. The Finite Element Model

Oxide spallation is taken to occur by the growth of an interfacial wedge crack [1] nucleated during cooling by the formation of a shear crack within the adherent surface oxide layer, as shown schematically in Fig. 1. The oxide layer is taken to be initially stress-free. Continued growth of this crack occurs during further cooling as a result of the wedging action of the adherent oxide layer to the right hand side of the shear crack, aided by sliding along the shear-crack surface.

The finite element model of this fracture process has been described in detail

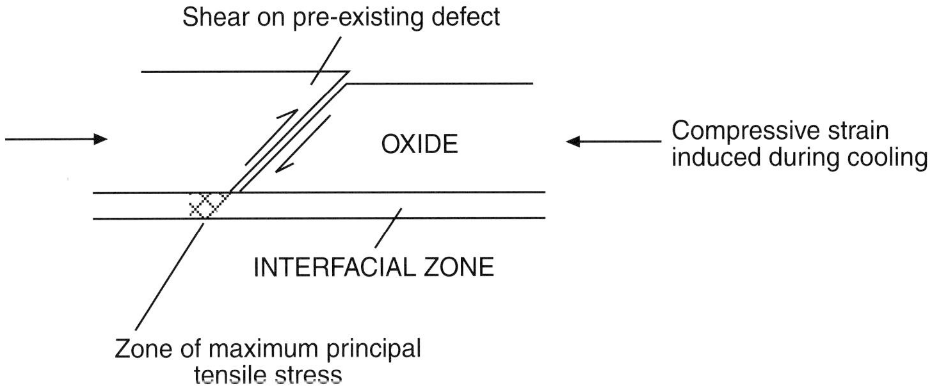

Fig. 1 *Shear within the oxide develops tensile stresses across the oxide/metal interface which can lead to the nucleation of a wedge crack.*

elsewhere [5]. It envisages localised spallation to occur over a circular zone of 8 μm radius centred within a spall cell of 80 μm radius, so ensuring that edge effects are negligible. The half-thickness of the underlying alloy is 190 μm and the results presented in this paper refer to a surface oxide thickness layer of 5 μm. Axisymmetric elements are used throughout and the mesh in the vicinity of the 8 μm spall zone is shown in Fig. 2. The left hand edge of this mesh is the centre line of the (axisymmetric) model and all lateral displacements for these edge nodes are constrained to be zero for reasons of symmetry. An important feature of the model is the presence of interfacial elements [16] of high aspect ratio representing the oxide/metal interfacial zone and the shear crack. Both sets of these elements are 0.1 μm thick. In general, they can have mechanical and physical properties different from the adjacent phases and, so, can represent discrete phases or cracked zones. For the stainless steel considered later, the interfacial zone is amorphous silica and is, in reality, of similar thickness to the interfacial elements. In this case, the interfacial zone is treated as a distinct deformable phase within the model. For the alumina-forming Haynes alloy, also to be considered, there will be a true oxide/metal interface and the 0.1 μm thick elements used in the model will be a less-good approximation to this. Unfortunately, thinning these elements can lead to instabilities and difficulties with convergence and, so, the compromise approach has been to limit within-element deformation and fracture to those elements within the spall zone. Failure of an interfacial element, i.e. it becomes non load-bearing, can occur when the maximum tensile stress within it exceeds a pre-set fracture value, σ_f. Thus, wedge-crack propagation along the oxide/metal interface will occur by 1 μm increments (the length of an interfacial element to the left of the shear crack in Fig. 2) whenever this failure condition is satisfied.

For the computations undertaken here, the shear crack lies at 45° to the oxide/metal interface and is pre-existing, i.e. it penetrates to the top of the oxide/metal interfacial zone (but not to the underlying alloy) even at the oxidation temperature. This assumption is intended to produce conservative estimates of ΔT_c but these will not be unduly pessimistic in view of the large, microns-long, lenticular voids known to form in some alumina layers [17]. Shear displacements on this crack surface are taken to be frictionless. All the oxide layers considered are essentially creep-rigid

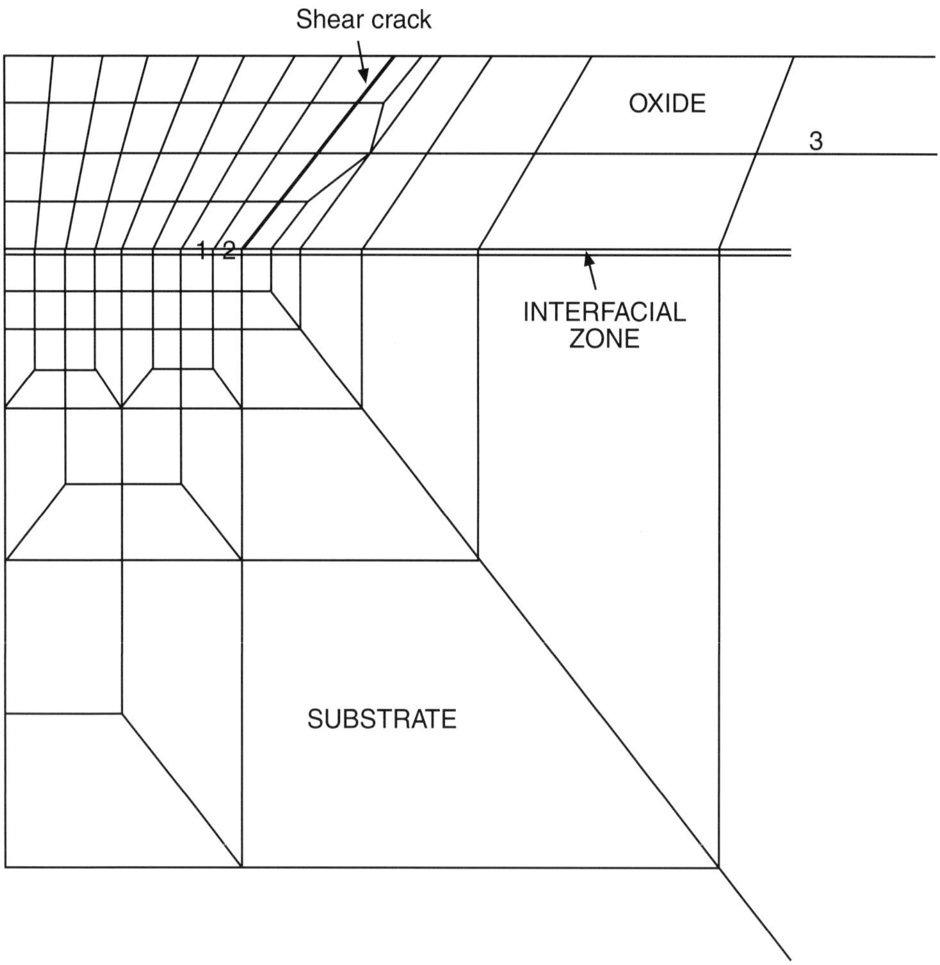

***Fig. 2** The finite element mesh in the vicinity of the spall zone.*

elastic solids but alloy creep is incorporated into the model. For all phases, the elastic properties are assumed to be temperature-invariant but this simplification has negligible effect on the predictions. The material parameters for the oxide phases and alloy substrates used in the computations are given in Table 1. The creep equations used have been obtained from the literature [18] for the austenitic steel or derived [7] from the manufacturers' data sheets for the Haynes alloy.

3. Crack Growth during Direct Cooling

Particular emphasis is given in this section to the behaviour of the Haynes 214 alloy. Some results on the kinetics of wedge-crack growth during cooling have been presented elsewhere [7] but are considered again in this paper since they provide a striking example of the influence of substrate creep on the growth kinetics of the

Table 1. *Material parameters used in the computations*

Material	E (GPa)	υ	$\alpha \times 10^6$ K^{-1}	σ_f (MPa)	Expression for creep rate, $\dot{\varepsilon}$, s^{-1} (σ = stress, Pa; T = temperature, K)
Alumina	367.0	0.27	7.9	1700	Creep rigid
Chromia	260.0	0.30	8.5	n.a.	Creep rigid
Silica	65.7	0.15	0.5	600	Creep rigid
20Cr25Ni steel	156.8	0.30	17.8	n.a.	$\dot{\varepsilon} = 2.10 \times 10^{-28}\sigma^{5.0} \exp(-42100/T)$
Haynes 214	177.0	0.30	20.0	n.a.	$\dot{\varepsilon} = 9.08 \times 10^{-31}\sigma^{5.24} \exp(-48450/T) + 1.65 \times 10^{-15}\sigma^{6.24} \exp(-106030/T)$

n.a. = not applicable.

wedge crack. They also provide a physical understanding for the success of the critical strain-energy criterion for predicting the temperature drop for spallation.

Haynes 214 is a Ni-based high-temperature alloy with nominal composition, in mass%, of Ni–16.0Cr–3.0Fe–4.5Al–Y. It is strengthened by a γ' dispersion at intermediate temperatures but, at temperatures much above 900°C (1173 K), this phase re-dissolves and creep strength drops rapidly with increasing temperature. This behaviour is represented in the model by a 2-term creep equation (Table 1) where the second term with high activation energy accounts for this weakening with increasing temperature. Indeed, at such temperatures this alloy is much weaker than the austenitic steel to be considered later. The protective oxide formed on the Haynes alloy at typical (high) operating temperatures or after pre heat treatment is alumina for which the oxide/metal interfacial zone in the model is also taken to have the same elastic properties, albeit with an intrinsic tensile fracture stress of 1700 MPa.

The computed kinetics of wedge crack growth for this alloy cooled from 1100°C at 10^2 and 10^4 °C/h are shown in Fig. 3. In both cases, wedge crack nucleation occurs at temperature drops less than 100°C to form a crack of 1 μm in length but, following this, there is an extended further temperature drop during which no further crack growth occurs. This is because creep processes in the alloy in the vicinity of the crack result in a reduction in the rate of stress build-up at the crack tip and, hence, a protracted period of cooling before the fracture stress of the adjacent interfacial element can be reached. The tensile stress normal to the oxide/metal interface at the location marked "1" in Fig. 2, i.e. 0.33 μm from the crack tip, is shown in Fig. 4 as a function of temperature drop. The lower rate of increase at the lower cooling rate is clear and is due to the longer time available for creep relaxation to occur. These broad trends have now been established for other alumina-forming alloys such as a NiCrAlY coating alloy [6] and, in unpublished work, in Fe20Cr5Al (mass%) steels as well as for the chromia-forming austenitic steel [5] to be considered below and Alloy 800.

Eventually, after this period of creep relaxation, the fracture condition of the adjacent interfacial element is reached and continued propagation of the wedge crack occurs. As can be seen from Fig. 3 for the present Haynes 214 alloy, this occurs at similar rates for both cooling conditions and suggest that creep processes have become

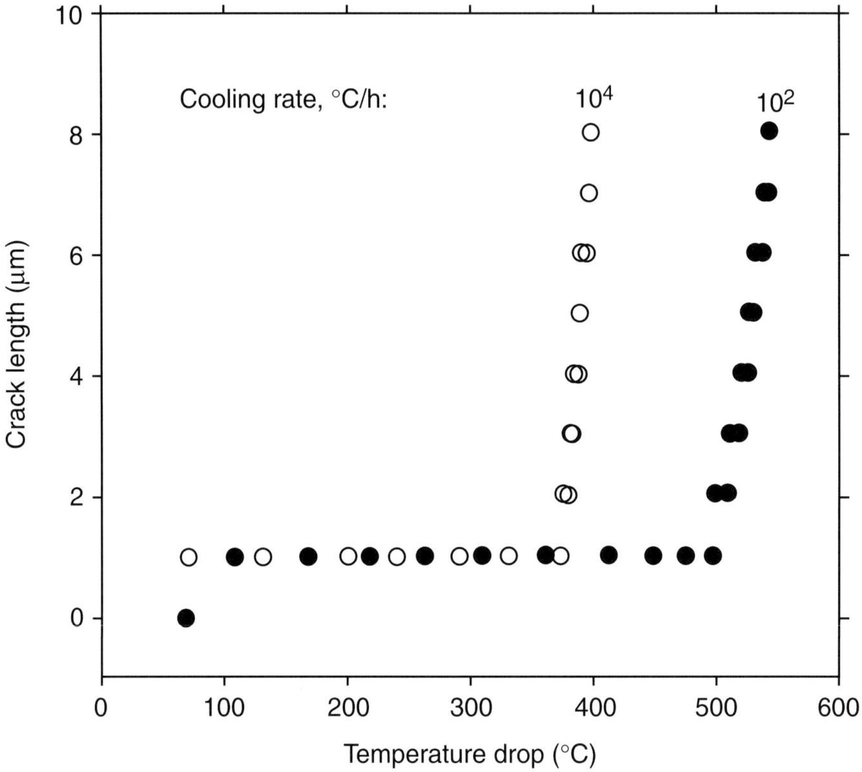

Fig. 3 *The kinetics of wedge crack growth for the Haynes 214 alloy with a 5 μm alumina layer cooled from 1100°C at two different cooling rates.*

unimportant at these temperatures. This is entirely consistent with the increased strengthening associated with γ' precipitation at intermediate temperatures during cooling. As a consequence of the limited stress relaxation available, subsequent crack growth then occurs rapidly over a narrow range of temperature and, to some level of approximation the entire oxide/metal interface can be considered to have failed at an unique value, ΔT_c, of temperature drop. As has been pointed out elsewhere [7], this is the condition required by the critical strain-energy criterion (CSEC) for spallation since this envisages all of the stored elastic energy within the oxide layer overlying the spall zone to be released instantaneously in interface fracture at a certain, critical temperature drop, ΔT_c [4]. Since this delay in crack propagation arises because of stress relaxation by substrate creep, a corollary is that the CSEC will only be a valid predictive tool under conditions in which substrate creep will occur.

One consequence of substrate creep is that energy must be dissipated from the fracture process into deforming the substrate. This means that the use of eqn (1) to calculate the fracture energy, γ_F, will produce an effective value which, for an alloy of low creep strength, will be dominated by substrate deformation. Thus for the two examples shown in Fig. 3, which give a ΔT_c of 543°C on cooling at 100°C/h and of 397°C at 10^4 °C/h, the corresponding effective fracture energies, from eqn (1), are 57.8 and 30.9 Jm^{-2}, respectively. Although no direct measurements appear to exist of

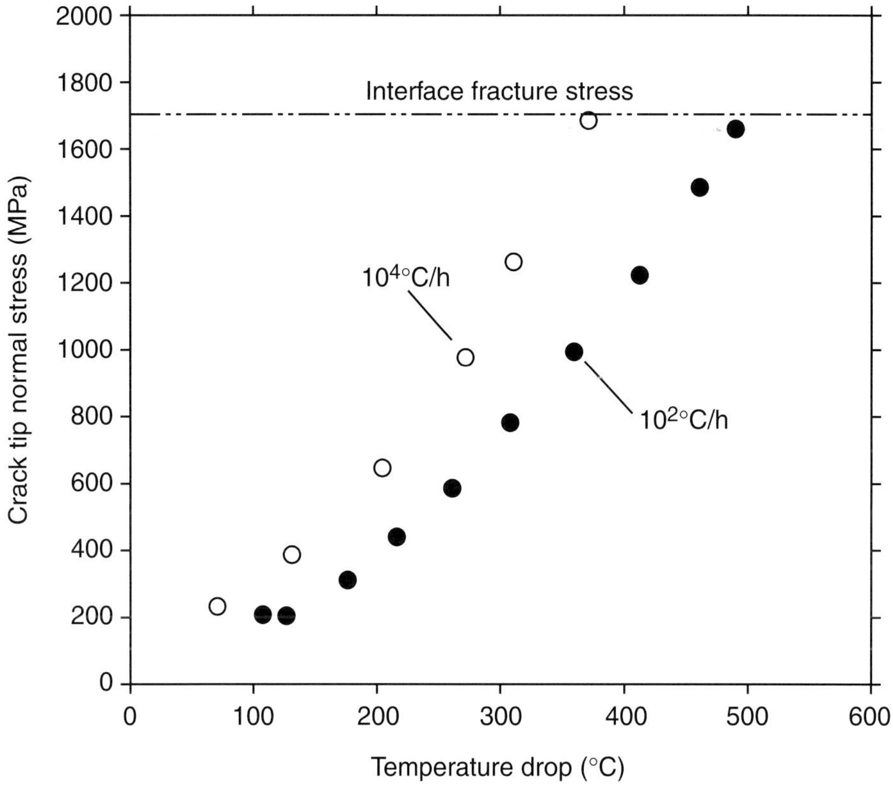

***Fig. 4** The development of normal stress at the tip of a 1 μm interfacial wedge crack in a Haynes 214 alloy cooled from 1100°C at two different rates.*

the elastic fracture properties of the oxide/metal interface, a true fracture energy for naturally-grown alumina layers of ≈ 2 Jm^{-2} would be expected from measurements [17,19] of the critical stress intensity factor, i.e. at least an order of magnitude lower than the values deduced here.

Creep deformation of the substrate alloy produces essentially a blunting of the crack tip, as can be appreciated from Fig. 5 which shows the displacement of the original oxide/metal interface, containing a 1 μm long wedge crack, after a temperature drop of 490°C at 100°C/h. It is clear that, in the vicinity of the crack, the substrate has moved away from the original location of the interface to form a void-like structure. This localised creep deformation also affects the stress distribution along the interface ahead of the crack as is shown in Fig. 6 for both the normal and shear stresses after the same temperature drop of 490°C. Whereas the normal stress is tensile with a value of 1650 MPa close to the crack tip, it rapidly reduces over a distance of 2 μm, presumably because the displacement of matter from the fracture zone imposes a compressive tendency across the interface. This type of behaviour is, of course, far removed from expectations under elastic deformation and shows the significance of substrate creep. The shear stress behaves in a similar manner (Fig. 5) but the peak value is approximately half that of the tensile stress.

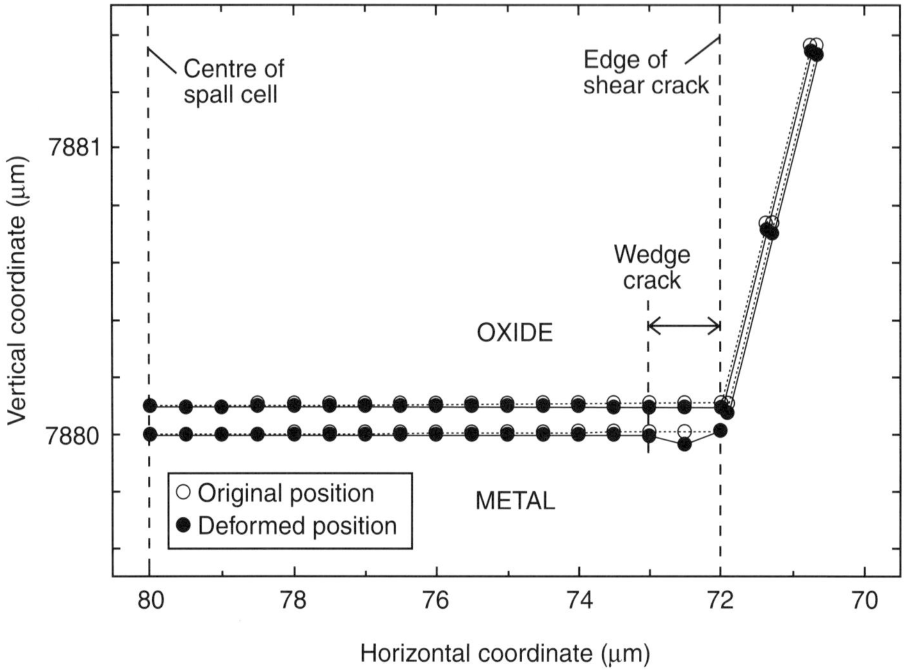

Fig. 5 *The displacement of the oxide/metal interface in the spall zone for the Haynes 214 alloy after cooling from 1100℃ to 610℃ at 100℃/h.*

4. Behaviour under Thermal Cycling

Although the intention is to undertake modelling of the Haynes 214 alloy under thermal cycling conditions, initial scoping tests are in progress on the 20Cr25Ni steel since the more limited creep in this alloy leads to more convenient processing times. As indicated in the Introduction, even this initial study is incomplete, however, and preliminary results only are reported here.

As with most stainless steels, this alloy contains approximately 0.6% Si and, during oxidation, will form a thin layer of amorphous silica between the outer chromia layer and the substrate [20]. As discussed earlier, this structure is modelled using the mesh of Fig. 2 by allocating materials properties for silica (Table 1) to the oxide/metal interfacial zone. The shear-crack interfacial elements have chromia properties but, as in the case of the alumina layer in the Haynes alloy, the shear modulus is set to zero to permit easy sliding over the crack surfaces. Whereas under simple cooling the oxide layer always experiences in-plane compressive stresses, stress relaxation during cooling can lead to tensile in-plane oxide stresses on the return to temperature during thermal cycling. This has been well demonstrated previously by Barnes *et al.* [21] on a Ni30Cr alloy but this current study is the first attempt to incorporate oxide and interface cracks as well as a creeping substrate into a finite element model. As will be shown, tensile stresses are also predicted in the present study but then the effect of these on the oxide shear crack must also be considered. Obviously, it is likely that this will open under tensile stressing but it is also likely that a critical

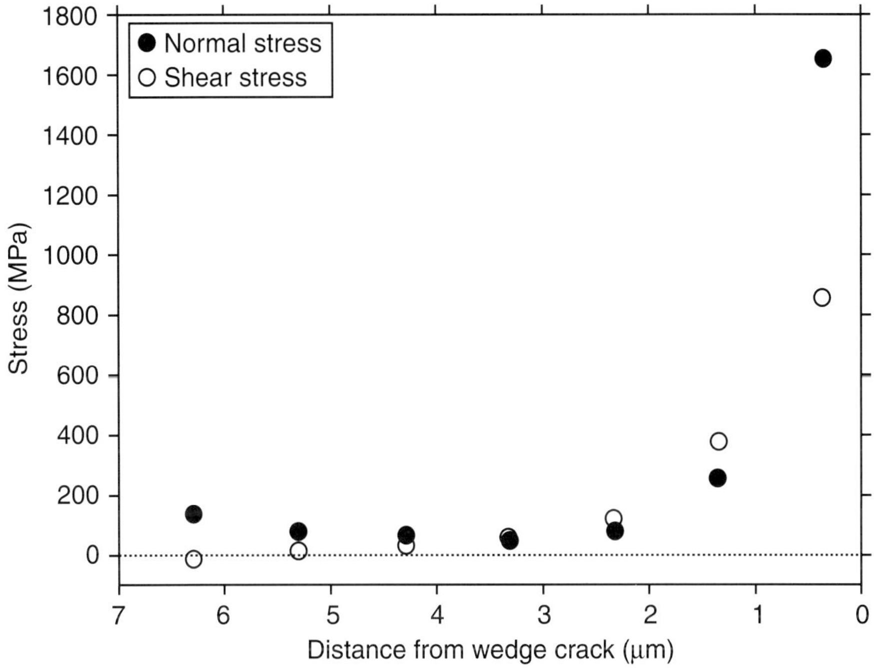

***Fig. 6** The normal and shear stress distribution over the interface in the spall zone for the Haynes 214 alloy cooled from 1100°C to 610°C at 100°C/h.*

stress, σ_c, will be needed to achieve this. In the work undertaken to date, a single value of σ_c of 100 MPa has been used since initial tests indicated that this would produce full opening of the shear crack after a limited number of cycles.

For the 5 μm chromia layer used in this work, the model predicts that spallation will occur during cooling at 100°C/h from the chosen oxidation temperature of 900°C (1173K) at a temperature drop of 290°C. This is consistent with published [1,2] experimental measurements. Accordingly, thermal cycles of smaller amplitude were used in the computations to ensure that no cracking at the oxide/metal interface occurred during the first cooling period. It then becomes possible to examine the effect of continued thermal cycling on stress development and crack initiation. The cyclic history chosen to be modelled is shown in Fig. 7 and consisted of ten cycles from 900°C → 720°C → 900°C at constant cooling and heating rates of 100°C/h and at a constant oxide thickness. After these ten cycles, the structure was allowed to cool until oxide spallation occurred.

The development of the normal stress across the interfacial zone during cycling was monitored at the base of the shear crack at the location marked "2" in Fig. 2. As indicated in Fig. 1, in the absence of wedge-crack nucleation, this would be the location of maximum tensile stress during cooling. The in-plane oxide stress was monitored at the location marked "3" in Fig. 2. This was chosen since it was a sensible distance away from the shear crack and the spall zone. The stress values reported should be considered as no more than representative of the oxide behaviour but, as will be seen, yield a consistent and understandable response to the cycling history.

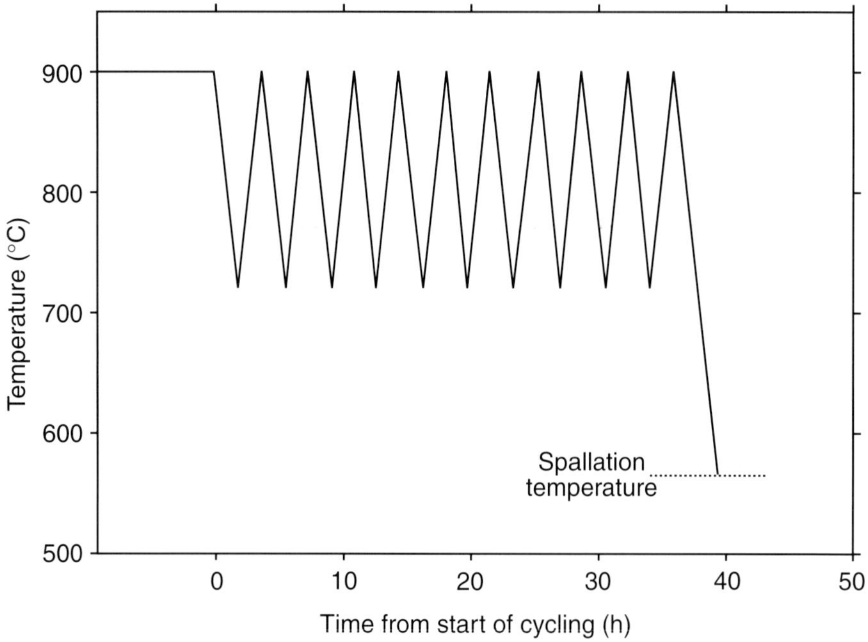

Fig. 7 *The thermal cycling history for the 20Cr25Ni steel used in the finite-element modelling.*

The variation of stress at both location "2" and "3" during the first complete thermal cycle is shown in Fig. 8; the arrows on these curves show the sense of temperature change. As expected (Fig. 1), the normal stress across the interface becomes tensile as temperature drops but, because of substrate creep, a hysteresis develops on the return to temperature resulting in residual compressive stress across the oxide/metal interface. A similar effect occurs with the in-plane oxide stress (Fig. 8) although, in this case, compressive stresses develop during cooling and then become tensile on the return to temperature. The influence of substrate creep in reducing the rate of development of these compressive thermal stresses can also be appreciated from the lower broken line in this figure. This represents the theoretical average in-plane oxide stress, σ_{ox}, calculated, using the parameters of Table 1, from [e.g. 22]:

$$\sigma_{ox} = -\frac{E_{ox}(\Delta\alpha)(\Delta T)}{(1-\upsilon_{ox})\left\{1+\frac{\xi}{h}\left[\frac{E_{ox}(1-\upsilon_m)}{E_m(1-\upsilon_{ox})}\right]\right\}} \tag{2}$$

where ξ is the oxide thickness, h is the half-thickness of the alloy substrate and the substrates m and ox refer to the substrate and oxide, respectively. This equation assumes elastic deformation only, i.e. no substrate creep. The trend shown in Fig. 8 in which the rate of increase of the compressive stress within the oxide layer is reduced in the presence of substrate creep is consistent with computations made for other alloy systems [21,23].

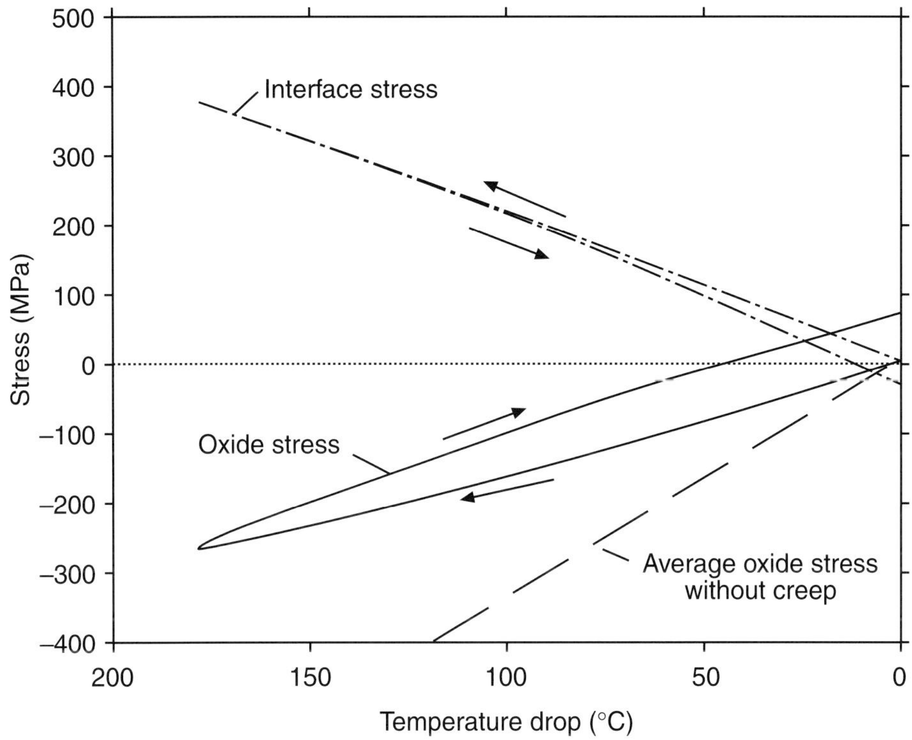

***Fig. 8** The variation of the normal interface stress at the base of the shear crack and of a representative in-plane oxide stress over the first temperature cycle from 900°C → 720°C → 900°C at 100°C/h.*

The variation with the number of complete thermal cycles of both the interface stress (location "2" in Fig. 2) and in-plane oxide stress (location "3") on return to 900°C is shown in Fig. 9. As is indicated in this figure, the residual stress in the oxide resulted in tensile failure of the shear crack after two complete cycles. Clearly, this had a marked effect in reducing the rate of development of residual normal stress across the interface but, even so, values of approximately –200 MPa were found after eight or so cycles. In contrast, the development of tensile stress within the oxide, at least as measured at location "3" remote from the shear crack, is little affected by shear-crack failure. This in-plane tensile stress seems to reach a shallow maximum of approximately 130 MPa after some six cycles.

Subsequent cooling after the ten thermal cycles results in oxide spallation at a temperature drop of 335°C, i.e. larger than the 290°C found by direct cooling in the absence of thermal cycling. This apparently beneficial influence of cycling is, however, consistent with the development of compressive stresses across the oxide/metal influence (Figs 8 and 9) since these need to be offset during cooling before wedge-crack initiation and propagation can occur. This feature seems inescapable, at least in the absence of cracks at the interface, since the tensile stresses developed across the entire length of the interface during initial cooling will be converted by creep hysteresis into compressive stresses on the return to temperature (Fig. 8). However,

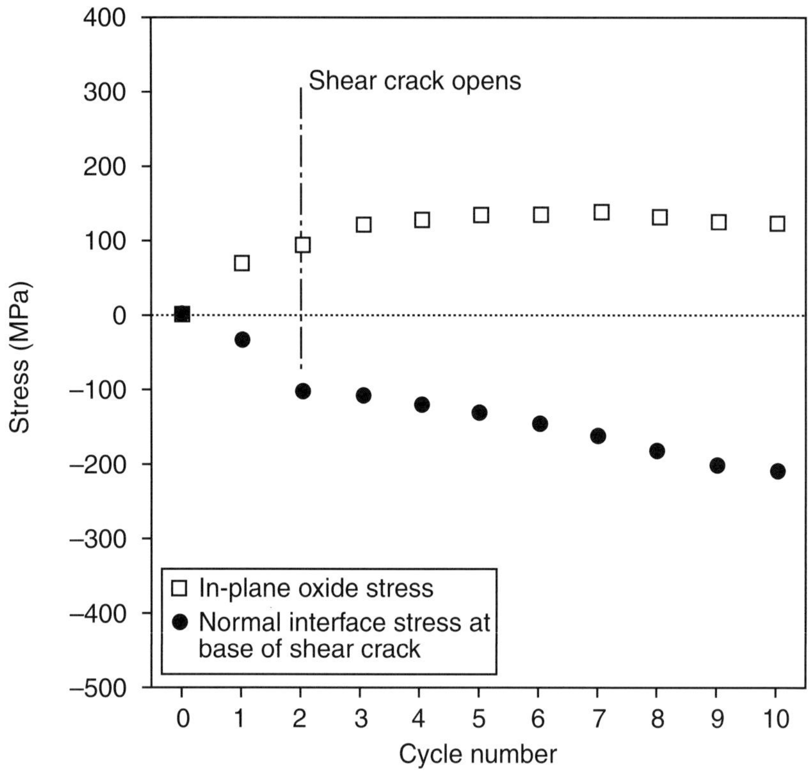

Fig. 9 *The variation of the normal interface stress and representative in-plane oxide stress with increasing number of complete thermal cycles.*

in future work this aspect will be confirmed for conditions where a wedge crack exists at the oxide/metal interface. Another aspect that requires further study will be the effect of the opening of the shear crack on the interfacial shear-stress distribution at its base. If particularly high values are produced then the interface may be susceptible to failure in a shear rather than tensile mode (these concepts are reviewed in [24]).

Even though the present study is at an early stage, it seems clear that the in-plane tensile stresses produced in the oxide layer on the return to temperature can produce direct-access paths for the oxidant. This is most likely to arise through the opening of pre-existing shear cracks although the propagation of new tensile cracks from oxide defects will occur if the residual stresses are sufficiently high. Such cracks may heal at temperature [25] (but still produce a mass gain) or, if substrate depletion is advanced, may lead to sub-surface oxidation and disruption of the protective oxide layer. The latter, clearly, would be one route whereby thermal cycling had a deleterious influence on oxidation resistance. Indeed, Quadakkers *et al.* [26] have shown how tensile cracking of the alumina scale under thermal cycling conditions increased the oxidation rate of MA956 alloys. It is not clear from their work, however, whether the cracks were caused by the present mechanism or were associated with the near-surface rumpling of the alloy substrate which also occurred.

5. Conclusions

Finite-element analyses have been performed of the oxide spallation process in which decohesion occurs by the growth of a wedge crack along the oxide/metal interface. It is shown, for the alumina-forming Haynes 214 alloy during direct cooling from 1100°C, that the growth of the crack is inhibited by stress relaxation at its tip associated with alloy creep but that the final stage of crack growth occurs rapidly at intermediate temperatures where creep is limited. These general cracking kinetics where an extended quiescent period is followed by rapid propagation gives confidence in the validity of the Critical Strain Energy Criterion for predicting spallation.

Initial results have also been presented of a related study on the influence of thermal cycling on a chromia-forming austenitic steel. It has been shown that alloy creep is again a dominant factor such that relaxation during the cycle will develop normal compressive stresses across the oxide/metal interface and in-plane tensile stresses in the oxide layer on the return to temperature. The latter will tend to open pre-existing shear cracks in the oxide and this process has also been modelled; these tensile stresses may also initiate new cracks from defects within the oxide layer. Such open cracks will provide direct gas access which will lead to an increase in the oxidation rate and may also increase the rate of spallation.

6. Acknowledgements

This work has been undertaken in association with the UK's National Physical Laboratory and forms part of the underpinning research in the programme "Degradation of Materials in Aggressive Environments", funded by the UK Department of Trade and Industry. I gratefully acknowledge the active support of Mr D. D. Gohil, Dr S. Osgerby and Dr S. R. J. Saunders of the NPL.

References

1. H. E. Evans and R. C. Lobb, *Corros. Sci.*, 1984, **24**, 209–222.
2. J. Asher *et al.*, *Werkst. Korros.*, 1987, **38**, 506–516.
3. K. Bouhanek, D. Oquab and B. Pieraggi, *Mater. Sci. Forum*, 1997, **251–254**, 33–40.
4. U. R. Evans, in *Internal Stresses in Metals and Alloys*, The Institute of Metals, London, 1948, pp.291–310.
5. H. E. Evans, G. P. Mitchell, R. C. Lobb and D. R. J. Owen, *Proc. Roy. Soc. London*, 1993, **A440**, 1–22.
6. H. E. Evans, A. Strawbridge, R. A. Carolan and C. B. Ponton, *Mater. Sci. Eng.*, 1997, **A225**, 1–8.
7. H. E. Evans and A. Strawbridge, in *Fundamentals of High-Temperature Corrosion VI* (D. A. Shores, P. Y. Hou and R. A. Rapp, eds). The Electrochem. Soc., Pennington, NJ, USA, 1997, pp. 1–15.
8. C. A. Barrett and C. E. Lowell, *Oxid. Met.*, 1975, **9**, 307–355.
9. R. Mevrel, *Mater. Sci. Technol.*, 1987, **3**, 531–535.
10. V. Ramakrishnan, J. A. McGurty and N. Jayaraman, *Oxid. Met.*, 1988, **30**, 185–201.
11. G. H. Meier, *Mater. Sci. Eng.*, 1989, **A120**, 1–11.
12. P. Y. Hou and J. Stringer, *Oxid. Met.*, 1990, **33**, 357–369.

13. C. E. Lowell *et al., Oxid. Met.*, 1991, **36**, 81–112.
14. Chang-Oh Moon and Soon-Bok Lee, *Oxid. Met.*, 1993, **39**, 1–13.
15. K. S. Chan, *Metall. Mater. Trans.*, 1997, **28A**, 411–422.
16. J. M. A. Cesar de Sa and D. R. J. Owen, "Computer Aided Modelling of Grain Boundary Failure Mechanisms", Report No. C/R 449/83, Dept. Civil Engineering, Univ. College Swansea, UK, 1983.
17. J. P. Banks, D. D., Gohil, H. E. Evans, D. J. Hall and S. R. J. Saunders, in *Materials for Advanced Power Engineering, II* (D. Coutsouradis *et al.*, eds). Kluwer Academic Pub., Dordrecht, The Netherlands, 1994, pp. 1543–1552.
18. B. D. Clay, in *Mechanical Behaviour and Nuclear Applications of Stainless Steels at Elevated Temperatures.* The Metals Soc., London, 1982, pp.122–128.
19. M. Schütze, *Int. J. Pressure Vessels Piping*, 1991, **47**, 293–315.
20. M. J. Bennett, J. A. Desport and P. A. Labun, *Proc. Roy. Soc. Lond.*, 1987, **A412**, 223–230.
21. J. J. Barnes, J. G. Goedjen and D. A. Shores, *Oxid. Met.*, 1989, **32**, 449–469.
22. C. Liu, A. M. Huntz and J. L. Lebrun, *Mater. Sci. Engng*, 1993, **A160**, 113–126.
23. S. J. Bull. *Oxid. Met.*, 1998, **49**, 1.
24. H. E. Evans, *Int. Mater. Rev.*, 1995, **40**, 1–40.
25. M. Schütze, *Oxid. Met.*, 1986, **25**, 409.
26. W. J. Quadakkers, K. Schmidt, H. Grübmeier and E. Wallura, *Mater. High Temp.*, 1992, **10**, 23–32.

2
The Influence of Surface Pre-treatment on the Integrity of Alumina Scales on the ODS Alloy MA 956

V. GUTTMANN, F. HUKELMANN, P.A. BEAVEN and G. BORCHARDT*
European Commission, Joint Research Centre, Institute for Advanced Materials, Petten, The Netherlands
*Institut für Allgemeine Metallurgie, TU Clausthal, Clausthal-Zellerfeld, Germany

ABSTRACT

Quasi-isothermal and cyclic oxidation experiments at 1100°C in air have been performed on the ODS alloy MA 956 in order to investigate the effects of prior surface treatments on the integrity of the alumina scales formed. The oxidation behaviour of samples with 800 grit SiC ground surfaces has been compared with that of samples exposed (i) in the as-received condition (pre-oxidised and Al_2O_3 grit blasted surfaces) and (ii) with lathe turned surfaces. Effects of these surface treatments on the growth kinetics of the scales, their morphologies, defect structures and spallation behaviour have been studied for exposure times up to 5000 h (quasi-isothermal) and 3500 h (cyclic). Assessment of the scale failure mechanisms revealed that deleterious effects resulted mainly from geometrical surface irregularities and that scale defects had no obvious influence. Defects at the scale–substrate interface appeared to play a more important role in the long term exposure regime.

1. Introduction

The technical applicability of materials for use as high-temperature components is to a large extent dependent on their ability to develop and maintain protective oxide scales such as Cr_2O_3 and Al_2O_3. This protection may diminish or even be lost completely as a result of stresses that can arise from various sources. Stresses may be generated during scale growth due to differences in specific volume between the oxide and the substrate and during cooling as a result of differences in thermal expansion coefficients. In addition, loading of components will involve the superposition of external mechanical stresses. Scale integrity depends not only on the global stress state but also on the presence of physical defects such as voids, pores and microcracks in the scale and at the scale–substrate interface. Since oxide scales generally exhibit brittle behaviour, linear elastic fracture mechanics (LEFM) are applicable and the size, number and distribution of defects play a vital role in determining the scale integrity [1–7].

Within the context of scale failure the influence of the surface condition of the alloy substrate prior to oxidation is of particular concern. For instance, geometrical surface irregularities generate additional stresses which facilitate defect formation. Theoretical [8–16] and experimental [1,14,17–19] aspects of this matter have been dealt with in the literature. Since technical components can be subjected to a variety of final surface treatments, it is of interest to study the influence of prior surface condition on scale development and the resulting degree of protection. The present work deals with the effect of different surface treatments on the integrity of the alumina scales formed on MA 956, an iron-based oxide dispersion strengthened (ODS) type alloy designed for service applications above about 1000°C. The high strength of MA 956 at elevated temperatures is achieved via Y-rich oxide dipersoids in a microstructure consisting of coarse, highly elongated grains. Good oxidation resistance results from the yttria addition.

2. Experimental

MA 956 material with a nominal alloy composition of Fe–20Cr–4.5Al–0.5Y_2O_3–0.4Ti (mass %) was supplied by Inco Alloys Ltd, (Hereford, UK). Experiments were carried out on 2 mm sheet and on rods machined to 6 mm dia. All oxidation exposures were performed at 1100°C in laboratory air using heating rates of 6°C/min and furnace cooling for both quasi-isothermal and interrupted tests. For the latter, cycle durations of 200–300 h were applied. Microstructural studies were carried out on cross sections and fractured samples using scanning electron microscopy (SEM). Substrate surfaces were also examined after descaling in order to facilitate the investigation of interfacial defects.

The basic reference condition for the oxidation tests was a smooth surface finish produced by grinding with 800 grit SiC paper. The second surface condition tested was the 'as-received' state in which the alloy exhibits an oxide scale formed during the final recrystallisation treatment at 1300°C which had subsequently been subjected to blasting with Al_2O_3 grit. The third condition studied was a rough surface produced by lathe turning of rod material. Surface roughness data are given in Table 1.

Table 1. *Roughness data for pre-treated MA 956 surfaces*

Surface condition	Roughness parameters*	
	Average roughness R_a (μm)	Maximum roughness R_m (μm)
800 grit SiC	0.28	1.7
As-received† Side 1:	1.56	16.54
As-received† Side 2:	1.38	10.16
Lathe turned	1.32	8.9

*Roughness parameters corresponding to peak-to-valley heights according to DIN ISO1302.
†The values given for this condition apply to the surface oxide and not to the MA 956 substrate. Sides 1, 2 refer to the strongly and weakly blasted sides of the sheet, respectively.

3. Results

Mass change vs exposure time data are shown in Fig. 1. The straight line fitted to the data for the 800 grit finish samples corresponds to subparabolic oxidation kinetics, consistent with previously reported results [20,21]. Scale adherence for these samples was found to be excellent; only after exposure times exceeding about 3200 h were a few small blank spots observed indicating the very beginning of spallation. Exposure up to 5000 h (quasi-isothermally with one interruption) also induced very limited spallation and no deviations in the oxidation kinetics. As a result of this very good scale behaviour the data for the 800 grit finish samples were taken as the standard reference for comparative purposes.

The mass change data for the as-received material condition showed consistently lower values than the standard data as a result of the retardation in oxidation caused by the presence of the pre-existing scale. First indications for spallation, characterised by a few, blank spots about 150 μm dia. were observed on one side of the sheet samples after about 250 h. After approximately 800 h, spallation was apparent on both sides. (Note: It will be demonstrated later that the scale morphology was different on the two sides of the sheets.) With extended exposure beyond about 2400 h, increasing amounts of both the original and newly formed scales were lost giving rise to mass decreases.

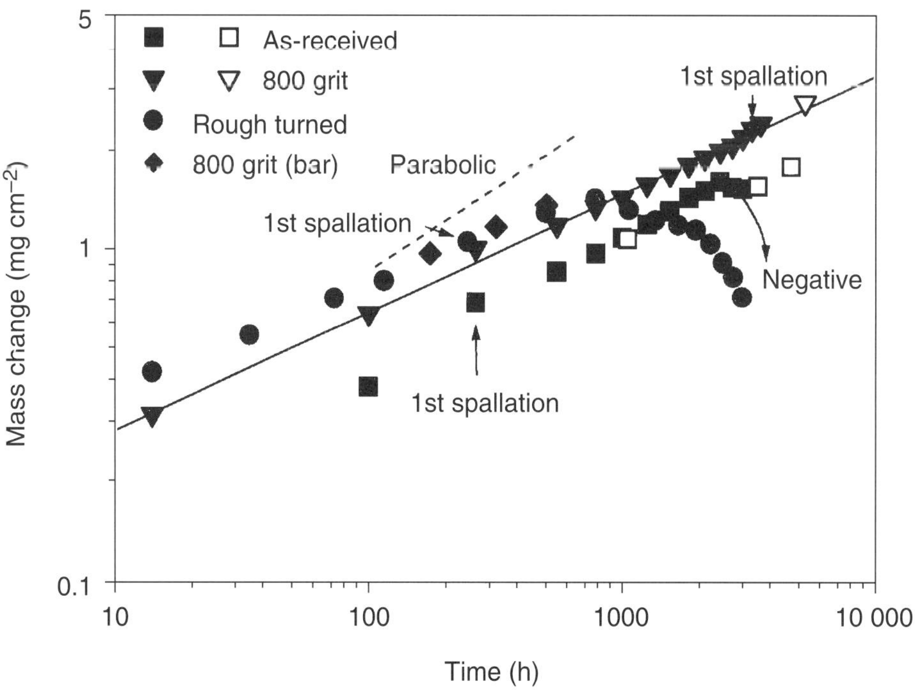

Fig. 1 *Mass change vs exposure time at 1100°C for MA 956 after different surface treatments. Full symbols represent cyclic oxidation, open symbols indicate fewer than 3 interruptions.*

Comparison of the mass changes for lathe turned samples with those of the 800 grit surface finish samples revealed somewhat more rapid oxidation kinetics. Tests performed on the same rod material prepared with an 800 grit finish also showed increased mass changes (Fig. 1). It is therefore concluded that the difference in kinetics probably results from batch to batch variations rather than from the severe surface pre-treatment. Lathe turned samples were found to be extremely prone to spallation, which began after about 100–150 h and was so extensive after about 800 h that mass decreases ensued.

For all three surface treatments, the assessment of scale integrity was based on observations of the major flat surfaces of sheet samples or the circumferential surfaces of rods. However, in all cases, repeated descaling and re-healing took place at the edges and corners of the samples from the very beginning of the exposures

4. Microstructural Studies

In order to elucidate the mechanisms governing scale spallation on the variously treated alloy surfaces, the morphology and defect structure of the scales were studied. The micrograph of a sample exposed for 5000 h shown in Fig. 2 serves to demonstrate the principle types of defects observed in Al_2O_3 scales formed on MA 956. Physical defects occur as pores of various sizes and compositional defects take the form of Y-rich and Ti-rich inclusions. Since the Y-rich oxides (bright contrast) are formed at grain boundaries, their morphologies give a first indication of the grain structure of

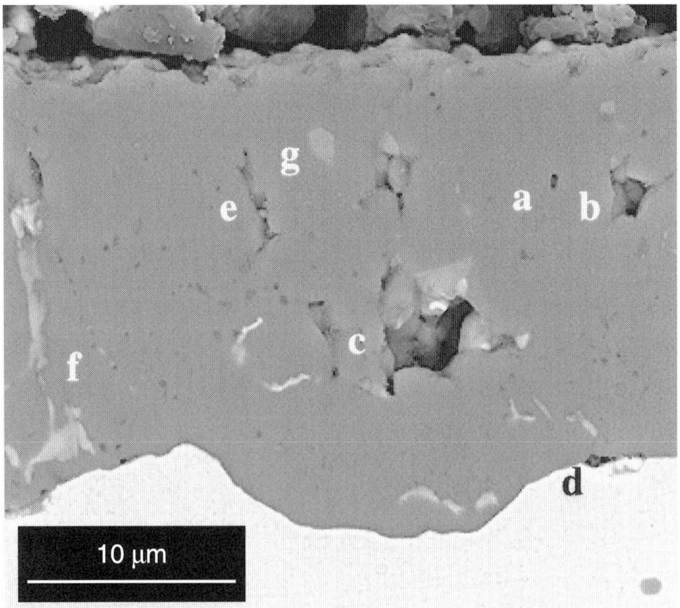

Fig. 2 SEM micrograph of the scale formed on an 800 grit finish sample of MA 956 after oxidation for 5000 h at 1100 °C. Defects are identified as: (a) micropores, (b) mid-size pores, (c) macropores, (d) interfacial pores, (e) chain-like defects, (f) Y-rich oxides and (g) Ti-rich oxides.

the scale, with the 'finger-like' morphology demonstrating that it is columnar. The Ti-rich particles (light grey contrast) are typically associated with pores in the scale, partially filling them.

The physical defects could be classified according to their sizes into micro, mid-size and macropores [21]. As shown in Fig. 3 there is a general tendency for the pore size in each category to increase with exposure time. The pores in the individual categories also differed in their morphology and spatial distribution. Micropores were roughly spherical in shape and occurred intergranularly, often forming chains along triple grain boundaries. Mid-size pores appeared for the first time at depths 2.5–3.5 μm below the scale surface. (Note: growth of the Al_2O_3 scale on MA 956 is known to be mainly inward [20,22].) These mid-size pores were usually crystallographically faceted and often located at grain boundary triple points. Their density could be relatively high in regions having equiaxed Al_2O_3 grain structures. A low density of mid-size pores was detected at the scale–substrate interface after short and intermediate exposure times. These initially small defects could develop into relatively large, hollow-like defects after extended exposure times, depending on the prior surface treatment. Macropores were first observed after the scale reached a thickness of about 4–5 μm corresponding to an exposure time of about 250 h. Their formation was clearly linked to localised agglomerations of mid-size pores associated with equiaxed Al_2O_3 grains. Generally, enhanced local oxidation took place in conjunction with macropores.

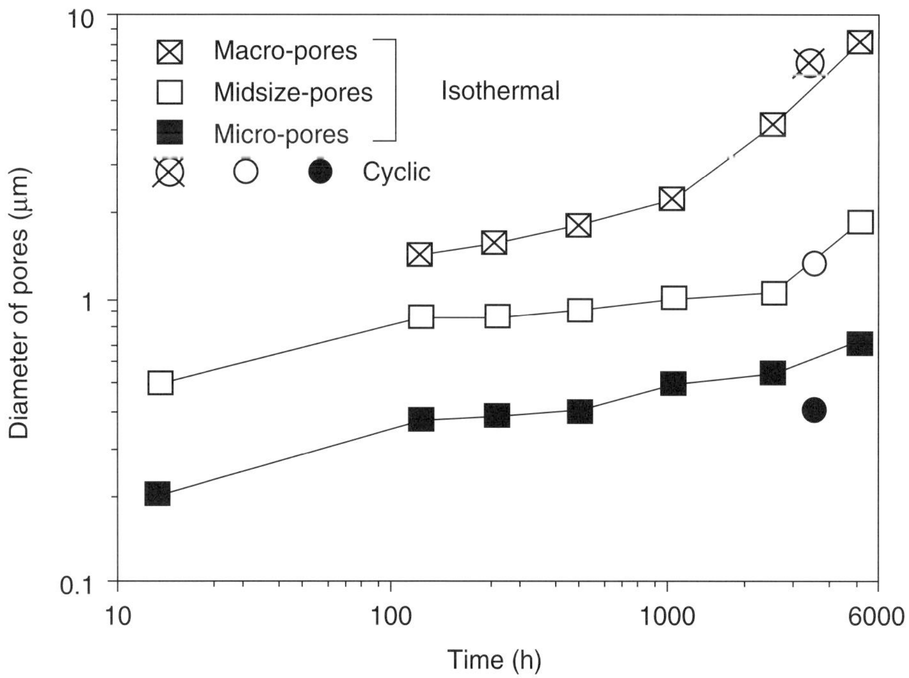

Fig. 3 *Time dependent evolution of the maximum diameters of pores in the scale during oxidation of 800 grit finish samples of MA 956 at 1100°C.*

4.1. 800 Grit Surface Finish

Studies of the scales formed under cyclic conditions on 800 grit treated surfaces did not reveal any specific, structural differences compared to samples exposed isothermally. Exposure for 250 h led to the formation of a compact scale of uniform thickness containing micropores and a few mid-size pores (Fig. 4). The agglomeration of mid-size pores seen in Fig. 4 might be considered as the initiation of a macropore. Figure 4 also shows Y-rich oxides with finger-like shapes indicating a columnar grain structure and small Ti-rich oxides typically associated with mid-size pores.

An example of the scale microstructure after 3500 h of cyclic exposure is shown in Fig. 5. The scale exhibited excellent adherence although the defect density was

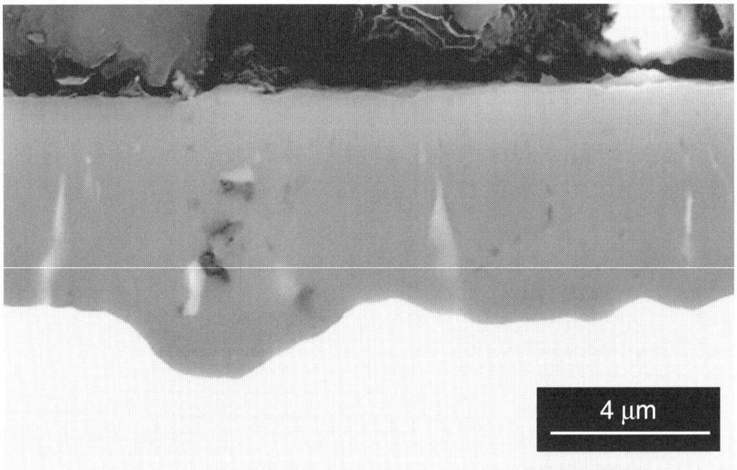

Fig. 4 *Scale formed on an 800 grit finish sample of MA 956 after exposure for 250 h at 1100°C.*

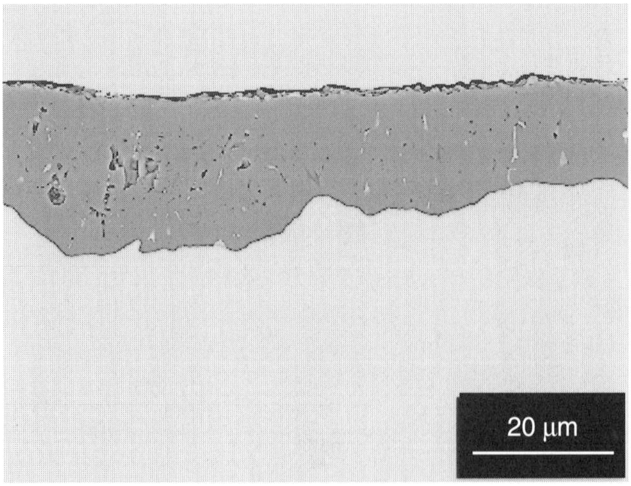

Fig. 5 *Scale formed on an 800 grit finish sample of MA 956 after cyclic exposure for 3500 h at 1100°C.*

higher than for the short term exposure. The pattern of defects appears very similar to that displayed in Fig. 2. The spatial distribution of defects along the scale was non-uniform with defect-rich, relatively thick regions of scale appearing directly adjacent to nearly defect-free ones. Details of a defect-rich scale section are illustrated in Fig. 6. The first formed, 4–6 μm thick outer zone of the scale is composed of columnar grains with lengths of 3–5 μm and thicknesses of 1–2 μm. Beneath this zone, equiaxed grains become more prevalent and the density of physical and chemical defects is relatively high. In scale sections consisting predominantly of columnar grains the defect density was always rather low.

A plan view of the substrate after descaling (Fig. 7), reveals the presence of hollow-like defects at the substrate–scale interface. (Note: descaling was achieved by rupture of the sample.) These hollows, which were only rarely observed after 3500 h, had maximum lateral extensions of about 25 μm. They could be distinguished from the imprints of the Al_2O_3 scale crystals by virtue of their larger sizes and their smooth surfaces. The latter is indicative of a loss of contact between scale and substrate during the growth process. Micropores can be seen at the traces of scale grain boundaries. In the upper part of Fig. 7, residual oxide scale in association with a macropore is also noted.

4.2. As-received Condition

The as-received condition of the MA 956 sheet material was characterised by significant differences in the roughness and the morphology of the pre-existing Al_2O_3

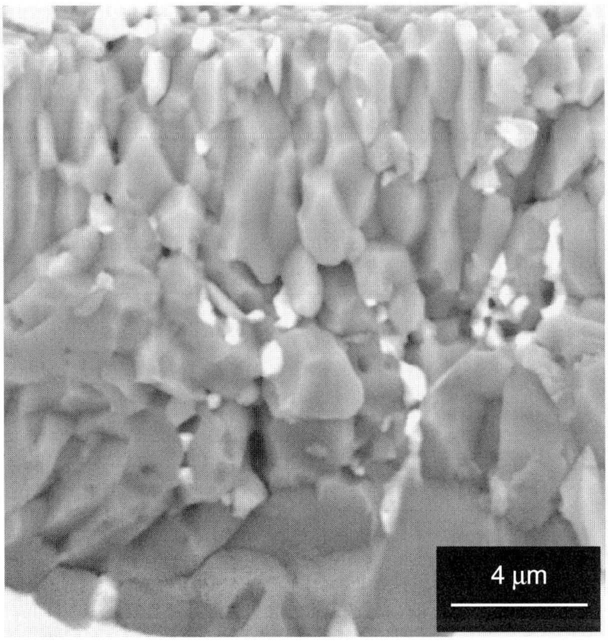

Fig. 6 *Fracture surface of the scale formed on an 800 grit finish sample of MA 956 after cyclic exposure for 3500 h at 1100°C.*

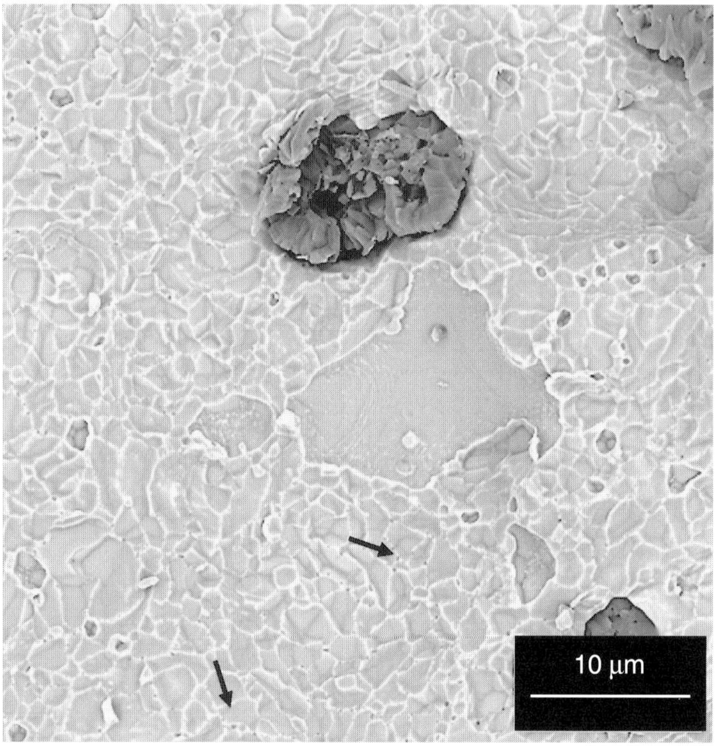

Fig. 7 Plan view of the alloy surface after descaling of an 800 grit finish sample of MA 956 after cyclic exposure for 3500 h at 1100°C. Arrows indicate micropores.

scale on the two sides (Fig. 8). One of the sides exhibited a high degree of roughness (see Table 1) with amplitudes of up to 15 μm and isolated, up to 25 μm deep craters with sharp edges and corners (Fig. 8a). These very probably resulted from the powerful impacts during the final Al_2O_3 grit blasting treatment. Some spallation of the original scale had also taken place. The scale remaining had a non-uniform thickness ranging up to 6 μm and a predominantly equiaxed grain structure with grains in the size range 1–3 μm. Partial intergranular decohesion of the grains in the scale was frequently observed. By contrast, the other side of the sheet exhibited a much smoother scale (see Table 1) and only very occasional spallation (Fig. 8b). This scale was 3–8 μm thick and was composed of equiaxed grains ranging in size up to 3 μm. Intergranular damage of the scale appeared to be limited to the outermost zone.

These initial differences in scale morphology on the two sides of the samples caused marked differences in the morphology, defect density and distribution, as well as the spallation sensitivity of the scales formed during intermediate and long term oxidation exposure. After intermediate exposure times under cyclic conditions the new scale on the more strongly blasted side of the samples generally had a lower defect density. On the other hand, it was much more prone to spallation. After an exposure of 370 h, the scales on both sides of the sheet exhibited alternating sections with high and low defect densities (Fig. 9). Defect-rich regions with their typical,

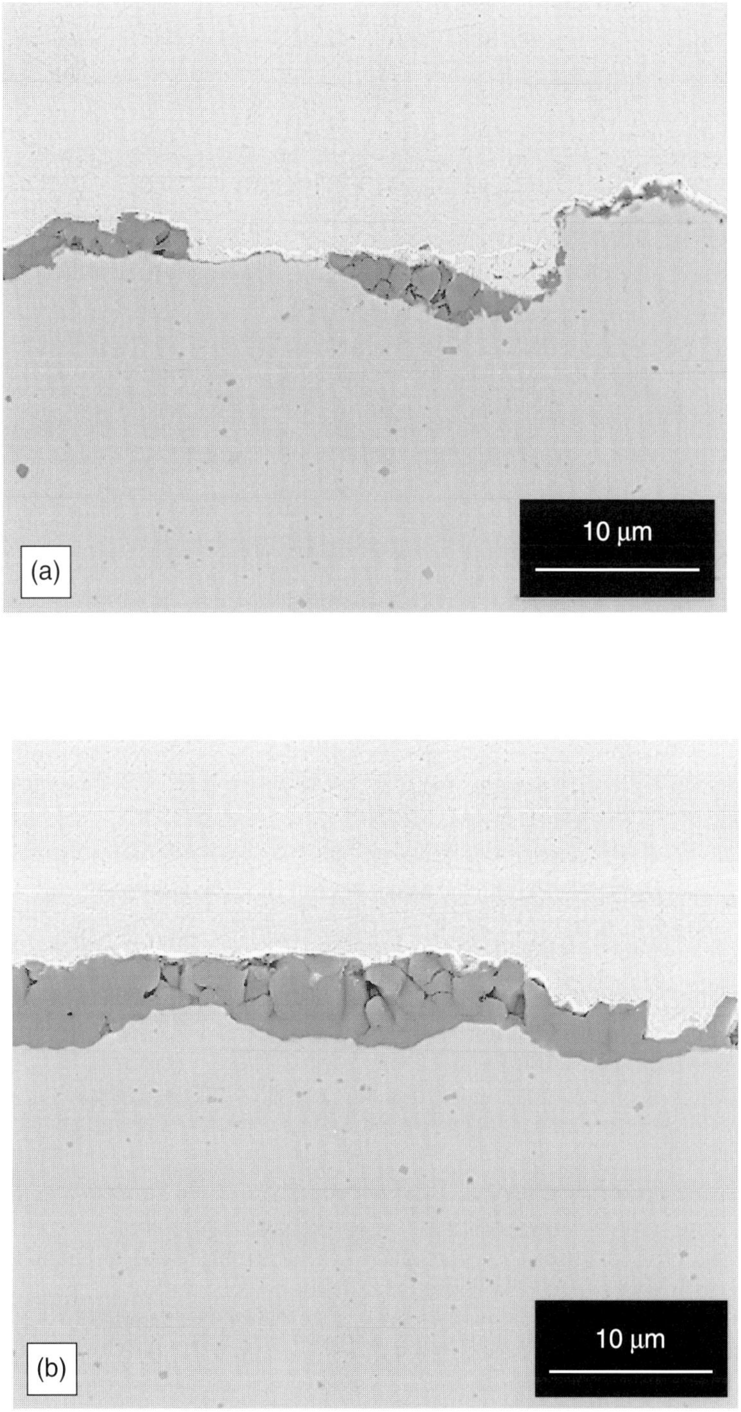

Fig. 8 *Oxide scales on (a) the strongly blasted and (b) weakly blasted sides of an MA 956 sheet in the as-received condition.*

equiaxed grain structures were clearly predominant for the scale on the weakly blasted side, but the scale adherence here was excellent (Fig. 9a). The defects observed were frequently quite large mid-size pores, partially or completely filled with Ti-rich oxides.

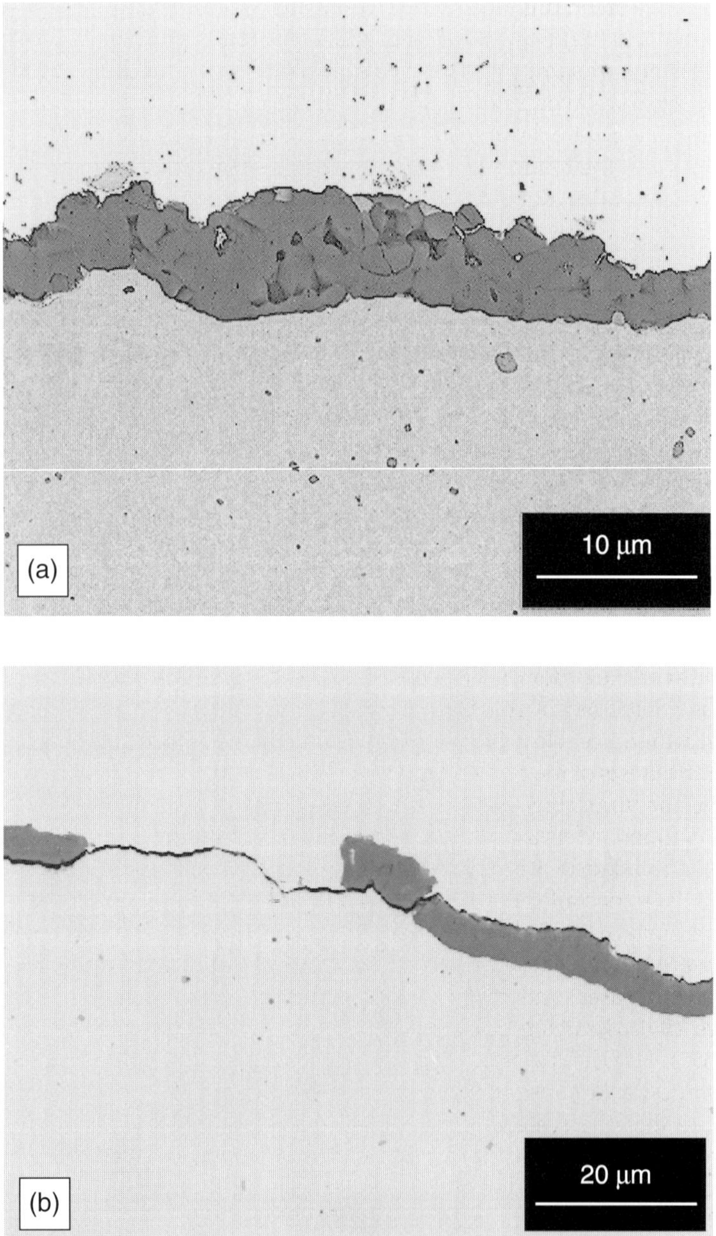

Fig. 9 *Oxide scale formed on (a) the weakly blasted and (b) strongly blasted sides of an as-received MA 956 sheet after 370 h at 1100°C: (b) documents the sensitivity to spallation.*

The presence of numerous Ti-rich particles at the scale–substrate interface suggests the existence of filled interfacial pores. At many places the scale reached total thicknesses of up to about 10 μm. The more strongly blasted side of the sheet (Fig. 9b), exhibited an approximately 5 μm thick scale with a much larger proportion of nearly defect-free regions with their typical, columnar grain structures. The sensitivity to scale spallation on this side of the sheet was high, especially on convex sections of the substrate (Fig. 9b). Scale adherence was particularly good on the concave sections.

Micrographs of samples with as-received surfaces exposed to 3500 h of cyclic oxidation are displayed in Figs 10 and 11. At this stage only occasional remnants of the original scale were detected as flakes on the surface. On the more strongly blasted side of the sheet the tendency to spallation was still greater for both the original and newly formed scales. Repeated descaling and re-healing caused marked differences in local scale thickness in the range 4–16 μm, with the thickest sections still retaining parts of the original scale (Fig. 10a). Double layer scales could form via growth under those scale sections which had experienced incomplete detachment (Fig. 10). The inner scales were either fully or partially separated from the outer layers. Whenever a new scale formed, the columnar grain morphology was preferred (Fig. 10b). Although the new scales were sometimes still rather thin, spallation could recur, especially at sharp, convex substrate locations (Fig. 11). However in concave substrate regions, e.g. at the crater seen in Fig. 11, scale sections with thicknesses up to 30 μm remained adherent even though highly defective.

4.3. Lathe Turned Condition

The surface topography of the lathe turned samples was characterised by a regular, wave-like distribution of ridges and valleys with amplitudes of about 20 μm and a wavelength of the order of 100 μm (see also Table 1). After short term exposures, the scale was composed of mainly columnar grains having a low defect density and excellent adherence. However, after exposures of only about 150 h, scale detachment began at the crests of the ridges. Figure 12 illustrates the typical finding that spallation took place at numerous convex sites whereas the scale remained adherent in the concave regions. After a few hundred hours, practically all the crests of ridges had lost their initial oxide. With increasing exposure time, spallation was also observed occasionally in the valley regions. Additional tests on samples with slightly less rough machined surfaces revealed that spallation was retarded. However, in this case, the valleys were often included in descaling and this led to spallation taking place over quite extensive areas rather than being localised.

Figure 13 illustrates the scale morphology on a lathe turned sample after cyclic exposure for 3000 h. The existence of scale sections with different thicknesses mirrors the spallation and subsequent re-healing behaviour. Figure 13 also demonstrates once more that the scale in a valley location can adhere well, even when highly defective. The chain-like arrangements of compositional defects associated with pores and microcracks oriented more or less parallel to the substrate surface reveal the former sites of incomplete detachment. The latter led to the formation of multiple oxide layers as illustrated in Fig. 14. Typically, the grain structure of the newly formed layers was predominantly columnar.

Details of the spallation process typical for convoluted scales are illustrated in

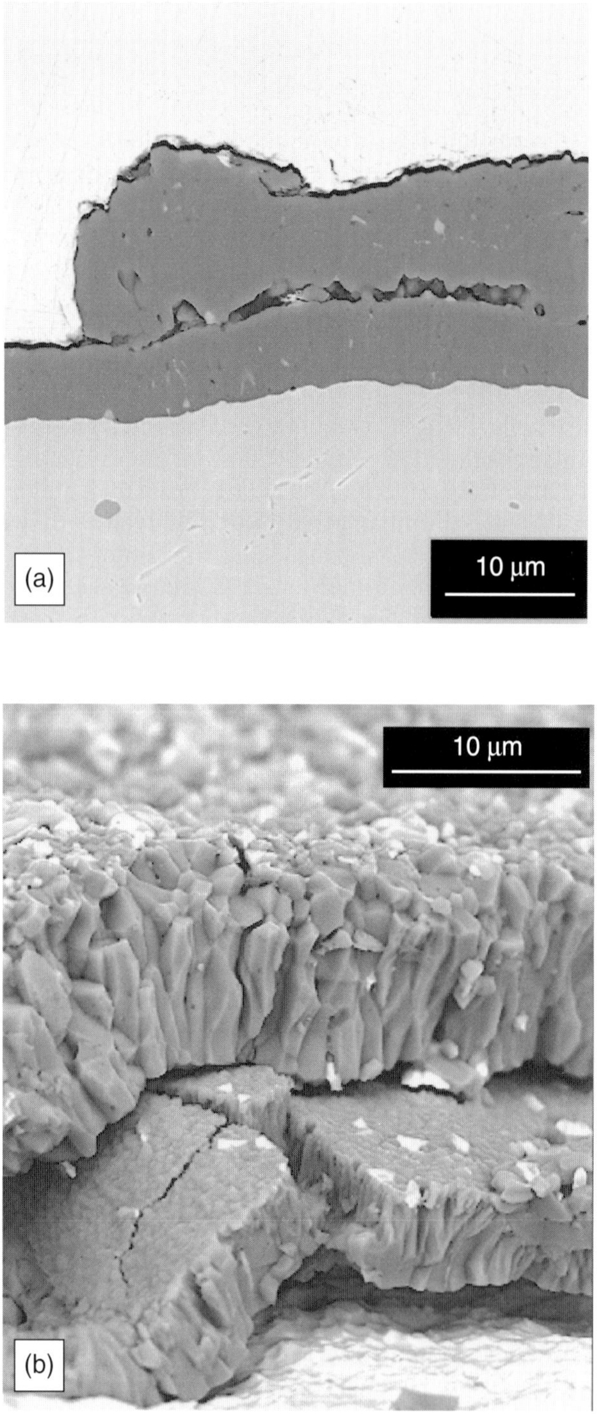

Fig. 10 *Multiple scales formed on as-received MA 956 samples after cyclic exposure for 3500 h at 1100°C. (a) Plan cross-section; (b) scale fracture surface.*

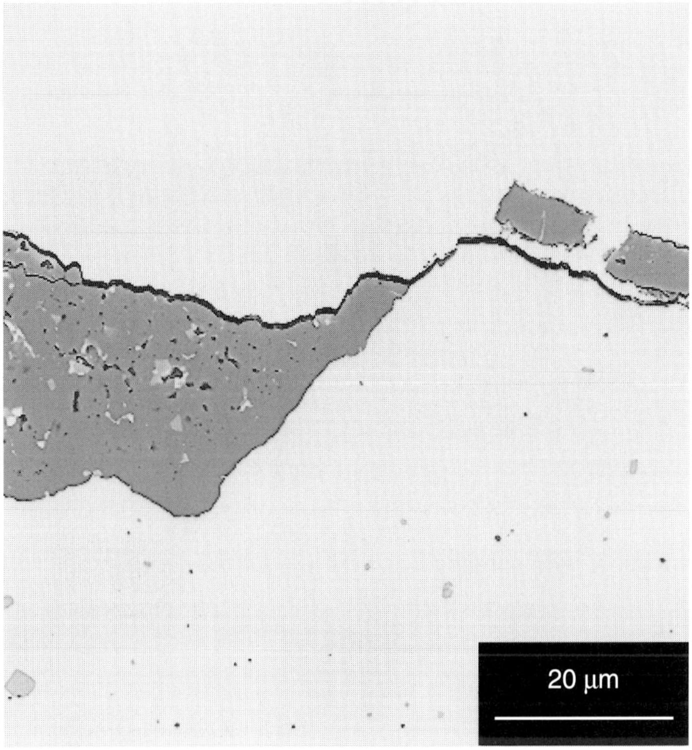

***Fig 11** Spallation of a newly formed scale from an as-received sample of MA 956 after cyclic exposure for 3500 h at 1100°C (right, convex region) and scale stability (left, concave region).*

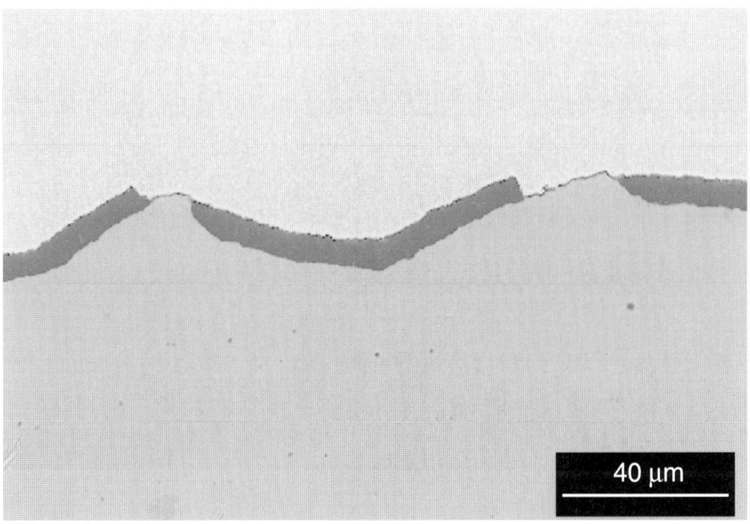

***Fig. 12** Scale formed on a lathe turned sample of MA 956 exposed for 250 h at 1100°C.*

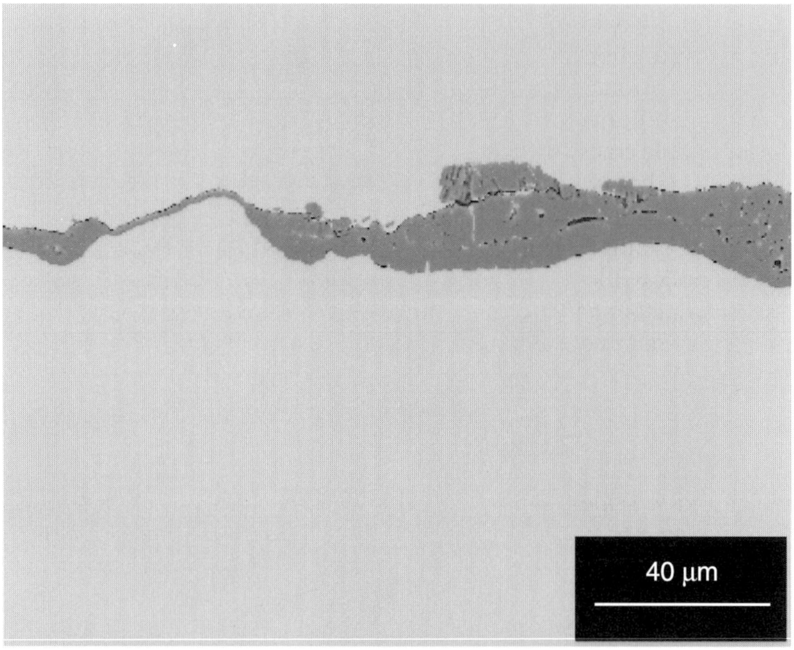

Fig. 13 *Scales formed on a lathe turned sample of MA 956 after cyclic exposure for 3000 h at 1100°C.*

Fig. 14 *Triple scale formed on lathe turned MA 956 after cyclic exposure for 3000 h at 1100°C. Note: the fracture plane of the uppermost layer is slightly inclined to the surface.*

Fig. 15. Lifting of the oxide layer was initiated on a convex section of the substrate, apparently in conjunction with interfacial pores (see arrows Fig. 15a). There was no evidence for prior through-scale cracking. At the apex of the scale a small, inwardly growing crack is indicated. A pronounced scale detachment stage is shown in Fig. 15(b). Through-scale cracking is observed at the crest of the ridge and on the flanks.

Examination of the substrate surfaces of the lathe turned samples after descaling revealed similar interfacial defects to those found for the other surface conditions investigated. The example shown in Fig. 16 demonstrates that at the crests of ridges the number of hollow-like defects can be rather high and their sizes relatively large, despite the fact that the spalled scale resulted from a short exposure time. In addition, a large number of relatively deep scale crystal imprints is observed.

5. Discussion

Spallation of oxide scales in connection with compressive (cooling) stresses is generally considered to take place either by a 'buckling' or a 'wedging' mechanism [5,7,23]. The buckling process is expected for relatively thin scales and moreover it requires an area of prior scale–substrate separation, e.g. in the form of interfacial pores. The critical in-plane compressive stress for scale buckling is given [8,24,25] as:

$$\sigma_c^b = \frac{1.22 \cdot E_{ox}}{1-\nu_{ox}^2}\left(\frac{s}{R}\right)^2 \tag{1}$$

where E_{ox} and ν_{ox} are the Young's modulus and Poisson's ratio of the oxide concerned (for alumina, 380 GPa [26] and 0.22 [27] respectively), s is the scale thickness and R the radius of the area of prior detachment. After buckling has initiated, spallation can take place by rupture of the scale as a consequence of the in-plane tensile stresses acting at the top and the periphery of the buckle. Disruption of the scale may be delayed until the buckle becomes sufficiently extended under the influence of the intensified stresses at the buckle front. The compressive stress in the scale needed to extend the buckle (unstable buckling) is given [8,25] as:

$$\sigma_c^{ub} \approx K_c^i\left[\frac{1+\nu_{ox}}{0.6}\right]^{1/2} \cdot \left(\frac{1}{s}\right)^{1/2} \tag{2}$$

where K^i_c is the critical stress intensity factor for the interface. (Note: K^i_c holds for mixed Mode I and II loading.) Buckling is facilitated on wavy surfaces because tensile stresses act perpendicular to the interface on the convex segments. For a sinusoidal geometry the maximum tensile stress at the top of a wave is approximately [8]:

$$\sigma_n^i \approx \frac{\Delta\alpha \cdot \Delta T}{z \cdot (1+r/s)} \tag{3}$$

with $z = (1+\nu_{ox})/2E_{ox} + (1-2\nu_s)/E_s$ and $\Delta\alpha = \alpha_s - \alpha_{ox}$.

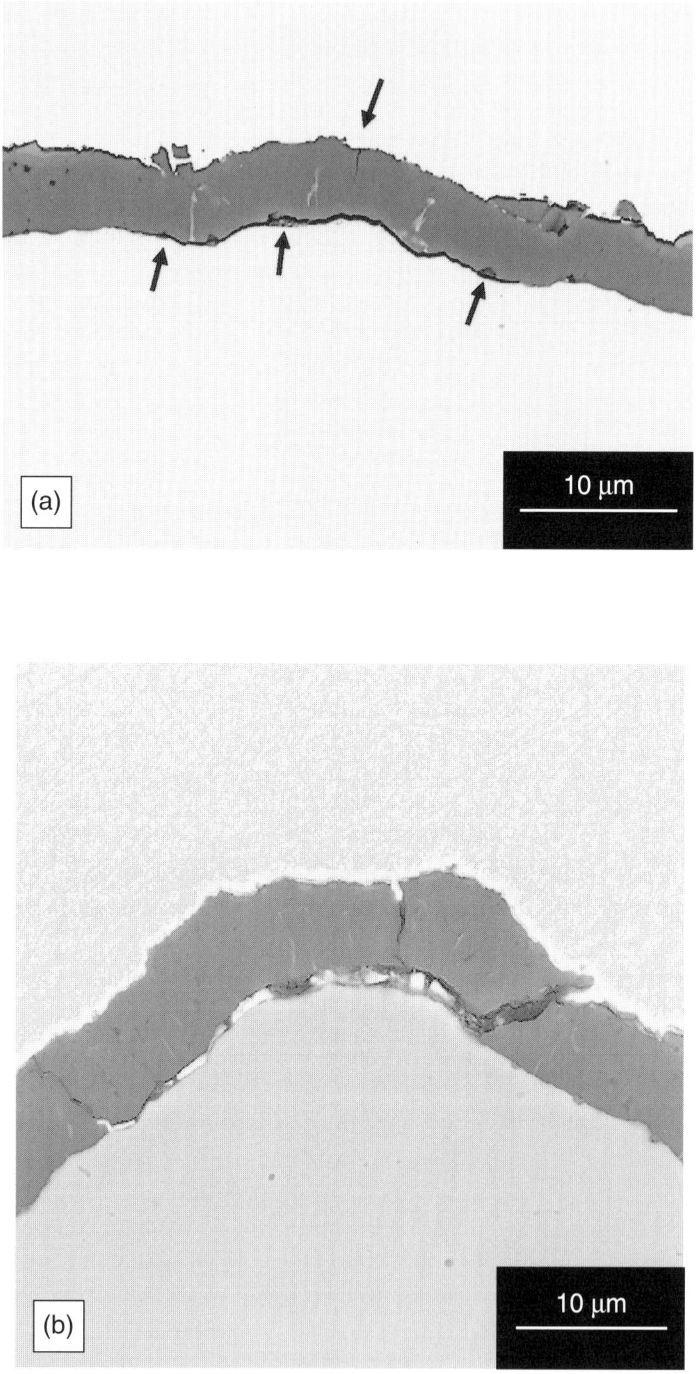

Fig. 15 *Evolution of spallation for lathe turned MA 956 after cyclic exposure for 3000 h at 1100°C. (a) Scale lift-off at the crest of a ridge; arrows indicate interfacial defects (bottom) and an arrested scale crack (top); (b) through-scale cracking and advanced detachment.*

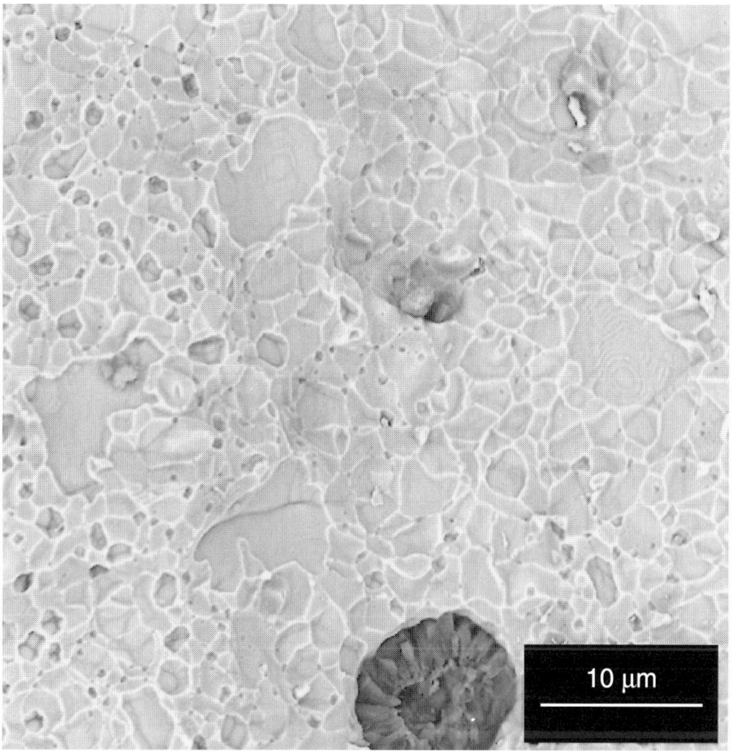

Fig. 16 *Plan view of the crest of a ridge after descaling of lathe turned MA 956 following cyclic exposure for 3000 h at 1100°C.*

The thermal expansion coefficients of the alloy substrate α_s and the alumina scale α_{ox} are taken as 15.5×10^{-6} K^{-1} [28] and 8.5×10^{-6} K^{-1} [27] respectively. ΔT corresponds to the temperature drop and E_s and ν_s refer to the Young's modulus and Poisson's ratio of the MA 956 substrate, having values of 270 GPa [28] and 0.3 respectively. The radius of the relevant curved area is denoted by r and the remaining terms s, E_{ox} and ν_{ox} are as defined in eqn (1).

Wedging, the second possible scale failure mechanism, is anticipated for relatively thick scales with rather defect-free interfaces. In this case, the compressive stresses in the scale must first generate through-scale shear cracks. These then trigger interfacial cracking and the scale segment concerned is pushed outwards. The critical compressive stress for wedging is given by [5,23,29] as:

$$\sigma_c^w = \left(\frac{E_{ox}}{1-\nu_{ox}}\right)^{1/2}\left(\frac{\gamma_F}{s}\right)^{1/2} \quad (4)$$

where γ_F is the energy needed to produce unit area of fracture surface. The other terms are already defined for eqn (1).

The present investigations have clearly demonstrated that buckling is the crucial scale failure mode for the wave-like surfaces of the as-received and lathe turned samples. It is very probable that the rarely observed scale failure for the 800 grit finish samples was also associated with buckling, although evaluation of the mechanism here was more difficult. Since the lathe turned samples exhibited a rather uniform, undulating surface topography the numerical analysis will be focused on this case.

On the ridges of the lathe turned samples, spallation generally began when the scale reached a thickness of about 5–6 μm regardless of the total exposure time, whereas in the valleys the scale remained adherent to larger scale thicknesses. Decohesion along the interface, in the absence of through-scale cracking, was the first step in the loss of scale integrity (Fig. 15a). Also visible in Fig. 15(a) is a thin scale crack, initiated by the in-plane tensile stress acting at the outer surface of the scale at the apex of the ridge, propagating towards the interface. Interfacial cracking started at or close to the crests of the ridges where the tensile stress acting perpendicular to the interface is a maximum. This cracking must have been initiated at defects such as the mid-size pores or the hollow-like craters. Although the observations appear to demonstrate that such defects play a role (e.g. Fig. 15a), it was difficult to identify a specific defect acting as a crack source. Interfacial cracks propagated downwards along the ridge flanks but beyond the curved region soon became arrested. This is consistent with the diminished tensile stresses acting normal to the interface as the curvature decreases. The interfacial cracks always remained very sharp and no evidence for crack blunting could be found. Moreover, during this first stage, the scale–substrate separation was rather small; even at the crest of the ridges it was < 1 μm. The process should therefore be characterised as slight decohesion rather than buckling in the usual sense. Similar cracking features have been observed at the onset of spallation of an Al_2O_3 scale formed on a NiCrAlY coating [18].

Taking this decohesion process to correspond to brittle crack propagation, as implied by the various observations, enables a critical stress intensity factor K^i_{Ic} for the interface to be calculated, according to [30]:

$$K^i_{Ic} = \sigma^i_n \sqrt{\pi \cdot r_p} \cdot \frac{M}{\sqrt{Q}} \tag{5}$$

where σ^i_n is the stress acting normal to the interface and r_p is the radius of an embedded interfacial defect. M and Q are the membrane stress correction factor and the defect shape parameter respectively; both may be taken as unity for the defect configuration here. With the observed values of scale thickness $s \approx 5$ μm and ridge curvature $r \approx 17$ μm, eqn (3) gives $\sigma^i_n \approx 550$ MPa. The appropriate defect size is that which is associated with the initial stages of the decohesion process. With $r_p \approx 2.0$ μm (see Fig. 15a), eqn (5) yields $K^i_{Ic} \approx 1.4$ MPa$\sqrt{m}$. A similar value of $K^i_{Ic} \approx 1$ MPa$\sqrt{m}$ has been reported for an Al_2O_3–NiCrAl interface [8]. For comparison, $K^i_{Ic} \approx 0.4$–1.0 MPa$\sqrt{m}$ was found for the fracture of alumina scales [21,31,32].

Following this initial process of slight scale decohesion, the next step involves crack formation in the scale in the vicinity of the interfacial crack tips, as a result of in-plane tensile stresses. The cracks in the scale were sometimes found to be arrested and this might be due to the stress gradient existing across the scale [16]. They may subsequently develop into through-scale cracks during reheating of the sample. Alternatively, incomplete scale detachment may lead to the formation of multiple oxide layers.

The propensity to spallation in the valley regions where compressive stresses act normal to the interface was much decreased (Figs 11 and 12). The large compressive stresses expected for highly convoluted surfaces may be the reason why the spallation tendency is lower for the latter compared to surfaces with intermediate roughness, as found both in the present work and in other investigations [14,19]. The thickness of the scale in valley regions was however limited by an oxide cut-off process involving cracks from neighbouring ridge flanks which, if oriented at small angles to the sample axis, could extend into the oxide in the valley. In contrast, at the scale–substrate interface within the valley region, spallation was very rarely observed.

For the rather flat, 800 grit finish samples spallation was very rare. In a few cases, decohesion over larger areas about 300 μm in diameter took place. Two observations suggest that the latter was also a result of a buckling process. First, the rupture surfaces of the through-scale cracks at the periphery of the spalled regions were only slightly inclined to the sample surface. Shear cracks would be expected to exhibit an orientation of about 45° corresponding to the plane of maximum shear stress. Secondly, cracks in the scale-substrate interface extended into the scale; no evidence was found for the initiation of interfacial cracking by through-scale cracks moving from the scale to the interface as required by wedging. On the other hand, for a typical scale thickness of 17 μm (= 5000 h exposure) and a characteristic interfacial hollow-like defect diameter of 30 μm, calculations indicate that buckling would require a much larger defect size or a higher scale in-plane compressive stress than are available. However, if the sample surface is not absolutely flat, tensile stresses will act perpendicular to the interface. The stress needed to cause extension of the defects present can be calculated by using the previously determined K^i_{Ic} value for the interface, (which should not differ substantially from that given above) and by taking a hollow-like interfacial defect with $r \approx 15$ μm as the source of spallation. From eqn (5), one obtains $\sigma^i_n \approx 175$ MPa. Equation (3) indicates that this stress could be generated at a surface with radius of curvature of about 220 μm, a value which does not seem unreasonable.

An essential feature of all the observations made is that buckling is most probably the operating scale failure mechanism. Cracking in the scale was clearly associated with tensile stresses, as indicated by the crack generation sites and the characteristic crack opening morphologies (Fig. 15b). Moreover, interfacial cracks formed at the apexes of the ridges without the assistance of through-scale cracks and propagated downwards along the flanks. For wedging, shear cracks in the scale should have initiated interfacial cracks at the ridge flanks which would have propagated along the ridge flanks in an upward direction [14,33]. On the other hand, spallation maps published for alumina scales on FeCrAlY [7] and NiCrAlY [33] substrates suggest that the conditions applying here would favour the wedging mechanism. It appears that it is the initial step necessary for wedging that is suppressed, i.e. the formation of shear cracks in the scale. These cracks have never been observed, possibly because the fracture toughness under shear conditions is too high.

6. Conclusions

Assessment of the role of surface pre-treatment on the integrity of alumina scales formed on MA 956 during oxidation in air at 1100°C leads to the following conclusions:

- For flat surfaces the scale exhibits very good adherence;
- Geometrical surface irregularities can cause a drastic deterioration in scale integrity;
- Scales on convex surfaces are prone to spallation whereas scale adherence in concave regions is good;
- Decohesion occurs preferentially along the scale–substrate interface;
- Loss of scale integrity is based on spallation via the buckling mechanism;
- The fracture toughness of the scale–substrate interface appears to be similar to that of the oxide scale; and
- For technical applications the use of controlled flat surfaces is recommended.

7. Acknowledgements

The authors would like to thank E. Conceicao, P. Frampton, J. Mendes, R. van Nieuwenhuyzen and K. Schuster of I. A. M. Petten for their technical assistance. This research was carried out within the European Commission's Research and Development Programme. One of the authors (F.H.) gratefully acknowledges the European Commission for the provision of a Research Training Grant.

References

1. M. Schütze, *Mater. Sci. Technol.*, 1988, **4**, 407–414.
2. J. Robertson and M. I. Manning, *Mater. Sci. Technol.*, 1990, **6**, 81–91.
3. P. Hancock and J. R. Nicholls, *Mater. High Temp.*, 1994, **12** (2/3), 209–218.
4. H. E. Evans, J. R. Nicholls and S. R. J. Saunders, *Solid State Phenomena*, 1995, **41**, 137–156.
5. H. E. Evans, *Int. Mater. Rev.*, 1995, **40**, 1–40.
6. J. P. Wilber, J. R. Nicholls and M. J. Bennett, Assessment of defects in alumina scales formed on ferritic ODS alloys, in *Microscopy of Oxidation III* (S. B. Newcomb and J. A. Little, eds). The Institute of Materials, London, UK, 1997, pp.207–220.
7. J. R. Nicholls, H. E. Evans and S. R. J. Saunders, *Mater. High Temp.*, 1992, **14**, 5–13.
8. A. G. Evans, G. B. Crumley and R. E. Demaray, *Oxid. Met.*, 1983, **20**, 193–216.
9. G. C. Chang, W. Phucharoen and R. A. Miller, *Surf. Coat. Technol.*, 1987, **32**, 307.
10. M. Schütze, Failure of oxide scales on advanced materials due to the presence of stresses, in *High Temperature Corrosion of Advanced Materials and Protective Coatings* (Y. Saito, B. Onay and T. Maruyama, eds). Elsevier Science, New York, USA, 1992, pp.39–49.

11. G. Petrus and B. L. Ferguson, in *Thermal Barrier Coatings Workshop*, NASA Conference Publication 3312, 251, 1995.
12. Z. Suo, *J. Mech. Phys. Solids*, 1995, **43**, 829–846.
13. R. L. Williamson, J. K. Wright and R. M. Cannon, in *Fundamental Aspects of High Temperature Corrosion*, ECS Proc. Vol. 96–26 (D. A. Shores, R. A. Rapp and P. Y. Hou, eds), p.16. The Electrochem. Soc., Pennington, NJ, USA, 1997.
14. J. Chao and J. L. Gonzalez-Carrasco, *Mater. Sci. Eng. A.*, 1997, **A230**, 39–48.
15. J. K. Wright, R. L. Williamson and R.M. Cannon, *Mater. Sci. Eng.* A., 1997, **A238**, 411–421.
16. J. K. Wright *et al.*, *Mater. Sci. Eng.A.*, 1999, **A262**, 246–255.
17. B. A. Pint, P. F. Tortorelli and I. G. Wright, *Mater. Corros.*, 1996, **47**, 663–674.
18. A. Strawbridge, H.E. Evans and C. B. Ponton, Modes of oxide spallation from MCrAlY overlay coatings, in *Microscopy of Oxidation — III* (S. B. Newcomb and J. A. Little, eds). The Institute of Materials, London, UK, 1997, pp.320–329.
19. P. Y. Hou, R. M. Cannon, H. Zhang and R. L. Williamson, Interface convolution and its effect on alumina scale spallation, in *Fundamental Aspects of High Temperature Corrosion*, ECS Proc. Vol. 96–26 (D. A. Shores, R. A. Rapp and P. Y. Hou, eds). The Electrochem. Soc., Pennington, NJ, USA, 1997, pp.28–40.
20. W. J. Quadakkers, *Werkst. Korros.*, 1990, **41**, 659–668.
21. V. Guttmann, F. Hukelmann and G. Borchardt, The role of defects on the mechanical integrity of alumina scales on MA 956, in *Proc. Conf. High Temperature Surface Engineering*, 23–25 Sept. 1997, Edinburgh, UK, in press.
22. T. Ramanarayanan, R. Ayer, R. Petkovic-Luton and D. P. Leta, *Oxid. Met.*, 1988, **29**, 445–472.
23. H. E. Evans, *Mater. High Temp.*, 1994, **12**, 219–227.
24. S. Timoshenko and J. M. Gere, *Theory of Elastic Stability*, 2nd Edn. McGraw Hill, New York, USA, 1961.
25. A. G. Evans and J. W. Hutchinson, *Int. J. Solid Struct.*, 1984, **20**, 455–466.
26. J. R. Nicholls, D. J. Hall and P. F. Tortorelli, *Mater. High Temp.*, 1994, **12**, 141–150.
27. R. Morell, *Handbook of Technical and Engineering Ceramics*, Part I, 95, HMSO, London, UK, 1985.
28. Data Sheet, Incoloy Alloy MA 956, Inco Alloys Ltd., Hereford, UK.
29. H. E. Evans and R. C. Lobb, *Corros. Sci.*, 1984, **24**, 209–222.
30. P. Hancock and J. R. Nicholls, *Mater. Sci. Technol.*, 1988, **4**, 398–406.
31. M. Schütze, *Mater. Sci. Technol.*, 1990, **6**, 32–38.
32. J. P. Banks, D. D. Gohil, H. E. Evans, D. J. Hall and S. R. J. Saunders, The room temperature mechanical properties of alumina scales, in *Materials for Advanced Power Generation 1994*, Part II (D. Coutsouradis *et al.*, eds). Kluwer Academic, Dordrecht, The Netherlands, 1994, pp.1543–1552.
33. H. E. Evans, A. Strawbridge, R. A. Carolan and C. B. Ponton, *Mater. Sci. Eng. A*, 1997, **A225**, 1–8.

3

The Effect of Water Vapour on the Oxidation of Alloys that Develop Alumina Scales for Protection

R. JANAKIRAMAN, G. H. MEIER and F. S. PETTIT
Department of Materials Science and Engineering, University of Pittsburgh, Pittsburgh, PA 15261, USA

ABSTRACT

Cyclic oxidation tests have been performed at 1100°C and 1000°C in wet and dry air on a number of alloys and coatings that form α-Al_2O_3 scales upon exposure to oxidising conditions. The alloys that were investigated included PWA 1480, PWA 1484, CMSX-4, diffusion aluminide coatings on PWA 1480 and PWA 1484, and Co–24Cr–10.5Al–0.3Y. In cases where some cracking and spalling of the alumina scales occurred in dry air, the presence of water vapour caused the degradation rate to be increased by a factor of two. When no cracking or spalling of the alumina occurred in dry air, as was the case for low sulfur alloys, water vapour had no effect on the oxidation behaviour. It is proposed that water vapour causes stress corrosion cracking at the Al_2O_3–alloy interface during cyclic oxidation.

1. Introduction

Water vapour is present in virtually every atmosphere to which high temperature materials are exposed. There are numerous examples that show water vapour affects the high temperature corrosion of metallic alloys and ceramics. Some effects that water vapour has been reported to cause during the oxidation of alloys are as follows:

- The presence of water vapour permits a H_2–H_2O mixture to be formed in voids which facilitates a rapid inward transport of oxygen across the pores via dissociative gas transport (an oxidation–reduction process) [1,2].
- Water vapour can affect the plasticity of oxide scales. Some investigators propose that the plasticity of oxide scales is decreased and, consequently, the spalling of scales is increased. [3] Other investigators have proposed that plasticity of oxide scales is increased [2,4] with increased oxidation rates [2] or improved adherence [4].
- The presence of water vapour adversely affects the selective oxidation of aluminium and chromium in iron base alloys [5] and in TiAl [6].
- Water vapour can affect transport through oxides such as SiO_2 by causing the network structure to be changed [7].

- Water vapour can cause the vapourisation of some oxides to be increased because of the formation of hydrated species with high vapour pressures [8].
- Water vapour causes the concentration of proton defects to be increased which can influence defect-dependent properties such as high temperature creep and diffusion [9,10].

Based upon data in the literature it is evident that, depending upon the alloy and the oxidising conditions, water vapour can affect oxidation in a number of different ways.

Most alloys and coatings that are developed for use above 1000°C rely upon the formation of an α-Al_2O_3 scale for oxidation protection. The influence of water vapour on the oxidation of alumina-forming alloys has not been extensively studied [5,11]. Kvernes *et al.* [5] did not observe any effect of water vapour on the isothermal oxidation of an Fe–13Cr–4Al* alloy at temperatures of 680, 820 and 980°C. At 980°C these authors showed that a continuous layer of α-Al_2O_3 was formed on this alloy. Buscail *et al.* [11] have found that water vapour decreases the isothermal oxidation rate of an Fe–21.5 Cr–5.6 Al alloy at 1000°C. The effect was not large. For example, k_p values of 2.4×10^{-13} and 2.7×10^{-13} $g^2\,cm^{-4}s^{-1}$ were obtained for water vapour pressures of 0.15 and 0.32 atm, respectively, vs 2.9×10^{-13} $g^2\,cm^{-4}s^{-1}$ for dry oxygen. These investigators proposed that the initial transient oxidation processes are affected by water vapour. Smialek [12] has shown that the scale adhesion of alumina is sensitive to water vapour and Sigler [13] has reported humidity induced spalling of alumina. Smith *et al.* [14] have also observed that humidity may play a role in the spalling of alumina from nickel base superalloys.

The results that are available indicate that water vapour does not have a substantial effect on the isothermal oxidation of alumina forming alloys and coatings, however, water vapour may have significant adverse effects on the cyclic oxidation of such alloys. This paper is concerned with the effect of water vapour on the cyclic oxidation resistance of alloys and coatings on which alumina scales are formed during oxidation.

2. Experimental

The cyclic oxidation experiments were performed on state of the art superalloys, namely, PWA 1480, PWA 1484 and CMSX-4. The compositions of these superalloys are presented in Table 1. All of these alloys have compositions such that α-Al_2O_3 scales are formed during oxidation at 1100°C after periods of some transient oxidation [15]. Aluminide coatings were also studied on the PWA 1480 and PWA 1484 alloys. One coating was formed using chemical vapour deposition (CVD) techniques to establish the aluminising conditions. The other involved a pack aluminising procedure. Scanning electron micrographs of these two coatings on PWA 1480 are presented in Fig. 1. The pack aluminide coating was thicker than that formed by the CVD technique if the alloy interdiffusion zones were neglected. The aluminium concentrations at the surface of the coatings were 61.8 and 42.8 at.% for the pack coating and CVD coating, respectively.

---

*All compositions are presented in mass percent unless noted otherwise.

Table 1. Compositions (mass%) of the alloys

Alloy Designation	Ni	Cr	W	Ta	Al	Co	Ti	Mo	Re	Hf	Y	Sulfur Concentration
PWA 1480	Bal.	10	4	12	5	5	1.5					5–8 ppm
PWA 1484	Bal.	5	6	8.7	5.6	10		2	3	0.1		5–8 ppm
PWA 1484 Desulfurised	Bal.	5	6	8.7	5.6	10		2	3	0.1		<1 ppm, sulfur removal by heating in hydrogen for 100 h at 1200°C
1 ppm sulfur PWA 1484 type (poly+cystaline)	Bal.	5	6	8.7	5.6	10		2	3	0.1		1 ppm prepared using special techniques to control sulfur concentration
CMSX-4	Bal.	6.4	6.3	6.5	5.7	9.5	1	0.6	2.9	0.1		3 ppm
CoCrA1Y		24			10.5	Bal.					0.3	

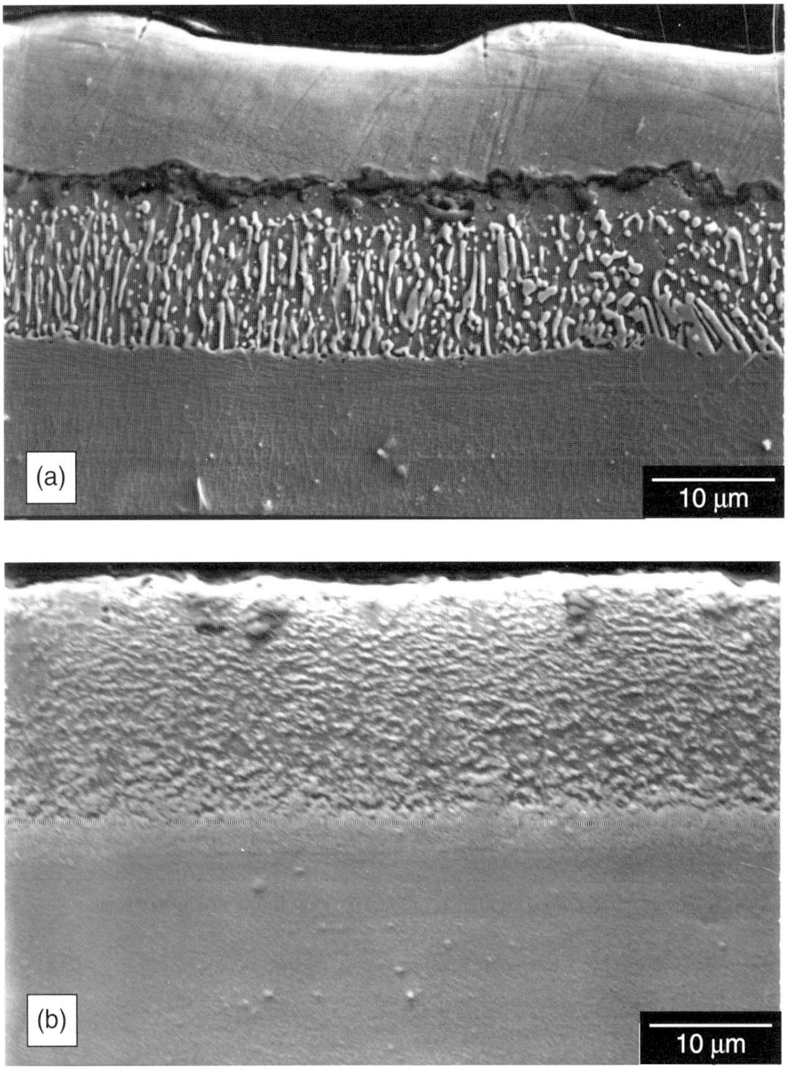

***Fig. 1** Scanning micrographs of PWA 1480 coated with aluminide by (a) chemical vapour deposition and (b) pack aluminising method.*

In cyclic oxidation tests the adherence of the alumina scales to the substrates is an important factor. The adherence of alumina scales on alloys is affected by the sulfur concentration in alloys as well as the concentrations of reactive elements such as yttrium [16–18]. The PWA 1480 and PWA 1484 alloys contained 5–8 ppm sulfur. The CMSX-4 alloy contained 3 ppm sulfur. In order to have alloys with very good adherence of alumina, a casting with essentially the same composition as PWA 1484 was prepared by Howmet using special techniques to control the sulfur concentration. This casting was not a single crystal, as PWA 1484 is usually fabricated, and the sulfur concentration was 1 ppm. This alloy is referred to as 1 ppm sulfur PWA 1484-type. In order to reduce the sulfur concentration in the PWA 1484 with 5–8 ppm sulfur, thin coupons of this alloy (~0.3 mm thick) were desulfurised by annealing in hydrogen for 100 h at 1200°C [19]. This treatment reduced the sulfur concentration to below 1 ppm. Finally, a CoCrAlY alloy with a nominal composition as indicated in Table 1 was prepared to investigate an alumina-forming alloy that contained the reactive element yttrium.

The cyclic oxidation specimens were coupons about 1 cm^2 and 3 mm thick except for the desulfurised specimens which were 1 cm^2 but 0.3 mm thick. The cyclic oxidation tests were performed in the apparatus shown in Fig. 2. The specimens were cycled in and out of the furnace once every hour, 45 min in the hot zone at 1100°C or 1000°C, 15 min in the cool zone at 100°C. The experiments were performed in dry air at 1 atm pressure with a dew point of –46°C ($p_{H_2O} = 6 \times 10^{-5}$ atm) and in wet air saturated with water at 46°C or 33°C (p_{H_2O} = 0.1 atm or p_{H_2O} = 0.05 atm). The water vapour pressure was achieved by passing dry air through water at a fixed temperature. At saturation the total pressure was one atmosphere with a partial pressure of H_2O fixed by the temperature of the water bath. The specimens were moved from the hot zone to the cool zone, or vice versa, in 250 s. The specimens were removed periodically from the apparatus and mass change measurements were determined along with visual examination of the specimens. On conclusion of the tests the specimens were examined in detail using optical metallography, scanning electron microscopy and X-ray diffraction.

Acoustic emission measurements were performed on some specimens in an

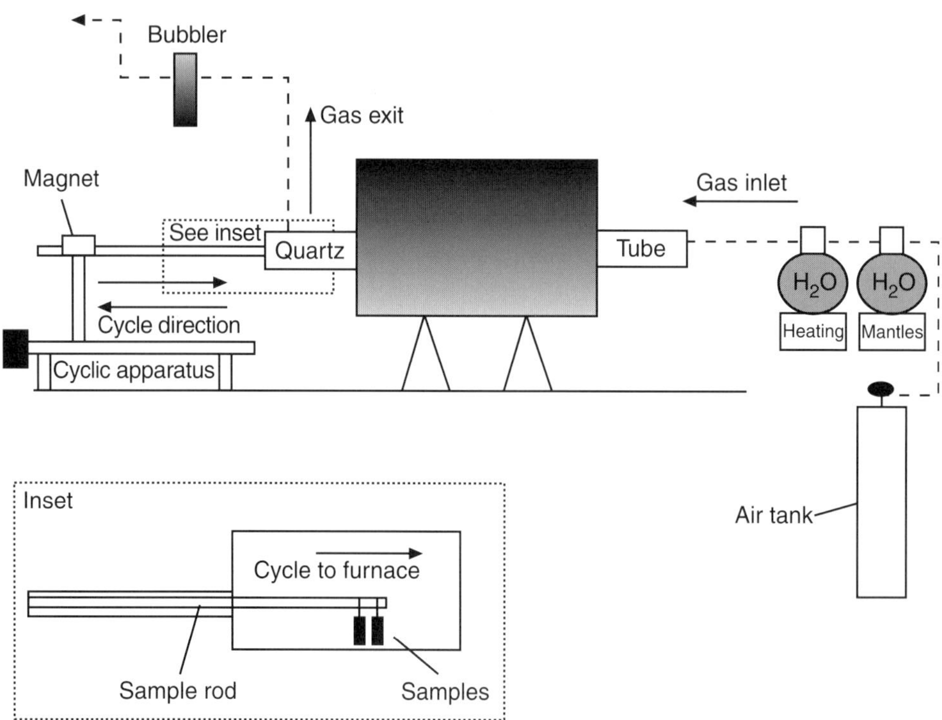

***Fig. 2** Schematic diagram to show the apparatus for cyclic oxidation of coupon specimens in gases with defined and controlled amounts of water vapour.*

attempt to determine if cracking of the oxide scales occurred during isothermal and cyclic oxidation. A platinum wire wave guide was attached to specimens to transmit acoustic signals from the specimens to a piezoelectric transducer which was part of a Dunegan/Endevco 3000 series monitoring system. These measurements were performed during isothermal oxidation and as specimens were removed from the furnace at a controlled speed.

Smialek [12,17] has shown that by repeated oxidation and polishing of NiCrAl specimens the sulfur content of the specimens can be lowered from 10 to 2 parts per million by mass since sulfur segregates to the alloy–oxide interface. This procedure was used to reduce the sulfur concentration of PWA 1484 by polishing from 320 grit through 600 grit silicon carbide papers followed by rinsing in methanol.

A water droplet test was also used to examine the effect of water on the cracking and spalling of oxide scales. This test consisted of adding a drop of water (~0.05 cm^3) to the surface of a specimen after it had cooled to room temperature. Surface photographs before and after adding the water droplet were used to document the amount of oxide spalling.

3. Results and Discussion

In cyclic oxidation tests spalling of the oxide scales occurs due to the development of growth and thermally induced stresses. As discussed by Evans [20] there are a number of factors, in addition to the environment, that can affect the oxide spalling process. The growth stresses are affected by the thickness of the oxide scale and by the thickness of the oxidation specimens as well as by creep relaxation in the specimen. The thermally induced stresses are affected by the differences in the coefficients of thermal expansion of the oxide scale and the metallic substrate, the total temperature change and the rate of temperature change. In this investigation all of these factors were held constant for a given alloy system and the results were used to compare alumina spalling in dry and wet conditions as a function of oxidation time or increasing oxide scale thickness.

3.1. Uncoated Superalloys Oxidised in Wet and Dry Air

Mass change (Δm) vs time data for the cyclic oxidation of PWA 1480 and PWA 1484 at 1100°C in wet (p_{H_2O}= 0.1 atm) and dry air are presented in Fig. 3. The mass losses are caused by the spalling of oxide and this spalling is more severe for both alloys in wet compared to dry air. To compare the effects of water vapour the times of cyclic oxidation to produce the same amount of mass loss were compared. It was not possible to use the same amount of mass loss for all of the alloys tested because the rates of mass loss varied substantially. A parameter $\phi_{\Delta m} = t_{dry}/t_{wet}$ was used to compare the effects of water vapour where Δm was used to indicate the value of the mass loss (mg cm^{-2}) at which the comparison was made. The results obtained for ϕ for PWA 1480 and PWA 1484 are presented in Table 2 and were 2.3 and 3.7, respectively, for Δm = –10 mg cm^{-2}. The time required to produce the same mass loss differed by about a factor of two, namely, it required about twice the time of cyclic oxidation in dry air to produce the same mass loss as that in wet air (p_{H_2O} = 0.1 atm).

Table 2. *The ratio of times to produce a given amount of oxide spalling,* $\phi_{\Delta m} = t_{dry}/t_{wet}$

Alloy	Δm	$\phi_{\Delta m}$
PWA 1480	$\Delta m = -10$ mg cm^{-2}	2.3
PWA 1484	$\Delta m = -10$ mg cm^{-2}	3.7
CMSX-4	$\Delta m = -10$ mg cm^{-2}	2.4
PWA 1480/CVD diffusion aluminide	$\Delta m = -10$ mg cm^{-2}	1.9

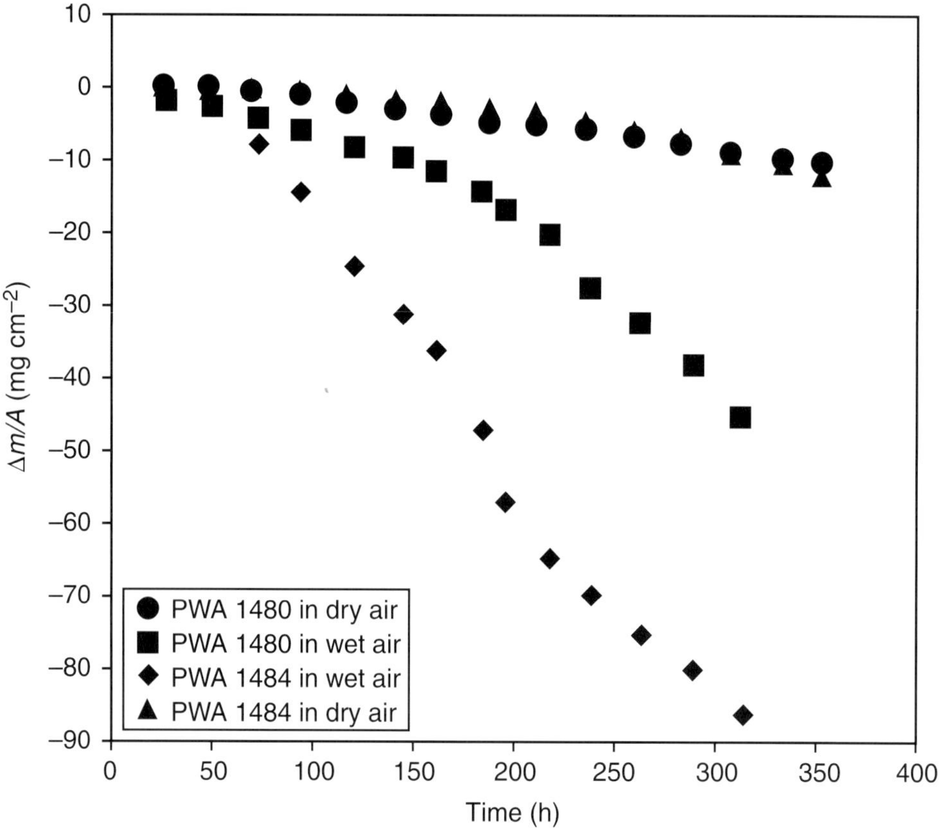

Fig. 3 *Mass change vs time data for the cyclic oxidation of PWA 1480 and PWA 1484 in wet* ($pH_2O = 0.1$ *atm) and dry air at 1100°C.*

The results obtained for cyclic oxidation of CMSX-4 at 1100°C are presented in Fig. 4. The mass losses for CMSX-4 in both wet ($p_{H_2O} = 0.1$ atm) and dry air were less than those observed for PWA 1480 and PWA 1484, Fig. 3. It is believed that these smaller mass losses were due to the lower sulfur concentration in CMSX-4 (3 ppm) which resulted in a more adherent oxide scale being formed on this alloy. Nevertheless, cracking and spalling of oxide did eventually occur on CMSX-4 (Fig. 4), and the mass losses were greater in wet than dry air. The value for $\phi_{\Delta m}$ was 2.4 where the value of Δm was –0.4 mg cm^{-2}.

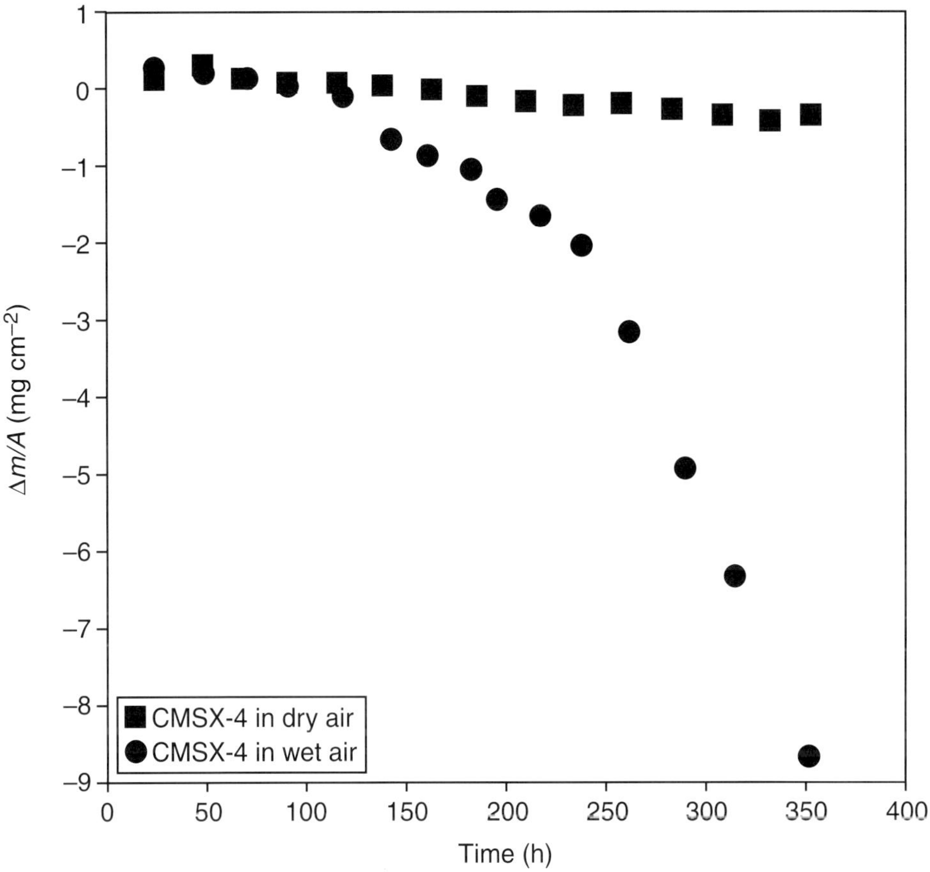

Fig. 4 Mass change vs time data for the cyclic oxidation of CMSX-4 in wet (p_{H_2O} = 0.1 atm) and dry air at 1100°C.

As mentioned previously, PWA 1480, PWA 1484 and CMSX-4 are alumina-formers. A continuous external scale of α-Al_2O_3 develops on these alloys after a transient period of oxidation during which other oxides such as NiO and Cr_2O_3 are formed. The presence of an α-Al_2O_3 scale on the alloys was confirmed using X-ray diffraction. In cyclic oxidation the oxide scales spall and the oxide formed on subsequent cycles becomes more enriched in Cr_2O_3 and NiO as an aluminium depleted zone develops in the alloys. Scanning electron micrographs comparing the depleted zones in CMSX -4 after 350 h at 1100°C of cyclic oxidation are presented in Fig. 5. The depleted zone that developed in wet air (p_{H_2O} = 0.1 atm), as indicated by the absence of the γ′ phase, is substantially larger than that for cyclic oxidation in dry air.

A limited amount of cyclic oxidation was performed at 1100°C using a water vapour pressure of 0.05 atm, and at 1000°C using a water vapour pressure of 0.1 atm. In Figs 6 and 7, results are presented to compare the cyclic oxidation of PWA 1484 and CMSX-4 at 1100°C with two different water vapour pressures. It can be seen that the cyclic oxidation of both alloys is affected by the water vapour pressure. However, the results are surprising, since in the case of PWA 1484 a water vapour pressure of 0.05 atm does not affect the cyclic oxidation process, whereas in the case of CMSX-4

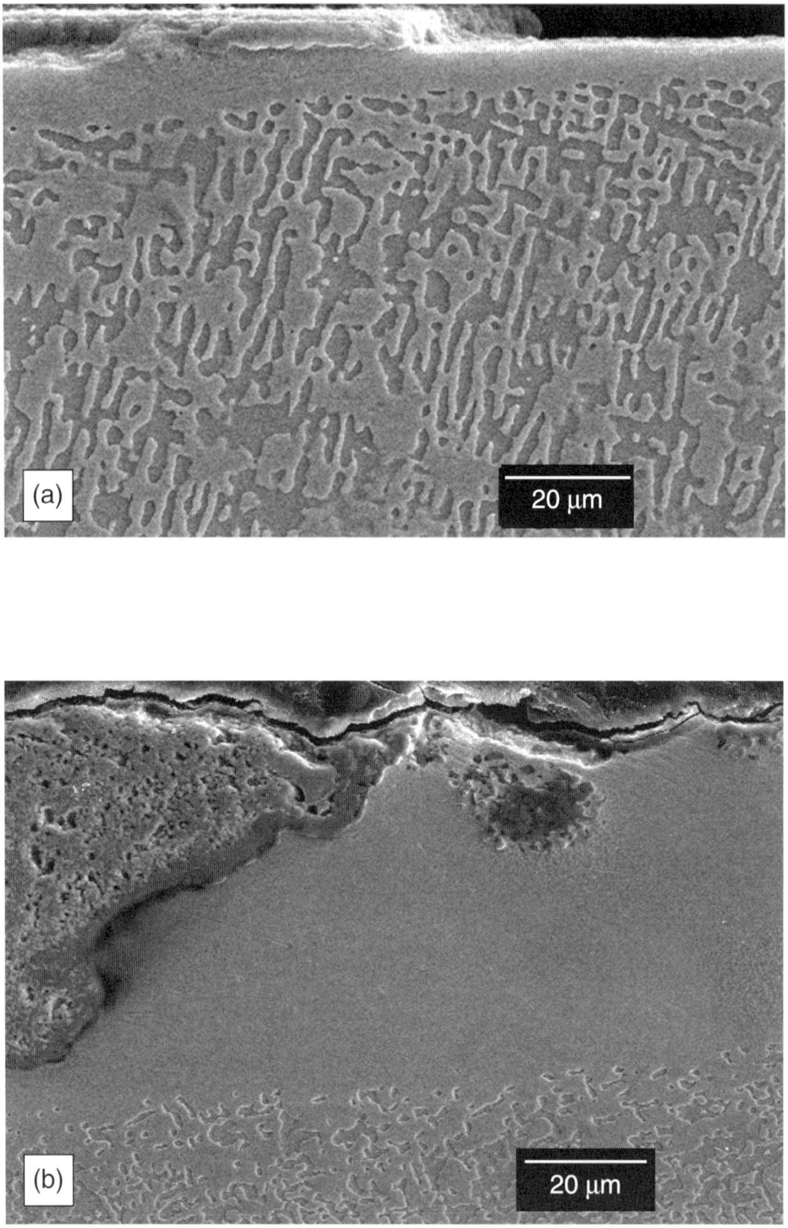

Fig. 5 *Scanning micrographs of CMSX-4 after 350 h of cyclic oxidation at 1100°C in (a) dry air and in (b) wet* (pH_2O *= 0.1 atm) air.*

a water vapour pressure of 0.05 atm increases the cyclic oxidation as much as a water vapour pressure of 0.1 atm. In view of the results obtained with a water vapour pressure of 0.05 atm, these experiments were repeated and extended to longer times. The results are presented in Fig. 8 and these results are in agreement with the data presented in Figs 6 and 7.

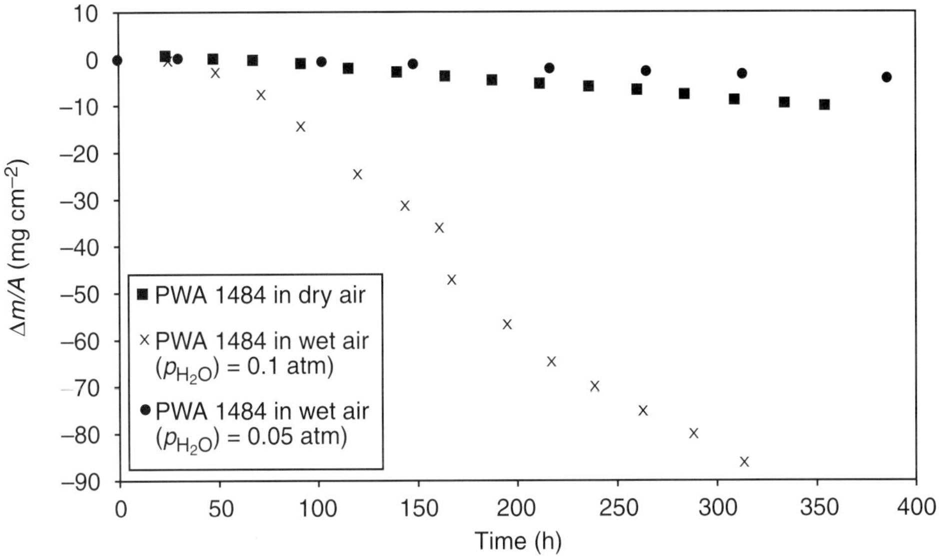

Fig. 6 Mass change vs time data for cyclic oxidation of PWA 1484 in dry air and in wet air at two different water vapour pressures at 1100 °C.

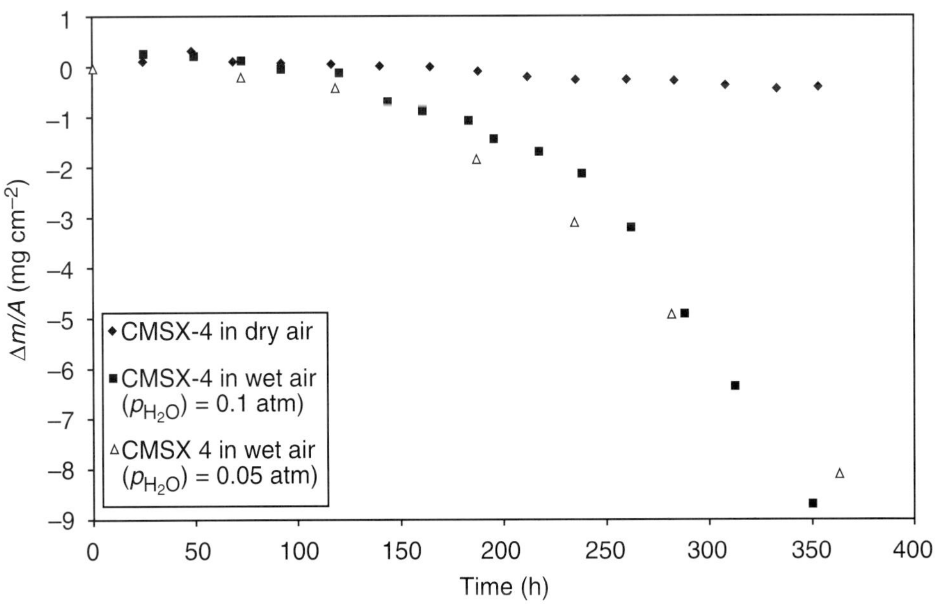

Fig. 7 Mass change vs time data for cyclic oxidation at 1100 °C of CMSX-4 in dry air and in wet air at two different water vapour pressures.

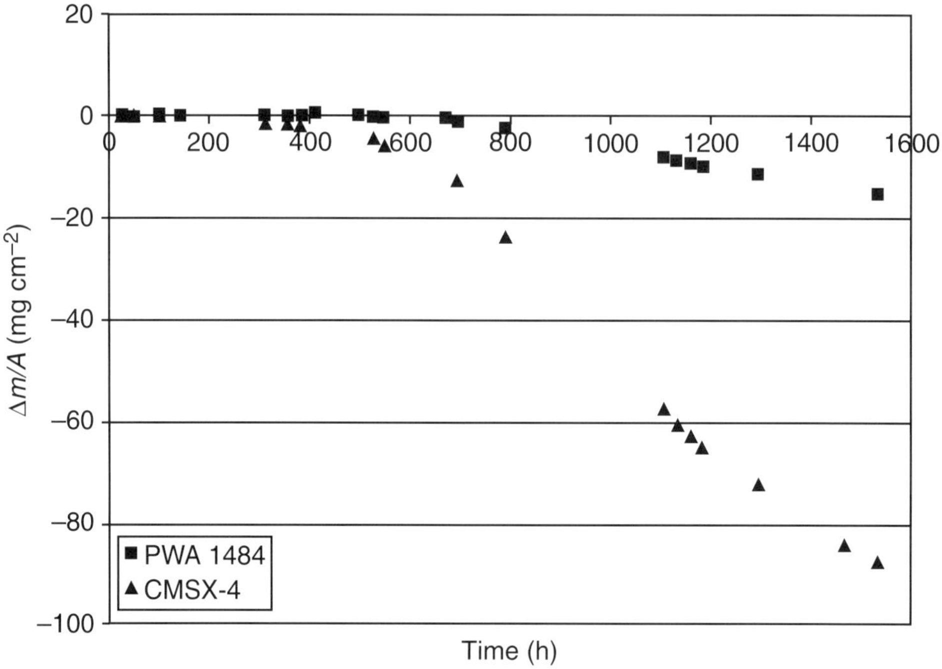

Fig. 8 *Cyclic oxidation data obtained for PWA 1484 and CMSX-4 at 1100°C in wet air (p_{H_2O} = 0.05 atm).*

Cyclic oxidation data obtained for CMSX-4 at 1000°C with p_{H_2O} = 0.1 atm are presented in Fig. 9. The mass changes are very small but do not show any evidence for substantial cracking and spalling of the oxide scale. In contrast to results obtained for this alloy at 1100°C, the water vapour does not appear to have influenced the cyclic oxidation behaviour.

3.2. Uncoated PWA 1484 with Low Sulfur Concentrations

In the case of cyclic oxidation of the desulfurised PWA 1484 and the 1 ppm sulfur PWA 1484 type alloys, no spalling of oxide occurred in either wet (p_{H_2O} = 0.1 atm) or dry air at 1100°C after 350 h of cyclic oxidation (Fig. 10). Some cyclic oxidation tests were performed for 3000 h (Fig. 11). No cracking or spalling was observed in wet (p_{H_2O} = 0.1 atm) or dry air. However, the mass gain was larger for the specimen tested in wet air after about 1500 h of cyclic oxidation (Fig. 11), and the γ' denuded zone was larger for the specimen tested in wet air (Fig. 12). In both cases the oxide adjacent to the alloy was α-Al_2O_3. Transient oxides of NiO, Cr_2O_3 as well as tantalum oxides were above the alumina scale. The reason for the lower mass gains of the desulfurised specimen in dry compared to wet air after 1500 h of cyclic oxidation is not known. The important point, as far as the present paper is concerned, is that no spalling of the alumina scales was evident in wet or dry air.

No isothermal oxidation experiments have been performed as yet to determine if water vapour affects the isothermal growth of alumina. In Fig. 13 mass changes

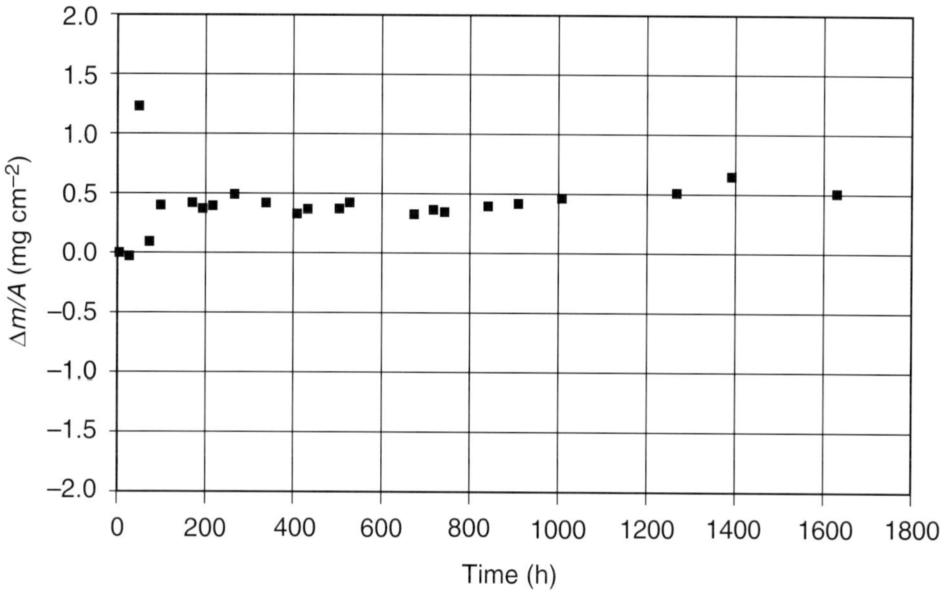

Fig. 9 *Mass change vs time for cyclic oxidation of CMSX-4 at 1000°C in wet air (p_{H_2O} = 0.1 atm).*

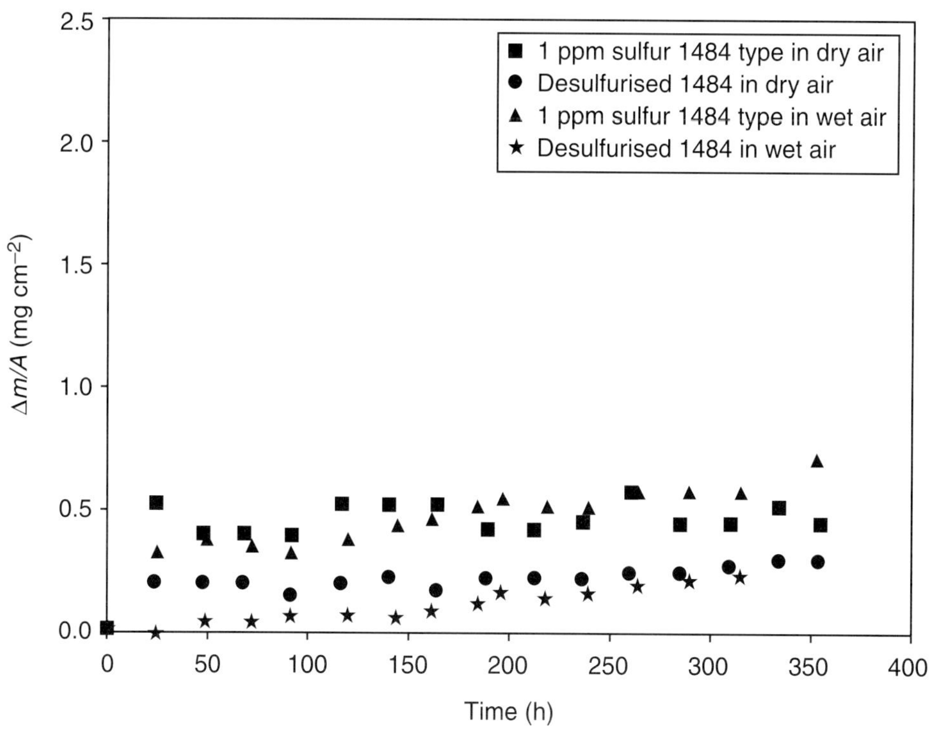

Fig. 10 *Mass change vs time data for the cyclic oxidation of desulfurised and low sulfur PWA 1484 in wet (p_{H_2O} = 0.1 atm) and dry air at 1100°C.*

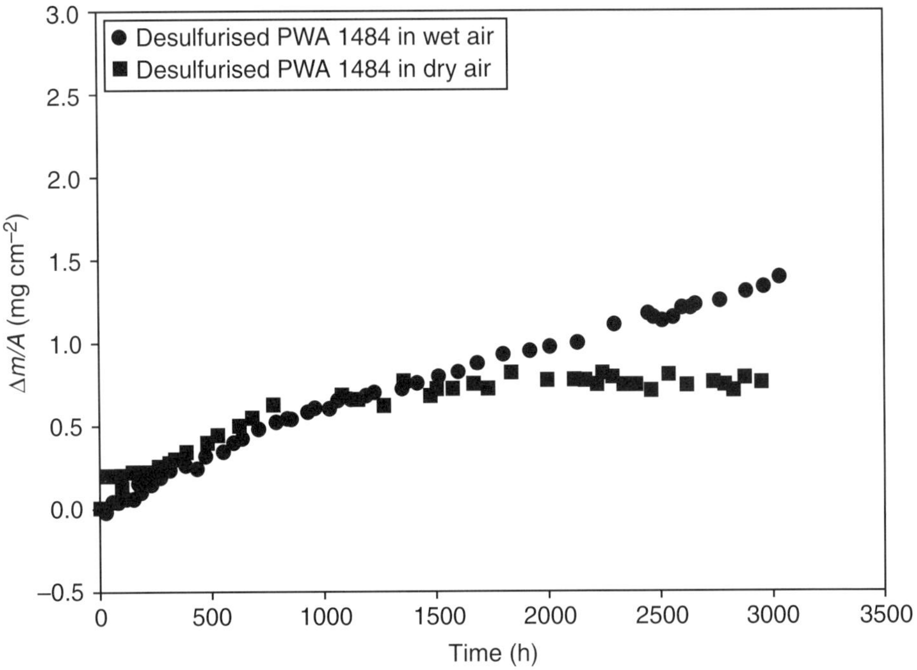

***Fig. 11** Mass change vs time data for the cyclic oxidation of desulfurised PWA 1484 alloy in wet (p_{H_2O} = 0.1 atm) and dry air at 1100°C.*

obtained for the two low sulfur alloys are plotted vs the square root of the oxidation time. Even though the test was cyclic and in water vapour no oxide spalling occurred. These data conform to a parabolic rate law with rate constants of 1.9 and $3.1 \times 10^{-13} g^2\ cm^{-4} s^{-1}$. These rate constants have values similar to that for isothermal oxidation. These results, which are consistent with data in the literature [5,21], indicate that water vapour does not have a substantial effect on the isothermal oxidation of alumina forming alloys.

3.3. Coated Superalloys

Mass change vs time measurements for cyclic oxidation of the two aluminide coatings on PWA 1480 at 1100°C in dry and wet air (p_{H_2O} = 0.1 atm) are presented in Fig. 14 and 15. Less cracking and spalling of oxide occurred in dry air for both the pack and CVD aluminides. Similar results were obtained for these two coatings on PWA 1484. The mass losses occurred at a smaller rate initially for the pack aluminide coatings on both PWA 1480 and PWA 1484. The larger coating thickness and the higher aluminium concentration of the aluminide coating formed via pack aluminising compared to the CVD coating may account for the observed differences in mass loss. After the coatings have been depleted of aluminium the rates of mass loss begin to approach one another.

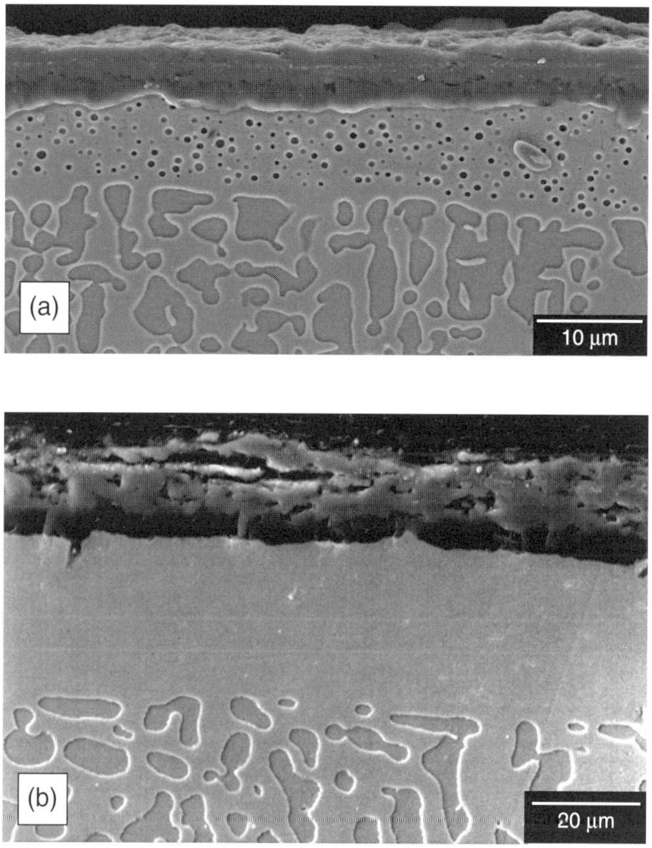

Fig. 12 Scanning electron micrographs of desulfurised PWA 1484 after 3000 h of cyclic oxidation in (a) dry and in (b) wet (p_{H_2O} = 0.1 atm) air at 1100°C.

3.4. CoCrAlY Specimens

Mass change measurements for the cyclic oxidation of the CoCrAlY alloy in wet (p_{H_2O} = 0.1 atm) and dry air at 1100°C are compared in Fig. 16. Cracking and spalling of the oxide scale was evident on the specimen tested in wet air after about 1000 h. The sample tested in dry air did not exhibit any mass loss, however, a very small amount of oxide cracking and spalling was evident on visual examination. This spalling was predominantly at the edges of specimens. Small amounts of yttrium in CoCrAl, NiCrAl and FeCrAl alloys normally result in extremely adherent alumina scales. While the cracking and spalling observed in the CoCrAlY specimen was not large, this alloy may not be representative of the optimum results that can be produced with yttrium. The yttrium concentration was 0.3mass%. Better results are obtained with lower values of 0.05 to 0.10mass%. At the higher levels of yttrium (e.g. 0.15–0.3mass%) yttrides are formed which preferentially oxidise to form stringers of yttria in the alumina scale and within the alloy itself.

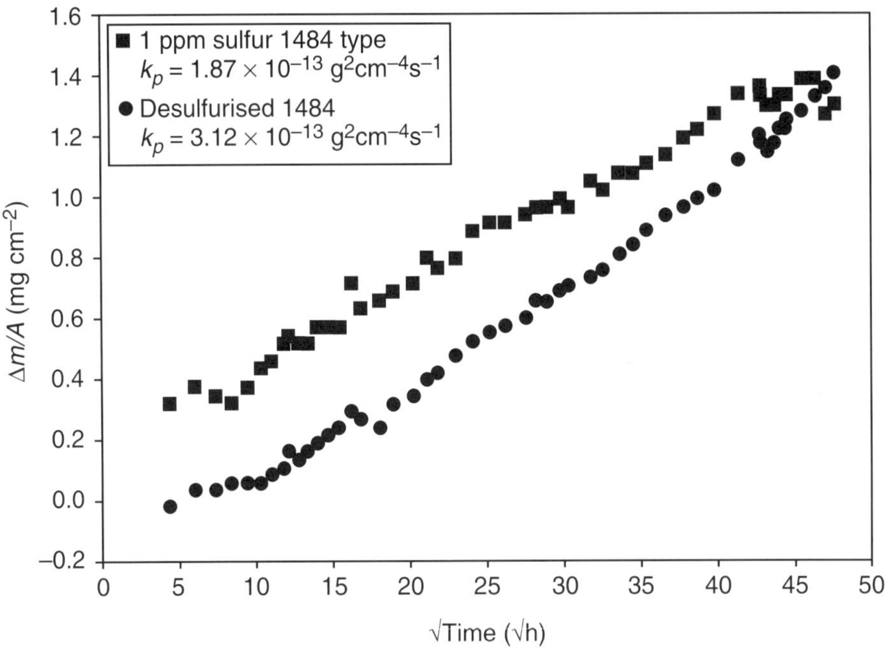

Fig. 13 *Mass change vs the square root of time for the cyclic oxidation of desulfurised and 1 ppm sulfur PWA 1484 type in wet air (p_{H_2O} = 0.1 atm) at 1100°C. The data conform to a parabolic rate law with rate constants as shown in the figure.*

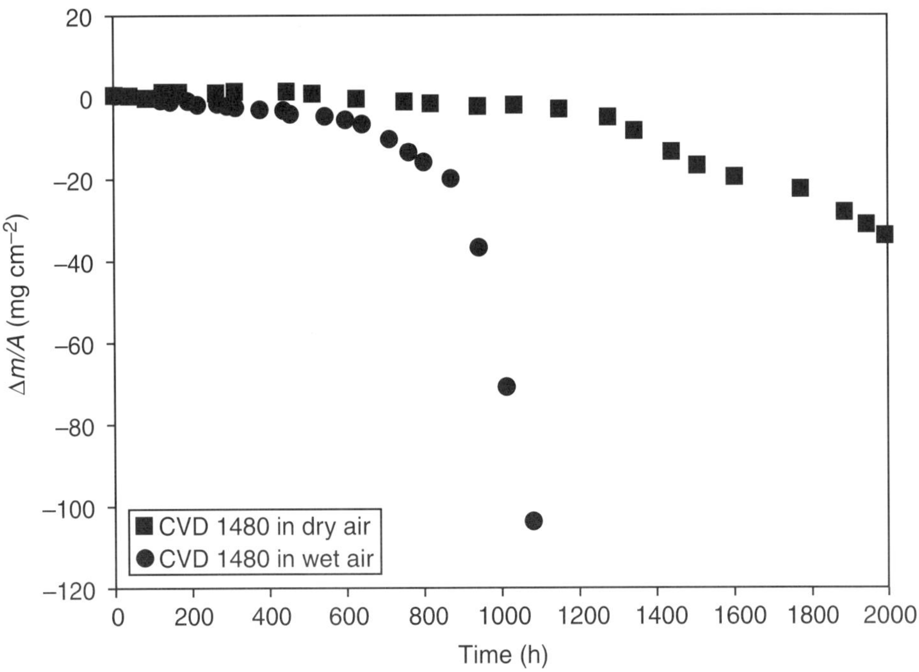

Fig. 14 *Mass change vs time data for the cyclic oxidation of aluminide coated 1480 (by CVD method) in wet (p_{H_2O} = 0.1 atm) and dry air at 1100°C.*

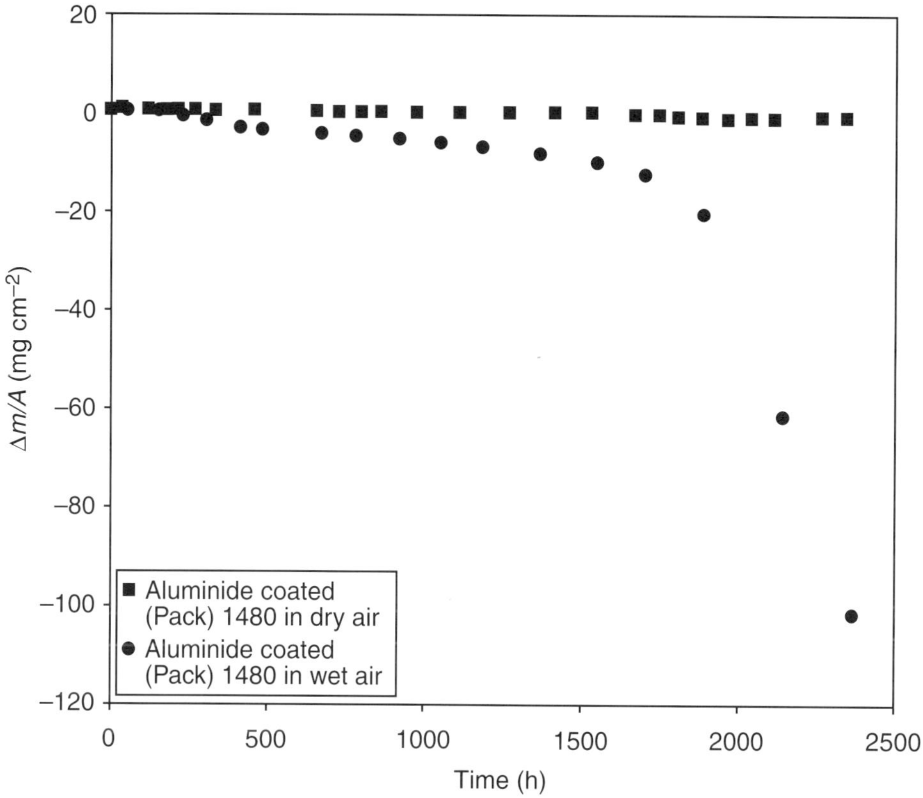

Fig. 15 *Mass change vs time data for the cyclic oxidation of aluminide coated (by Pack Al method) PWA 1480 in wet (p_{H_2O} = 0.1 atm) and dry air at 1100°C.*

3.5. The Effects Produced by Water Vapour

The results clearly show that water vapour does cause the cracking and spalling of α-Al_2O_3 to be more severe than in dry air. Water vapour (p_{H_2O} = 0.1 atm) causes mass losses to be increased by a factor of 2 to 4 compared to dry air. However, such a condition was not observed for extremely adherent alumina scales. It appears that when cracking and spalling occur in dry air, then water vapour causes the cracking and spalling to be more severe. When cracking does not occur in dry air, water vapour does not appear to have any influence on cracking and spalling of the alumina scales. Since the oxidation resistance of alloys is related to the times over which α-Al_2O_3 scales are developed on their surfaces, water vapour causes the lives of such alloys and coatings to be decreased. Since no effects of water vapour have been observed as yet in the low sulfur alloys, the importance of such alloys is great. Reactive elements such as yttrium may also be capable of negating the effects of water vapour, but the reactive element will have to be present at concentrations and distributions which are optimised to produce adherent alumina scales with no cracking or spalling.

In order to determine if water vapour is affecting the stresses in alumina scales, the residual stresses in α-Al_2O_3 scales formed on desulfurised PWA 1484 were

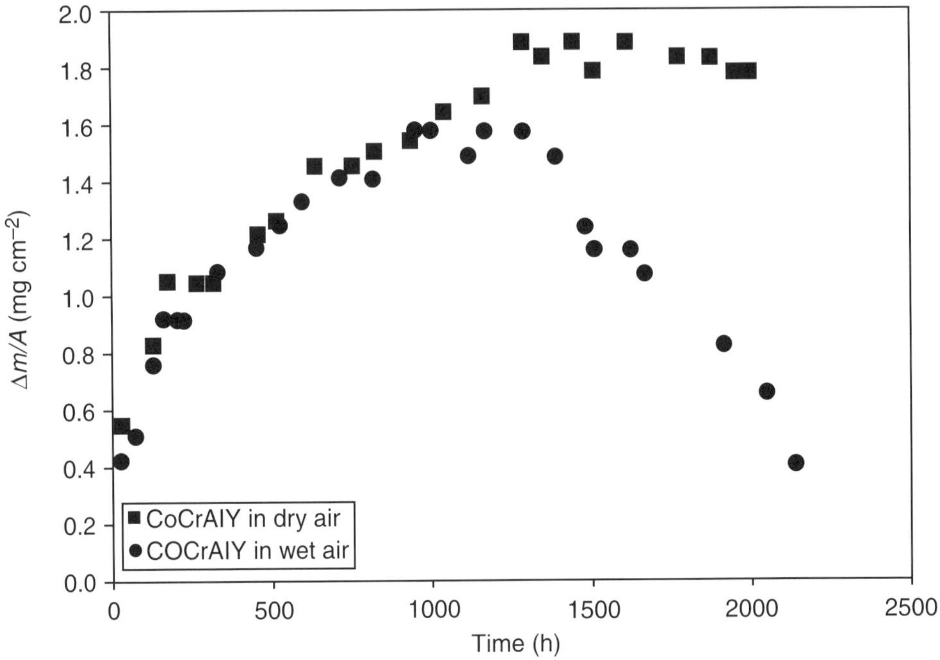

Fig. 16 *Mass change vs time data for the cyclic oxidation of CoCrAlY in wet (p_{H_2O} = 0.1 atm) and dry air at 1100°C.*

determined by X-ray diffraction. [22]. The stresses were compressive and values are presented in Table 3 where it can be seen that no differences in residual stresses were detected after 350 h of oxidation in wet and dry air at 1100°C. The residual stress levels did not change significantly when the oxidation time was extended to 3000 h at 1100°C in both dry and wet air.

In the case of the adherent Al_2O_3 scales an attempt was made to determine by acoustic emission measurements whether cracks formed in the Al_2O_3 during cyclic oxidation. Acoustic emission counts measured during cooling of a low sulfur PWA 1484 alloy after 350 h of oxidation are presented in Fig. 17. No counts were recorded during isothermal oxidation, and the fact that counts were evident during cooling of the specimen indicates that some cracking, but no spalling, of the oxide scales occurred during cyclic oxidation. Therefore, it appears that water vapour did have access

Table 3. *Residual stress measured by XRD in desulfurised PWA 1484 after oxidation at 1100°C*

Environment	**Time (h)**	**Residual compressive stress (MPa)**
Dry air	350	–3890
Wet air (p_{H_2O} = 0.1 atm)	350	–3760
Dry air	3000	–3480
Wet air (p_{H_2O} = 0.1 atm)	3000	–3710

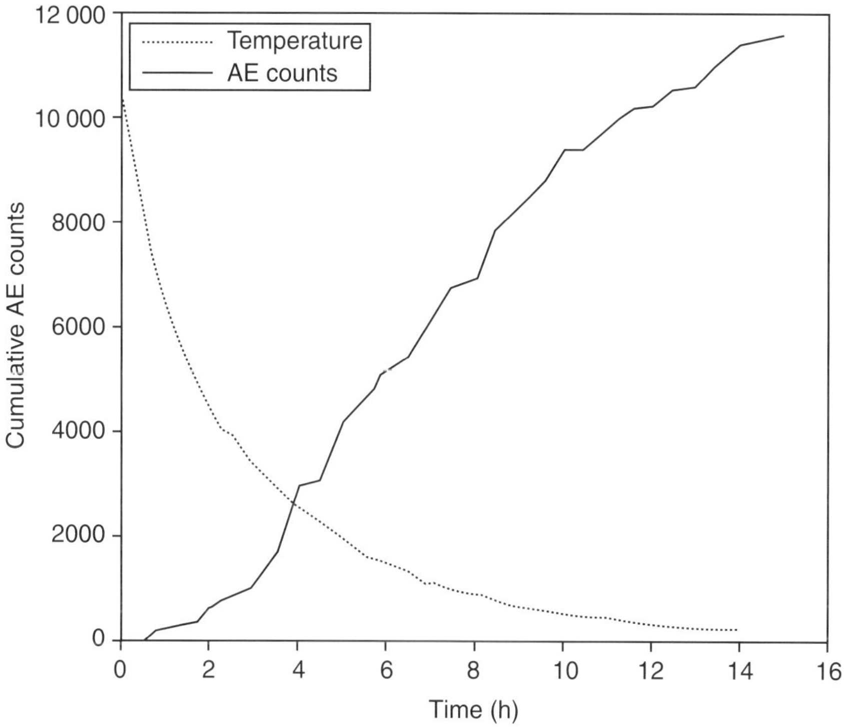

Fig. 17 *Acoustic emission counts recorded during cooling for a low sulfur PWA 1484 alloy isothermally oxidised at 1100 °C for 350 h and furnace cooled.*

during cooling to the α-Al_2O_3–alloy interfaces during cyclic oxidation of the low sulfur alloys.

To attempt to show conclusively that water vapour had access to the α-Al_2O_3–alloy interfaces in the cases where no spalling was observed during cyclic oxidation, repolishing of specimens followed by oxidation experiments, and water droplet tests were performed. The results from the repolishing–oxidation experiments are presented in Fig. 18. It can be seen that repolished specimens of PWA 1484 exhibited less oxide spalling than observed in dry air even when the samples were oxidised in wet air. In fact, as shown in Fig. 19, the PWA 1484 specimen which initially contained 5 ppm sulfur behaves similarly to 3 ppm CMSX-4. These tests indicate that water vapour may have access to the Al_2O_3–alloy interface but cannot induce substantial spalling if the interfacial toughness is sufficiently high.

The results obtained in the water droplet test are presented in Fig. 20 and show that as the interfacial toughness at the α-Al_2O_3–alloy interface is increased the amount of spalling caused by water vapour at this interface decreases. In Fig. 20 surface photographs of cyclicly oxidised PWA 1484 are presented after repolishing after every 24 h oxidation. These photographs also show the alloy surface after a droplet of water was added to the specimen. The water droplet caused spalling after 24, 48, 72 and 96 h oxidation but the amount of oxide spalling became progressively less as the sulfur was removed via the repolishing–oxidation procedure. After 120 h the sulfur has been reduced to levels such that the water droplet caused spalling only at sharp

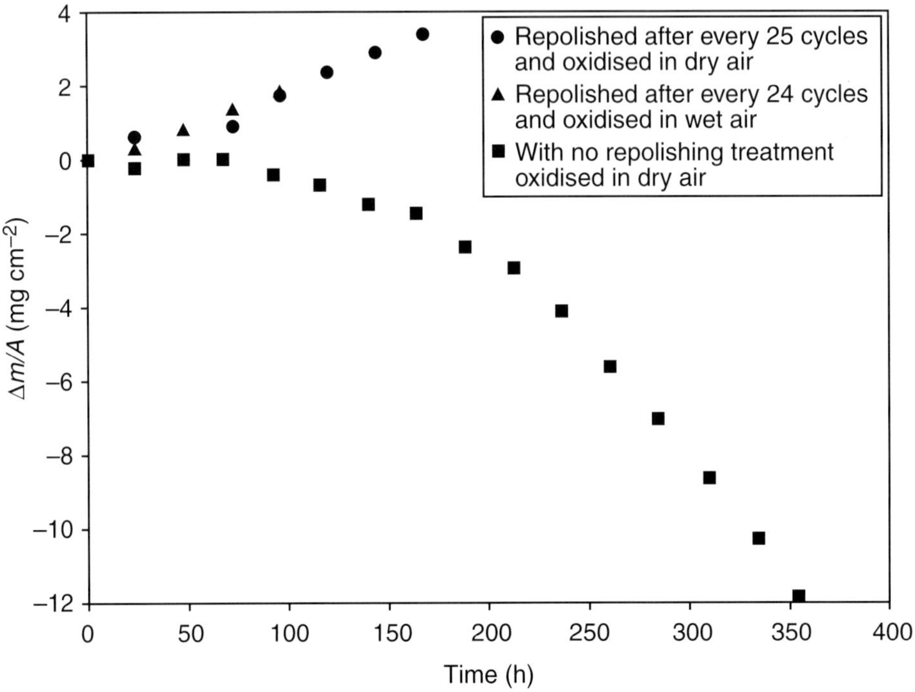

Fig. 18 *Effect of repolishing treatment on the cyclic oxidation of PWA 1484.*

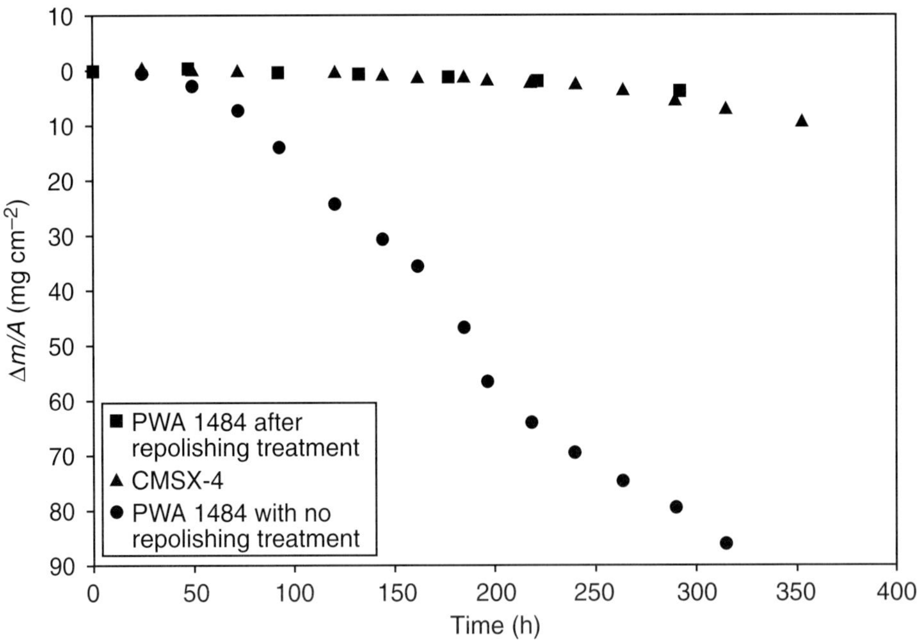

Fig. 19 *Mass change vs time data for the cyclic oxidation of an oxidised/repolished PWA 1484, CMSX-4 and PWA 1484 in wet (p_{H_2O} = 0.1 atm) at 1100°C.*

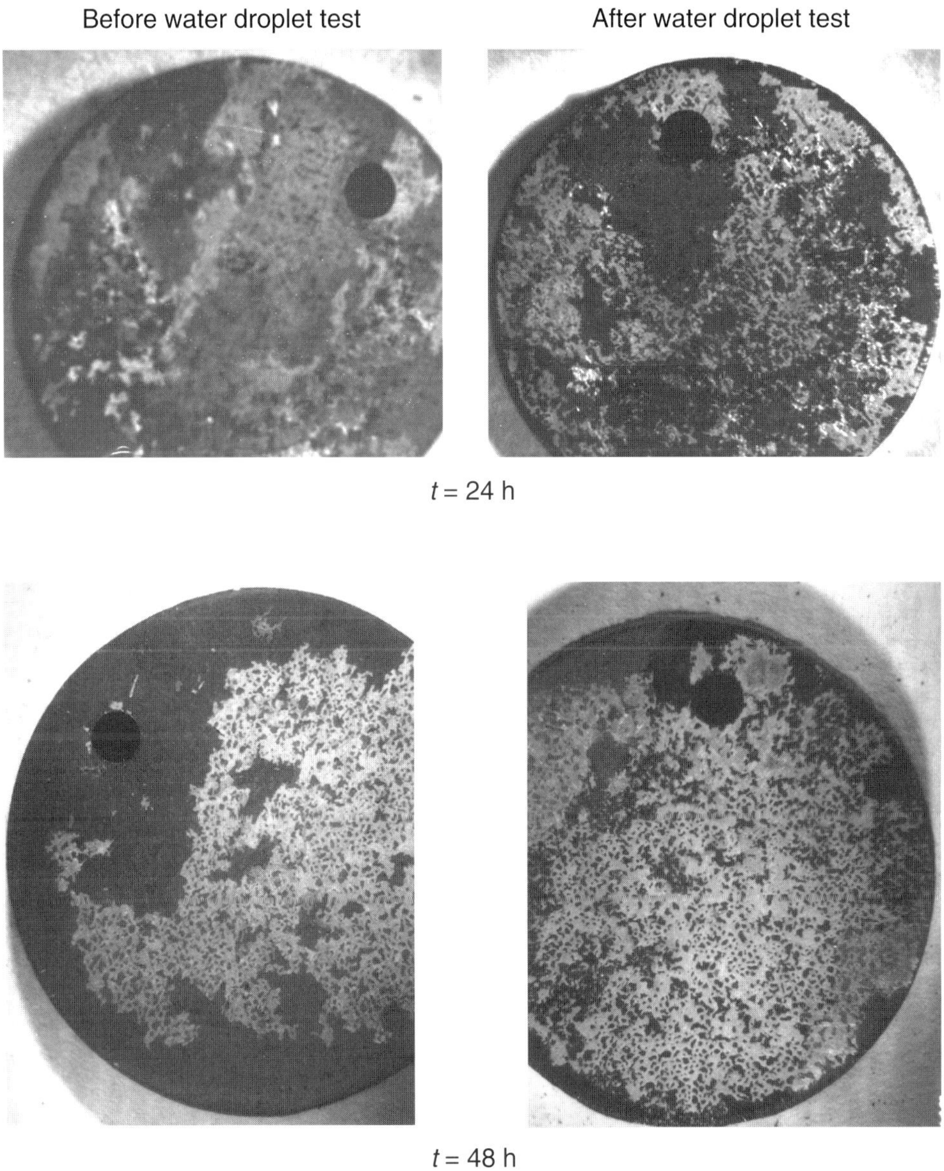

***Fig. 20 (a)** Effect of water droplet test on the oxidised (1100 °C)/repolished PWA 1484 specimen.*

edges. Hence, these experiments show that water vapour can have access to the interface but does not cause spalling if the interfacial toughness is high.

The results that have been obtained show that when water vapour has access to the α-Al_2O_3–alloy interface, it causes cracking to occur along this interface if the interfacial toughness is not high. Reimanis *et al.* [23] have investigated the interfacial toughness of gold–alumina specimens and found a large increase in the interfacial toughness in dry nitrogen compared to moist air. These investigators have proposed that this effect is due to interfacial stress corrosion cracking induced by water vapour

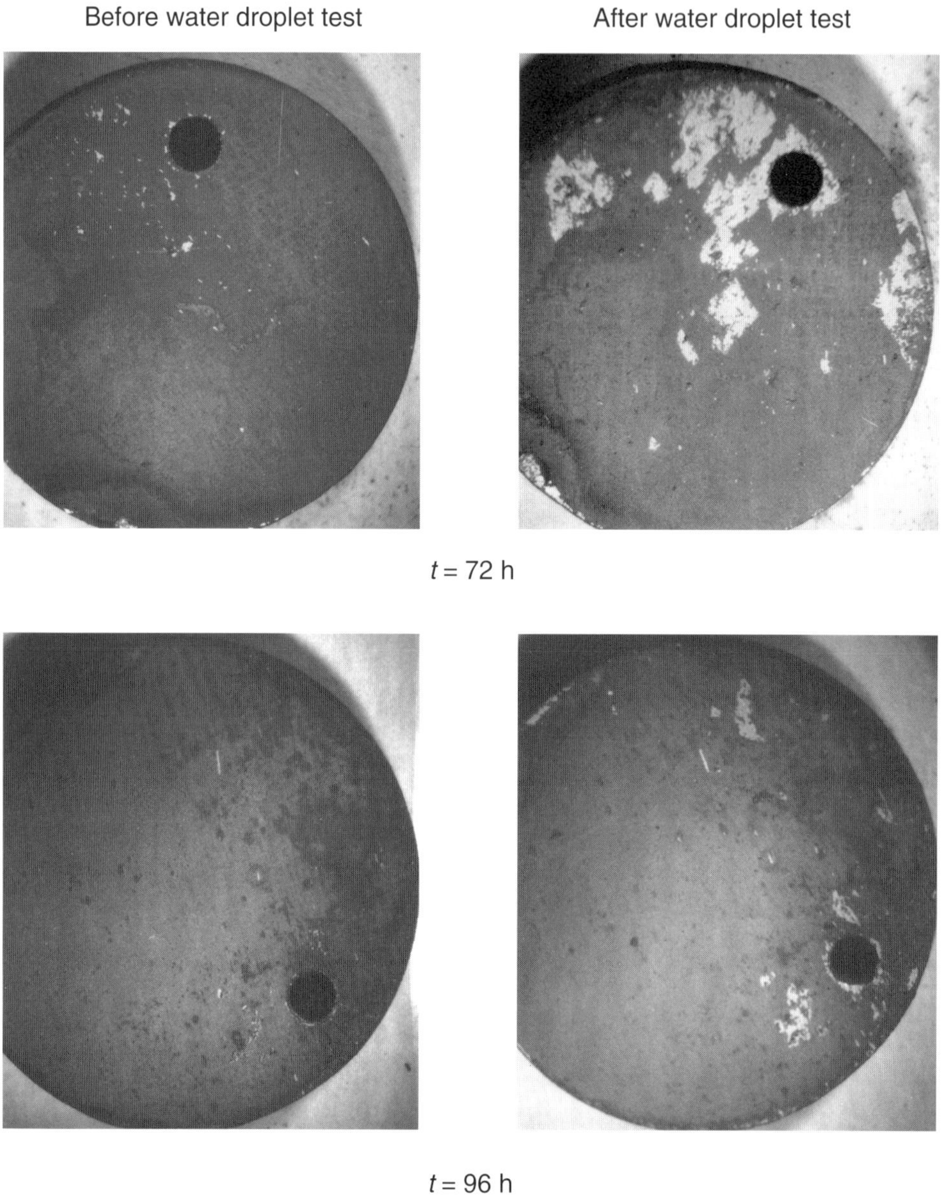

Fig. 20 (b) *Effect of water droplet test on the oxidised (1100 °C)/repolished PWA 1484 specimen.*

adsorption at the crack tip. Wiederhorn [24,25] has studied the influence of water vapour on crack propagation in soda-lime glass and in sapphire. The measured crack velocities were shown to be a complex function of stress and water vapour concentration in the environment. Portions of the data were adequately explained by the stress corrosion theory of Hillig and Charles [26] where crack motion was limited by the rate of reaction of water with the glass at the crack tip, or by the rate of transport of water vapour to the crack tip. The results were similar for both

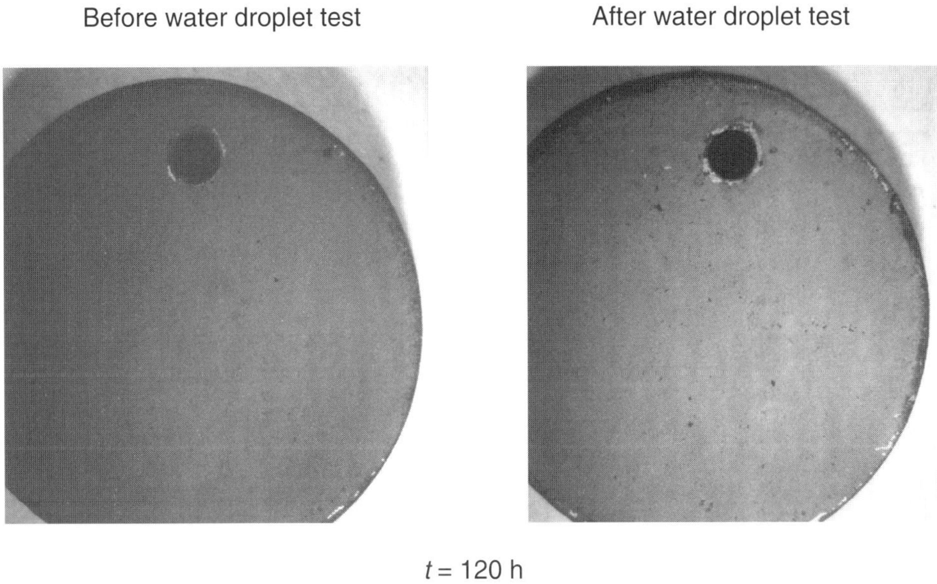

***Fig. 20 (c)** Effect of water droplet test on the oxidised (1100 °C)/repolished PWA 1484 specimen.*

soda-lime glass and sapphire but more experimental scatter was evident in the case of sapphire.

Michalski and Freiman [27] have proposed that the loss of strength in vitreous silica is associated with slow crack growth of pre-existing surface flaws due to stress corrosion by water from the surrounding environment. Michalski *et al.* [28] showed that single crystal sapphire had slow fracture behaviour similar to silica when exposed to nonaqueous environments. In this model it is the bridging Si–O bond which supports stress in the material and its rupture is important to the fracture process. The resulting stress on the bridging Si–O bond can be discussed in terms of decreasing bond overlap and, consequently, increased availability of Si and O atoms to bond with other species. The structure of the water molecule is well documented in the literature and consists of proton donor sites at one end of the molecule and lone pair orbitals at the other. Michalski and Freiman [27] propose a water molecule attaches to a strained bridging Si–O–Si bond at the crack tip by the formation of the hydrogen bond with the O(bridging), and interaction with the lone-pair orbitals from water with the Si atom. Two new bonds are formed, one between O(water) and Si, and one between hydrogen and O(bridging). The original bridging bond is destroyed. The rupture of the hydrogen bond with O(water) yields Si–O–H groups on each fracture surface. The hydrogen bond is weak and, therefore, this process occurs immediately after the proton transfer to the bridging oxygen atom. The results obtained by Michalski *et al.* [28] with single crystal sapphire were predicted since the chemistry of the Si–O bond and Al–O bond are quite similar, the nonaqueous medium possessed structural bonding features similar to those of water, namely, proton donor sites at one end of the molecule and lone pair orbitals on the other, and the molecule was of comparable size with the Si–O bond.

In view of the results that have been obtained in this paper, and the data in the literature concerning water vapour and stress corrosion cracking, it is reasonable to propose that water vapour causes increased spalling of α-Al_2O_3 scales from alloys by a stress corrosion process. Water vapour must affect the interfacial toughness of the α-Al_2O_3–alloy interface by altering the nature of the bonds at this interface. It is known that H_2O attacks the bridging oxygen atoms of the SiO_2 network [27,29,30] and H_2O probably causes similar effects with Al_2O_3 [28]. The bonds at the α-Al_2O_3–alloy interface have been proposed to be dominated by the Al^{3+} cation to nickel bond in the case of nickel base substrates [31] and sulfur adversely affects such bonds. Cracking and spalling at the Al_2O_3–alloy interface is determined by the toughness of this interface and also by the elastic energy available to drive the spallation failure. Increases in the elastic energy driving spallation can be caused by increases in the oxide thickness, residual stresses or stiffness. A decrease in toughness of the interface can be caused by a decrease in the adhesive strength of the interface due to sulfur segregation and/or water interactions, or by a decrease in the plastic deformation occurring in the oxide or alloy during crack extension. The fact that all of these factors have been constant except sulfur concentration and the presence of water vapour shows that the interfacial toughness is the critical factor in the current experiments. When sulfur is present at the interface, water vapour lowers the fracture toughness even more, hence spalling of oxide is increased. When sulfur is not present, either water vapour has no effect, or the interfacial toughness is so high that even with water interaction the interfacial toughness is still sufficient to inhibit spalling.

The absence of an observed water vapour effect at 1000°C must be due to the small thickness of the α-Al_2O_3 at this temperature. As the oxide scale increases in thickness, the elastic energy driving spallation should increase and eventually, for sufficiently long oxidation times, spalling should occur for CMSX-4 at 1000°C (Fig. 9), and for the low sulfur alloys oxidised at 1100°C, (Fig. 10 and 11). As spalling occurs water vapour can be expected to cause more severe degradation even in the case of an alumina-forming alloy with sulfur concentrations below 1 ppm. Experiments must be performed to confirm this proposed result.

There are other important questions that must be investigated. The results that have been obtained indicate that the interfacial toughness of the alumina on CMSX-4 is greater than that on PWA 1484 due to the lower sulfur concentration in the former. It appears that water vapour at both p_{H_2O} = 0.1 atm and 0.05 atm can alter this toughness in CMSX-4 but only a pressure of 0.1 atm affected the toughness of PWA 1484. It is necessary to examine in more detail the initial stages of oxide spalling on alloys with sulfur contents such as those in PWA 1480, PWA 1484 and CMSX-4, and to compare cracking and spalling with and without water vapour present. As indicated schematically in Fig. 21, water vapour is proposed to reduce the interfacial toughness of the alumina–alloy interface. For systems where the stored elastic energy approaches that of the interfacial toughness, water vapour therefore causes spalling to occur sooner.

4. Concluding Remarks

The influence of water vapour on the oxidation of alloys has been difficult to describe because the effects of water vapour depend profoundly on the alloy system being

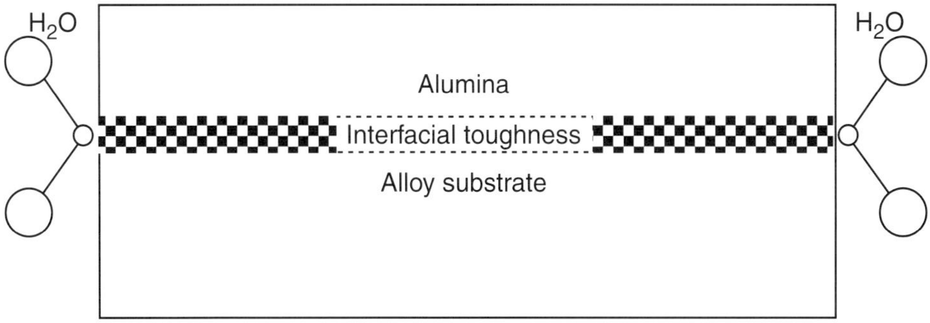

Fig. 21 *Schematic diagram illustrating a possible mechanism for water vapour effects on alumina scales formed on alloy substrates.*

considered. Different effects are probable for iron oxide scales, chromia or α-Al_2O_3. In the case of those alloys that develop oxidation resistance via the formation of α-Al_2O_3 scales, it is clear that water vapour causes more profuse spalling of such scales and, consequently, more rapid degradation of such systems. The mechanism by which water vapour causes spalling of α-Al_2O_3 is proposed to consist of stress-corrosion cracking whereby water vapour causes the bonding across the α-Al_2O_3–alloy interface to be altered.

5. Acknowledgements

The authors are grateful for financial support from the Office of Naval Research (Contract # NOOO14-94-1-0085, A. John Sedriks, Program Officer). The desulfurised specimens were prepared by Dr J. L. Smialek, NASA Lewis Research Center.

References

1. A. Rahmel and J. Tobolski, *Corros. Sci.*, 1965, **5**, 333–346.
2. C. W. Tuck, M. Odgers and K. Sachs, *Corros. Sci.*, 1969, **9**, 271–285.
3. R. L. McCarron and J. W. Schulz, The effects of water vapour on the oxidation behaviour of some heat resistant alloys, in *Proc. Symp. on High Temperature Gas–Metal Reactions in Mixed Environments*. AIME, New York, 1973, p.360.
4. H. Bouaouine, F. Armanet and C. Coddet, *9th Int. Congr. on Metallic Corrosion,* Toronto, Canada, 1989, pp.379–381.
5. I. Kvernes, M. Oliveira and P. Kofstad, *Corros. Sci.*, 1977, **17**, 237–252.
6. R. Kremer and W. Auer, *Mater. Corros.*, 1997, **48**, 35.
7. E. A. Irene, *J. Electrochem. Soc.*, 1974, **121**, 1614–1616.
8. J. F. Cullinan, 'The Oxidation of Carbon–Carbon Composites between 300°C and 900°C in Oxygen and Oxygen/Water vapour Atmospheres', M. S. Thesis, University of Pittsburgh, 1989.
9. A. J. Sedriks, *Corrosion of Stainless Steels*, 2nd Edn. John Wiley and Sons, Inc., New York, 1996.

10. P. Kofstad, in *Microscopy of Oxidation* (M. J. Bennett and G. W. Lorimer, eds). The Institute of Metals, London, 1991, p.2.
11. H. Buscail, S. Heinze, Ph. Dufour and J. P. Larpin, *Oxid. Met.*, 1997, **47**, 445–464.
12. J. L. Smialek, *Met. Trans.*, 1991, **22A**, 739–752.
13. D. R. Sigler, *Oxid. Met.*, 1993, **40**, 555–583.
14. M. A. Smith, W. E. Frazier and B. A. Pregger, *Mater. Sci. Eng.*, 1995, **A203**, 388–398.
15. M. Levy, P. Farrell and F. Pettit, *Corrosion*, 1986, **42**, 708–717.
16. J. G. Smeggil, A. W. Funkenbusch and N. S. Bornstein, *Met. Trans.*, 1986, **17A**, 923–932.
17. J. L. Smialek and B. K. Tubbs, *Met. Trans.*, 1995, **26A**, 427–435.
18. G. H. Meier, F. S. Pettit and J. L. Smialek, *Werks. Korros.*, 1995, **46**, 232–240.
19. J. L. Smialek, *Met. Trans.*, 1991, **22A**, 739–752.
20. H. E. Evans and M. P. Taylor, Creep relaxation and the spallation of oxide layers, *Surf. Coat. Technol.*, 1997, **94–95**, 27–33.
21. F. S Pettit, *Trans. Met. Soc. AIME*, 1967, **239**, 1296–1305.
22. C. Sarioglu, J. R. Blacheré, F. S. Pettit and G. H. Meier, Room temperature and in-situ high temperature strain or stress measurements by XRD techniques, in *Microscopy of Oxidation — 3* (S. B. Newcomb and J. A. Little, eds). The Institute of Materials , London, 1997, p.41.
23. I. E. Reimanis, B. J. Dalgleish, M. Brahy, M. Rühle and A. G. Evans, *Acta Metall.*, 1990, **38**, 2645–2652.
24. S. M. Weiderhorn, *J. Am. Ceram. Soc.*, 1967, **50**, 407–414.
25. S. M. Weiderhorn, *Int. J. Fract. Mech.*, 1968, **4**, 171–177.
26. W. B. Hillig and R. J. Charles, in *High Strength Materials* (V. F. Zackay, ed.). John Wiley & Sons, New York, 1965.
27. T. A. Michalski and S. W. Freiman, *J. Am. Ceram. Soc.*, 1983, **66**, 284–288.
28. T. A. Michalski, S. W. Freiman and B. Bunker, *Am. Ceram. Soc. Bull.*, 1982, **61**, 414.
29. R. H. Doremues, *J. Phys. Chem.*, 1971, **75**, 3147–3152.
30. R. J. Bruckner, *J. Non-Crystalline Solids*, 1971, **5**, 177–183.
31. S. Y. Hong, A. B. Anderson and J. F. Smialek, *Surf. Sci.*, 1990, **230**, 75–183.

4

The Influence of Thermal Cycling on Internal Oxidation and Nitridation of Nickel-Base Alloys

S. Y. CHANG, U. KRUPP and H.-J. CHRIST
Institut für Werkstofftechnik, Universität Siegen, Germany

ABSTRACT

The oxidation behaviour of the three single-crystal nickel-base superalloys CMSX-4, CMSX-6 and SRR99, and of several model alloys of the system Ni–Cr–Al–Ti, was investigated in air at 1000°C under isothermal and thermal cyclic conditions. The study was performed using a continuous thermogravimetric method which permits mass changes to be measured in different atmospheres and under cyclic conditions. It is shown that even under isothermal conditions the commercial Ni-base superalloys suffered internal oxidation of Al (Al_2O_3) as well as internal nitridation of Ti and Al (TiN and AlN, respectively), particularly at specimen edges where geometrically-induced growth stresses reach high values. Under temperature-cycling conditions the extent of internal corrosion increased clearly. Microstructural examinations of NiCr model alloys showed that under these conditions low Cr concentrations (≤10 mass%) led to excessive oxide spalling of the superficial NiO layer followed by massive internal oxidation of Cr_2O_3. All the model alloys containing Ti suffered internal nitridation even during isothermal exposure to air. If in addition to Ti a small Al concentration (2 mass%) was present, a combined process of internal Al oxidation and internal Ti nitridation took place, which was also clearly enhanced by cyclically varying temperatures. In the present paper, results on the kinetics of isothermal and cyclic oxidation are reported and the mutual interaction between oxide scale failure and the formation of corrosion products and microstructural changes in the base metal are illustrated by means of analytical electron microscopy. It is intended to incorporate the experimental results into a computer model which allows a realistic simulation of the material behaviour during cyclic high-temperature exposure.

1. Introduction

The applicability of high-temperature alloys depends strongly on the formation of slow growing protective oxide scales, such as chromia and alumina which are responsible for the resistance to high-temperature corrosion. However, since most high-temperature components, e.g. gas turbine blades, are used under conditions that involve thermal cycles, failure of the oxide scales cannot be avoided completely. The cyclic oxidation behaviour of the alloys is determined by the adherence of the oxide scale, which is related to the morphologies and compositions of the oxides formed. Repeated spalling of the oxide scale during cyclic loading leads to a massive depletion of the oxide-forming elements inside the substrate resulting in the loss of the healing effect and an excessive growth of less-protective oxides of the base metal or alloying elements (e.g. Ni and Ti). Finally, a partial destruction of the component takes place due to breakaway oxidation [1].

The failure process starts in the form of a transition from protective superficial oxide-scale formation to internal oxidation when the content of one of the less noble elements falls below a critical value. In many cases, this mechanism, which is well-described by Wagner's classical theory [2], is responsible for the premature failure of high-temperature alloys because of a significant near-surface degradation of the material properties. In the case of nickel-base superalloys, the Ti and Al consumption due to external oxide scale formation as well as to the internal corrosion processes — particularly internal oxidation and internal nitridation, contributes to a dissolution of the strengthening γ'-phase ($Ni_3(Al,Ti)$). Furthermore, as a consequence of the higher specific volume of the internal corrosion products compared to that of the base material, near-surface embrittlement takes place with deterioration of the high-temperature corrosion and strength properties [3].

In order to quantify the influence of different alloying elements or the conditions under which failure of a protective oxide scale leads to internal oxidation and nitridation, the three commercial single-crystal superalloys CMSX-4, CMSX-6 and SRR99 and various model alloys of the Ni–Cr–Al–Ti system were investigated thermogravimetrically under isothermal and thermally cyclic conditions in air at 1000°C. Microstructural examinations were performed in order to obtain a physical foundation of the thermogravimetrical data and to improve the understanding of the mechanisms leading to material failure. It is planned to incorporate the experimental results into a computer simulation which contains a numerical diffusion calculation for all the species involved in the corrosion process using the finite-difference technique, the commercial thermodynamic equilibrium calculation ChemApp and a model which defines the conditions under which cracking or spalling of a protective oxide scale occur.

2. Experimental Details

The chemical compositions of the single-crystalline Ni-base superalloys CMSX-4, CMSX-6 and SRR99, as well as of the Ni-base model alloys which were used in this study are given in Table 1. All the materials were obtained as cylindrical rods of 10–12 mm dia. Specimens of <001> orientation (with respect to the rod axis) single-crystal superalloys were fully heat treated according to the specifications given in Table 2 in order to dissolve the γ-Ni/γ'-$Ni_3(Al,Ti)$ eutectics and to reduce the degree of segregation resulting from the dendritic growth during the solidification process.

For the thermogravimetric experiments, discs of 1mm thickness were cut from the rods by a slow-turning saw using a diamond-coated disc. The specimens were wet ground with SiC abrasive paper down to 1200-grit and polished with 3 μm and finally 1 μm diamond suspension. Prior to the corrosion tests the specimens were cleaned ultrasonically in ethanol and weighed by means of a microbalance.

The isothermal and cyclic oxidation studies were performed in a thermobalance

Table 1. *Chemical composition of the single-crystal Ni-base superalloys and the model alloys used in the present and in earlier studies [3,17] (in mass%)*

	Ni	**Cr**	**Al**	**Ti**	**Co**	**Ta**	**Mo**	**Re**	**Hf**	**W**	**C**
CMSX-6	(71.6)	9.6	4.7	4.6	4.8	1.8	2.8	—	—	0.1	—
CMSX-4	(61.7)	6.0	5.6	1.0	10.0	6.0	0.6	3.0	0.1	6.0	—
SRR 99	(65.6)	8.5	5.6	2.2	5.2	2.8	0.6	—	—	9.5	0.014

	Ni	Cr	Ti		Ni	Cr	Al	Ti
NiTi2	98	—	2	NiCr20Ti2	78	20	—	2
NiTi6	94	—	6	NiCr30Ti2	68	30	—	2
NiCr10	90	10	—	NiCr20Ti6	74	20	—	6
NiCr20	80	20	—	NiCr20Al2	78	20	2	—
NiCr30	70	30	—	NiCr10Al2Ti2	86	10	2	2
NiCr5Ti2	93	5	2	NiCr20Al2Ti2	76	20	2	2
NiCr10Ti2	88	10	2					

Table 2. *Heat treatment of the single-crystal Ni-base superalloys*

Alloy	**Solution annealing**	**Ageing**
CMSX-4	1305°C (6 h)	1140°C (5 h)/870°C (16 h)
CMSX-6	1204°C (1 h)/1238°C (3 h)/1267°C (6 h)	1080°C (4 h)/870°C (16 h)
SRR 99	1280°C (6 h)/1290°C (2 h)/1300°C (0.5 h)	1080°C (4 h)/870°C (16 h)

that was designed and built especially for this purpose. A computer-controlled lift was used to move the specimens periodically into and out of the hot zone of the furnace within a few seconds to allow rapid heating and cooling. Each temperature cycle of the cyclic oxidation experiments consisted of 5 h during which the temperature is held constant at 1000°C and a 15 min period of cooling down to room temperature. The repeated spalling of the oxide scale as a consequence of thermal cycling leads to a depletion of scale-forming alloying elements in the base metal. This process was accelerated for a number of samples by mechanically removing the oxide scale formed after 24 h periods of high-temperature exposure.

The composition, phases and morphology of the corrosion products formed within the external scale and subsurface region of the alloy were investigated applying optical and analytical scanning electron microscopy as well as X-ray diffraction to the corroded surfaces and cross-sections of the exposed specimens.

3. Results and Discussion

3.1. Oxidation of the Single-Crystal Nickel-Base Superalloys

The mass-change data obtained during cyclic thermogravimetry represent integral information that reflects a combination of the oxygen and nitrogen consumption due to external and internal oxidation and nitridation, respectively, and the mass loss caused by scale spalling. Thermal cycling can often lead to a step-wise mass loss in thermogravimetric mass-change curves. The isothermal and cyclic oxidation kinetics of the commercial superalloys at 1000°C are shown in Fig. 1.

The mass change for cyclic exposure of CMSX-6 (Fig. 1a) increases during the second cycle due to scale cracking. After the second cycle the scale adhesion is not sufficient to avoid massive scale spalling which is repeated periodically after the subsequent cycles. The isothermal results for the superalloys CMSX-4 and SRR99 represented in Figs 1(b) and 1(c) demonstrate the better corrosion behaviour of these alloys compared to CMSX-6 by lower mass gain values. This difference can be attributed to the lower Al and significantly higher Ti contents in CMSX-6 resulting in a comparatively less-protective oxide scale system containing a higher amount of porous Ti oxides. Although stepwise spalling of the oxide scale (see Fig. 1(a) for the example of CMSX-6) could not be observed for the superalloys SRR99 and CMSX-4, thermal cycling leads especially for SRR99 to a decreased mass gain which indicates

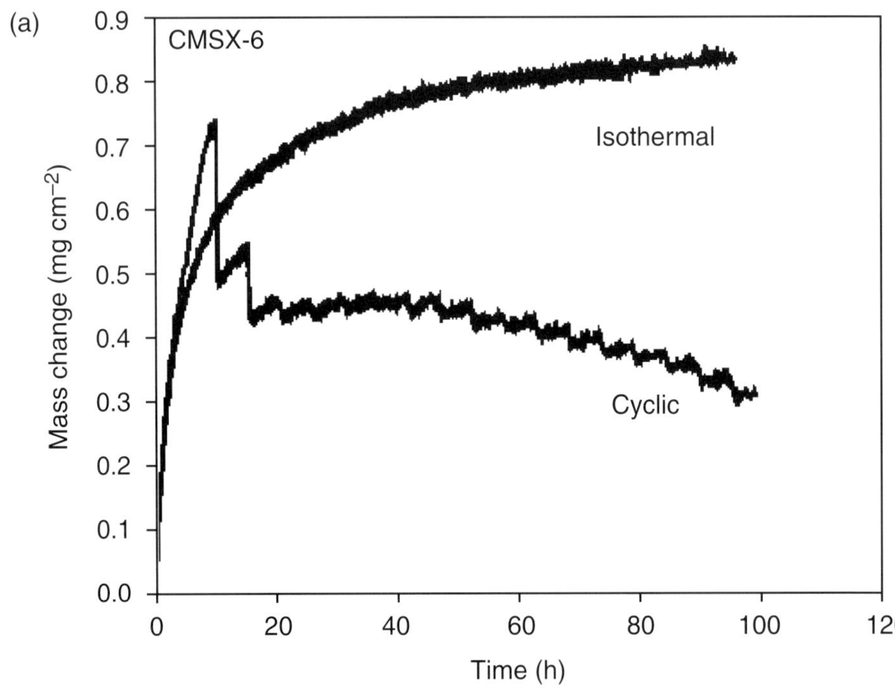

Fig. 1 *(a) Comparison of the thermogravimetrically measured mass gain vs time for the superalloy CMSX-6 exposed isothermally and under cyclic thermal conditions to air for 100 h at 1000°C.*

(b)

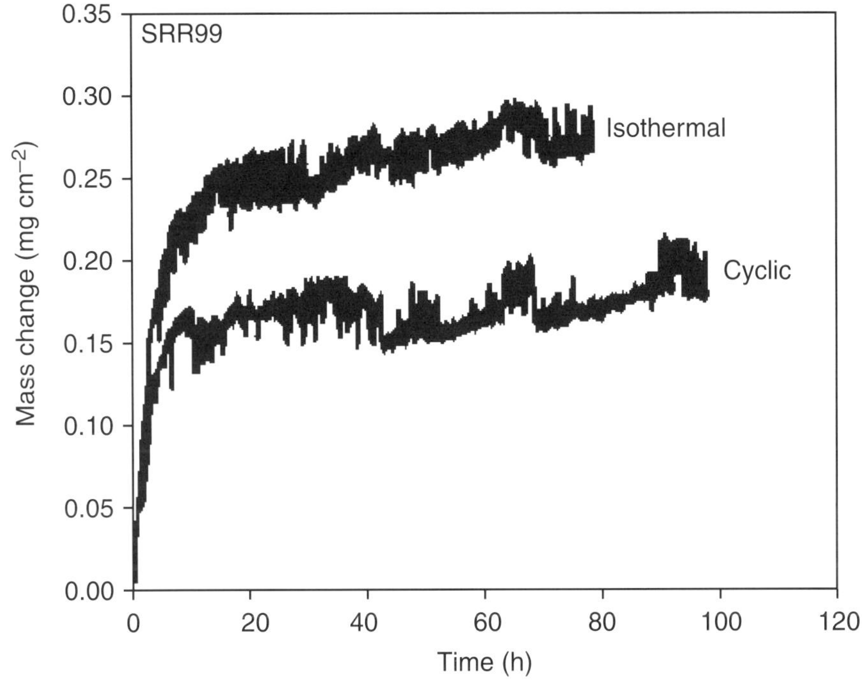

(c)

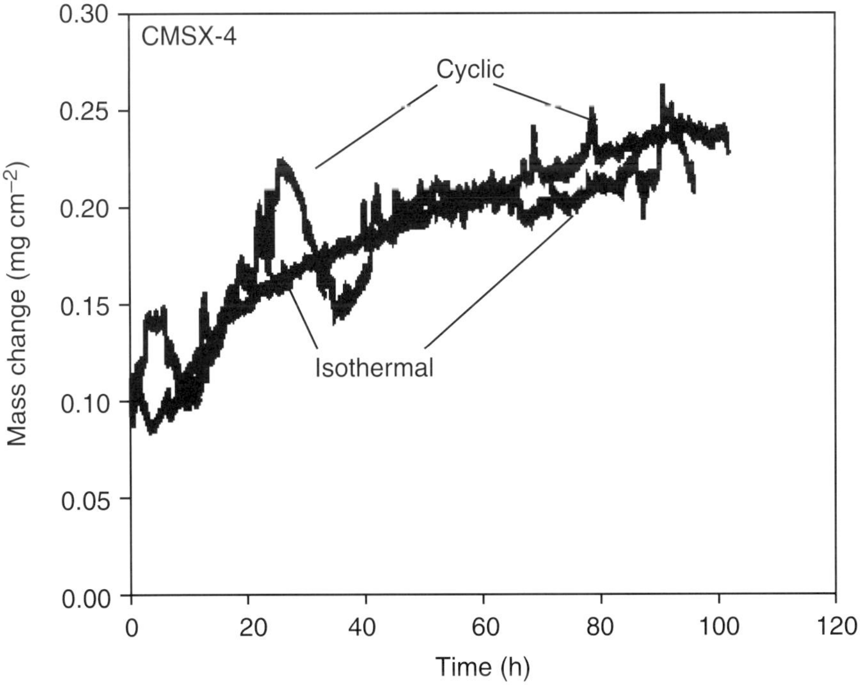

Fig. 1 *(b,c) Comparison of the thermogravimetrically measured mass gain vs time for the superalloys SRR99 and CMSX-4 exposed isothermally and under cyclic thermal conditions to air for 100 h at 1000°C.*

locally limited cracking and spalling processes [4]. The excellent isothermal and cyclic oxidation behaviour of the superalloy CMSX-4 can be attributed to small additions of the reactive element Hf [5].

The dendritic segregation which occurs in single-crystal nickel-base superalloys during the solidification process affects clearly both the isothermal and the cyclic oxidation behaviour of these alloys [6]. After exposure to air an enrichment of Al could be identified within the fine-grained oxide scale that was formed on interdendritic areas, while a coarse-grained oxide scale enriched with Ni developed on the dendritic areas. This is shown in Fig. 2(a) for the example of SRR99 after isothermal exposure at 1000°C for 100 h to air. Figure 2(b) represents the results of an Energy Dispersive Spectroscopic (EDS) element analysis along the marked line between dendritic and interdendritic areas. Similar results were reported by Göbel *et al.* [7].

Figure 3 shows an oxidised surface region of CMSX-6 after cyclic testing in which parts of the oxide cracked off in the interdendritic area. In contrast, the surface of CMSX-4 after oxidation revealed a large number of nodules mainly in the interdendritic regions, as shown in Fig. 4. These nodules were found to be composed of Ta_2O_5 and TiO_2 particles.

The microstructural examination of cross-sections, especially of the cyclically exposed superalloys, showed that locally massive internal oxidation and nitridation had occurred underneath the oxide scale. The internal oxidation process beneath dense external scale is typically controlled by the diffusion of dissolved oxygen [2]. The transition from external to internal oxidation occurs in this case if the concentration of the less noble, scale-forming alloying element (e.g. Al and Cr) falls

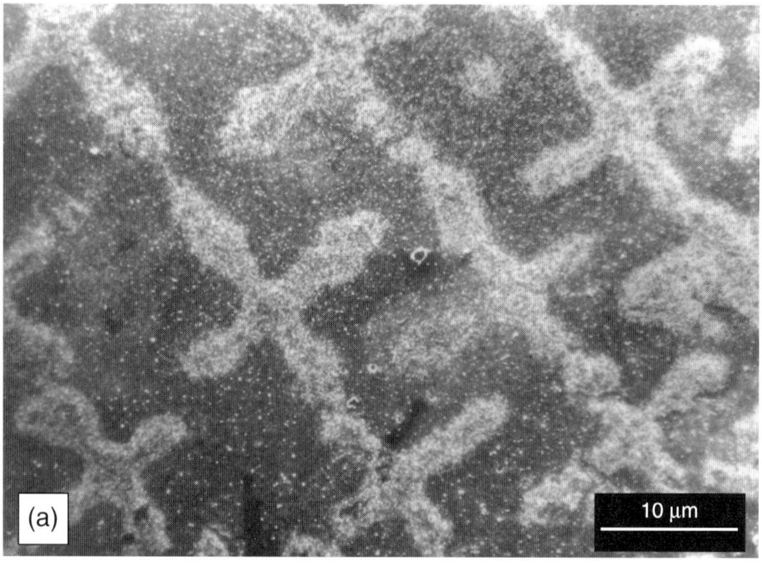

Fig. 2 *(a) Preferential oxidation of the dendritic area of SRR99 after 80 h exposure to air at 1000°C (surface normal to the <001> zone axis) and (b) (opposite) EDS–line scan for the elements Al, Ti, Cr, and Ni between the dendritic and interdendritic areas.*

below a critical concentration [8]. The depth of the internal oxidation zone depends on the concentration of dissolved oxygen at the interface of the compact oxide scale and the alloy [2]. The value of this concentration is determined by the local thermodynamic equilibrium between oxygen, oxide-forming alloying element and the corresponding oxide. Furthermore, atmospheric nitrogen may penetrate into the base metal either via diffusion or through cracks and pores in the oxide scale. Although

(b)

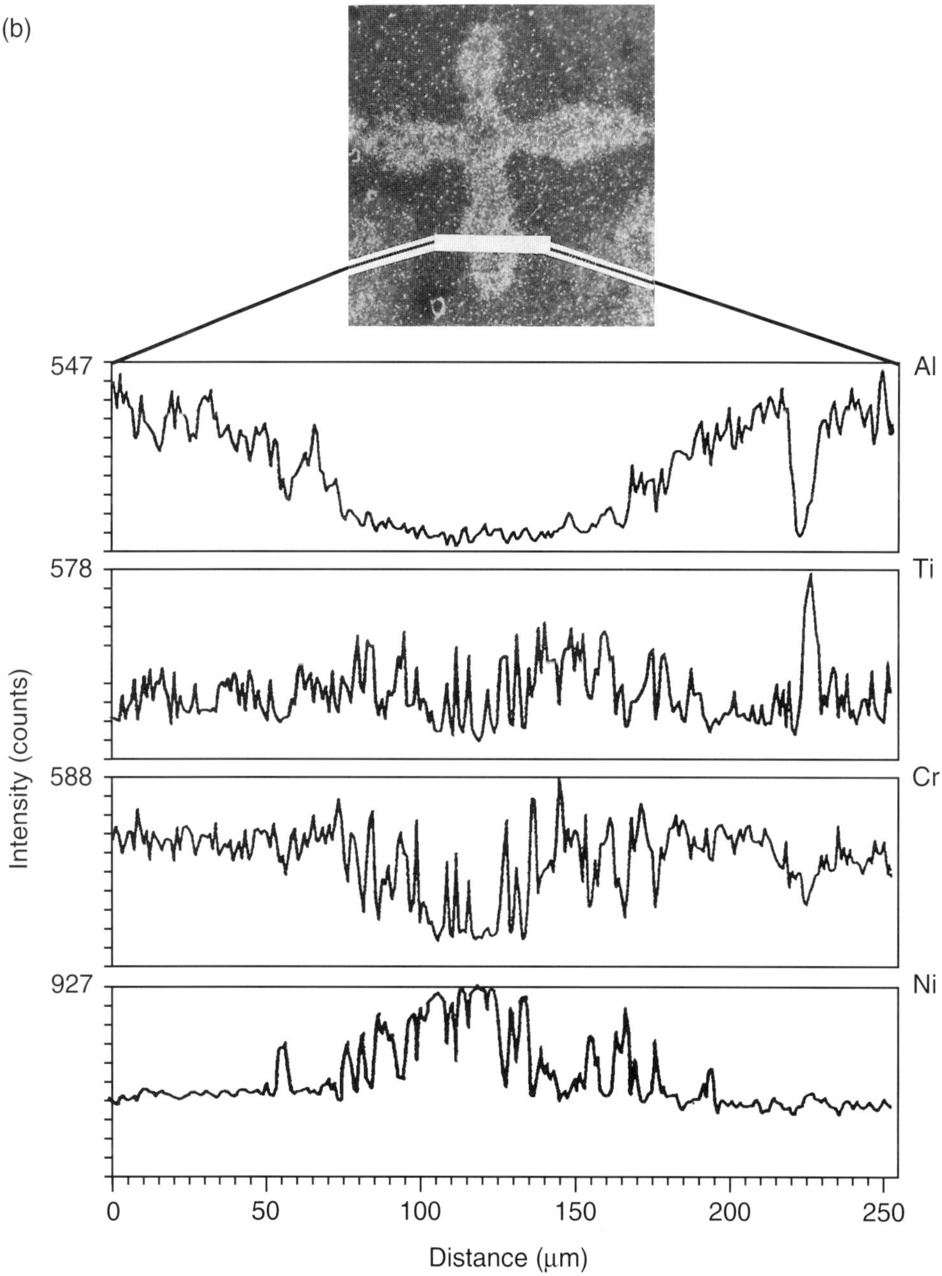

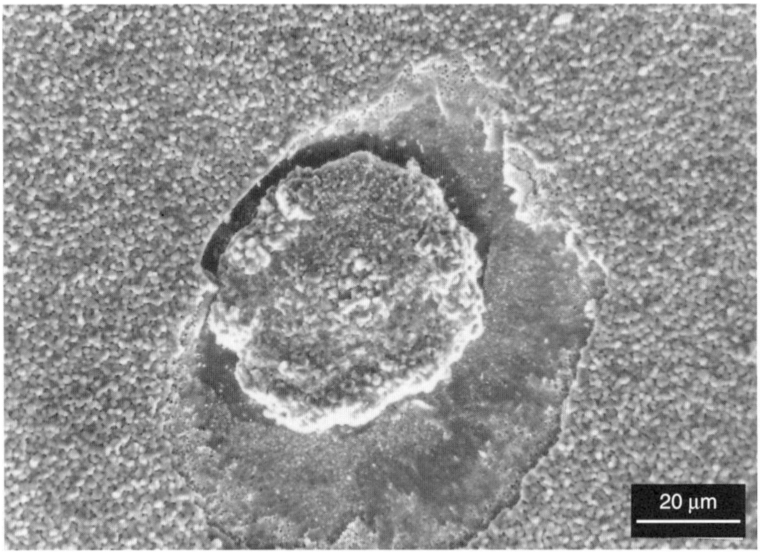

Fig. 3 *Morphology of the oxidised surface of the superalloy CMSX-6 after 100 h cyclic exposure to air at 1000°C.*

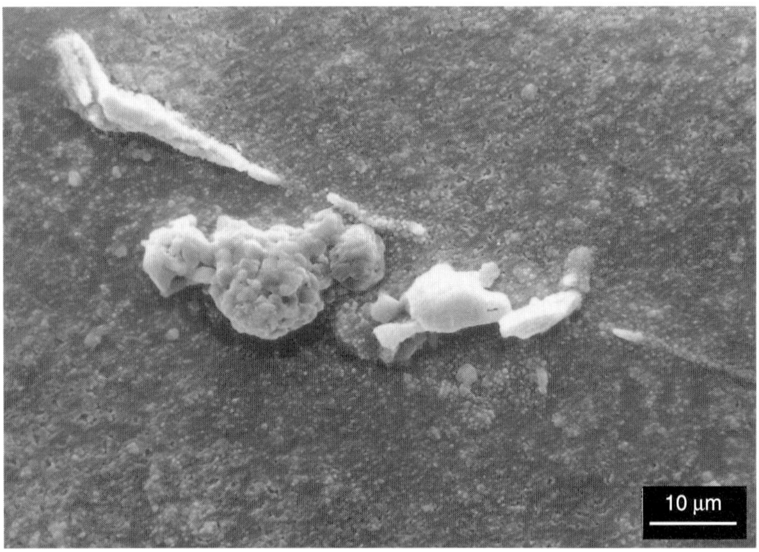

Fig. 4 *Ta_2O_5 and TiO_2 particles on the oxidised surface of CMSX-4 after 100 h cyclic exposure to air at 1000°C.*

near to and at the surface of the alloy the formation of oxides is more probable than the formation of nitrides because of the higher stability of most oxides, the formation of nitrides becomes more likely deeper in the alloy where the oxygen activity has reached low values [9]. Since the diffusion of nitrogen is faster than that of oxygen in the Ni-base matrix [10,11] a relatively deep internal nitridation occurs in superalloys. A typical example of an oxidation process which is combined with internal nitridation of Ti is

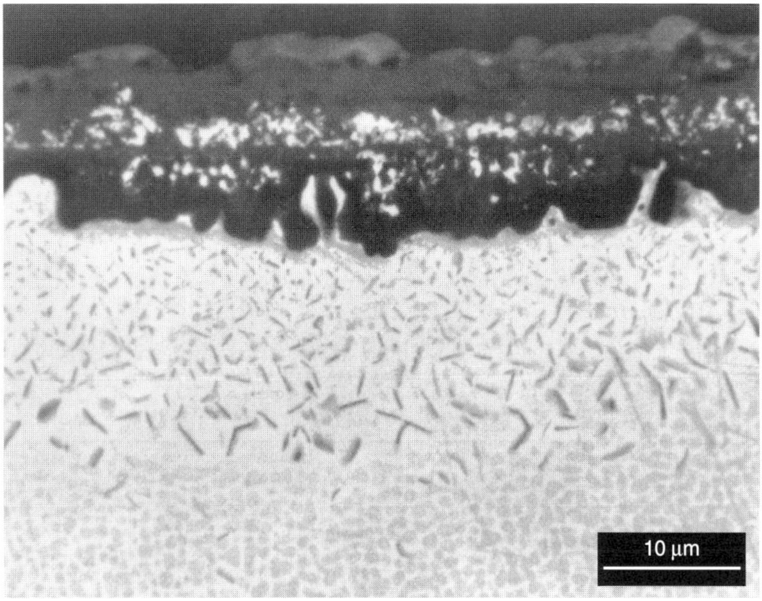

Fig. 5 *Cross-section through the oxide scale and the internal corrosion zone with TiN precipitates for the superalloy SRR99 exposed isothermally to air at 1000°C for 100 h.*

shown in Fig. 5. Below the external oxide scale, which consists of a superficial layer of NiO, TiO_2 and Cr_2O_3 and an Al_2O_3 subscale, needle-shaped TiN precipitates were formed. The γ'-depleted zone which reaches almost the TiN precipitation front indicates the consumption of Al and Ti due to oxidation and nitridation.

For cyclic thermal conditions the thickness and composition of the oxide scale varied locally among the tested alloys, and this can be partly attributed to the influence of spalling [12]. It was generally observed that cyclic conditions drastically increased the extent of the internal corrosion attack, particularly nitridation. Table 3 represents a comparison of the measured internal corrosion depths of the superalloys SRR99 and CMSX-6 and the model alloy Ni–20Cr–2Ti–2Al after isothermal and cyclic exposure for 100 h at 1000°C to air.

Table 3. *Comparison of the internal corrosion depths of the superalloys CMSX-6 and SRR99 and the model alloy Ni–20Cr–2Al–2Ti*

Alloy	**Condition**	**Thickness of the internal Al_2O_3 zone (µm)**	**Thickness of the internal TiN zone (µm)**
CMSX-6	Isothermal	1.1	1.9
	Thermal cycling	2.0	6.0
SRR99	Isothermal	1.1	2.8
	Thermal cycling	1.1	4.0
Ni–20Cr–2Al–2Ti	Isothermal	32.1	58.7
	Thermal cycling	34.3	81.1

3.2. Oxidation of the Model Alloys

The isothermal oxidation kinetics for the model alloys of the system Ni–Cr–Ti are represented in Fig. 6 as mass change vs exposure time at 1000°C in air. All the model alloys studied are characterised by a higher mass gain than any of the commercial nickel-base superalloys but generally this mass gain obeys a parabolic rate law.

The results of the cyclic thermogravimetric measurements for the model alloys of the system Ni–Cr–Ti in air at 1000°C are summarised in Fig. 7. In all cases a step-wise mass loss during the cycles due to spalling and/or cracking of the oxide scale could be observed.

During isothermal exposure of the model alloys of Cr content less than 20 mass% the oxidation behaviour is determined by the growth of an external NiO scale. Underneath this scale internal oxidation takes place that leads to internal Cr_2O_3 and/or TiO_2 precipitates, depending on the alloy composition. Increasing the Cr content to 20 mass% or higher changes the high-temperature oxidation behaviour to the formation of a continuous external Cr_2O_3 scale. Underneath the chromia scale internal oxidation (Al_2O_3) and nitridation (TiN) occur within alloys which contain small amounts (2 mass%) of Al and Ti, repectively. If, in addition to the increased Cr content, the Ti concentration is raised to 6 mass%, a thin external TiO_2 layer above the Cr_2O_3 scale can be established. The greater stability of the TiO_2 compared to Cr_2O_3 suggests that the TiO_2 layer is discontinuous. The content of dissolved oxygen in the base

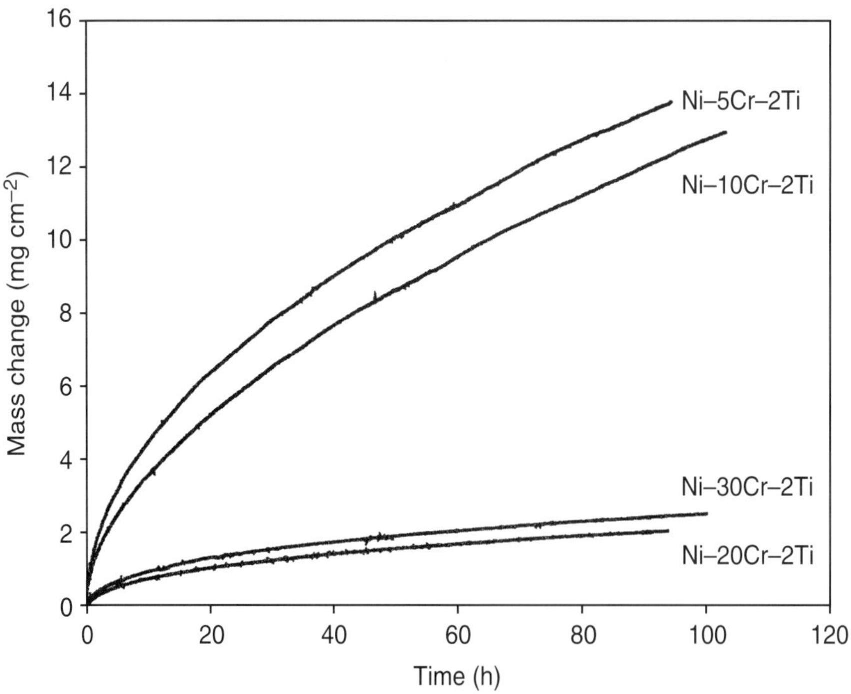

Fig. 6 *Thermogravimetrically measured mass gain vs time for the isothermal exposure of the model alloys of the system Ni–Cr–Al–Ti to air for 100 h at 1000°C.*

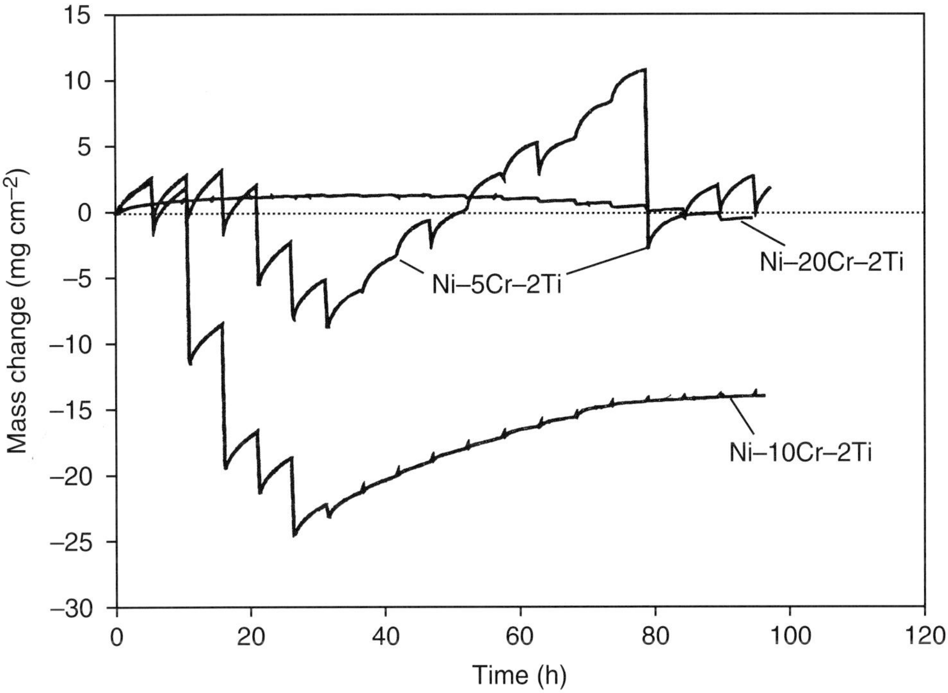

Fig. 7 *Thermogravimetrically measured mass gain vs time for thermally cyclic exposure of the model alloys of the system Ni–Cr–Al–Ti to air for 100 h at 1000°C.*

metal depends on the thermodynamic equilibrium conditions in the metal/oxide scale interface. Beneath a continuous external Cr_2O_3 scale only TiO_2 and Al_2O_3 can be formed as internal oxides [13].

Even isothermal exposure of the model alloys can result in the formation of microcracks, as is shown in Fig. 8(a) for the example of the model alloy Ni–5Cr–2Ti. However, during cyclic oxidation, significant spalling of the oxide scale takes place, in addition to cracking. This is exhibited in Fig. 8(b), which shows the surface of the model alloy Ni–5Cr–2Ti after thermal cycling.

The formation of a protective oxide scale is generally connected with a near-surface depletion of the scale-forming alloying element in the base metal. Since thermal cycling leads to a periodic oxide loss caused by cracking and spalling, the concentration of the scale-forming element drops below the critical value necessary for superficial scale formation. Then a transition to internal oxidation takes place and the protective effect of the oxide layer diminishes. In such cases the rapid oxidation of the base-element (Ni) accelerates the failure process [14].

Spalling and cracking of a protective oxide scale followed by a local formation of a non-protective scale during cyclic exposure allow for an increased oxygen penetration at the areas where the protective scale forming element such as Al has been depleted. Then the dissolved oxygen diffuses inward through the zone of internal oxidation. Hence, if the oxygen activity increases within an area of internal nitridation the nitrides can be oxidised. This process releases nitrogen which diffuses deeper into the alloy, where internal nitridation takes place again [15]. The comparison

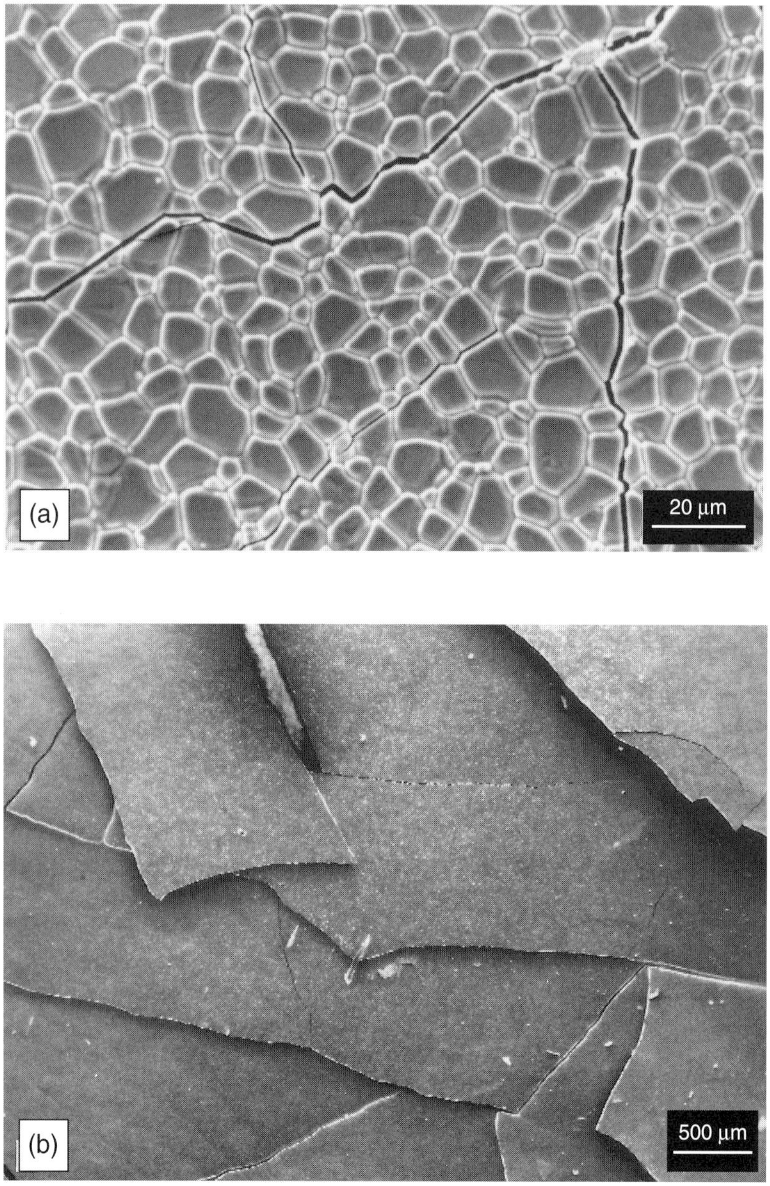

Fig. 8 (a) Microcracks within the NiO scale on the model alloy Ni–5Cr–2Ti exposed isothermally to air for 100 h at 1000°C, and (b) spalling of the NiO scale on the same alloy after cyclic thermal exposure to air for 100 h at 1000°C.

of the subsurface regions of the model alloy Ni–20Cr–2Al–2Ti after isothermal and cyclic exposure for 100 h at 1000°C to air in Fig. 9 confirms the effects mentioned above.

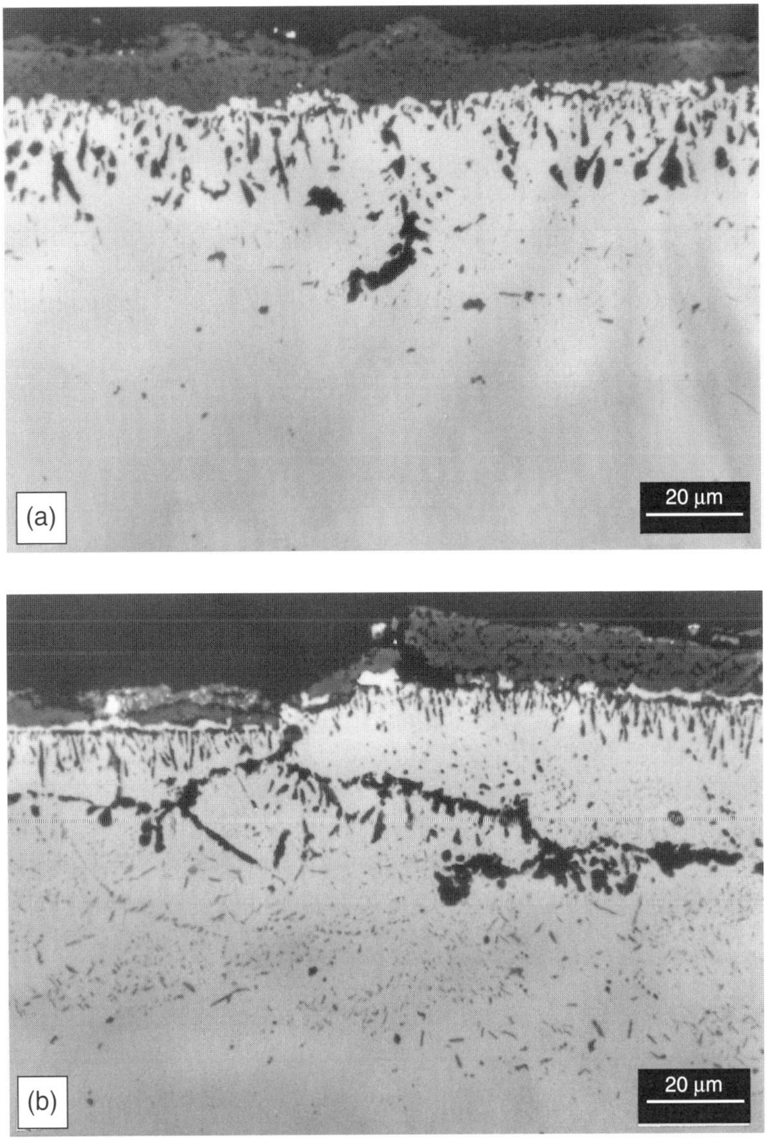

Fig. 9 *Comparison of cross-sections of the model alloy Ni–20Cr–2Al–2Ti after (a) isothermal and (b) thermal-cycling exposure to air at 1000 °C for 100 h.*

3.3. Compositional Changes in the Near-Surface Area During Oxidation

In order to obtain a better understanding of subscale depletion effects and to quantify them, several specimens were exposed to air and the oxide scale formed was polished off after every 24 h. Figure 10 represents the development of the surface oxides for the commercial superalloy CMSX-6 (Fig. 10a) and the model alloy Ni–20Cr–6Ti (Fig. 10b). During this specific cyclic oxidation experiment, a gradual transition from a

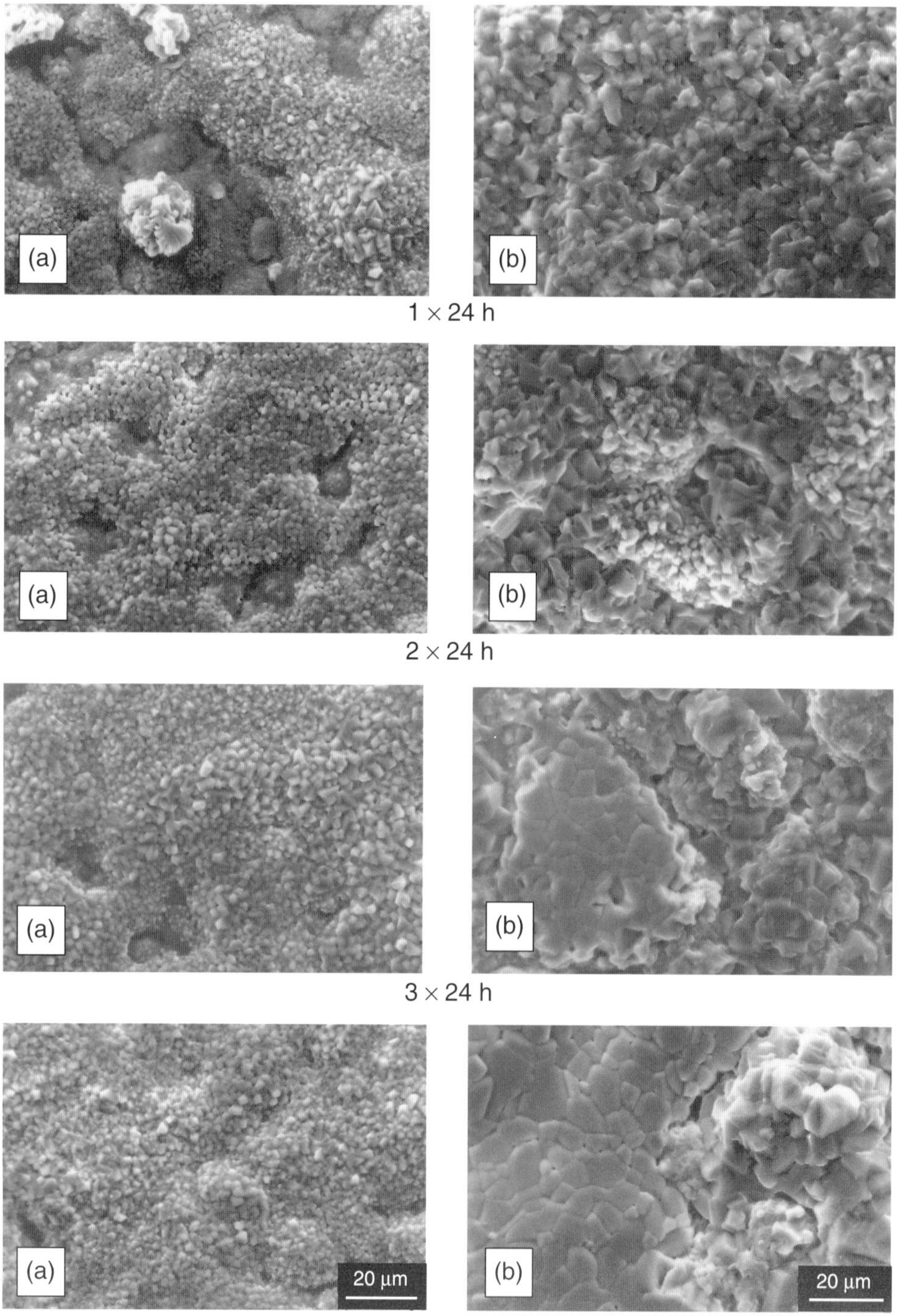

Fig. 10 (a) *Development of the oxidised surface of the superalloy CMSX-6 and (b) the model alloy Ni–20Cr–6Ti which were periodically exposed for 24 h at 1000°C to air and polished after each oxidation cycle.*

protective Cr_2O_3 layer to the formation of a NiO surface scale could be observed. The subsurface Cr depletion could be detected by EDS line profiles of chromium, as shown in Fig. 11 for cross-sections of the specimens represented in Fig. 10. These profiles show clearly a decrease in the Cr concentration at the metal/Cr_2O_3 scale interface.The concentration must be lower than that necessary for selective Cr_2O_3 formation, consequently resulting in enhanced Ni oxidation [16].

3.4. Development of a Computer Simulation for Cyclic Oxidation Processes

The observations reported so far serve as the experimental basis to expand a computer model, previously developed for internal corrosion phenomena [17], in order to simulate high-temperature corrosion processes during thermal cycling. The main part of this program is formed by a two-dimensional numerical diffusion calculation for all the species involved in the corrosion process by means of the finite-difference technique. The formation of corrosion products (e.g. internal and external oxides) is considered by the commercial thermodynamic software module ChemApp which calculates the precipitated and dissolved concentrations of all species, participated in the corrosion process, according to local thermodynamic equilibrium. Spalling and cracking of the scale has been integrated into the program in a first approximation applying both existing models from literature (e.g. a modified version of the computer model of cyclic oxidation (COSP) by Lowell *et al.* [18]) and FEM calculations of the stresses resulting from the different thermal expansion coefficients of oxide scale and base metal. Figure 12 shows, as an example, the two-dimensional N diffusion, which is assumed to be caused by a small crack in the oxide scale of the model alloy Ni–2Ti, leading to internal precipitation of TiN.

A further development of the computer program and the experimental verification of the results supplemented by creep experiments in different atmospheres is the subject of ongoing work.

4. Conclusions

It was shown that under both isothermal and cyclic thermal conditions the high-temperature corrosion behaviour of the three commercial single-crystal Ni-base superalloys CMSX-4, CMSX-6, and SRR99 is more favourable for technical applications than any one of the model alloys of the system Ni–Cr–Al–Ti studied. The high Al concentration in the alloy SRR99 and particularly the Re and Hf addition to the alloy CMSX-4 lead to both a small mass gain during isothermal exposure and reduced cracking and spalling during cyclic exposure to air at 1000°C. The isothermal oxidation of all the model alloys studied obeys the parabolic rate law while cyclic oxidation leads to a step-wise mass loss similar to the cyclic behaviour of the superalloy CMSX-6. Especially under cyclic thermal conditions all the superalloys are susceptible to internal oxidation of Al as well as internal nitridation of Ti. These internal corrosion phenomena contribute to a deterioration of the material properties due to the near-surface dissolution of the strengthening γ'-phase. Cyclic oxidation experiments which were intensified by periodic oxide scale removal documented the transition process that leads to the formation of nickel-rich oxides and finally to

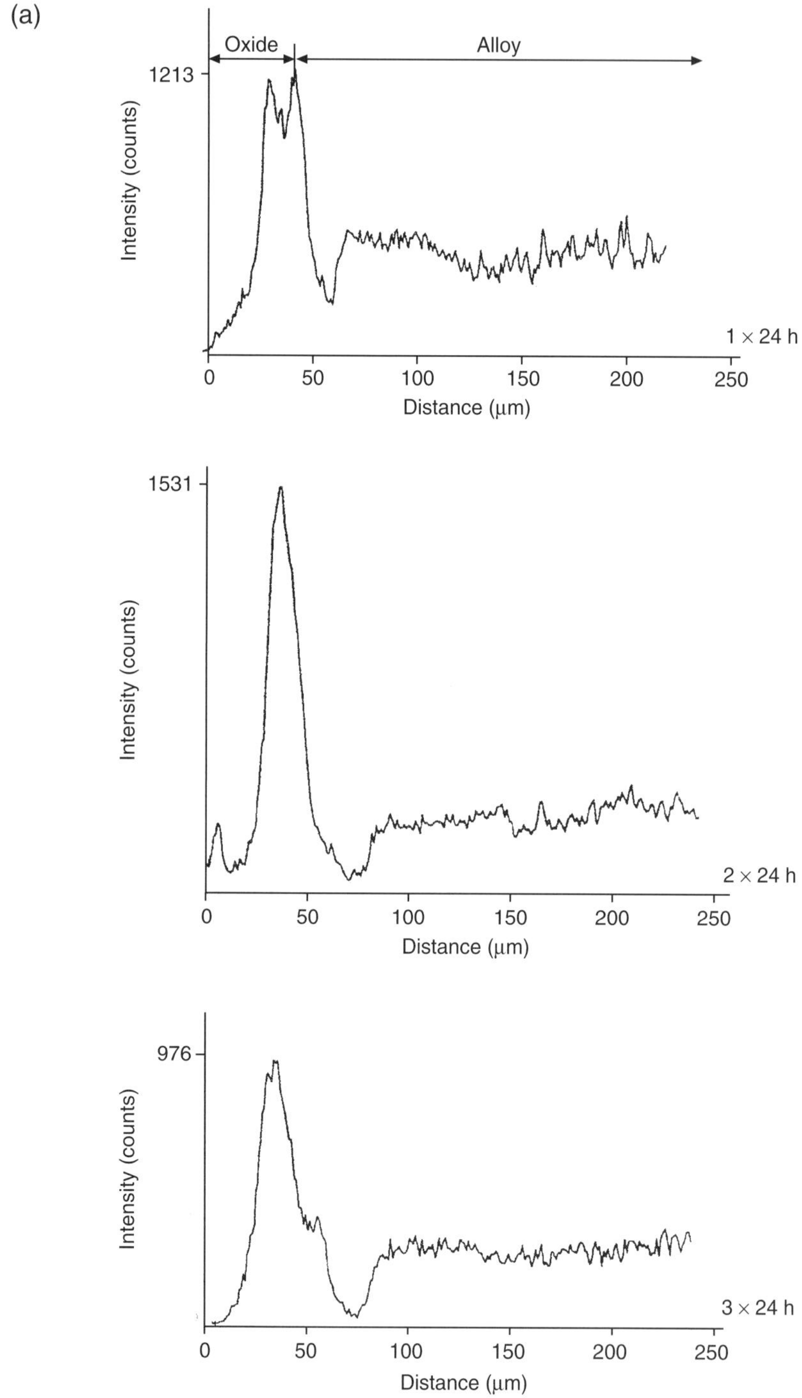

Fig. 11 (a) *EDS-line scans of the near-surface Cr concentration after cyclic oxidation of the superalloy CMSX-6 at 1000°C in air.*

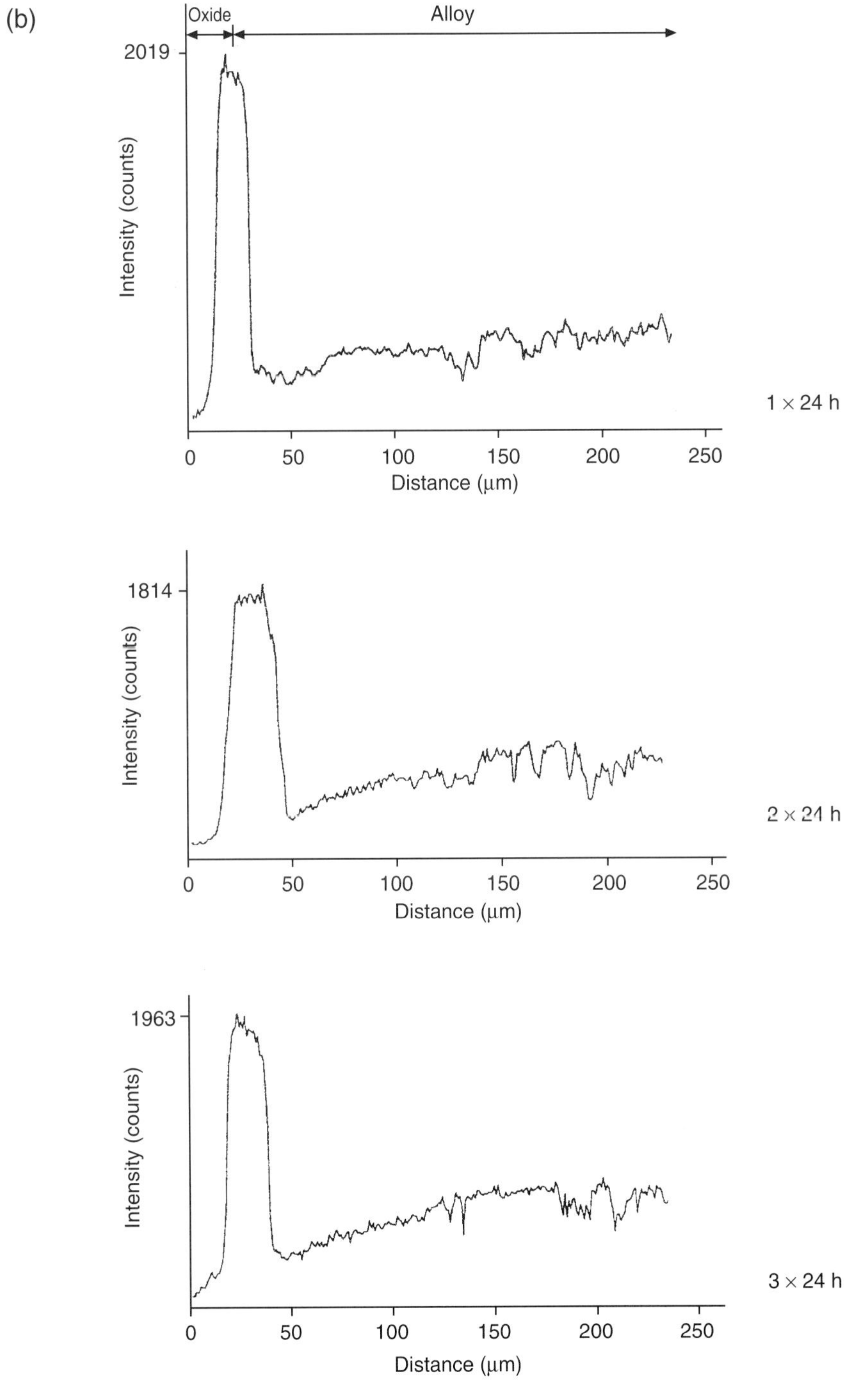

Fig. 11 (b) *EDS-line scans of the near-surface Cr concentration after cyclic oxidation of the model alloy Ni–20Cr–6Ti at 1000°C in air.*

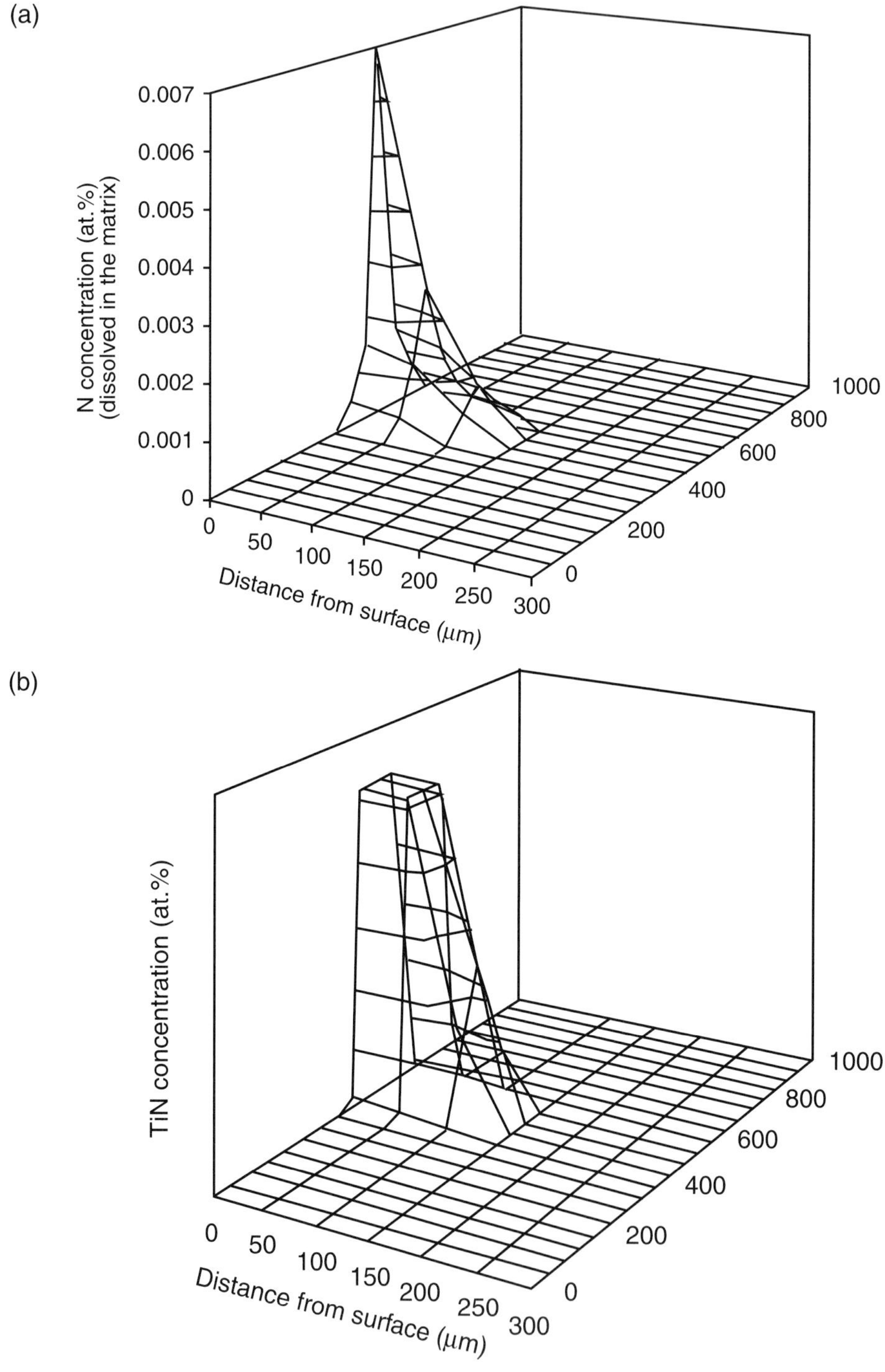

Fig. 12 *Calculated concentration profiles of (a) N and (b) TiN after cracking of the oxide scale on the model alloy Ni–2Ti exposed to air for 100 h at 1000°C.*

breakaway oxidation. On the basis of these results a computer simulation was developed which can be applied to technical cyclic corrosion problems.

References

1. F. H. Stott, *Mater. Charact.*, 1992, **28**, 311.
2. C. Wagner, *Z. Elektrochemie*, 1959, **63**, 772.
3. U. Krupp and H.-J. Christ, *Metall. Mater. Trans. A* (in press).
4. F.-R. Chien and R. Brown, *J. Mater. Sci.*, 1992, **27**, 1514.
5. B. A. Pint, *Oxid. Met.*, 1996, **45**, 1
6. S. Y. Chang, U. Krupp and H.-J. Christ, *Metall.*, 1998, **52**, 355.
7. M. Göbel, A. Rahmel and M. Schütze, *Oxid. Met.*, 1993, **39**, 231.
8. F. H. Stott, Y. Shida, D. P. Whittle, G. C. Wood and B. D. Bastow, *Oxid. Met.*, 1982, **18**, 3/4, 127.
9. G. H. Meier, *Mater. Sci. Eng.*, 1989, **120A**, 1.
10. J. H. Chen, P. M. Rogers and J. A. Little, *Oxid. Met.*, 1997, **47**, 5/6, 381.
11. D. L. Douglass, *J. Mater.*, 1991, **11**, 74.
12. D. P. Whittle, *Proc. High Temperature Alloys for Gas Turbines*, Liège, Belgium, 1978, p.109.
13. R. Mevrel, *Mater. Sci. Technol.*, 1987, **3**, 531.
14. M. Lery, P. Farrell and F. Pettit, *Corrosion*, 1986, **42**, 708.
15. G. H. Meier, N. Birks and F. S. Pettit, *Corrosion '81*, NACE-6, San Diego, California, 1983, p.327.
16. B. D. Bastow, D. P. Whittle and G. C. Wood, *Oxid Met.*, 1978, **12**, 413.
17. U. Krupp and H.-J. Christ, *Oxid. Met.*, 1999, **52** (in press).
18. C. E. Lowell, C. A. Barrett, R. W. Palmer, J. V. Auping and H. B. Probst, *Oxid. Met.*, 1991, **36**, 81.

5
Cyclic Oxidation of Thin FeCrAl(RE) Foils in Air: SEM and EPMA Investigations

G. STREHL, H. AL-BADAIRY*, L. M. RODRIGUEZ, J. KLÖWER§, G. BORCHARDT, G. TATLOCK* and A. J. CRIADO¶
Institut für Allgemeine Metallurgie, TU Clausthal, D-38678 Clausthal-Zellerfeld, Germany
*Department of Materials Science and Engineering, University of Liverpool, Liverpool L69 3BX, UK
§Krupp VDM GmbH, D-58778 Werdohl, Germany
¶Departamento de Ciencia de los Materiales e Ingeniería Metalúrgica, Facultad de C.C. Químicas, Universidad Complutense de Madrid, ES-28040 Madrid, Spain

ABSTRACT

Research has been carried out on the cyclic oxidation behaviour of thin commercial FeCrAl(RE) foils in air. The influence of cycle time (5-, 20-, 50-h) and temperature (1100, 1200, 1300°C) on the mass change has been investigated. The mechanism of oxidation and the morphology of the scale have been examined by SEM and EDX/WDX at different stages of the oxidation process. Preliminary conclusions can be drawn on the breakaway process in thin foils. The results show that breakaway appears in different stages: formation of a chromia layer underneath the still adherent alumina scale, slight changes in the growth law and spallation of the scale followed by rapid oxidation of the iron-rich matrix. This is different from the case of thicker sheet material, where breakaway can be established by rapid mass change solely.

The main cause of breakaway — the depletion of aluminium in the foil — has been examined by EPMA measurements of the remaining aluminium level in the alloy. These results are consistent with aluminium levels calculated from the mass change.

1. Introduction

The FeCrAl foils investigated are candidate substrate materials for modern automotive catalyst converters. With their properties (weight, thickness, lower exhaust pressure head) they help to meet future European regulations for emissions. Because automotive catalyst converters experience many thermal cycles during their lifetime the influences of cycle time and temperature on the life expectancy has been investigated. Special attention has been paid to the aluminium content in the alloy.

2. Materials, Sample Preparation and Oxidation Experiments

Table 1 shows the composition of the industrial alloys that have been tested and Table 2 the test conditions. All samples were cleaned in a reflux condensation apparatus with *iso*propanol after a degreasing procedure. The oxidation was carried out in laboratory air. The heating and cooling rates were *ca.* 11 Ks^{-1}. Apart from the

Table 1. *Alloy composion in mass%*

Alloy	Fabrication route	Fe	Cr	Al	Y	Y_2O_3	Zr	Si	Ti	Mn
Aluchrom I SE	Wrought	Bal.	20.5	5.2	—	—	0.002	0.39	0.010	0.25
Aluchrom YHf	Wrought	Bal.	20.25	5.6	0.05	—	0.04	0.26	<0.010	0.2
Kanthal AF	Wrought	Bal.	21.1	5.27	0.08	—	0.6	0.22	0.08	0.1
PM 2000	ODS	Bal.	18.83	5.83	—	0.5	—	0.02	0.49	0.05

Table 2. *Materials investigated, cycle times and temperature for oxidation in air*

Alloy	Thickness	Cycle time at 1100°C	Cycle time at 1200°C	Cycle time at 1300°C
Aluchrom I SE	50 μm	5-h, 20-h, 50-h	5-h, 20-h, 50-h	5-h, 20-h
Aluchrom YHf	50 μm		20-h	20-h
Kanthal AF	50 μm		20-h	
Kanthal AF	70 μm		20-h	
PM 2000	125 μm			20-h

Aluchrom I SE foils, all samples were coupons (10 mm × 20 mm) cut from sheet. To simulate a fixed structure in an automotive catalytic converter and to prevent additional effects by sample bending during oxidation, the Aluchrom I SE foils were electrically spot welded to rings with a diameter of approximately 30 mm. Oxidation was carried out until the first signs of breakaway oxidation were detected. These were:

- changes in colour, showing the formation of chromia, and
- rapid changes in the mass gain, showing classical breakaway.

After the exposure, scanning electron microscopy (SEM) pictures were taken and the composition of the remaining substrate was determined by EPMA (electron probe microanalysis) WDX/EDX (wavelength/energy dispersive X-ray spectroscopy).

3. The Influence of Cycle Time on the Lifetime

For every cycle time and temperature three specimens were oxidised. The mass gain was reproducible and has been averaged. With the assumption that the mass gain can be associated with the oxidation of aluminium only, the point where all the aluminium was consumed ($\Delta m(t_B)$ in gm^{-2}) can be calculated for given specimen dimensions (d = foil thickness in μm), alloy density (ρ in gcm^{-3}) and alloy composition (c_{Al} aluminium content in mass%):

$$\Delta m(t_B) = \frac{\rho d c_{Al}}{2 \times 1.12443} \qquad (1)$$

For Aluchrom I SE Δm (t_B) was 8.3 gm^{-2}. This value has been marked in the mass gain diagrams (Figs 1–3). The first evidence of a colour change is shown by filled symbols. This can be related closely to the start of chromia formation after the total consumption of aluminium in the alloy. Slight differences can be traced back to small variations in foil thickness.

A closer look on the $\Delta m(t)$ curves shows that the growth law for all cycles seems to be nearly the same. Thermal cycling could influence the growth law by the induction of additional stresses from the difference in the thermal expansion coefficient of the alloy and the oxide [1], resulting in microcracks in the oxide. Following this train of thought would imply faster oxygen transport through a cracked scale and thus a shorter lifetime for samples exposed to short cycles. This would appear to be substantiated on comparing the 20-h and 5-h cycles. However, the specimens cycled in 50-h periods show lifetimes similar to the 5-h cycles. So cycling also has a positive effect on the growth law. It enables the stress in the oxide metal system to relax and to heal the microcracks in the oxide.

For the evaluation of the growth law the data have been fitted to $\Delta m = kt^{1/n}$ [2]. The results are represented in Table 3. Also the k_p value for a parabolic growth law [3] $\Delta m = (k_p t)^{1/2}$ is shown.

The overall growth law changes with temperature from subparabolic to hyperparabolic, which might indicate a change in the growth mechanism. On the assumption that the parabolic constant k_p is proportional to a diffusion coefficient,

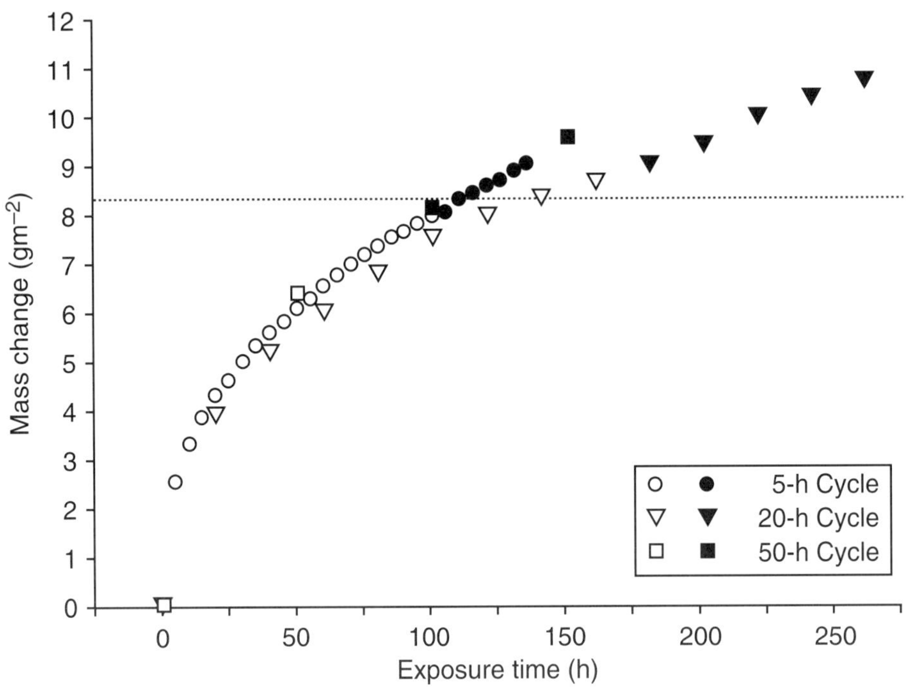

Fig. 1 Cyclic oxidation of Aluchrom I SE at 1100°C. The line at 8.3 gm^{-2} marks the total consumption of aluminium in the alloy. The filled symbols indicate the beginning of the colour change of the sample to green — an indication of chromia formation.

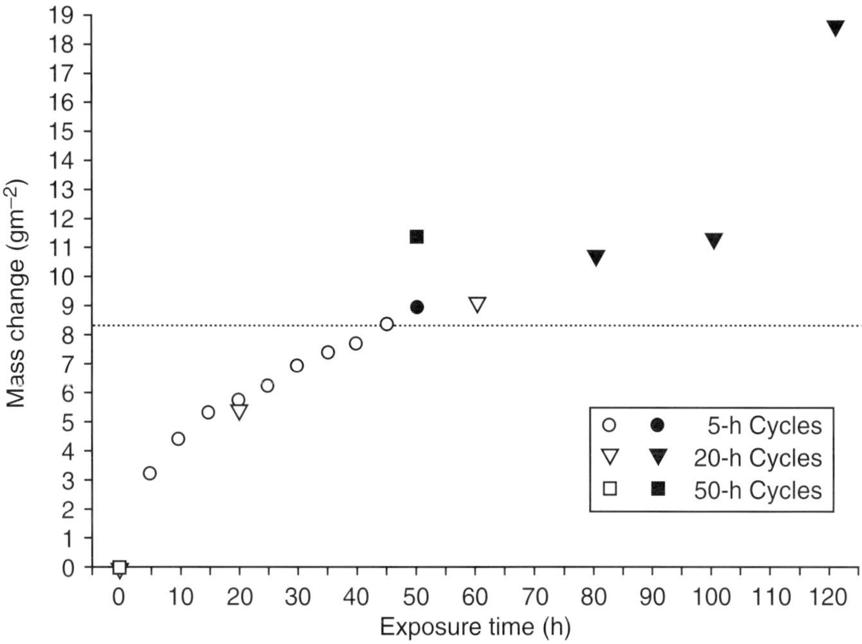

Fig. 2 *Cyclic oxidation of Aluchrom I SE at 1200°C.*

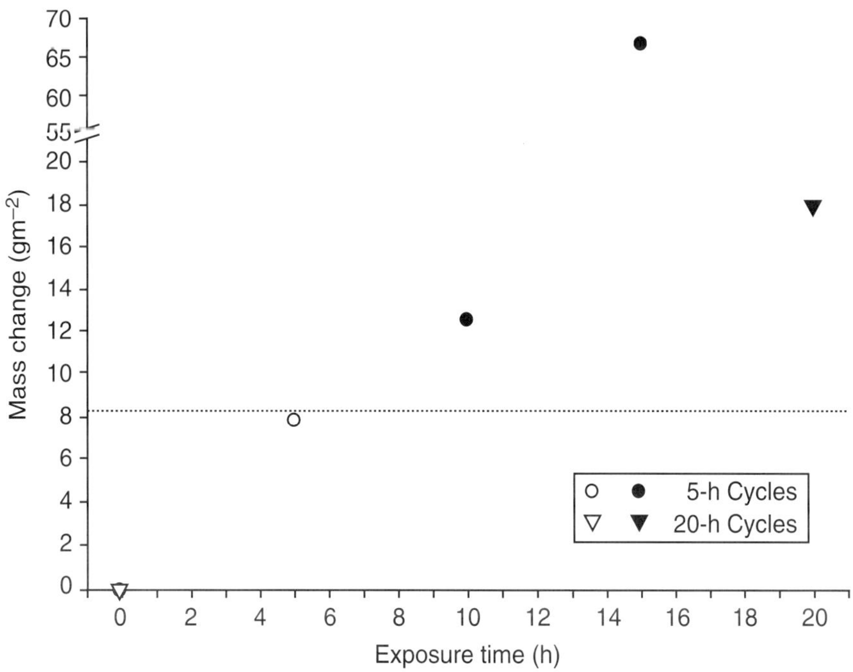

Fig. 3 *Cyclic oxidation of Aluchrom I SE at 1300°C.*

***Table 3.** Evaluated constants of the growth law for Aluchrom I SE. The relative errors are 4% for* k *and 1% for* n *and* k_p

Cycle	n	k ($gm^{-2}\,s^{-1/n}$)	k_p ($g^2m^{-4}s^{-1}$)
1100°C 5-h/50-h	2.535	0.05136	0.00018
1100°C 20-h	2.595	0.05336	0.00014
1200°C 5-h/20-h	2.362	0.05117	0.00041
1300°C 5-h/20-h	1.707	0.0262	0.00437

the activation energy for an effective diffusion process (combined diffusion through the alumina grains and along the grain boundaries) is calculated to be 284 $kJmol^{-1}$ with an error of at least 17%. This value still falls within the range of activation energies measured for similar materials [4].

The scale growth law for the other materials has still to be evaluated because data currently exist only for a single temperature. The gravimetric data for the remaining materials are shown in Fig. 4, revealing a similar behaviour compared to Aluchrom I SE. Breakaway starts only after complete aluminium depletion.

4. SEM Observations

The Scanning Electron Microscope used for these investigation was a Hitachi model S-2460N (operating voltage 25 kV) with an analytical EDX system. An SEM study of the surface morphology showed that the alumina scale on alloy Aluchrom YHf was

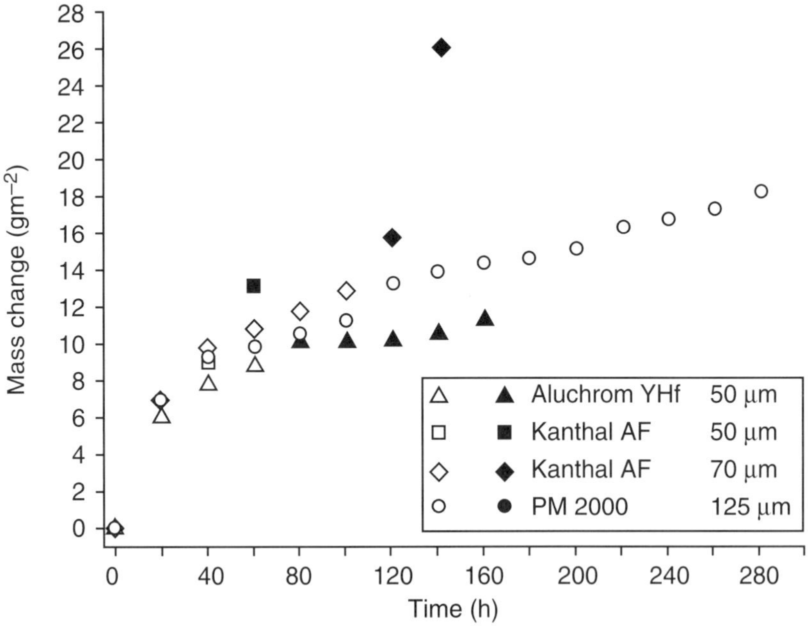

***Fig. 4** Oxidation of the remaining alloys at 1200°C in 20-h cycles.*

convoluted after 40 h at 1300°C. However, examination of the scale in cross-section showed that the scale was well adhered to the metal, although whenever there was a dip in the metal, whose origin is unknown as yet, cracks formed in the scale above (Fig. 5). These features could trigger breakaway oxidation because of poor adhesion of the scale. By comparison the investigation of a cross-sectional sample of alloy PM 2000, showed that after the third cycle the scale was well adhered to the metal substrate in the regions away from where breakaway oxidation had occurred and detached in the regions next to the breakaway location.

Initial observations indicated that breakaway oxidation started after 60 h oxidation at 1300°C for PM 2000 at all the corners of the sample (Fig. 6a). This can be explained by the geometry of the sample. From experiments with wedge samples [4,5], it is

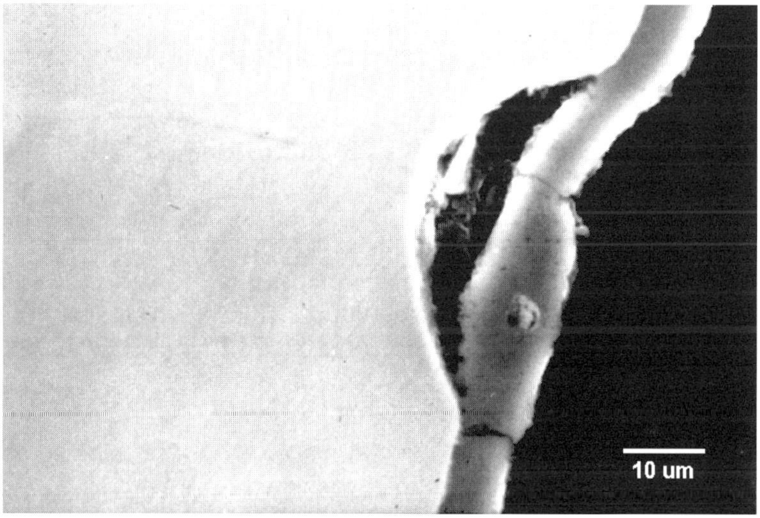

Fig. 5 *Cross-section of Aluchrom YHf oxidised at 1300°C for 40 h, where the still protective scale shows cracks where there is a dip in the metal.*

Fig. 6(a) *Photograph of a PM 2000 sample (10 mm × 20 mm × 125 µm) oxidised at 1300°C for 60 h. The white line marks the region of the cross-section in Fig. 6(b).*

known that growth stresses lead to cracks in the scales, especially when the materials are creep resistant like the ODS materials. These cracks in the scale lead to a faster consumption of aluminium in the corners and an earlier breakaway. Analysis of cross-sections (Fig. 6b) indicate that the substrate in the corners contains little aluminium and that chromia forms instead of alumina.

Another sample of Aluchrom YHf oxidised at 1200°C for 160 h also showed convolution and cracking of the scale (Fig. 7a). The same cracks can be seen in the cross-section (Fig. 7b). Additionally a chromia layer was formed underneath the alumina. Because Al_2O_3 and Cr_2O_3 have the same crystal structure and nearly equal lattice constants (0.513 nm and 0.535 nm respectively) [6] it is obvious that after aluminium depletion in the substrate chromium takes over the role of the aluminium as the main oxide former. Phase diagrams [7] suggest that alumina and chromia form a solid solution. Only the low cation diffusivities in the oxides can be the reason why two different oxide layers coexist on the substrate.

On thicker material the protective alumina layer is damaged by spallation and cracks. After the depletion of alumina in the substrate there are always some locations where oxygen has easy access to the substrate. The oxidation of the chromium is no longer slowed down by the limited oxygen transport through the alumina. The oxygen partial pressure increases and rapidly growing iron oxides are immediately formed as no substantial chromia formation takes place. The formation of iron oxide rods mechanically breaks up the scale and destroys the alloy.

Figure 8 shows an Aluchrom I SE sample which has been oxidised until aluminium depletion and chromia formation are about to start. The metal–oxide interface is different on both sides. The upper one is very straight and shows some voids in the metal. The lower one has no voids but convolution can be detected. This is a strong hint that the convolution and the void formation are two possible creep processes relaxing the stress at the oxide metal interface.

Fig. 6(b) Cross-section of a corner where breakaway occurred and chromia formed.

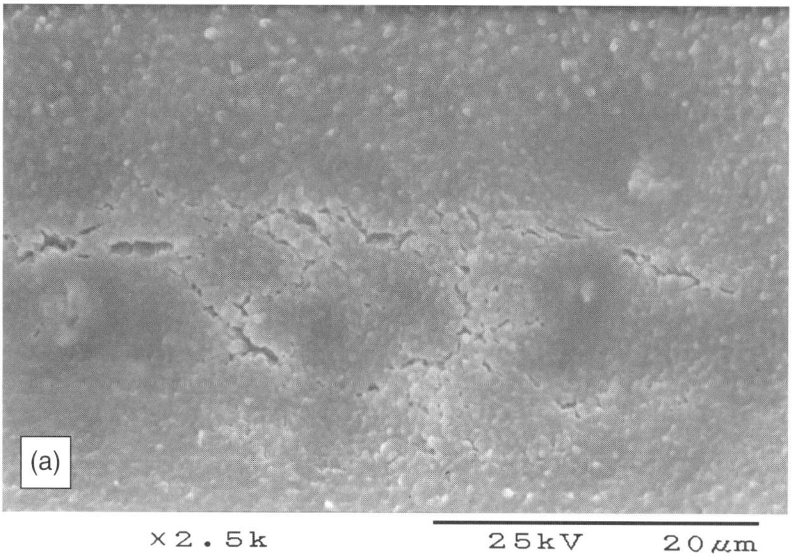

Fig. 7(a) *Surface of Aluchrom YHf after 160 h oxidation at 1200°C, showing convolution and cracks within the scale.*

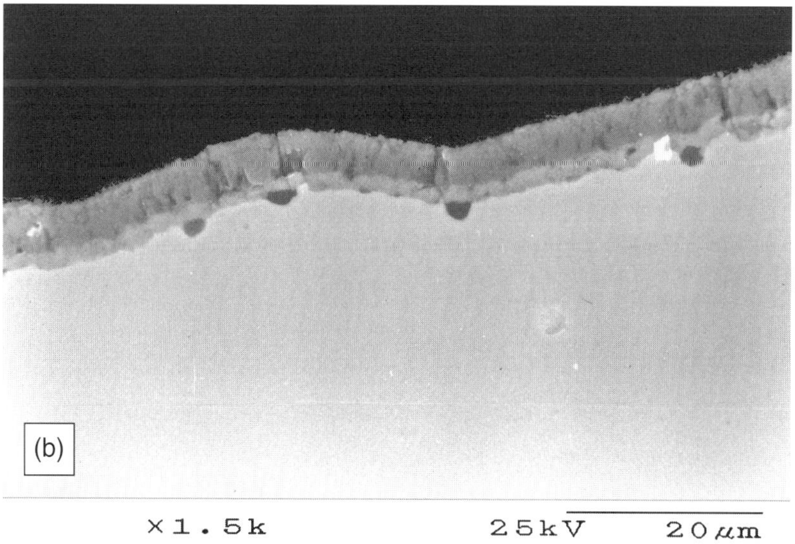

Fig. 7(b) *Cross-section of the same Aluchrom YHf sample, showing cracks in the oxide layer and the formation of a chromia scale underneath the alumina.*

5. The Aluminium Content near Breakaway

From the mass gain the remaining Al content in the alloy can easily be calculated based on the assumption that the oxygen reacts only with aluminium to form Al_2O_3. This approximation leads to negative aluminium contents after total aluminium depletion in the alloy and the start of chromia and iron oxide formation.

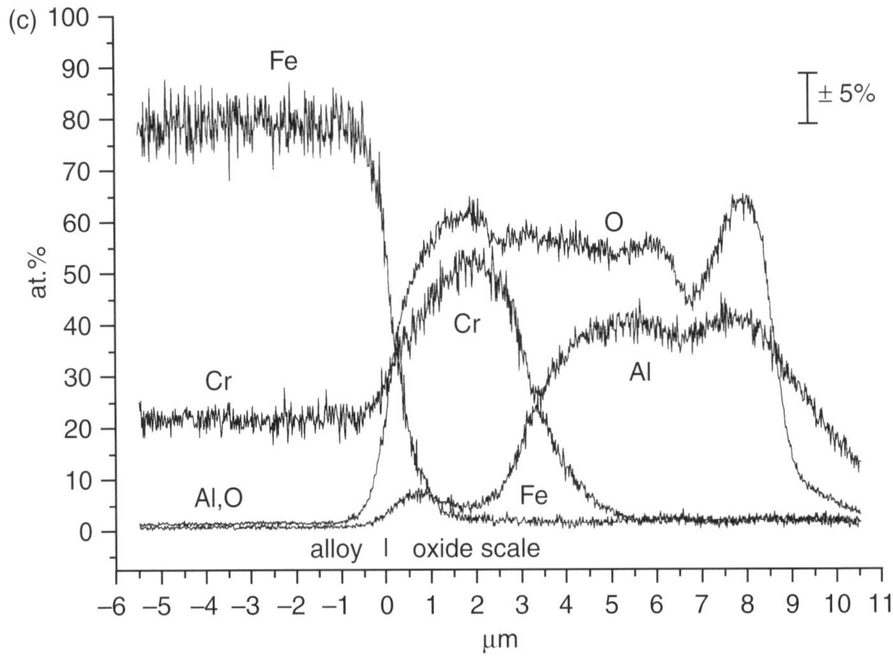

Fig. 7(c) EPMA profile of the same Aluchrom YHf sample, showing the depleted alloy, the inner chromia and the outer alumina layers.

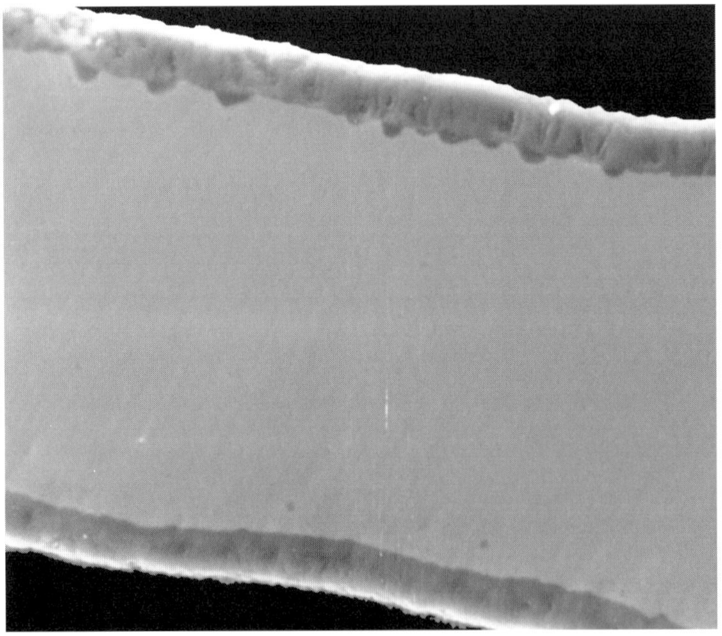

Fig. 8 Cross-section of Aluchrom I SE sample after 60 h oxidation at 1200°C. The aluminium in the alloy is totally depleted and the chromia formation just about to begin.

Figure 9 shows that measured and calculated values are close if the sample is only growing an alumina scale — that is, the assumption is fulfilled. In the region of negative calculated Al contents one observes that the remaining aluminium content after breakaway is not always zero. This means that the formation of chromia starts before all of the aluminium in the substrate is consumed. It is also clearly evident that the critical aluminium content is much lower than those contents published in recent years (1.2–2%)[8], (1.5%)[9], (2%)[10]. The greater consumption of the available aluminium may occur only with thin (<150 μm) foils and due to the fact that during the whole lifetime of these foils the alumina scale is adherent.

A closer look at the formation process of the chromia reveals that after aluminium depletion silicon first segregates at the metal–oxide interface. From the Gibbs energy of formation it is possible that silicon forms an oxide before chromium. This is very likely to happen because the silica layer can be observed even after total oxidation of the sample.

An interesting aspect with chromia formation is that it seems to take place within the alumina (see Figs 11 and 7c). A thin alumina layer between the metal and the chromia was often observed. A possible explanation might be that chromium ions have a diffusivity close to that of oxygen, so that the reaction takes place within the alumina.

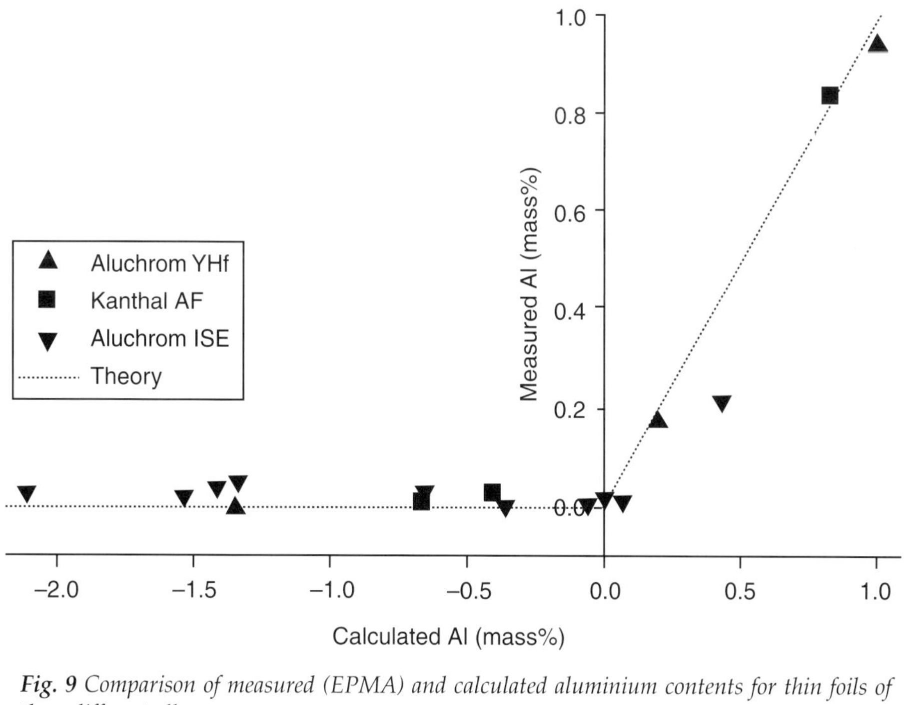

Fig. 9 *Comparison of measured (EPMA) and calculated aluminium contents for thin foils of three different alloys.*

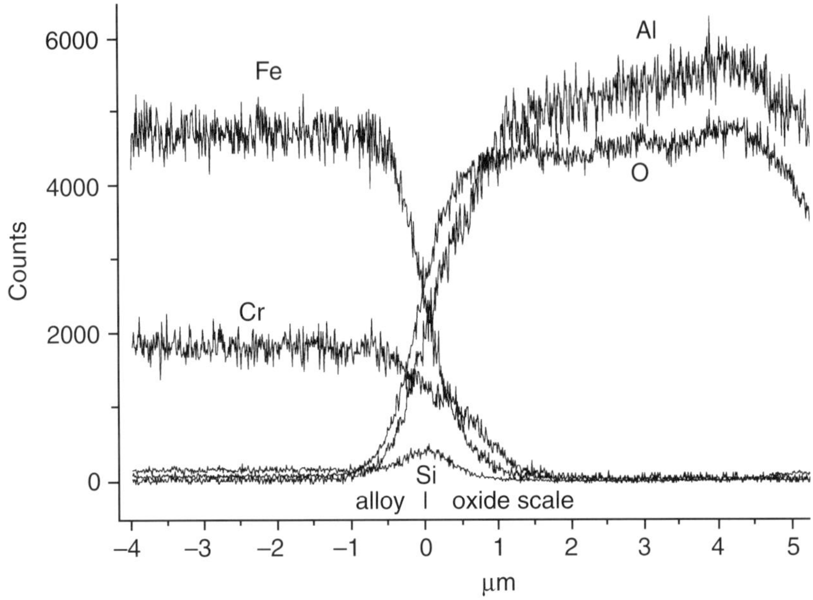

Fig. 10 *EPMA profile of Aluchrom I SE after oxidation at 1100°C for 135 h (i.e. just before breakaway), showing silicon segregation at the metal–oxide interface.*

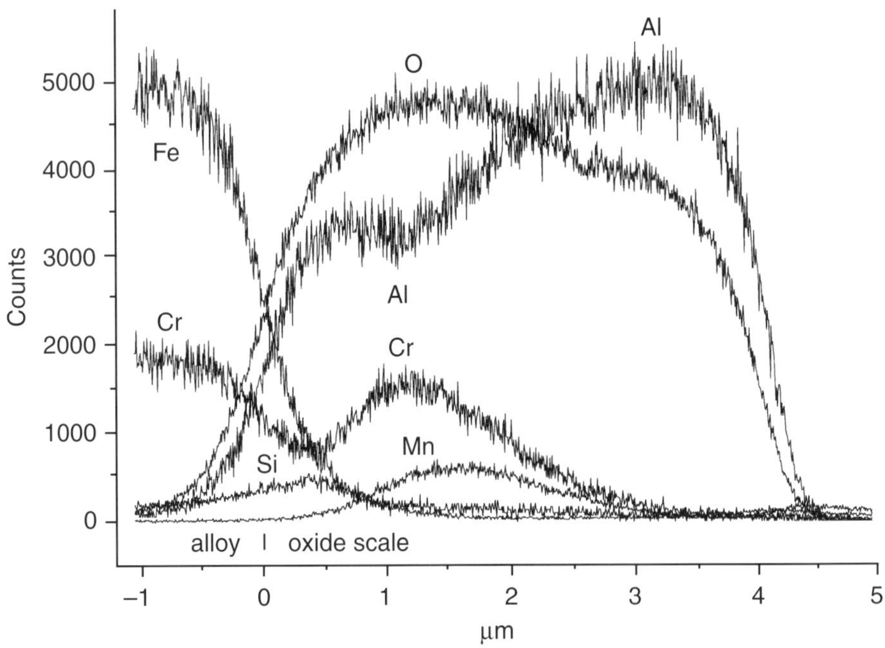

Fig. 11 *EPMA profile of Aluchrom I SE after breakaway (oxidation at 1200°C for 45 h), showing the segregation of silicon at the metal–oxide interface, the formation of an inner chromia layer in which manganese was enriched and an outer alumina layer.*

6. Discussion

For thicker FeCrAl(RE) alloys the lifetime governing process is the oxidation of the aluminium, which forms an adherent scale until a critical thickness is reached. Subsequent spallation and reformation of the scale consumes the aluminium in the alloy until a critical aluminium content is reached. The alloy has now lost its capability to form a protective alumina scale so that rapid oxidation of chromium and iron (called breakaway) ensues and is observed as a sudden increase of the mass.

The behaviour of thin foils is different. Their aluminium reservoir is much smaller and the alumina scale does not reach the thickness that leads to spallation. Underneath the alumina, chromia starts to form. From the literature [4,11] it is known that oxygen inward diffusion is the growth rate controlling process for this type of alloy. The chromia underneath the alumina cannot grow faster than the alumina did before. Its growth will be controlled by the oxygen supply through the alumina as the diffusivity of chromium is much higher than that of oxygen in chromia [12]. So the change in the chemical reaction at the metal–oxide interface can only be determined by the colour change of the sample (or sensitive *in situ* optical spectroscopy).

7. Acknowledgements

The research presented here was possible by the financial assistance of the following European programmes:

Brite Euram Project BE–97–4491
TMR (COPES/Technische Universität Clausthal)
TMR (COSI/University of Liverpool)
(TMR = Training and Mobility of Researchers Programme).

References

1. M. Schütze and W. Przybilla, The role of stresses in oxidation and shape stability, in *Metal-Supported Automotive Catalytic Converters,* October, 1997, Wuppertal, Germany, p. 163–177 (Ed. H. Bode). Werkstoff-Informationsgesellschaft mbH, Frankfurt. 1997.
2. W. J. Quadakkers and K. Bongartz, *Mater. Corros.,* 1994, **45**, 232–241.
3. N. Birks and G. H. Meier, *Introduction to High Temperature Oxidation of Metals.* Edward Arnold Ltd., London, UK, 1983.
4. M. Göbel *et al.*, Growth of α-alumina scales on Fe–Cr–Al alloys, in *Metal-Supported Automotive Catalytic Converters,* October, 1997, Wuppertal, Germany, pp. 191–205 (H. Bode, ed.). Werkstoff-Informationsgesellschaft mbH, Frankfurt, 1997.
5. H. Al-Badairy and G. J. Tatlock, The Application of a Wedge-Shaped-Sample Technique for the Study of Breakaway Oxidation in Fe–20Cr–5Al Base Alloys. Accepted for publication in *Oxid. Met.*
6. W. B. Pearson, *A Handbook of Lattice Spacings and Structures of Metals and Alloys*. Pergamon Press, Oxford, 1964.
7. E. Levin, C. Robbins and H. McMurdie, *Phase Diagrams for Ceramists,* p. 121. The American Ceramic Society, Columbus, Ohio, 1964.

8. W. J. Quadakkers, K. Bongartz and F. Schubert, The prediction of oxidation limited life of thin walled ODS heat exchangers for high temperature applications, in *Materials for Advanced Power Engineering, Part II,* p. 1533–1542, COST 501, Liège, October 1994 (D. Coutsouradis *et al.,* eds). Kluwer Academic Publishers, 1994.
9. J. Klöwer, *Mater. Corros.,* 1998, **49**, 758–763.
10. H. Al-Badairy and G. J. Tatlock, An electron probe microanalysis study of the breakaway oxidation of Fe–20Cr–5Al based Alloys, in *Electron Microscopy,* paper presented at *ICEM14,* Cancun, Mexico, Symposium U, Volume II, 1998.
11. C. Mennicke, E. Schuhmann, M. Rühle, R. J. Hussey, G. I. Sproule *et al., Oxid. Met.,* 1998, **49**, 455–466.
12. S. C. Tsai, A. M. Huntz and J. Philibert, *Defect and Diffusion Forum,* 1997, **143–147**, 1195–1200.

6
Cyclic Oxidation of PM 1000

H.-P. MARTINZ, F. E. H. MÜLLER* and K. PRANDINI
Plansee AG, Reutte, Austria
*Plansee GmbH, Lechbruck, Germany

ABSTRACT

The Ni-based ODS-Superalloy PM 1000† (Ni–20Cr–3Fe–0.5Ti–0.3Al–0.6Y_2O_3) because of its excellent high temperature strength and, presumably, adequate short term oxidation resistance represents a candidate material for future high temperature aerospace applications like hot metallic structural components, fasteners or control surfaces.

In order to simulate re-entries of reuseable space vehicles which create high thermal loads at various temperature levels, samples of PM 1000 with various surface states (sand blasted with iron grit or alumina; ground) were submitted to cyclic oxidation and thermal shock tests. The test equipments used consisted of air flooded furnaces and time controlled manipulation systems for up to six samples between 1150–1350°C and room temperature. Discontinuous weighing of the samples and the spalled–off scales during 10 to 100 cycles were used for the evaluation of the oxidation resistance of the various PM 1000 samples. Roughness measurements, SEM, EDX and XRD were applied before and after the tests for the clarification of oxidation mechanisms.

Iron-grit blasted samples showed higher mass gain and lower spallation of the scale mainly because of their homogeneously roughened surface and, presumably, to a minor extent, because of surface alloying effects by the grit. The much smoother surfaces of the ground samples on the other hand were oxidised most strongly with subsequent severe spallation effects. It can be assumed that this was due to the almost complete absence of anchoring sites for the growing oxide. The alumina blasted samples showed a behaviour slightly worse than that of the iron-grit blasted samples.

1. Introduction

The Ni-based (Oxide Dispersion Strengthened) ODS-Superalloy PM 1000 (Ni–20Cr–3Fe–0.5Ti–0.3Al–0.6Y_2O_3) as a result of its excellent high temperature strength and, presumably, adequate short term oxidation resistance is one of the candidate materials for future high temperature aerospace applications like hot metallic structural components, fasteners or control surfaces.

Much work has already been done on the oxidation behaviour of PM 1000 and related ODS-superalloys, but test modes in most cases were isothermal, and temperatures (in cyclic tests) were without exception below 1200°C [1–22]. Furthermore, surface states of the test specimens were not sufficiently characterised.

Taking into account the planned applications of the material the aim of this work was to examine the air oxidation behaviour of PM 1000 samples with various technical

†Trademark of PLANSEE GmbH Lechbruck.

surface conditions (blasted and ground) under thermocyclic and thermal shock conditions with maximum temperatures up to 1350°C.

2. Experimental Procedures

Samples of PM 1000 ($3 \times 18 \times 29$ mm^3 for cyclic oxidation tests and $8 \times 14 \times 80$ mm^3 for thermal shock tests) with three technically relevant surface states (blasted with iron grit of about 0.5 mm dia. at 4–5 bar; blasted with alumina grit of 0.25 to 0.5 mm dia. at 4–5 bar and ground with an alumina belt with 0.25 mm grit) were submitted to cyclic oxidation and thermal shock tests.

The cyclic oxidation test equipment consisted of a tubular furnace at constant temperature and with constant air flow. Samples (on a sledge) were moved into the hot zone and removed to a water cooled section at a controlled speed. The thermal shock test equipment comprised a chamber furnace at constant temperature. Samples (also on a sledge) were moved into the hot zone and removed to an air stream from a perforated spiral tube pneumatically.

The following temperature–time dependencies were used:

Cyclic oxidation testing
Type 1: within 6 min from 20°C to 1300 or 1350°C, 13 min dwell time, cooling to 20°C within about 10 min.

Type 2: within 6 min from 20°C to 1150°C, 60 min dwell time, cooling to 200°C within 9 min.

Thermal shock testing
Heating from 20°C to 1200°C within 5 min and immediately cooling to 200°C within about 6 min.

Discontinuous weighing of the samples and the spalled-off scales after 1 or 10 cycles (up to maximum 100 cycles) were used for the evaluation of the oxidation resistance of the various PM 1000 samples. Roughness measurements, metallographic sectioning, light microscopy (LiMi), SEM (scanning electron microscopy), EDX (energy dispersive X-ray analysis) and XRD (X-ray diffraction) were applied before and after the tests for the clarification of oxidation mechanisms.

3. Results and Discussion

Figure 1 shows SEM-pictures of the surfaces of the PM 1000 samples before exposure. The iron-blasted specimens were the most strongly roughened, Fe-contamination was detected by EDX. The alumina blasted surfaces were slightly, but significantly, less rough and enriched with alumina. The ground specimens were ten times smoother and much cleaner.

Starting conditions:

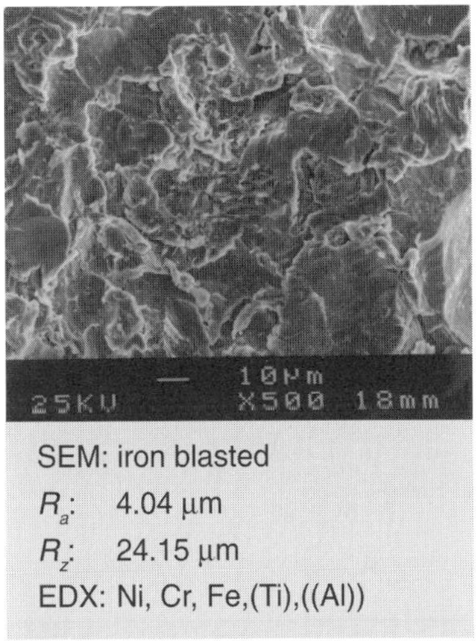

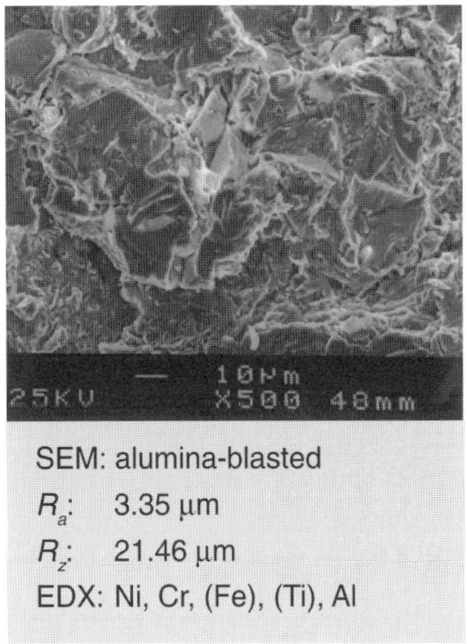

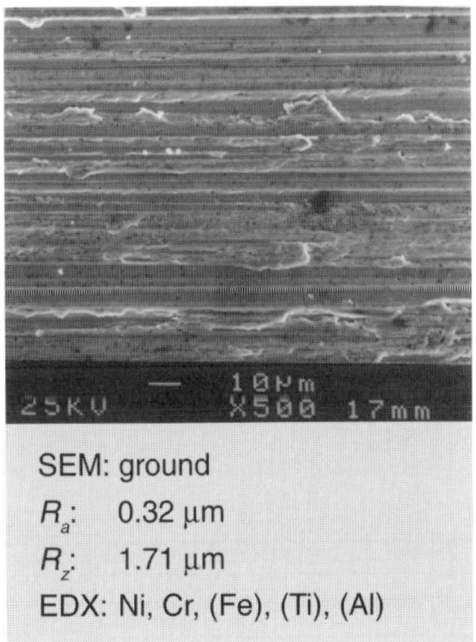

Fig. 1 *SEM-pictures, roughnesses and EDX analyses of the PM 1000 specimens — starting conditions.*

Figure 2 shows the gravimetric results of cyclic oxidation in air between 200°C and 1150°C over 100 cycles (total duration of one cycle: 75 min). Samples (a) and (b) behaved almost identically up to about 50 cycles, but then in case (a) a denser oxide scale seemed to flatten the mass gain curve. In case (c) the mass increases were lower presumably because of the smoother surface (smaller surface area) and at about 50 cycles the curves declined because of spallation effects. This is confirmed by surface SEM pictures (Fig. 3) and cross-sections (Fig. 4) of the samples after 100 cycles.

XRD measurements show that Cr_2O_3 (escolaite) is formed in all cases as the main oxide phase.

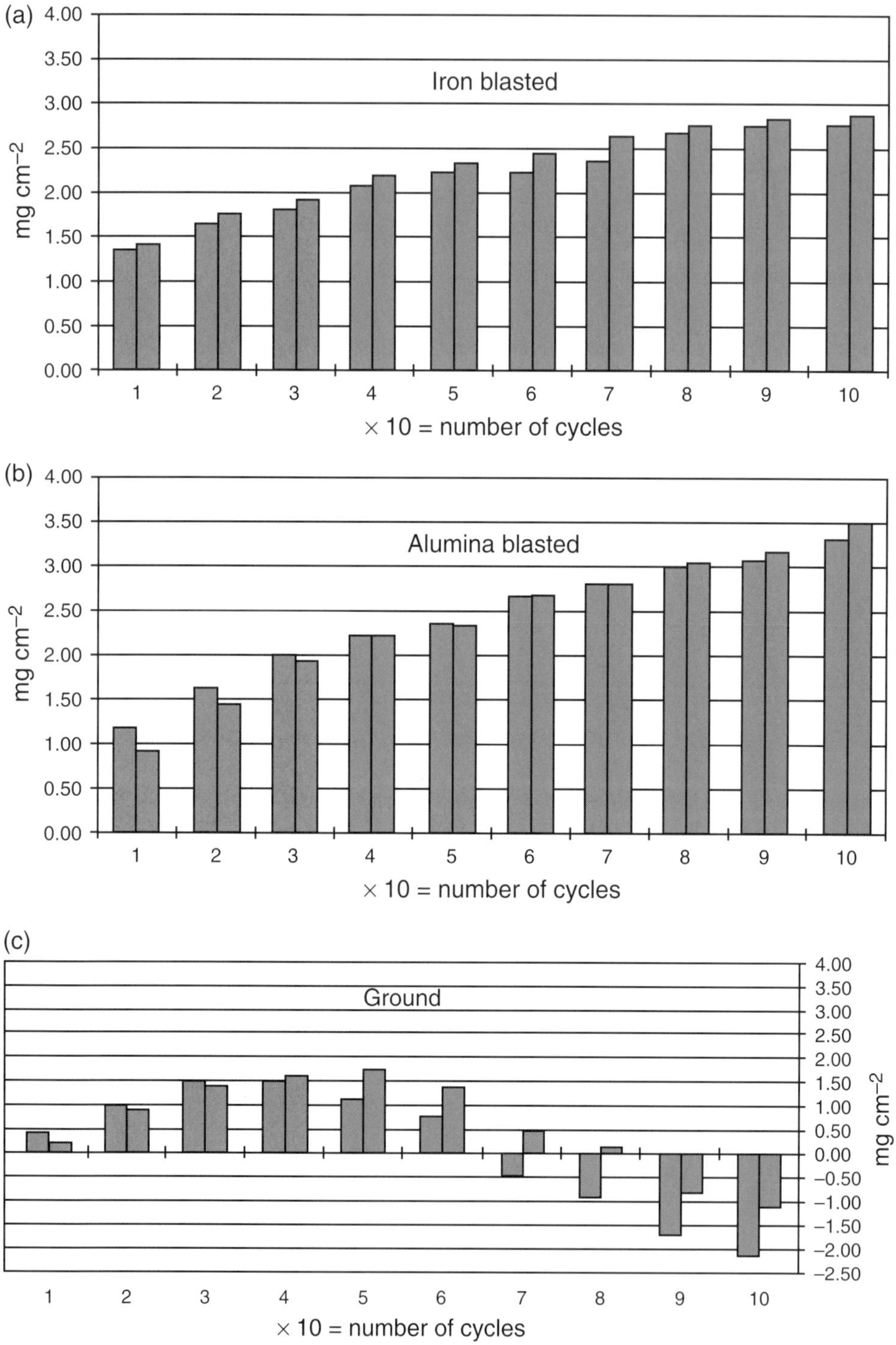

Fig. 2 *Gravimetric results of cyclic oxidation test between 200 and 1150°C (a–c); duplicate samples represented by two adjacent bars.*

Structure and composition: surface — 1150°C

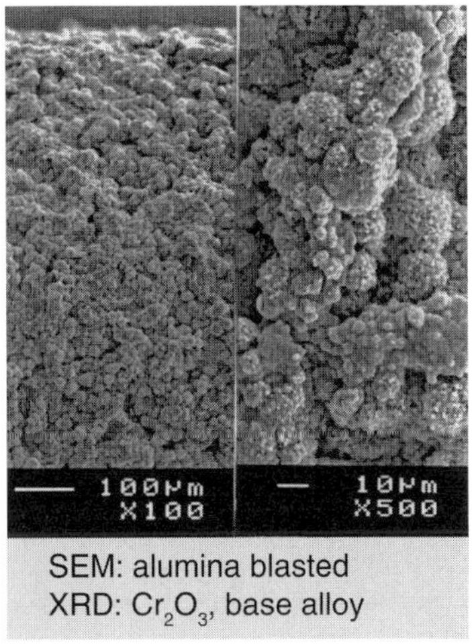

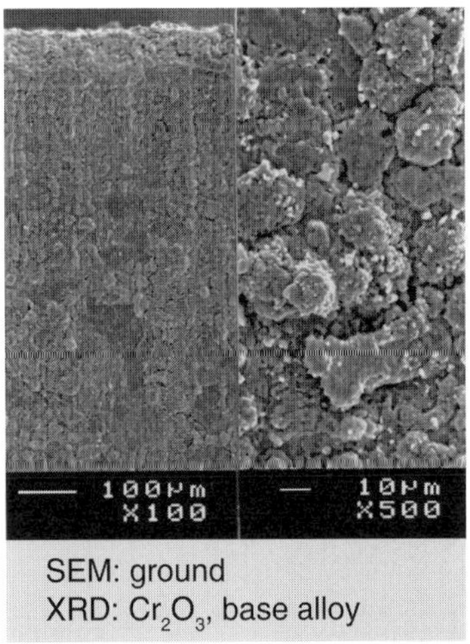

***Fig.* 3** *PM 1000 Surface — SEM and XRD after 100 cycles between 200 and 1150°C.*

Light micrographs show that porosity is formed in all cases, but that there is the highest pore density in the subsurface region of the ground specimens. This is caused most probably by the high chromium loss in case (c) because of spallation. The SEM-pictures confirm the rougher surfaces of the blasted samples with thicker oxides pinned between the roughness peaks. In all cases (a–c) a mass loss (because of evaporation of Cr–O–H species) [22] of about 0.5 $mgcm^{-2}$ over the whole test duration is probable (our own investigations on Cr_2O_3-oxidation, unpublished).

Gravimetric results of cyclic tests between 20 and 1350°C are presented in Fig. 5. Again, 2 parallel samples were investigated in cases (a), (b) and (c); the bar between the duplicate results represents the mean specific spalled-off mass of the pair of specimens which could be collected in the crucible. The iron-grit blasted samples show mass gains over the complete period of 10 cycles with some minor spallation

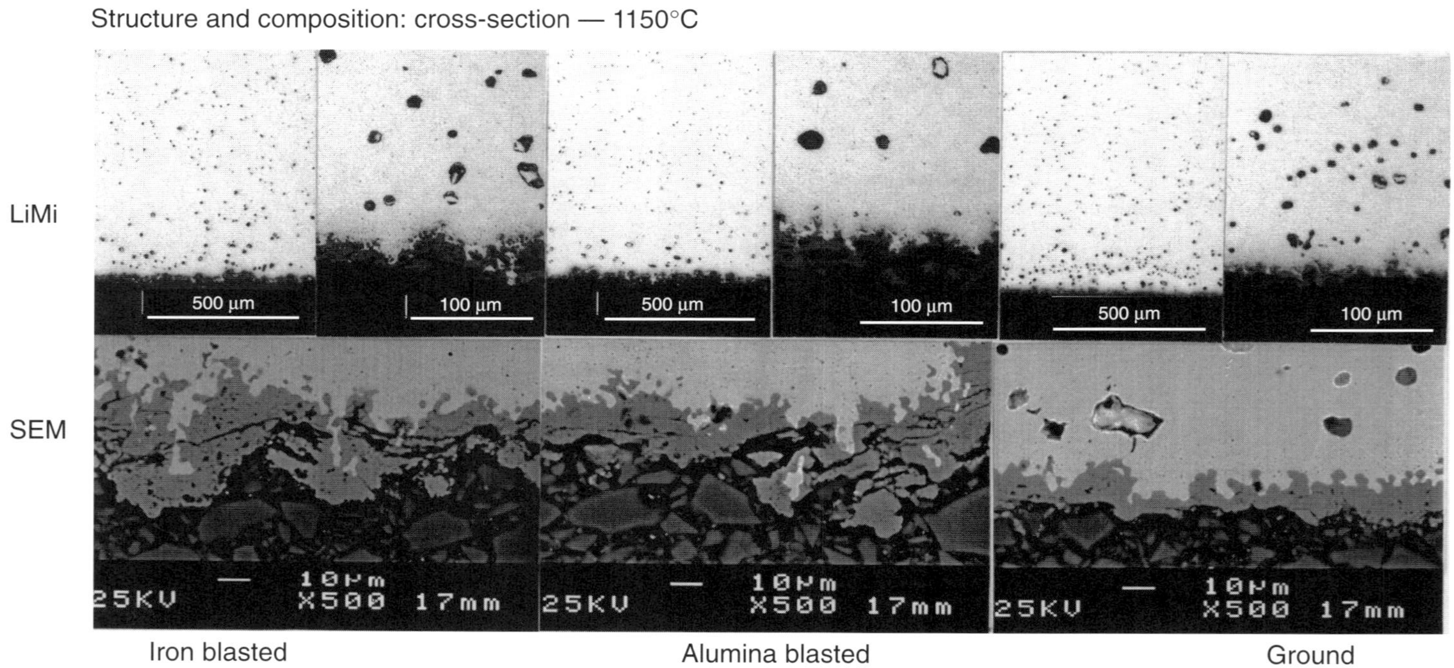

Fig. 4 *Cross sections (LiMi, SEM) after 100 cycles between 200 and 1150°C.*

(a)

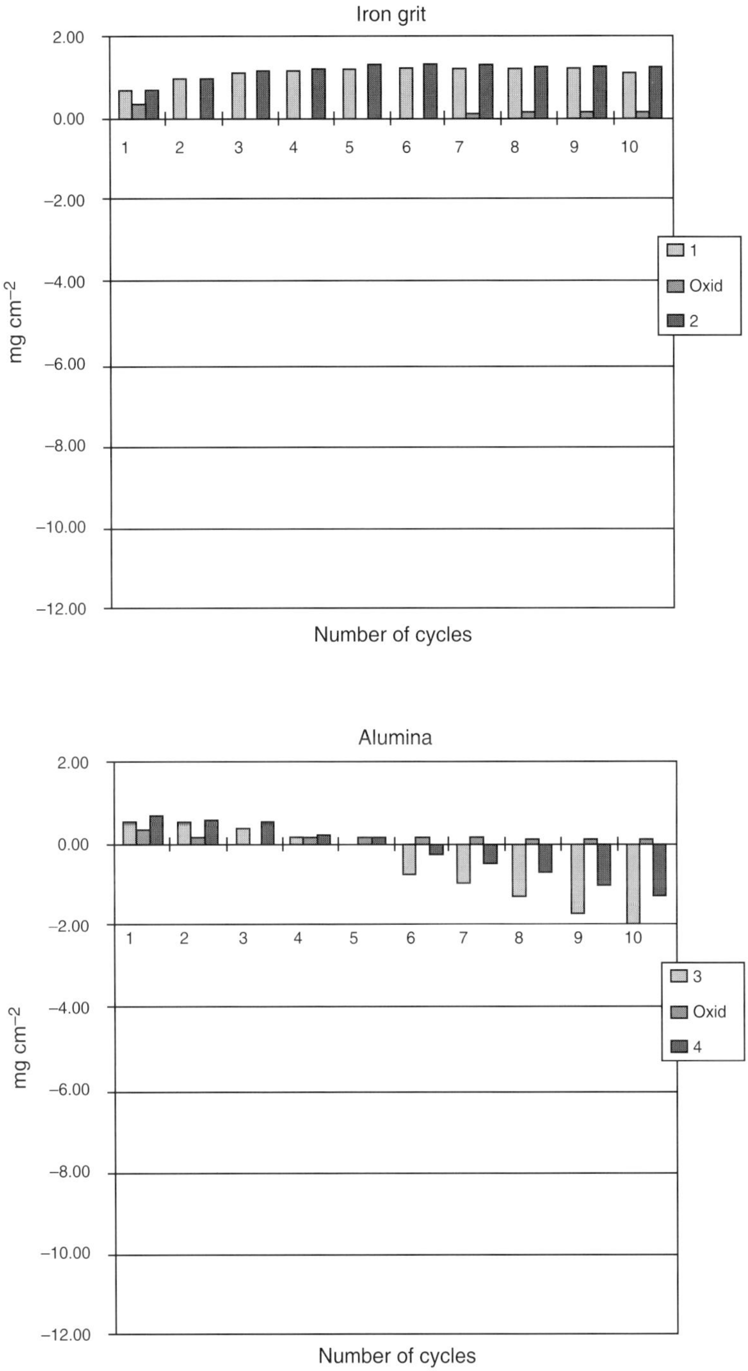

(b)

Fig. 5 *Gravimetric results of cyclic oxidation test between 20 and 1350°C (a–c); duplicate samples represented by 1, 2; 3, 4; 5, 6 respectively (the vertical bar between the duplicate results is the mean spalled-off mass of the duplicate pair collected in the crucible).*

(c)

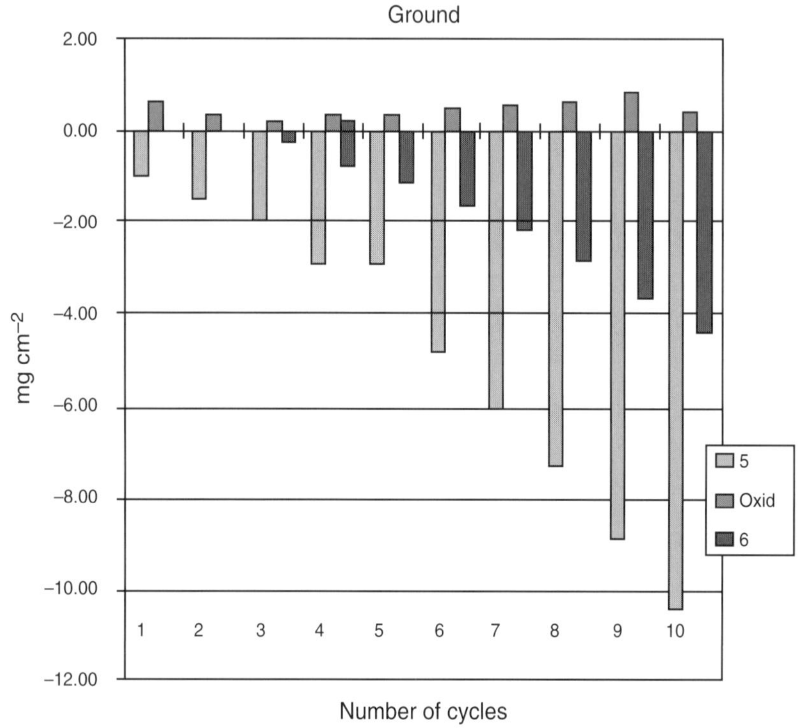

Fig. 5 (Continued) Gravimetric results of cyclic oxidation test between 20 and 1350°C (a–c); duplicate samples represented by 1, 2; 3, 4; 5, 6 respectively (the vertical bar between the duplicate results is the mean spalled-off mass of the duplicate pair collected in the crucible).

effects especially at the beginning and the end of the experiment. Evaporation of Cr–O–H species may contribute with a loss of about 0.25 mg cm^{-2} over the whole test period to the measured curves (our own unpublished investigations on Cr_2O_3 oxidation, unpublished). In cases (b) and (c) reproducibility between the pair of samples is lower, but nevertheless the following can be derived: in the case of alumina blasted samples mass loss occurs after about 5 cycles, in the case of the ground specimens strong spallation leads to mass loss from the beginning.

Surface investigations of the samples after 10 cycles shown in Fig. 6 reveal rough surfaces with strong spallation in the case of the alumina blasted and ground surfaces and a rough but better adherent scale on the iron-grit blasted specimen. EDX reveals Cr, Ti, Ni and traces of Ca in all cases with some additional Fe in case (a). XRD shows Cr_2O_3, $(Cr,Fe)_2O_3$ and the base alloy as detectable phases.

Light micrographs of the cross-sections (Fig. 7) show some porosity in the near-surface region in the case of the iron-grit blasted sample and a more uniform distribution of pores over the whole cross section in cases (b) and (c). This can again be explained by a much higher loss of chromium in the latter cases due to spallation, which creates vacancies within the bulk materials. These then coagulate to form pores.

Figure 8 shows the gravimetric results of a thermal shock test between 200 and 1200°C over 100 cycles. Iron-grit blasted, alumina blasted and ground samples behave

Structure and composition: surface — 1350°C

SEM: iron blasted
EDX: Cr, Ti, (Fe), (Ni), ((Ca))
XRD: Cr_2O_3, $((Cr,Fe)_2O_3)$, base alloy

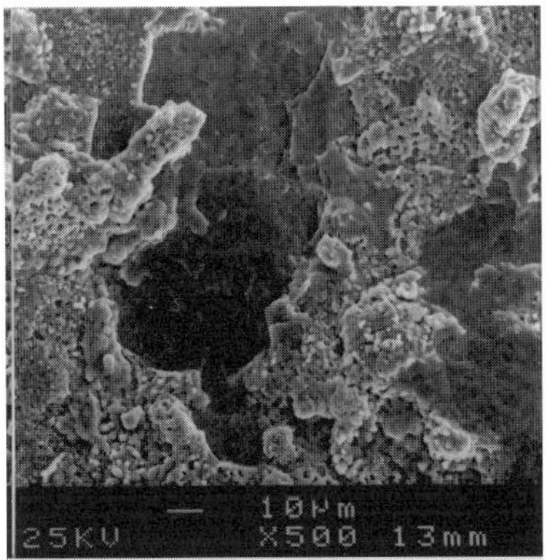

SEM: alumina blasted
EDX: Cr, Ti, (Ni), ((Ca))
XRD: Cr_2O_3, $((Cr,Fe)_2O_3)$, base alloy

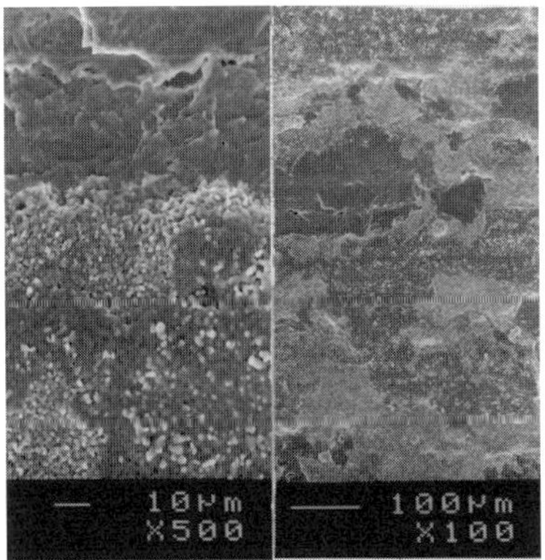

SEM: ground
EDX: Cr, Ti, (Ni), ((Ca))
XRD: (Cr_2O_3), $(Cr, Fe)_2O_3$, base alloy

Fig. 6 *SEM-surface pictures (incl. EDX and XRD) of samples after 10 cycles between 20 and 1350°C.*

similarly over the complete duration i.e. parabolic mass gain within the first 30 cycles, then a slight decline of the curves due to spallation (scale not collected in these cases) and evaporation (presumably about 0.5 mg cm^{-2} loss over the whole experiment according to our own unpublished investigations.) Obviously the shorter exposure to the high temperature in this test mode compared to 60 min in the case of the cyclic test up to 1150°C leads to slower growth of a consequently better adherent oxide on all types of surfaces. This similarity is also supported by the pictures of Fig. 9 and Fig. 10. The oxides mainly consist of Cr_2O_3 and only on the iron-grit blasted sample could a nickel-chromium spinel be additionally detected (XRD) in splash-like shapes

Structure and composition: cross-section — 1350°C

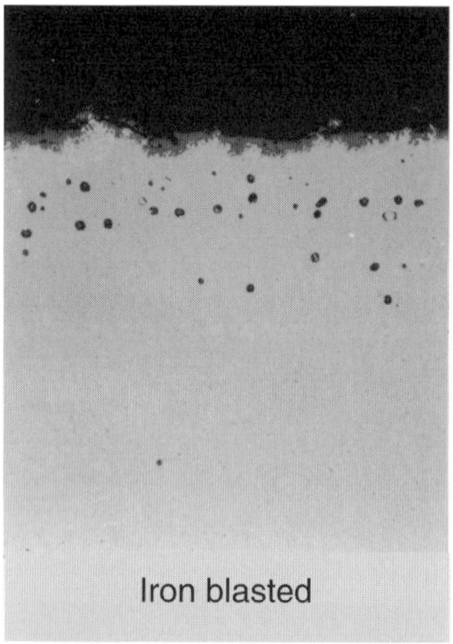

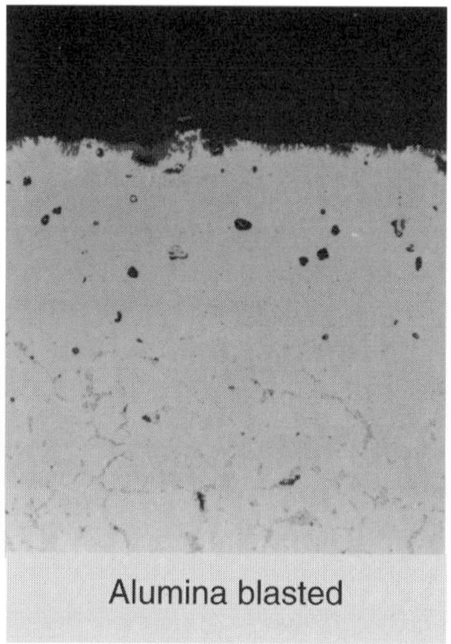

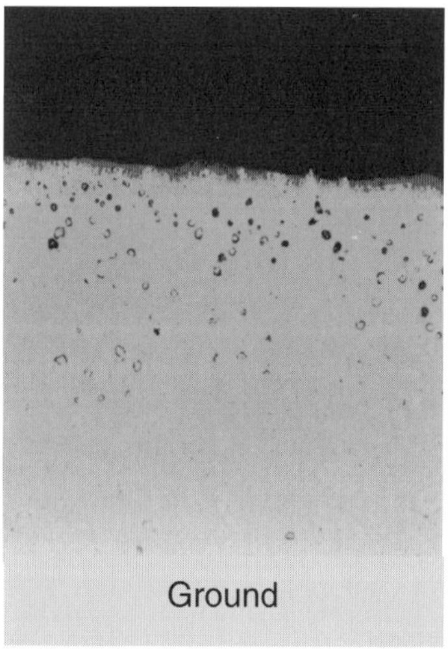

***Fig. 7** Light micrographs of cross-sections of samples after 10 cycles between 20 and 1350°C.*

(SEM-EDX). The distribution and density of pores is also comparable in all 3 cases (Fig. 10) with these being restricted to the surface regions because of relatively low consumption of chromium.

The following findings can be derived from the experimental work reported above:

- Fine oxide intrusions lead to a generally improved oxide adhesion in case of PM 1000 compared to alloys without a 'reactive metal' such as Y.

- Roughness leads to an additionally more wavy interface between oxide and metal (higher 'amplitudes') and thus stronger anchoring of the scale.

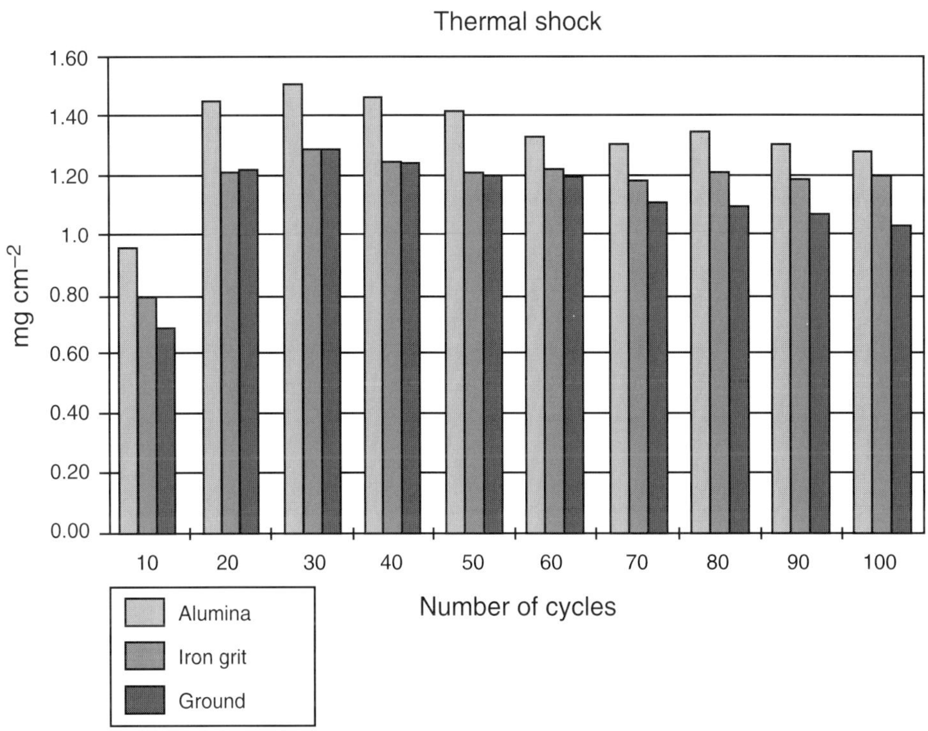

Fig. 8 Gravimetric results of a thermal shock test between 200 and 1200°C.

- The porosity of the base alloy which arises and increases during oxidation originates from the consumption of Cr for scale formation and may cause mechanical weakening of the material.
- Evaporation of Cr–O–H species is significant, in all cases, but spallation causes the differences between the various types of surfaces.
- Grit materials — such as iron — buried in the surface may, to a minor extent, chemically influence the growth and adhesion mechanism of the scale.

4. Conclusions and Outlook

- Iron-grit blasted PM 1000 is slightly rougher than alumina blasted material. Both blasted surfaces are distinctly rougher than ground surfaces.
- In cyclic oxidation and thermal shock tests the roughest samples generally show — as expected from higher surface areas — the highest specific mass gain; but, on the other hand, also the least spallation of oxide scale, this most probably being associated with anchoring effects. The ground sample with the smoothest surface consequently exhibits the severest oxide loss. In the

Structure and composition – surface – 1200°C

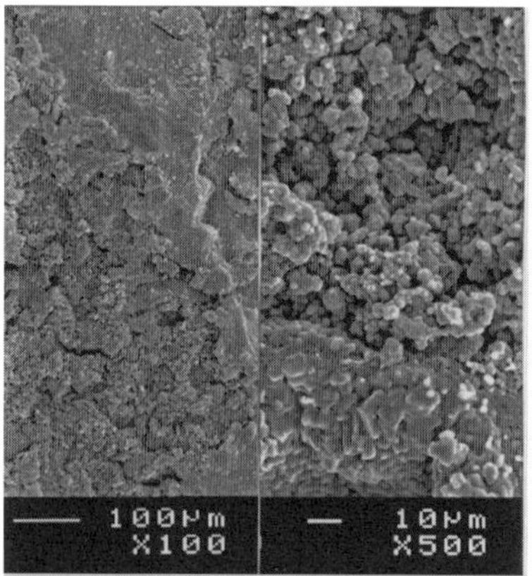

SEM: iron blasted
EDX: Cr, Ni, Fe, (Ti), drops: more Ni
XRD: Cr_2O_3, ($NiCr_2O_3$), base alloy

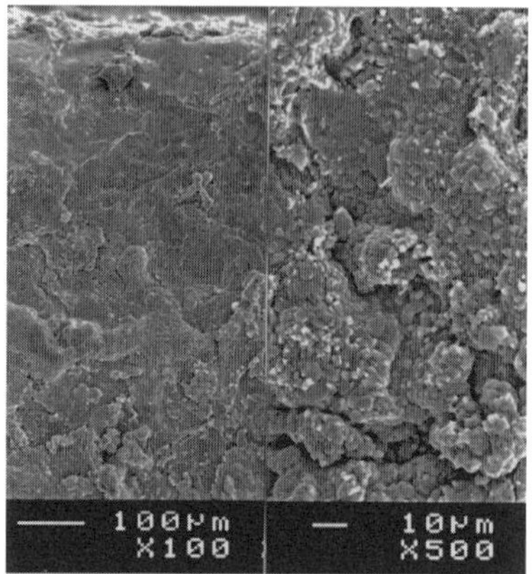

SEM: alumina blasted
XRD: Cr_2O_3, base alloy

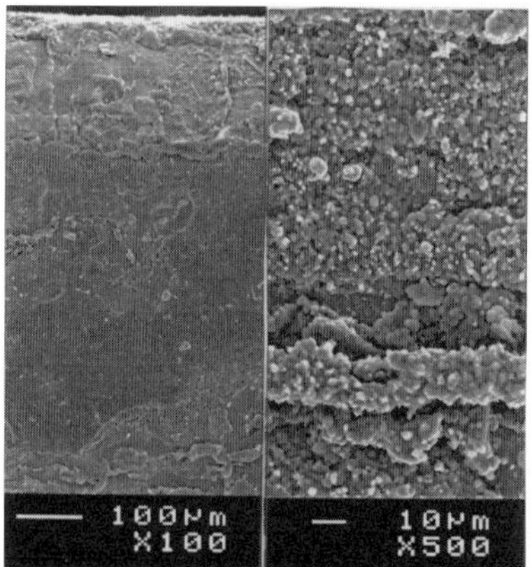

SEM: ground
XRD: (Cr_2O_3), base alloy/textured

Fig. 9 *SEM-surface pictures of samples after 100 thermal shocks between 200 and 1200°C; XRD and EDX of the surfaces.*

cases of thermal shocks between 200 and 1200°C which have very short exposure to the maximum temperature the various types of samples behave similarly. Evapouration of Cr–O–H species is significant in all cases, but different spallation behaviour is dominant.

- Long exposure times to maximum temperatures and high mass loss lead to higher porosity within the base material.

- In all cases Cr_2O_3 (escolaite) is the major phase in the scale, but there are indications of additional phases such as $(Cr,Fe)_2O_3$ on all samples and $NiCr_2O_4$

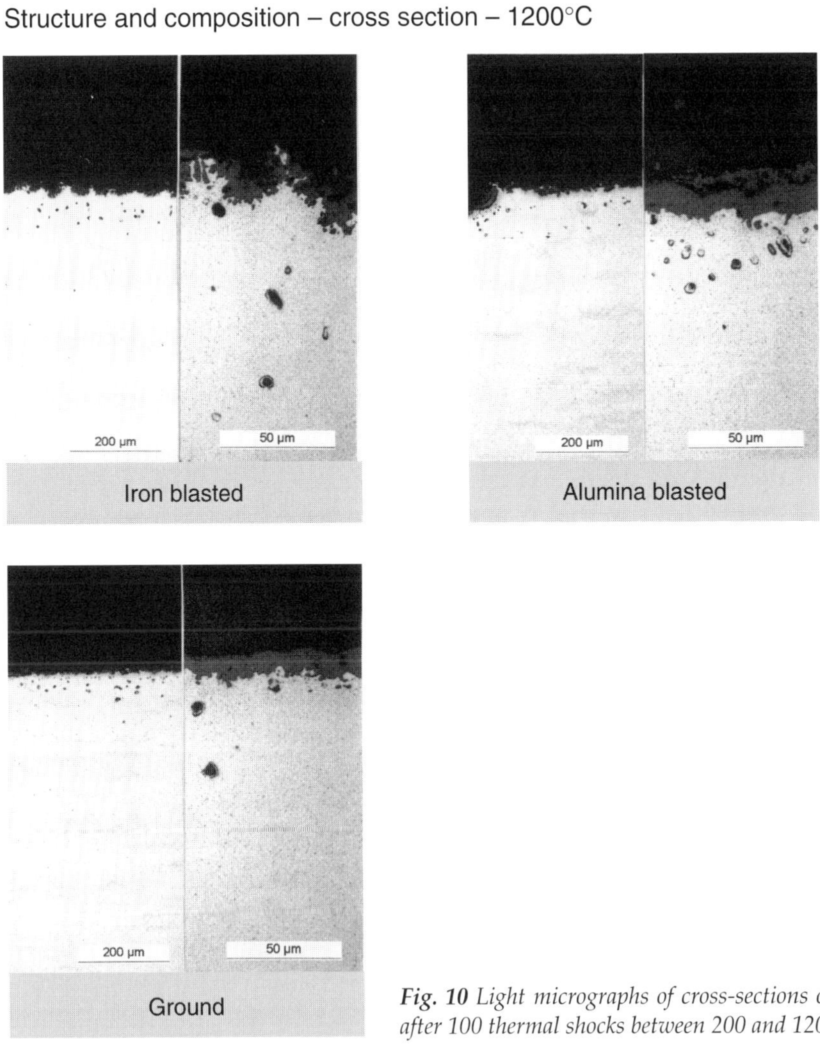

Fig. 10 *Light micrographs of cross-sections of samples after 100 thermal shocks between 200 and 1200°C.*

on iron-grit blasted specimens. By element analysis Cr, Ti, Fe, Ni and Ca could be detected within the scale in various concentrations.

For deeper clarification more sophisticated test and surface analytical methods need to be applied. The influence of surface treatment and oxidation on the mechanical properties of PM 1000 should also be examined.

References

1. H.-H. Angermann *et al.*, *Oxid. Met.*, 1997, **48**, (1–2), 1–39.
2. L. Brückner, E. Okorn, Oxidation resistant ODS-superalloys for HIP-equipments, *Conf. on Hot Isostatic Pressing: Theory and Applications*, Osaka, Japan, 1991, pp.533–547. Elsevier, Barking, Essex, UK, 1992.

3. G. D. Smith, J. J. Barbadillo and J. J. Fischer, High temperature oxidation resistance of MA ODS nickel-based alloys, *Conf. on Advances in Powder Metallurgy & Particulate Materials,* 1992, **7**, *Novel Powder Processing*. San Francisco, USA, 1992, pp.221–232. Metal Powder Industries Federation, Princeton, NJ, USA, 1992.
4. R. A. Testin, B. A. Ewing and J. A. Spees, A high performance austenitic ODS superalloy sheet for advanced gas turbine applications, *Superalloys 1992*, Champion, Pa, USA, 1992. The Minerals, Metals & Materials Society, Warrendale, Pa 15086, USA, 1992, pp.83–92.
5. W. J. Quadakkers, W. Speier, H. Holzbrecher and H. Nickel, SIMS Investigations of the transport phenomena in chromia and alumina scales on ODS alloys, *Microscopy of Oxidation,* Cambridge, UK, 1990, The Institute of Metals, Terrace, London, UK, 1991, pp.149–160.
6. C. S. Tassen, J. J. Fischer, G. D. Smith and M. J. Shaw, High temperature service experience and corrosion resistance for mechanically alloyed ODS alloys, Heat-Resistant Materials, Fontana, Wi, USA, 1991, ASM International, Materials Park, Ohio, USA, 1991, pp.105–114.
7. F. H. Stott, S. Berg, M. Sang and N. Karim, The oxidation performance of superalloys in gaseous environments at very high temperatures, *High Temperature Materials for Powder Engineering,* Liège, Belgium, 1990, Kluwer Academic Publishers, Dordrecht, NL, 1990, pp.213–226.
8. G. D. Smith and P. Ganesan, High temperature corrosion resistance of heat resistant mechanically alloyed products, *Structural Applications of Mechanical Alloying,* Myrtle Beach, SC, USA, 1990, ASM International, Materials Park, Oh, USA, 1990, pp.109–118.
9. H. Beske, W. J. Quadakkers, H. Holzbrecher, H. Schuster and H. Nickel, *Mikrochim. Acta,* 1990, **2** (1–6), 109–119.
10. K. Tsukuta, T. Iikubo and S. Isobe, Application of mechanically alloyed nickel- and iron-base alloys to skid rails for walking-beam-type reheating furnaces, in *Structural Applications of Mechanical Alloying,* Myrtle Beach, SC, USA, 1990, ASM International, Materials Park, Oh, USA, 1990, pp.99–108
11. H. E. Evans, Cavity formation and metallurgical changes induced by growth of oxide scale, *Scale Growth,* London, UK, 1988, *Mater. Sci. Technol.,* 1988, **4**, (12), pp.1089–1098.
12. T. A. Ramanarayanan, R. Ayer, R. Petkovic-Luton and D. P. Leta, *Oxid. Met.*, 1988, **29**, (5–6), 445–472.
13. K. Przybylski and G. J. Yurek, *Proc. Electrochem. Soc.*, 1988, **88–5** (*Proc. Symp. High Temp. Mater. Chem.* — 4, 1987), 306–313.
14. G. J. Yurek, K. Przybylski and A. J. Garratt-Reed, *J. Electrochem. Soc.,* 1987, **134**, 2643–2644.
15. T. A. Ramanarayanan and R. Petkovic-Luton, Growth mechanisms of Cr_2O_3 scales on Y_2O_3 -dispersed alloys, *Proc. Symp. on Fundamental Aspects of High Temperature Corrosion,* Boston, MA, USA, 1986, pp.75–102. The Electrochemical Society, Pennington, NJ, USA.
16. T. A. Ramanarayanan and R. Petkovics-Luton, *Ber. Bunsenges. Phys. Chem.,* 1985, **89**, (4), 402–409.
17. R. D. McIntyre, *Mater. Eng.,* 1982, **95**, (2), 34–39.
18, L. Hsu and A. R. Stetson, *Thin Solid Films,* 1980, **73**, (2), 419–428.
19. H.-P. Martinz, W. Köck and T. Sakaki, *Journal de Physique IV,* (Colloque C), supplement to *Journal de Physique* III, 1993, **3**, 205–213.
20. H.-P. Martinz, G. Leichtfried, W. Köck, O. Lang and E. Kotz, *Glasteknisk Tidskrift,* 1997, **52**, (1), 27–33.
21. S. Weinbruch, A. Anastassiadis, H. M. Ortner, H.-P. Martinz and P. Wilhartitz, *Oxid. Met.,* 1999, **51**, (1/2), 111–128.
22. E. J. Opila and N. S. Jacobson, Volatile hydroxide species of common protective oxides and their role in high temperature corrosion, *Proc. Symp. on Fundamental Aspects of High Temperature Corrosion*, 1997, pp.269–80. The Electrochemical Society, Pennington, NJ, USA.

Part 2

Test Procedures and Parameters Influencing Cyclic Oxidation Behaviour

7

Effect of Cycle Frequency on High-Temperature Oxidation Behaviour of Alumina- and Chromia-Forming Alloys

B. A. PINT, P. F. TORTORELLI and I. G. WRIGHT
Oak Ridge National Laboratory, P. O. Box 2008, Oak Ridge, TN 37831-6156, USA

ABSTRACT

Cycle frequency affects both high-temperature oxidation performance and the method in which the cyclic test is conducted. Several factors are discussed using results taken from Ni-base and Fe-base alumina- and chromia-forming alloys. For the most adherent scales, cycle frequency has little effect on results over extended test times (≥500 h). When the scale is less adherent, reducing the cycle frequency typically has the expected effect of increasing the mass loss per unit exposure time; however, the opposite effect is observed in other cases. Low-cycle frequency experiments performed in alumina crucibles have the important benefit of collecting the spalled oxide and, for chromia-formers, volatility products. The test method and cycle frequency ultimately have a strong effect on lifetime predictions.

1. Introduction

Cyclic oxidation testing is one of the fundamental ways to assess high-temperature environmental resistance of materials. Thermal strains that occur during heating and, especially, cooling are often a primary cause of degradation of the external oxide scale that affords protection against further attack [1–4]. The practical reason for this type of test is to simulate application cycles in a controlled manner. Ideally, the cyclic test is designed to mimic the expected duty cycle. However, in a broader sense, it is attractive to use standardised tests that allow direct performance comparisons of different materials and that may provide valuable information for the understanding of cracking and spallation mechanisms [5–11].

Because cyclic oxidation is such a broad topic, this paper attempts to focus on one aspect, cycle frequency or the length of time at temperature for each cycle. Surprisingly, this variable is rarely discussed in any detail [12,13]. Longer cycle times are of significant importance to projects such as the U S Department of Energy's Advanced Turbine Systems Program, which is attempting to transfer state-of-the-art materials technology from aircraft engines to land-based gas turbine engines for power generation [14,15]. The duty cycle for aircraft is quite different than that for power generation. Consequently, there has been concern that materials and, in particular, coatings which are well suited for aircraft applications will perform differently in the longer-term cycle experienced in land-based gas turbines, heat exchangers or other power generation technologies.

A second aspect of cycle frequency relates to data collection for lifetime predictions. If more oxidation-resistant alloys (which are generally more expensive) are to be implemented in commercial applications, it is essential to clearly define the benefit to life extension that these materials are expected to achieve. Depending on the cycle frequency and the procedures used in conducting the experiment, very different lifetime predictions are possible. Therefore, it is essential to correctly select appropriate test conditions and methods for collecting data in order to develop a reliable prediction.

This paper draws together a wide range of experimental data from a variety of alumina-and chromia-forming alloys to illustrate some basic issues regarding cycle frequency and test procedures. In several cases, the testing is still in progress and the results are still evolving. Two forms of testing are described: uncontained cycling, which is more appropriate for high-frequency tests, and contained cycling, where the specimen is held in an individual alumina crucible which collects spalled and evaporated material (i.e. CrO_3 vapour condenses on the crucible walls). It is demonstrated that, for lifetime predictions, the total mass gain data from contained cycling is more useful than the specimen mass change data alone. However, contained cycling tests are more labour and equipment intensive and are not appropriate for high-cycle frequencies and certain exposure conditions.

2. Experimental Procedure

A variety of alloys were used in this study. Chemical compositions of Ni-base superalloys [16], NiAl variations [17], Pt-containing alloys [18], Fe-base castings [19], ODS FeCrAl [20,21], ODS Fe_3Al [22], and ODS NiCr [23] are given in detail elsewhere. Unless otherwise specified, specimens were polished to a 0.3 μm finish and ultrasonically cleaned in acetone and methanol prior to oxidation.

Short-term (high-frequency) cycles (1–10 h) were performed in a computer-controlled cyclic rig in dry flowing O_2, with seven specimens hanging in a vertical furnace on individual alumina rods and connected to the rod by Pt–Rh wires. Specimens up to 3 g and 2 cm × 2 cm can be accomodated. Cooling was for 10 min to approximately 25°C. Long-term (low-frequency) cycles (100–500 h) were conducted in box or horizontal tube furnaces in laboratory air with the specimen contained in a pre-annealed alumina crucible including an alumina lid. Specimens were cooled for at least 24 h prior to weighing. Specimens were weighed on a Mettler model AG245 balance. Isothermal kinetics were measured using a Cahn model 1000 microbalance.

Gravimetric results are used as the principal indicator of the influence of cycle frequency on performance. Little microstructural information is provided, although some conclusions are based on characterisations by metallography, X-ray diffraction, scanning electron microscopy, energy dispersive X-ray spectroscopy (EDS) and transmission electron microscopy.

3. Results

The experimental results are divided by alloy type in order to demonstrate several important points about the role of cycle frequency.

3.1. Ni-base Superalloys: Sulfur Effects

A series of single-crystal Ni-base superalloys similar to General Electric alloy René N5 were cast in order to investigate the role of yttrium and sulfur on their oxidation performance [16]. Castings without Y were de-sulfurised both in the melt (N5B, N5C and N5D) and by hydrogen annealing (N5AH). Sulfur contents were measured by glow-discharge mass spectrometry. While it has been widely observed that de-sulfurisation improves performance in high frequency cycles [24–29], less work has been performed in low frequency cycles [16].

The specimen mass change data from exposures at different cycle times at 1100°C are shown in Fig. 1. The 1- and 10-h cycle tests were conducted in an uncontained test while the 100-h cycles were conducted in a contained cyclic test. Also, the 10-h specimens were only polished to 600 grit SiC finish. A number of alloys are shown in order to illustrate that both hydrogen annealing and melt de-sulfurisation result in an improvement in scale adhesion relative to a Y-free, nominal sulfur content (4–5 ppma) casting (N5A). However, the de-sulfurised alloys differentiated somewhat in the various tests. In particular, N5B (melt de-sulfurised, 2.1 ppma S) showed little spallation in 1-h cycles but showed progressively higher mass losses with increasing cycle time (Fig. 2a). Increasing the cycle time to 100-h also resulted in significant spallation for N5C (melt de-sulfurised, 1.8 ppma S). The onset of spallation in 100-h cycles increased with decreasing sulfur. The standard N5 and N5AH both accumulated more than 3000 h of exposure time with only minor spallation.

When the mass losses were large, it was clear from examination of the specimen that spallation was to the metal interface. However, some spallation appeared to include only the outer transient scale, which was typically blue for these alloys. In both 1-h and 10-h cycles, the specimens occasionally experienced minor but distinct drops in mass (arrows in Fig. 1a and 1b) but then continued to increase in mass. This was most evident for the castings with more adherent scales, N5 and N5AH, but not for NiAl + Hf (Fig. 2a). The result was noticeably lower specimen mass values for 1- and 10-h cycles compared to 100-h cycles (Fig. 2b). These drops were attributed to the spallation of the transient Ni-, Co- and Cr-rich scale (e.g. Fig. 3), which can spall and still leave behind the underlying, rate-controlling, alumina layer.

These relatively small deviations in specimen mass change make performance comparisons difficult among alloys which do not have large amounts of spallation. One way to try to resolve these differences is to calculate rate constants from the various specimen masses, k_{ps} (Table 1). These values can be compared to those from isothermal thermo-gravimetric tests [16], k_{pi}, and to a rate constant calculated from the total mass gain measured for the 100-h cycles, k_{pt}. In general, the substrates with more adherent scales (N5, N5AH) also exhibited lower rate constants (a factor of 4–10 compared to N5B and N5C) suggesting that there may be decreased transient scale formation or some type of reactive element (RE) effect [31–33] occurring in those alloys. While the low-S alloys do not contain Y, they do contain Hf, which has

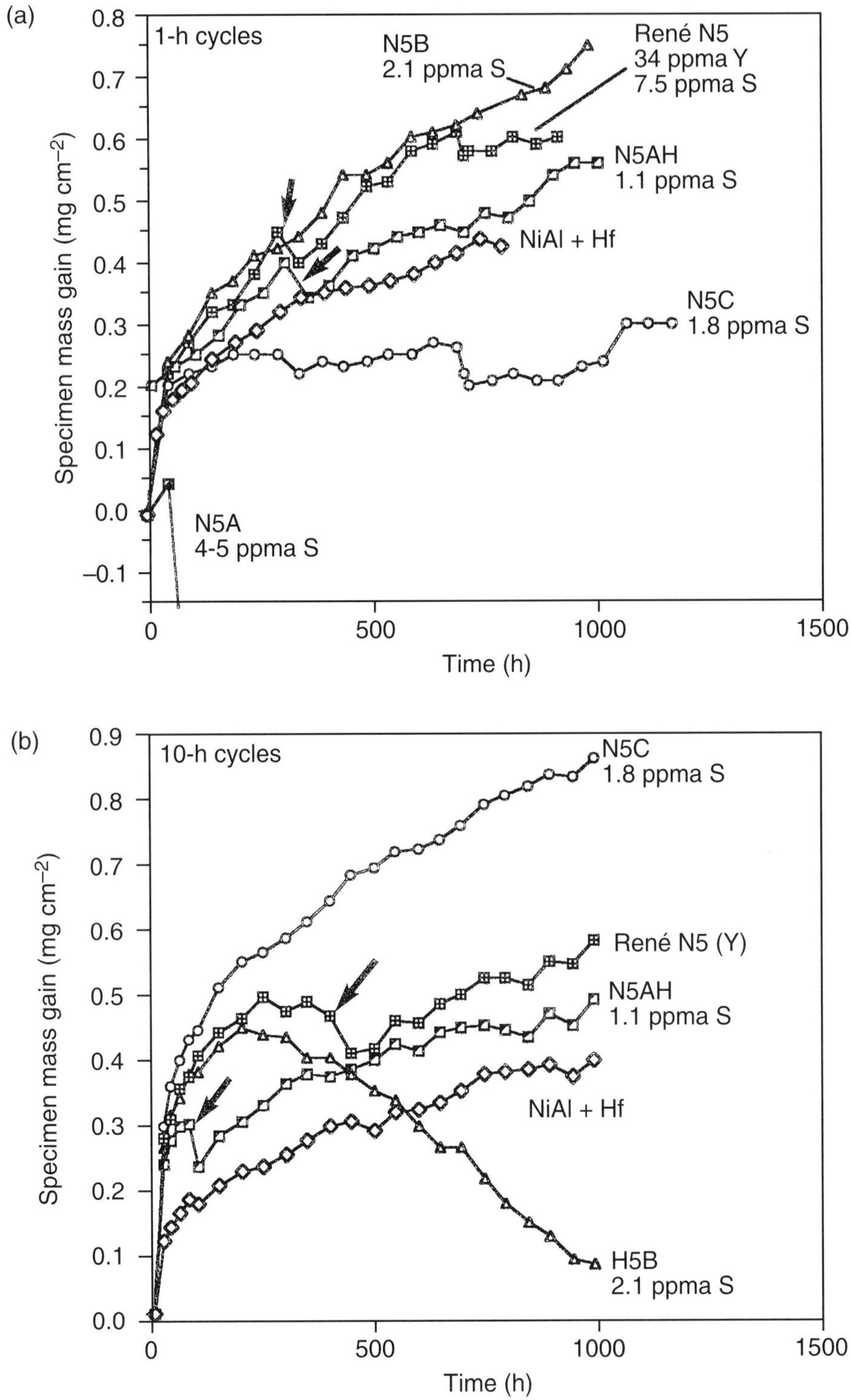

Fig. 1 *Mass gain for several single crystal Ni-base superalloys oxidised at 1100°C in (a) 1-h, (b) 10-h and (c) (opposite) 100-h cycles.*

(c)

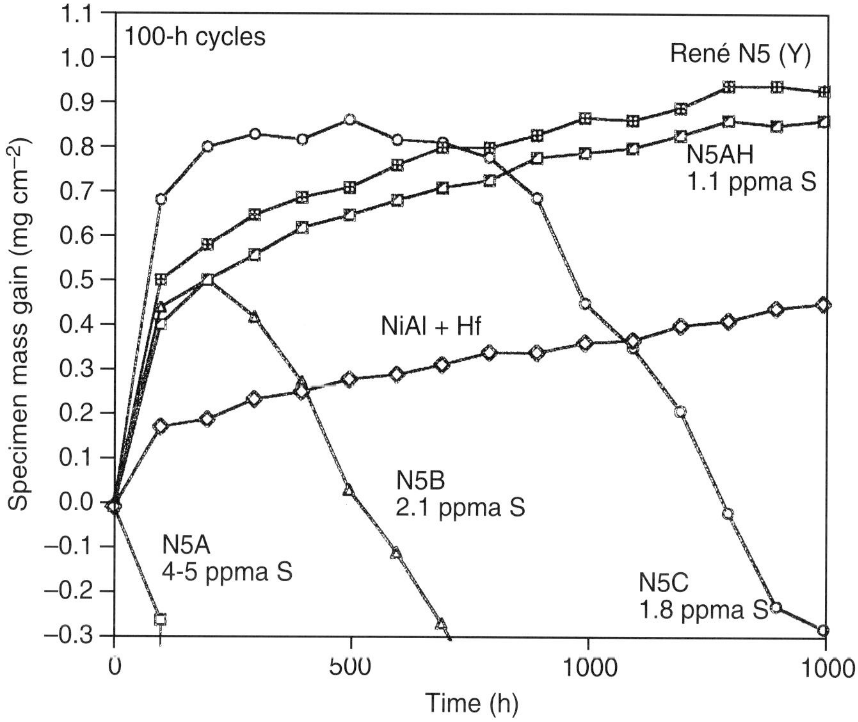

been shown to have a dramatic effect on the scale growth rate in β–NiAl [17,18]. It may be an important factor that H_2-annealing removed C as well as S from the superalloy [16], which may affect the Hf activity in the substrate [29]. Also, the higher isothermal rates for N5B and N5C suggest that a thicker alumina scale was forming. This would create higher strains on cooling than alloys with slower scale growth rates.

Overall, there is some benefit associated with comparing these types of calculated rate constants (Table 1). However, they provide only qualitative information about relative rates of spallation and are not well suited to incorporation into a quantitative spallation model. One limitation in comparing rates measured under isothermal and cyclic conditions is that the former experiments were run for 100–200 h while the latter generally were of 1000 h duration. Such a comparison requires a large extrapolation or much longer isothermal tests. The fact that k_{pi}>k_{pt} for the substrates with the most adherent scales suggests that the rate drops with time.

3.2. NiAl

When properly doped with a RE such as Zr or Hf, NiAl is one of the most oxidation-resistant alloys. This is clearly illustrated in Fig. 2(a) where NiAl + Hf is observed to show very similar mass gains up to 1000 h, independent of cycle frequency. The mass gains were extremely low because of the optimal Hf doping, which has been found to reduce the parabolic rate constant by a factor of 10 or more at this temperature [17,18]. It is likely that these curves for different cycle frequencies would differentiate but only after much longer times.

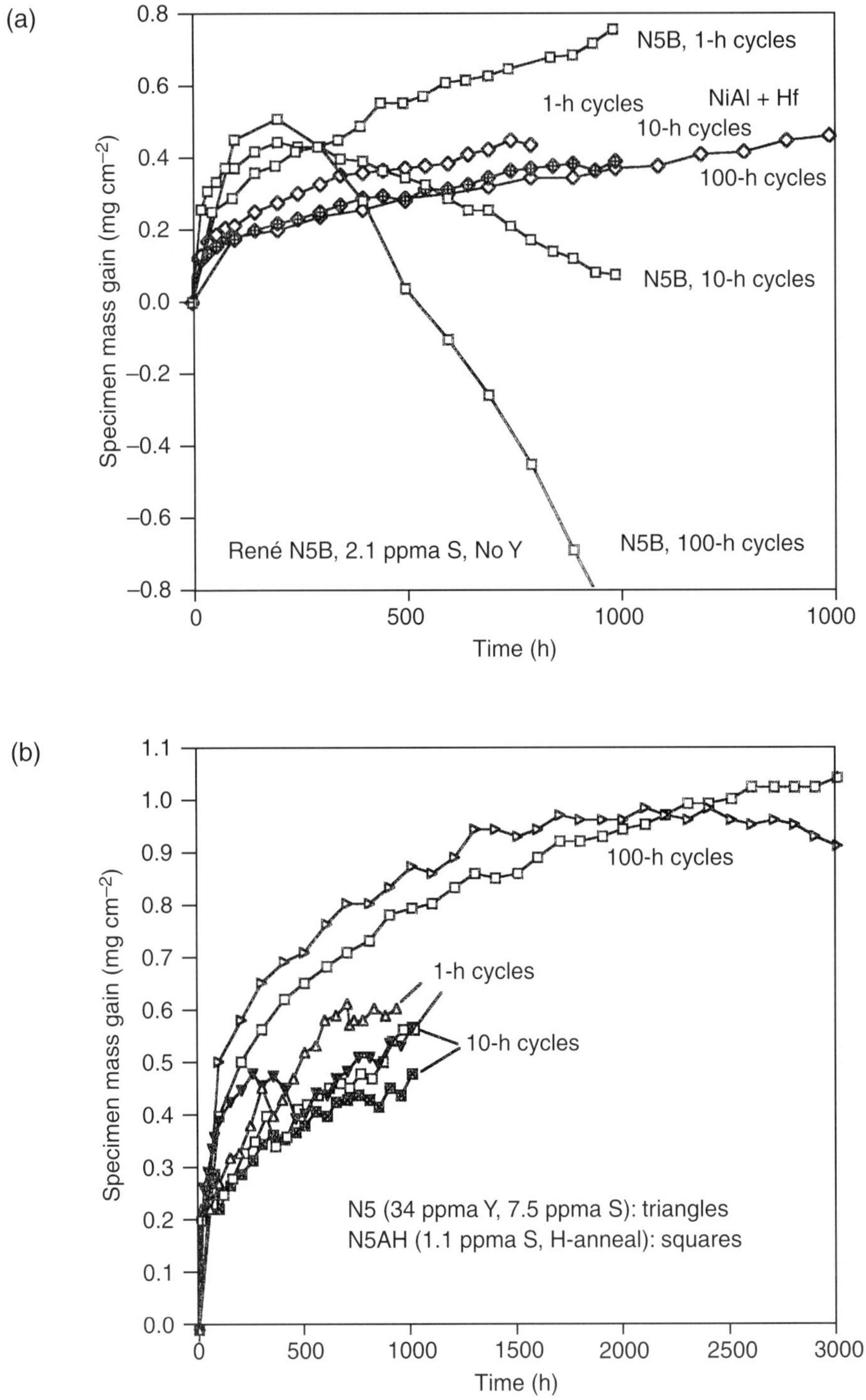

Fig. 2 *Mass gain for various cycle times at 1100°C for (a) N5B and NiAl + Hf and (b) standard N5 (with Y) and hydrogen annealed N5, N5AH.*

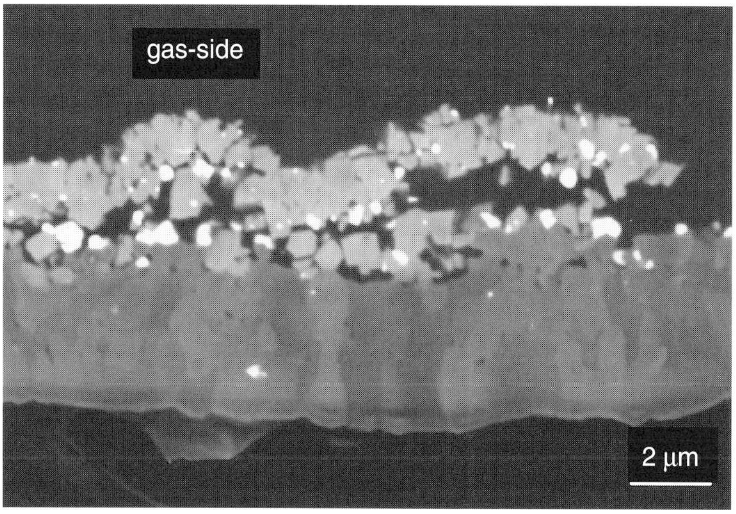

***Fig. 3** SEM back-scattered electron image of the scale formed on N5AH after 100 h at 1200°C in a polished cross-sectional mount. (The very bright particles are rich in Ta as determined by EDS analysis.) [30].*

***Table 1.** Parabolic rate constants, $g^2cm^{-4}s^{-1}$ ($\times 10^{-14}$) at 1100°C for variations of René N5, a single crystal Ni-base superalloy, for different cycle tmes and isothermally*

Alloy	Comments	Calculated from Specimen ΔW (k_{ps}) 1 h	10 h	100 h*	Total ΔW (k_{pt}) 100 h	Isothermal ΔW (k_{pi}) 100 h
β-NiAl-Hf	Cast 0.05 at.% Hf	4.5	3.3	2.9 (6)	2.7	11.4
N5	Standard 34 ppma Y, 7.5 ppma S	10.4	7.1	6.8 (42)	14.1	31.6
N5A	No Y, 4–5 ppma S	—	—	—	11 800	132
N5B	No Y, 2.1 ppma S	13.1	—	(40)	64.5	44.7
N5C	No Y, 1.8 ppma S	—	13.6	(72)	58.5	62.1
N5AH	H_2 annealed No Y, 1.1 ppma S	5.9	3.2	4.9 (32)	5.9	35.7

* Values in parentheses estimated from first 200–300 h data.

The same types of exposures were performed at 1200°C in order to form a thicker oxide (Fig. 4). However, there still appeared to be little effect of cycle time up to 1000 h due to the excellent scale adhesion on NiAl + Hf. The observed scatter in the results may be explained by the quality of the NiAl casting. An extreme example is shown with two specimens for 100-h cycles, the specimen with the higher mass gain

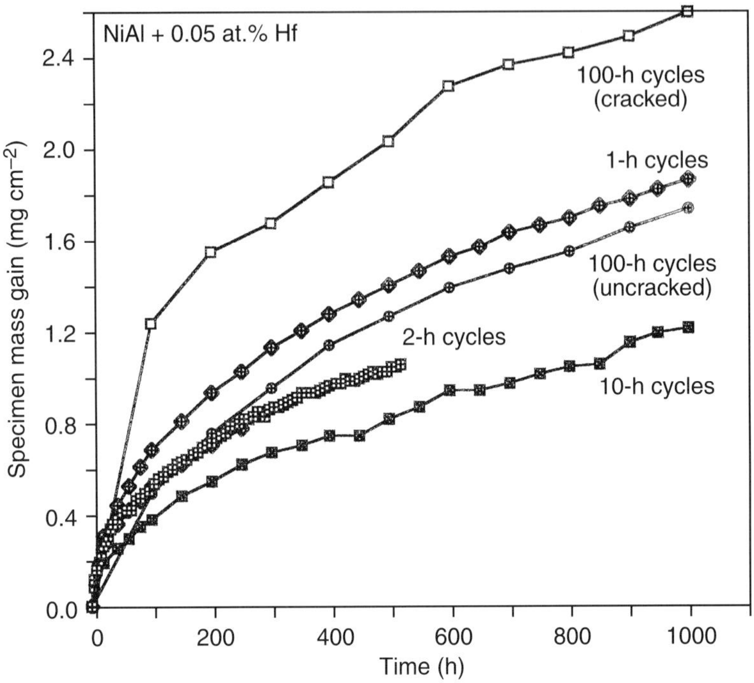

Fig. 4 *Specimen mass gain for NiAl + 0.05at.%Hf for various cycle times at 1200°C.*

contained cracks in the alloy (which became evident after exposure) while the specimen with the lower mass gain showed no cracks. The higher mass gain including some spallation was attributed to internal oxidation along cracks in the alloy.

The effect of cycle frequency was also evaluated on undoped NiAl (Fig. 5). On the first casting (NiAl-1, Al = 50.2at.%, 27 ppma S), switching from 100-h to 1-h cycles caused a significant increase in the linear rate of mass loss. After one thousand, 1 h cycles the specimen had cracked and was forming large amounts of Ni-rich oxide and spinel. A second casting (NiAl-2, Al = 51.2at.%, 3 ppma S), showed only a slight increase in the linear spallation rates as the cycle time was increased from 100-h to 1-h. The difference in performance between the two castings could be an effect of aluminium content, sulfur content or casting defects. However, it is surprising how little cycle frequency affected the performance of NiAl-2.

3.3. Pt-containing Alloys

To study the role of Pt in aluminides, a series of Pt-containing alloys were cast. Previous papers have reported that these cast (Ni,Pt)Al and PtAl alloys showed little spallation in 1-h cyclic testing up to 500 h at 1150°C and 100 h at 1200°C [18,34]. This is a considerable improvement over cast, undoped NiAl (e.g. Fig. 5) but not as good as that observed for Hf-doped NiAl (Fig. 4). However, as illustrated in Fig. 6, in 100-h cycles at a lower temperature (1100°C) there was no improvement in the total mass gain associated with PtAl or Pt-doped NiAl (2.3at.%,10mass%Pt). After the first 500 h, the total mass gains increased nearly linearly, indicating a large amount of spallation similar to that observed for undoped NiAl (50.1at.%Al, < 4 ppma S). The specimen

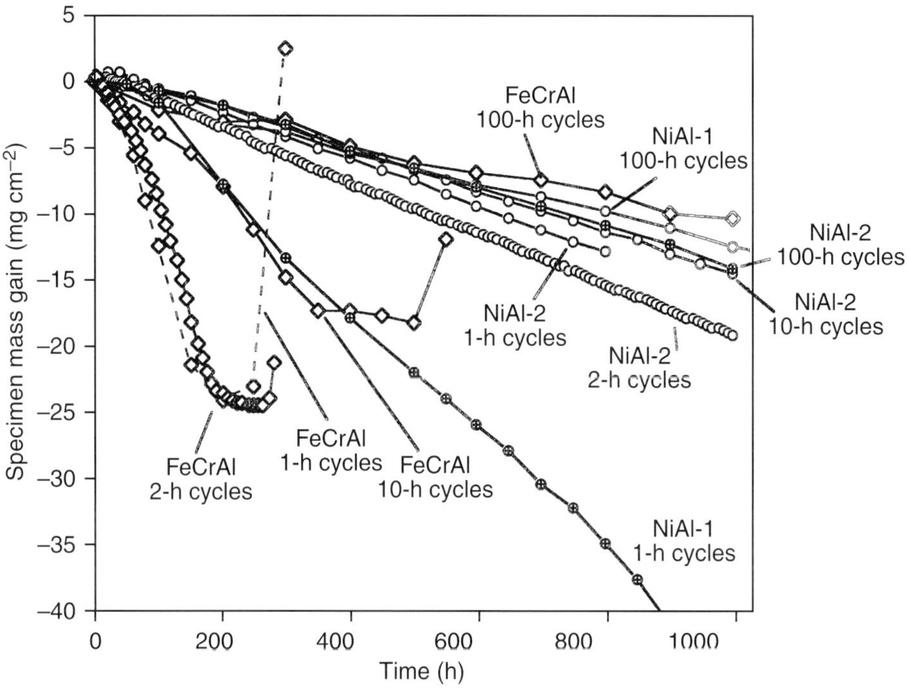

***Fig. 5** Specimen mass loss for undoped FeCrAl and undoped NiAl as a function of cycle time at 1200°C.*

mass gains (dashed lines) also reflect the spallation from these substrates, with PtAl showing slightly less spallation than NiAl or (Ni,Pt)Al. However, this reduction in specimen mass loss was mainly a result of cracking and internal oxidation of PtAl [34] and not due to a more adherent alumina scale.

Total mass gains were also similar for NiAl, PtAl and (Ni,Pt)Al at 1200°C. However, this is not surprising as the beneficial effect of Pt was only observed for 100, 1-h cycles at this temperature, which is only comparable to one 100-h cycle.

Similar to the N5B alloy, Pt-containing aluminides also showed a reverse effect of cycle frequency, with spallation increasing as frequency decreased. If the role of Pt is to slow the growth of interfacial voids [34], this may explain the negative effect of longer cycle time. With a low cycle frequency, the extended time at temperature allows for larger voids to grow at the metal–scale interface [35,36]. Smaller voids formed during shorter cycles may limit the damage to the scale allowing for a healing alumina layer to quickly form, whereas large voids may result in major scale losses. This result is particularly disturbing for Pt-modified aluminide coatings being considered for land-based gas turbine engines. For this long-cycle time application, the benefit of Pt additions may be substantially reduced.

3.4. Fe-base Alloys — Difficulty with Lifetime Predictions

Iron-base alloys show a somewhat different response to cycling from that of Ni-base alloys. For example, with undoped FeCrAl (20.1at.%Cr, 9.6%Al, 85 ppma S) in 100 h

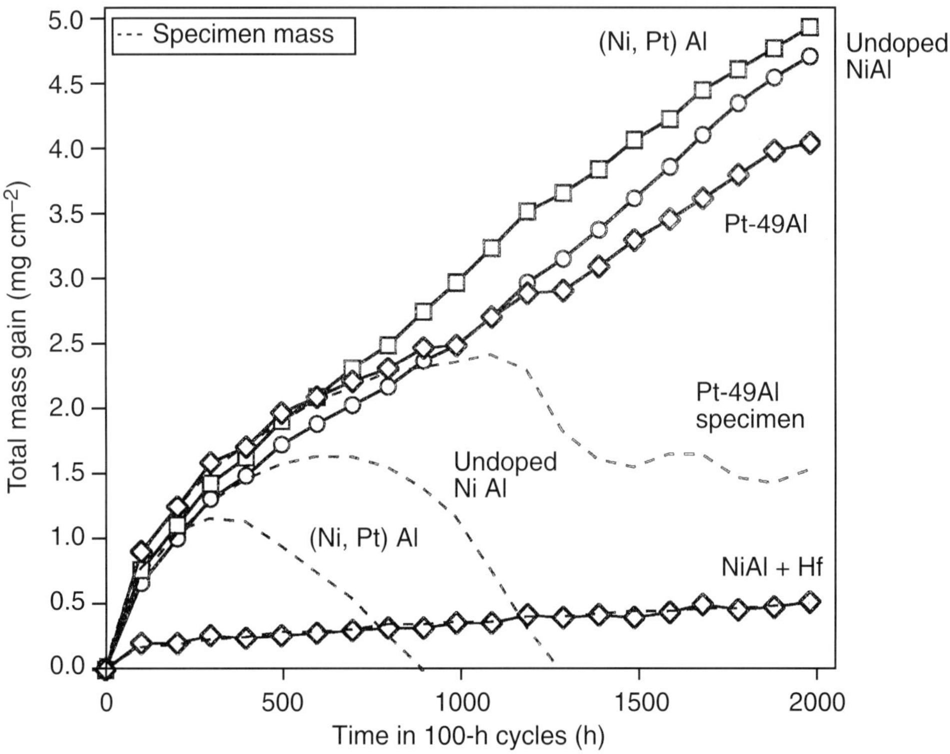

***Fig. 6** Specimen (dashed line) and total mass gains (symbols) for several aluminides during 100-h cycles in crucibles at 1100°C in air.*

cycles, the performance was nearly identical to NiAl-1 and -2 (Fig. 5). This is expected as the scale growth rate is similar on both alloys [33] and essentially all of the oxide spalls after each cycle. When the cycle frequency is increased, the mass loss increases significantly and breakaway oxidation occurs after 550 h in 10-h cycles, 280 h in 2-h cycles and 300 h in 1-h cycles. (All of these specimens were within 5% of a 1.5 mm thickness to make this comparison realistic.) The lower Al content relative to NiAl is a major reason for the different behaviour of FeCrAl. Particularly in higher frequency cycles, scale spallation from FeCrAl quickly leads to Al depletion and the formation of Fe-rich spinel. This increases the rate of oxide growth and thus higher rates of spallation.

When FeCrAl is doped with Y, there is a well-known strong improvement in scale adhesion [37,38]. As a result, there is little effect of cycle frequency up to ≈500 h at 1200°C (Fig. 7). In 100-h cycles, the specimen was run to 5000 h with virtually no spalled oxide and only a few cracks observed in the oxide. However, in 1-h cycles, the specimen began to crack and deform much earlier and began a continuous mass loss. In both cases the specimens were 1.5 mm thick. In 10-h cycles, some spallation was observed after 1000 h and, in 2-h cycles, spallation began after 600 h of total exposure.

For iron aluminides, the situation is somewhat different. Two examples are given: a cast alloy, FAL (Fe–28Al–5Cr–0.1Zr) and an oxide-dispersion strengthened substrate, Fe–27Al–2Cr–0.2Y (as Y_2O_3) (Figs 8 and 9, respectively). Several papers have pointed

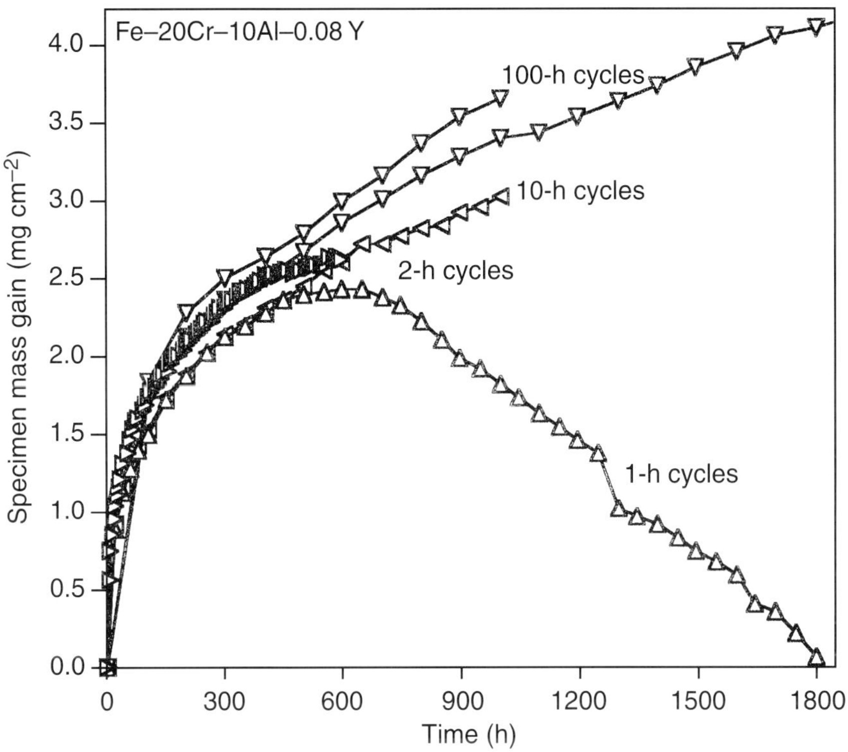

Fig. 7 *Specimen mass gain for 1.5 mm thick FeCrAlY as a function of cycle time at 1200°C.*

out the increased amount of scale spallation for iron aluminides compared to RE-doped FeCrAl or NiAl [19,22]. In both cases, increasing cycle frequency initially results in more scale spallation. However, unlike Ni-base alloys, the mass loss does not continue linearly. This has been observed both in 1-h cycles for FAL and 100-h cycles for ODS Fe_3Al.

The observed non-linearity in specimen mass loss is associated with several different phenomena, represented by four different stages in the specimen mass change curve. A general example for Fe-base alloys is given in Fig. 10. The first two stages and last stage also are common to Ni-base alloys: the adherent (mass gain) stage followed by linear spallation (mass loss) and finishing with breakaway. For Fe-base alloys, however, the rate of linear mass loss often decreases to nearly zero when the alloy begins to form spinel oxide and usually starts to deform. The deformation effect is reflected in the mass loss data for FAL (Fig. 8). When a thinner (1 mm) specimen was used, the onset of specimen deformation (≈1000 h) was earlier and the mass loss began to level off sooner. A thicker specimen (1.3 mm) continued to lose mass after 1700 h, although the rate of loss had decreased. Similarly, mass loss was arrested for ODS Fe_3Al during 1-h cycles (Fig. 9) when the specimen began to deform and form spinel. Another occurrence unique to iron aluminides was the formation of an FeO nodule along the specimen edge which subsequently spalled (arrow in Fig. 8). On FeCrAl specimens the formation of FeO nodules quickly leads to breakaway oxidation.

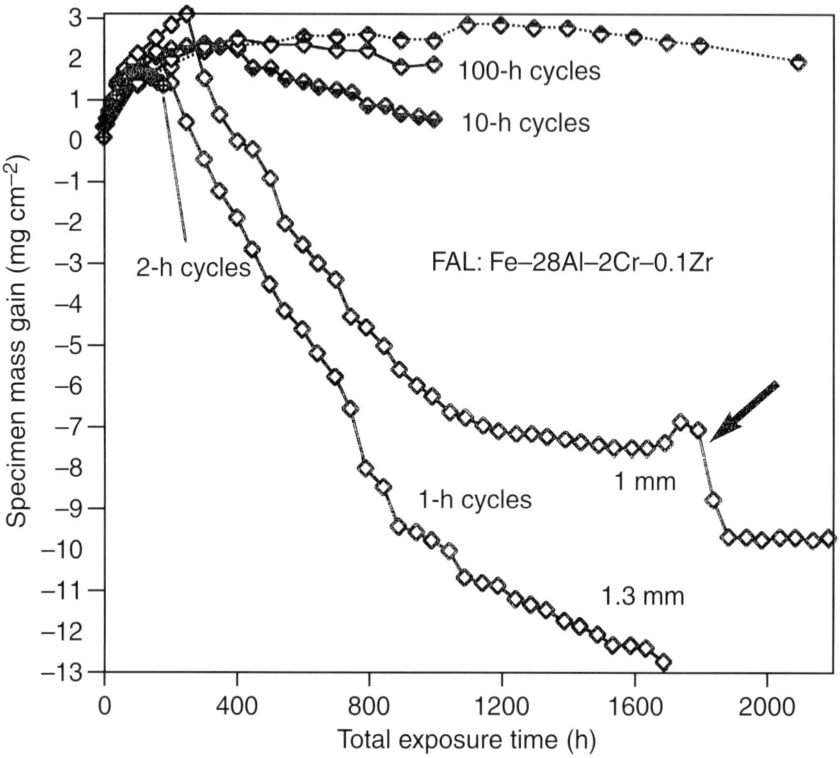

Fig. 8 *Specimen mass gain for cast/extruded FAL as a function of cycle time at 1200°C.*

Thus, unlike Ni-base alloys such as NiAl and NiCrAl, specimen mass losses in thermal cycling are less well-behaved for Fe-base alloys. This problem is illustrated in Fig.11 for several ODS FeCrAl alloys [21] with a 1.5 mm specimen thickness. If the data are examined after only 1000 h (dashed line), it is difficult to accurately rank the long-term performance of the different materials. The commercial Kanthal alloy APM and Y_2O_3-dispersed FeCrAl appear to be performing similarly, and yet their subsequent behaviour is radically different. Likewise, FeCrAl co-doped with Y_2O_3 and Ta_2O_5 appears to be performing very poorly with the largest amount of spallation, however, it has a longer life than Y_2O_3-dispersed FeCrAl and performs almost as well as FeCrAl co-doped with TiO_2. With no general pattern followed by the various specimens, it is difficult to see how the specimen mass change data could be used in a lifetime prediction model. As a result, the only important number generated by this type of experiment is the time to breakaway (Table 2), which, for the best alloys, is much longer than two thousand, 1-h cycles and therefore involves a large amount of experimental time.

An alternative way to collect similar data is to use a contained cyclic test which yields the total mass gain (Fig. 12). The specimen mass change is also obtained (dashed lines) but whereas the specimen masses follow different patterns the total mass gains are reasonably well-behaved. Because the total mass gain is easily correlated with metal wastage, it is much more useful in calculating lifetimes [19,39–41].

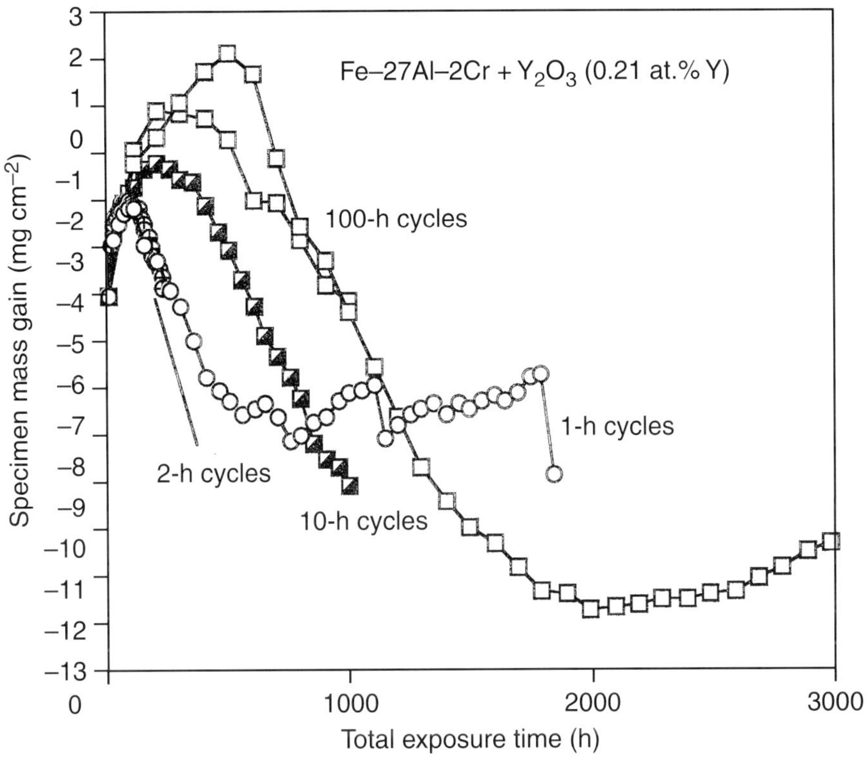

Fig. 9 Specimen mass gain for 1.5 mm thick ODS Fe_3Al as a function of cycle time at 1200°C.

3.5. Chromia-formers

Two examples are given here to illustrate some of the issues with chromia-formers and effects at lower temperatures. Figure 13 shows the effect of cycle frequency on an experimental Ni–26at.%Cr + 0.24Y (as Y_2O_3) alloy at 1000°C. After an initial mass gain for both specimens, the rate of specimen mass loss is greater in 1-h cycles. However, no simple conclusion can be reached because of CrO_3 evaporation. The 100-h cycles were conducted in alumina crucibles, while the 1-h cycles were performed in flowing O_2. Therefore, the greater specimen mass loss may be a result of increased evaporation in the dynamic environment, independent of scale spallation.

Containing a chromia-forming specimen during thermal cycling has a number of benefits. Figure 13 also shows the total mass gain for NiCr + Y during 100-h cycles. No spallation was observed in the bottom of the crucible but the inside was green due to the deposited CrO_3 vapour forming Cr_2O_3. While in principle some CrO_3 vapour may be lost during the experiment, as the crucible lid is not sealed, no mass losses were recorded during this test. Thus, the difference between the specimen mass change and the total mass gain measured the mass lost due to evaporation.

Total mass gains are, therefore, more reliable for comparing performance at this temperature. The total mass gain for Ni–26Cr + Y is very similar to that for two commercial Y_2O_3–dispersed NiCr alloys (Inco alloy MA754 Ni–21at.%Cr and MA758

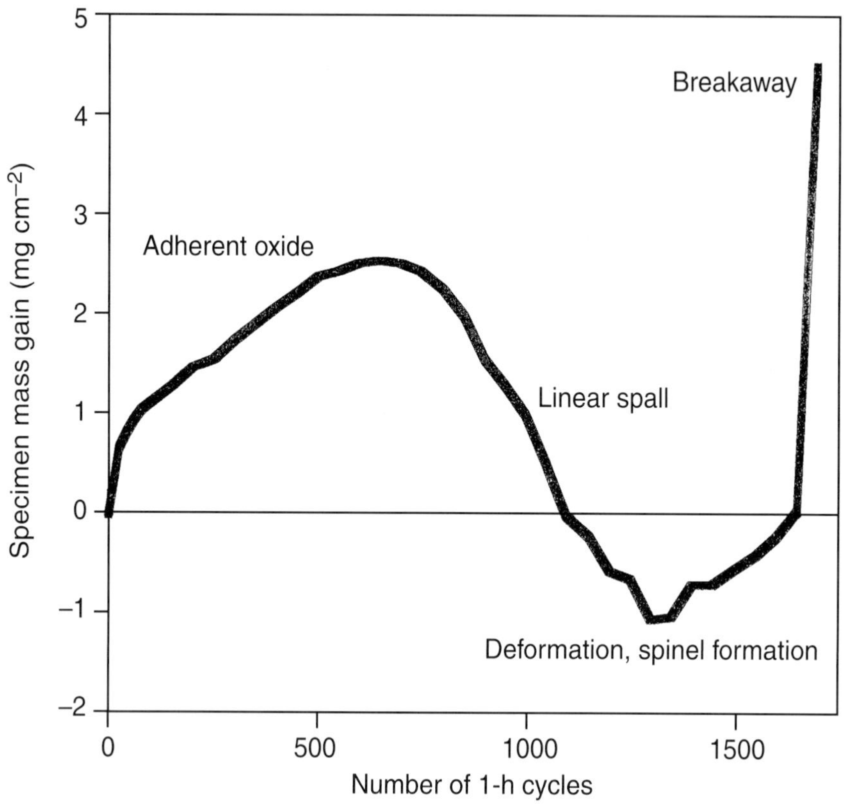

Fig. 10 Representative example of specimen mass change for Fe-base alumina-forming alloys in 1-h cycles at 1200°C.

Ni–33at.%Cr) shown as dashed lines in Fig. 13. When the dispersion of Y_2O_3 was replaced with La_2O_3 there was a significant drop in the chromia growth rate [23] (Fig.13). Comparing only the specimen mass change data, it is difficult to reach this same conclusion for Ni–26Cr–0.07at.%La (as La_2O_3); instead, it could be incorrectly concluded that this alloy spalls more readily than those with Y_2O_3.

A final example is for foil of Nb-modified Fe–20Cr–25Ni (Fig. 14). In this case, contained cycling was used with a cycle time of 500 h at various temperatures. The data are plotted against the square root of time to emphasise the deviations from parabolic behaviour. At the higher temperatures, there was an acceleration in the rate of oxidation but spallation and failure were observed only at 900°C. The acceleration at 800°C was not due to scale spallation (owing to the coincidence of the specimen and total masses) but due to curling and slumping of the foil at a temperature where Nb-modified Fe–20Cr–25Ni has limited strength. One problem with this type of test is noted at 650°C, where the specimen mass gain (dashed line) was higher than the total mass gain (diamond symbols). This was due to the extremely low specimen mass gain and the large difference between the mass of the crucible (≈12 g) and the foil specimen (≈0.2 g). A slight change in the crucible mass due to, for example, outgassing of sintering aid impurities, masks the small specimen mass

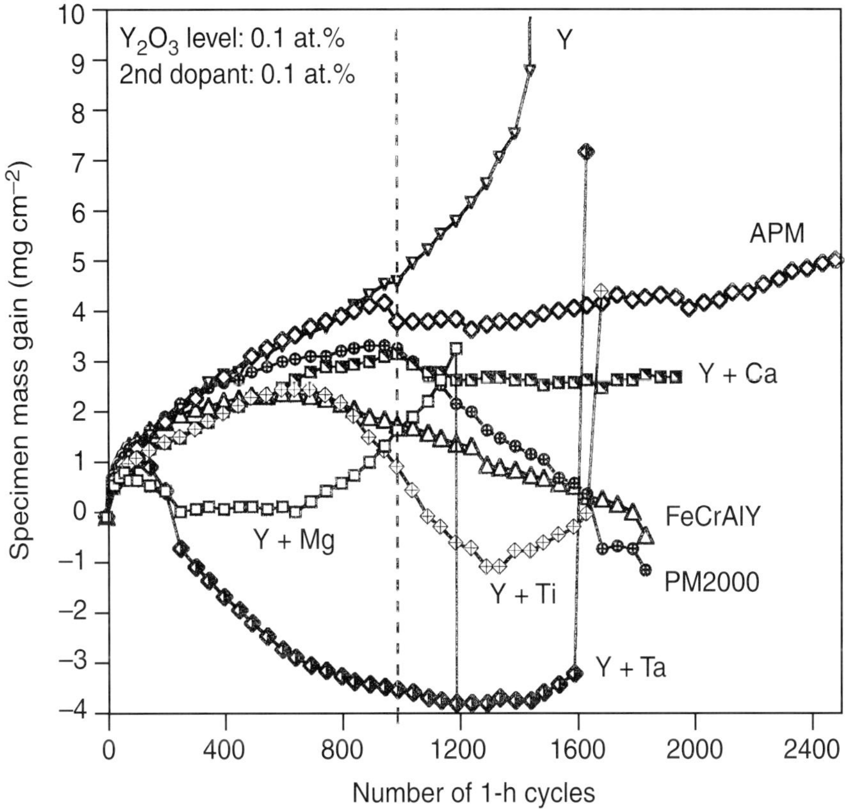

Fig. 11 Specimen mass gain for ≈1.5 mm thick FeCrAl alloys in 1-h cycles at 1200°C.

change in this test. Overall, however, the total mass gain gives a higher confidence to the test data, which are being used to predict foil lifetime [42].

4. Discussion

4.1. Effect of Cycle Frequency

The effect of cycle frequency on oxidation performance is a very important variable which is rarely discussed in the oxidation literature. It is shown above that the effect of cycle frequency is not consistent for all alloy systems. In some cases, higher cycle frequencies increase oxidation losses, while in others, the opposite effect is observed. A standard explanation would be that there are two broad competing factors when cycle frequency is decreased:

(1) The frequency of thermal shocking decreases, so that scales that contain defects capable of concentrating the stresses generated in each cooling cycle or acting together to cause fracture are subjected to fewer such events. Also, any defects nucleated by thermal cycling would accumulate at a slower rate.

Table 2. *Normalised life in 1 h cycles at 1200°C for various Fe-base alloys extrapolated to a uniform 1.5 mm specimen thickness. Many of the tests are continuing to completion*

Alloy	**Actual thickness (mm)**	**Normalised life 1 h cycles, 1200°C 1.5 mm thickness**
Fe–28Al–2Cr	1.0	2500
FAL	1.0	3450
FAL	1.3	>2100
FAP	0.9	1270
PMWY2A	1.5	>2280
FeCrAl	1.5	310
FeCrAlY	1.5	>2370
PM 2000	1.7	>2040
APM	1.2	>4500
FeCrAl + Y_2O_3	1.4	1570
FeCrAl + La_2O_3	1.4	2400
FeCrAl + HfO_2	1.5	>2970
FeCrAl + SrO	1.5	>3100

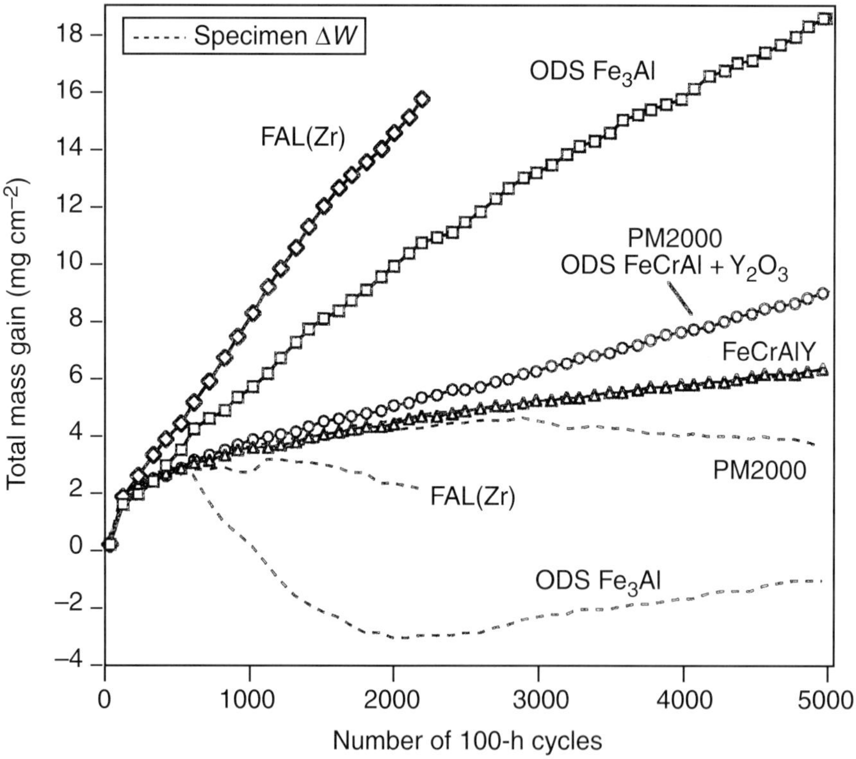

Fig. 12 *Total and specimen mass gains for several Fe-base alloys in 100-h cycles at 1200°C.*

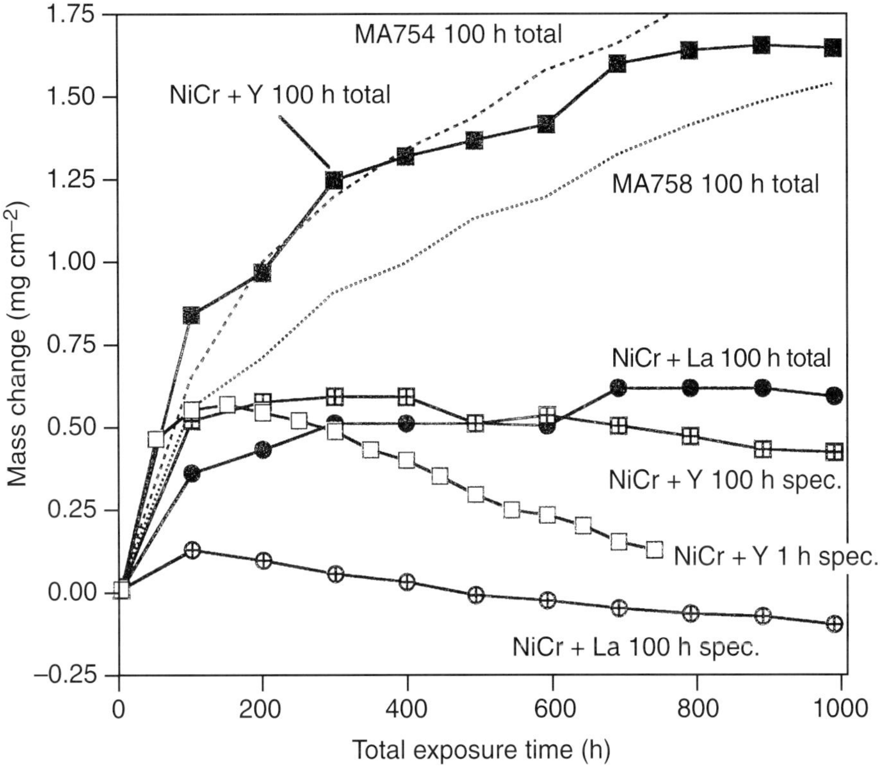

Fig. 13 *Total and specimen mass gains for several ODS NiCr alloys at 1000°C.*

(2) The scale and any defects that develop at temperature (such as interfacial voids) have additional time to grow between thermal shocks. A thicker scale increases the strain that must be accommodated in the cooling cycle and larger defects may increase the amount of spallation.

Thus, decreasing the cycle frequency could yield a positive or negative effect on the rate of spallation, and empirically both positive (e.g. NiAl-1, FeCrAl, FeCrAlY, iron aluminides) and negative (e.g. N5B and (Ni,Pt)Al) effects have been observed. However, the more expected result is that increasing the cycle frequency (shorter cycles) should cause far more damage to the scale than low frequency cycles.

Spallation models tend to emphasise the accumulation of damage or defects such as cracks in the scale which lead to spallation. It is commonly assumed that these defects result from thermal cycling. For example, Lowell *et al.* [8,9] considered the effects of cycle duration on oxide spallation behaviour in the development of an oxidation life-prediction model for superalloys in aircraft engine service. The model assumed that the loss of oxide occurred by spallation within the oxide scale (not to bare metal) in a random fashion over the alloy surface, with the fraction of scale lost in a given segment of the surface being described by one of several statistical distributions. The amount of oxide lost depended on the time available for scale growth between successive spallation events, hence the cycle dwell time was an important parameter. The model predicted that decreasing the cycle frequency would increase maximum mass change and decrease the number of cycles to net scale spallation. Their data for Ni–30Cr showed this effect when mass change in cycles from 5–300 min against the *number of cycles* was plotted. However, when the data are replotted against *time at temperature*, cycles from 5–60 min showed no difference, and only for 5-h cycles was there less spallation. This is similar to the small effect observed in this study between 1-h and 2-h cycles (Figs. 5, 8 and 9).

The negative effect of decreasing cycle frequency (i.e. more spallation with longer time cycles) perhaps requires a different type of mechanism than what is described by Lowell *et al.* There are processes which occur isothermally, such as oxidation-induced deformation of the substrate and growth of interfacial voids and dopant-rich oxides (such as Ta-rich oxides on superalloys [16]) which may have different effects when the cycle frequency is changed. Allowing these defects to increase in size (by increasing the length of the cycle) may change the spallation event from a localised minor event, such as that nucleated by a small defect, to a much larger event affecting a large fraction of the scale. The specific mechanisms by which each type of material reacts to a change in the cycle frequency is beyond the scope of this paper. Future work in this area should include extensive characterisation work, including a time-series of specimens.

The time at which the effect of cycle frequency begins to differentiate appears to be a strong function of substrate scale adhesion. For NiAl + Hf, that time exceeds 1000 h at 1200°C, compared to about 500 h for FeCrAlY, 100 h for iron aluminides and very short times for most undoped alumina-formers. This result is not surprising, because until the scale starts to spall it grows at a rate essentially equal to that under isothermal conditions. This critical time is an important factor in terms of experimental design. Only by exceeding the critical time can the effect of cycle frequency be evaluated. For example, the effect of cycle frequency is not known for NiAl + Hf because the experiments have not been run for a sufficiently long time.

Another point to emphasise is the shape of the specimen mass gain curve. In general, the Ni-base alloys exhibit more regular behaviour, with a protective period followed by a nearly-linear mass loss period when the alloy begins to spall, (Figs 1, 2, 5 and 6). This type of behaviour has been successfully modelled by workers at the NASA Lewis Research Center [8,9,43–45] and others [46]. However, this type of behaviour is not as clearly observed in Fe-base alloys which were not included in these previous modelling efforts. As illustrated in Fig. 10, the spallation curves often have a nearly zero rate loss period not observed in Ni-base alloys. Even this figure is an oversimplification, as there are a wide variety of behaviours, as shown in Fig. 11. This complication makes it difficult to apply models such as those developed for Ni-base alloys to Fe-base alloys. As suggested by the data in Fig.12, it may be easier to model lifetime if the total mass gain is collected in a contained cyclic test, rather than only the specimen mass change in an uncontained test. The total mass gain curves are easily fitted to an exponential or other function to predict lifetimes.

Thus, an important underlying issue on the effect of cycle frequency concerns the method by which the test is conducted. There are two general methods; both have strengths and weaknesses, depending on the goals of the test.

4.2. Comparison of Test Methods

4.2.1. Uncontained testing

This type of experiment is characterised by specimens hanging in a furnace or otherwise not self-contained in a crucible and is more widely reported in the oxidation literature. One of the benefits of this type of test is that it is more flexible and easier to conduct. This method is also preferred for complex environments, such as mixed gas or water vapour, where it is necessary to ensure that the specimen experiences a constant gas composition and gas velocity may be a variable. With a computer-controlled cyclic rig, essentially any cycle time can be used and cooling can be easily modified by the rate of withdrawal or by using fan cooling on removal from the furnace. In any case, heating and cooling times are relatively short. However, one of the important failings of this type of test is that it is very difficult to know the total mass gain; when the specimen spalls its scale only the specimen mass change can be accurately measured. This appears to be less of a problem for Ni-base alloys which appear to be more well-behaved than Fe-base alloys (see above).

Sources of error for this test are the evaporation of the Pt–Rh attachment wire (for a 2 cm long, 0.5 mm dia. wire this can vary from 0.4 μgh^{-1} at 1000°C to 12 μgh^{-1} at 1200°C in flowing O_2) and the uncertainty regarding chromia volatilisation vs spallation for mass losses of chromia formers (e.g. Fig. 13). Also, depending on the friability of the scale, there may be more or less scale lost from the specimen depending on how it is handled, particularly while being weighed.

4.2.2. Contained testing

This type of test is characterised by individual crucibles with lids to contain any spalled or volatile CrO_3 products. Without the lid, the validity of the test is far more questionable. The primary benefit of this test is that it yields the total mass gain (including spallation products) and appears to retain most evaporated product on the crucible walls and lid. This allows better assessments of relative metal consumption rates than are possible from the specimen mass change alone. The amount of spallation or evaporation is directly quantified without any required assumptions and is collected in the crucible if analysis is required.

There are several limitations of contained testing. First, because of the large thermal mass and problem of thermal shock of alumina, short cycles are impractical. Also, when a set of multiple crucibles are inserted into a tube furnace, the furnace temperature is often altered for 30 min or more. For 1-h or 2-h cycles this would cause a significant difference in the thermal history of the specimen compared to uncontained tests or longer cycles. This test is best suited for 10 h or longer to avoid such problems.

Because the specimen is isolated in a crucible, it is difficult to use environments other than air with this test. With increased emphasis on the effect of mixed gas environments and particularly water vapour in combustion environments, this is a significant limitation. For iron aluminides, which can exhibit accelerated oxidation in air compared to O_2 [19,47] this is also a concern. Another drawback of this test is that weighing the crucible, the specimen and then the crucible without the specimen (as a check) is much more time consuming than merely weighing the specimen. The alumina crucibles are also an added expense, although they rarely need to be replaced. The specimen size is also restricted by the size of the crucible.

The sources of error in this test are more varied. One factor is that an alumina crucible mass (10–13 g) is often an order of magnitude larger than the specimen (1–2 g). As pointed out for low specimen mass gains (650°C in Fig. 14), slight changes in the crucible mass can distort the total. This problem can be corrected in a similar manner to the evaporation of Pt, and is most strongly encountered after the first cycle. For well-used crucibles, mass changes (mainly slight losses) are typically 0.1 mg per 1000 h of usage. In general, for mass changes like those observed at 1200°C (Fig. 12) this error is extremely small.

Additional precautions are the use of gloves when handling crucibles to prevent transfer of contaminants onto the crucible surface. Also, if the crucible contents are spilled, re-weighing and corrections may be required. The total mass gain only decreases in value in extremely rare circumstances. Irrespective of specimen and cycle conditions, the total mass gain should increase after every cycle. Finally, the assumption that no CrO_3 vapour is lost from a crucible with a lid has not been tested.

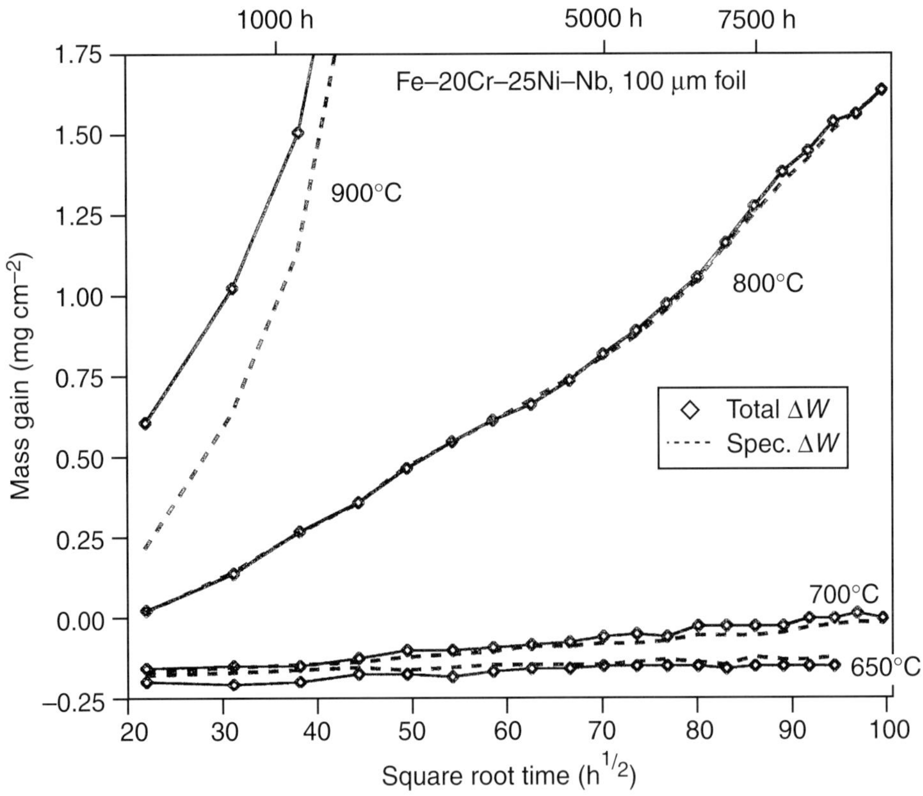

Fig. 14 *Mass gain for Fe–20Cr–25Ni + Nb 100 mm foil at several temperatures plotted against the square root of time.*

5. Conclusions

As is readily apparent in this paper, the issue of cycle frequency is far from resolved but a number of important points can be emphasised. First, the effect of cycle frequency is not the same in every case. It is highly dependent on the scale adherence to the substrate, the type of defects controlling spallation, the total test time and (likely) the test temperature. Cycle frequency also affects how the test is performed, as longer term cycles (100 h) are better suited for contained testing in alumina crucibles and shorter cycles (1–2 h) require uncontained testing. In general, the objectives of the testing strongly influence the test procedures. For life prediction, total mass gain from contained testing appears to be very valuable and reliable, particularly for Fe-base alloys which perform somewhat differently from Ni-base alloys. Further work will be required to understand fully the issue for cycle frequency and its implications for high-temperature oxidation performance and life prediction.

6. Acknowledgements

The authors are indebted to M. J. Bennett (formerly of Harwell Laboratory, UK) who started us on our programme on contained cyclic testing. The authors also would like to thank M. Howell, L. D. Chitwood, G. Garner, N. Bonwit and C. Roberts at ORNL for many hours of assistance with the experimental work and J. R. DiStefano, D. F. Wilson, and J. A. Haynes at ORNL for their comments on the manuscript. This research was sponsored by the U.S. Department of Energy, Fossil Energy AR&TD Materials Program and the Advanced Turbine Systems program under contract DE-AC05-96OR22464 with Lockheed Martin Energy Research Corporation.

References

1. C. A. Barrett and E. B. Evans, *Am. Ceram. Soc. Bull.*, 1973, **52**, 353–360.
2. C. E. Lowell and D. L. Deadmore, *Oxid. Met.*, 1980, **14**, 325–336.
3. R. Mevrel, *Mater. Sci. Tech.*, 1987, **3**, 531–535.
4. M. Schütze, *Protective Oxides Scales and Their Breakdown.* J. Wiley & Sons, Chichester, UK, 1997.
5. H. Jonas and J. A. Golczewski, *J. Nucl. Mater.*, 1984, **120**, 272–277.
6. H. E. Evans, *Mater. Sci. Tech.*, 1988, **4**, 415–420.
7. P. Hancock and J. R. Nicholls, *Mater. Sci. Tech.*, 1988, **4**, 398–406.
8. M. J. Bennett, H. E. Evans and D. A. Shores, *Mater. High Temp.*, 1994, **12**, 127–133.
9. J. Jedlinski, M. J. Bennett and H. E. Evans, *Mater. High Temp.*, 1994, **12**, 169–175.
10. H. E. Evans, J. R. Nicholls and S. R. J. Saunders, *Solid State Phenom.*, 1995, **41**, 137–156.
11. J. R. Nicholls, H. E. Evans and S. R. J. Saunders, *Mater. High Temp.*, 1997, **14**, 5–13.
12. C. E. Lowell, S. R. Levine and S. J. Grisaffe, in *Proc. Conf. on Gas Turbine Materials in the Marine Environment,* MCIC Report No. 75-27, pp.535–554. Metals and Ceramics Information Center, Battelle, Columbus, Oh, USA, 1975.
13. C. E. Lowell, J. M. Smialek and C. A. Barrett, in *High-Temperature Corrosion,* pp. 219–226. (R. A. Rapp, ed.). NACE, Houston, TX, 1983.

14. Comprehensive Program Plan for Advanced Turbine Systems, Office of Fossil Energy and Office of Energy Efficiency and Renewable Energy, Department of Energy, Report to Congress No. DOE/FE-0279, July 1993.
15. W. P. Parks, E. E. Hoffman, W. Y. Lee, and I. G. Wright, *J. Thermal Spray*, 1997, **6**, 187–192.
16. I. G. Wright, B. A. Pint, W. Y. Lee, K. B. Alexander and K. Prüßner, in *High Temperature Surface Engineering* (in press) (J. Nicholls, ed). The Institute of Materials, London, UK, 1999.
17. B. A. Pint, *Oxid. Met.*, 1998, **49**, 531–560.
18. B. A. Pint, I. G. Wright, W. Y. Lee, Y. Zhang, K. Prüßner and K. B. Alexander, *Mater. Sci. Eng.*, 1998, **A245**, 201–211.
19. B. A. Pint, P. F. Tortorelli and I .G. Wright, *Mater. High Temp.*, 1999, **16**, 1–13.
20. B. A. Pint and K. B. Alexander, *J. Electrochem. Soc.*, 1998, **145**, 1819–1829.
21. B. A. Pint, A Quantification of Dopant Effects in ODS FeCrAl, for submission to *Oxid. Met.*
22. B. A. Pint, P. F. Tortorelli and I. G. Wright, *Mater. Corr.*, 1996, **47**, 663–674.
23. B. A. Pint, C. R. Roberts and I. G. Wright, The Effect of Dopant Ion Size on the Oxidation Behaviour of Oxide-Dispersed NiCr Alloys, manuscript in progress.
24. A. W. Funkenbush, J. G. Smeggil and N. S. Bornstein, *Met. Trans.*, 1985, **16A**, 1164–1166.
25. J. L. Smialek, D. T. Jayne, J. C. Schaeffer and W. H. Murphy, *Thin Solid Films*, 1994, **253**, 285–292.
26. J. L. Smialek and B. K. Tubbs, *Met. Trans.*, 1995, **26A**, 427–435.
27. G. H. Meier, F. S. Pettit and J. L. Smialek, *Werkst. Korros.*, 1995, **46**, 232–240.
28. M. A. Smith, W. E. Frazier and B. A. Pregger, *Mater. Sci. Eng.*, 1995, **A203**, 388–398.
29. C. Sarioglu, C. Stinner, J. R. Blachere, N. Birks, F. S. Pettit and G. H. Meier, in *Superalloys 1996* (R. D. Kissinger, D. J. Deye, D. L. Anton, A. D. Cetel, M. V. Nathal, T. M. Pollack and D. A. Woodford, eds). TMS, Warrendale, PA, 1996, pp.71–80.
30. K. Prüßner and I. G. Wright, unpublished research, 1998.
31. D. P. Moon, *Mater. Sci. Tech.*, 1989, **5**, 754–764.
32. A. Strawbridge and P. Y. Hou, *Mater. High Temp.*, 1994, **12**, 177–181.
33. B. A. Pint, *Oxid. Met.*, 1996, **45**, 1–37.
34. E. C. Dickey, B. A. Pint, K. B. Alexander and I. G. Wright, submitted to *J. Mater. Res.*, 1999.
35. J. L. Smialek, *Met. Trans.*, 1978, **9A**, 309–320.
36. B. A. Pint, *Oxid. Met.*, 1997, **48**, 303–328.
37. C. S. Wukusick and J. F. Collins, *Mater. Res. Stand.*, 1964, **4**, 637–646.
38. F. A. Golightly, F. H. Stott and G. C. Wood, *J. Electrochem. Soc.*, 1979, **126**, 1035–1042.
39. W. J. Quadakkers and M. J. Bennett, *Mater. Sci. Tech.*, 1994, **10**, 126–131.
40. W. J. Quadakkers and K. Bongartz, *Werkst. Korros.*, 1994, **45**, 232–241.
41. I. G. Wright, B. A. Pint, C. S. Simpson and P. F. Tortorelli, *Mater. Sci. Forum*, 1997, **251–4**, 195–202.
42. B. A. Pint, P. F. Tortorelli and I. G. Wright, unpublished research, 1998.
43. J. A. Nesbitt, *J. Electrochem. Soc.*, 1989, **136**, 1511–1517 & 1518–1527.
44. C. E. Lowell, C. A. Barrett, R. W., Palmer, J. V. Auping and H. B. Probst, *Oxid. Met.*, 1991, **36**, 81–99.
45. J. A. Nesbitt, E. J. Vinarcik, C. A. Barrett and J. Doychak, *Mater. Sci. Eng.*, 1992, **A153**, 561–566.
46. K. S. Chan, N. S. Cheruvu and G. R. Levant, presented at the *42nd ASME Gas Turbine and Aeroengine Congress*, Orlando FL, June 1997, Paper ASME 97-GT-389.
47. B. A. Pint, J. Regina, K. Prüßner, L. D. Chitwood, K. B. Alexander and P. F. Tortorelli, submitted to *Intermetallics*, 1999.

8
The Effect of Thermal Cycling on the Mechanical Failure of Alumina Scales formed on Commercial FeCrAl-RE Alloys

J. P. WILBER, M. J. BENNETT* and J. R. NICHOLLS
School of Industrial and Manufacturing Science, Cranfield University, Cranfield, Bedford MK43 0AL, UK
*Materials Research Consultant, Three Chimneys, South Moreton, Oxon OX11 9AH, UK

ABSTRACT
Two crucial aspects, namely oxidation cycle frequency and cooling rate from the oxidation temperature, could influence the effect of thermal cycling upon the mechanical failure of alumina scales formed on commercial FeCrAl-RE alloys. Four alloys (PM2000, MA956, Kanthal APM and Fecralloy steel JA13) were oxidised in air at 1300°C in three series of experiments. These involved, respectively: (i) isothermal (i.e. single cycle) exposures followed by furnace cooling to ambient temperature, (i.e. cooling rate of 10^2 °C h^{-1}), (ii) discontinuous (i.e. multi-cycle) long term exposures up to 7500 h comprising 20 h to >300 h dwell times at temperature followed by furnace cooling, and (iii) up to *ca.* 180 rapid thermal cycling exposures, each involving 50 min at temperature, followed by 10 min of forced air cooling to below 100°C at a rate of 10^4 °C h^{-1}. The extents of oxidation and spallation were measured gravimetrically.

The initial (i.e. up to ~100 h) rates of alumina scale growth on all alloys, in all three series of exposures, were comparable. No scale spallation was measured after the single cycle exposures. However, spallation was detected during the discontinuous tests once the scales exceeded critical mass gains, which varied from alloy to alloy: MA956 and PM2000 ~2–3 mg cm^{-2}; Kanthal APM ~7 mg cm^{-2}; Fecralloy steel JA13 >14 mg cm^{-2} during slow furnace cooling. The increased cycle frequency and more rapid cooling rate in the series (iii) experiments had no effect on the mechanical failure of the scales during the early stages of oxidation (i.e. for mass gains less than the critical mass gains for the onset of significant spallation).

1. Introduction

Over the past decade three successive major European programmes (COST 501/2 WP4 [Heat Exchangers], Brite/Euram 'Improve Al_2O_3' (BE7972) and Brite/Euram 'LEAFA' (BE-97-4491)) have focused on producing life-time prediction models of commercial FeCrAl-RE alloys for high temperature industrial applications. During the COST 501 programme work concentrated on the performance of model and commercial oxide dispersion strengthened (ODS) FeCrAl-type alloys over long-term exposures at temperatures between 1050 and 1300°C, e.g. 2.7×10^4 h at 1050°C, which are typical of some industrial applications for these alloys [1–3]. The Brite/Euram programme 'Improve Al_2O_3' focused on understanding and improving the mechanical performance of the protective alumina scales that form in oxidising atmospheres above circa 1000°C on a range of these alloys fabricated by powder metallurgy, casting and mechanical alloying processes [4–6]. This, in turn, has led to the on-going Brite/Euram programme 'LEAFA' ('Life Extension of Alumina Forming Alloys in High Temperature Corrosion Environments'), in which the parameters affecting the time to chemical failure of these alumina-forming alloys are being examined. The overall aim of these programmes has been to understand the many factors that affect the life-time of protective scales in order to develop life-time prediction models for commercial FeCrAl-RE type alloys. This work complements similar studies in the USA aimed at understanding the cyclic oxidation behaviour of alumina forming alloys and intermetallics [7–13]. An overall picture has emerged following a review of these studies, indicating that there are three key stages governing the life-time of the protective alumina scales formed, essentially comprising:

'Birth' — During the initial exposure to an oxidising atmosphere at over 1000°C a protective alumina scale becomes established over the surface of the alloy.

'Life' — With continued exposure this alumina scale grows and thickens. As it matures, however, the scale becomes prone to mechanical failure, via cracking and spallation. The onset and extent of this mechanical failure depends primarily on the interplay between three factors:

(i) Defects, such as voids, pores and micro-cracks, that develop in the scale as it grows;

(ii) Stresses that act on the scale, generated for example by growth and particularly during thermal transients due to the thermal expansion mis-match between the oxide and the alloy. The stresses are locally concentrated by the defects in the scale;

(iii) Extent of stress relief within the scale, particularly by the plastic deformation of the alloy substrate while at temperature and during the early, high temperature part of the cooling stage [6,14].

The interaction between these three factors will determine at what stage in the growth of the scale mechanical failure is likely to occur and to what extent it will continue. It also illustrates that the mechanical failure of the scales cannot be predicted by examining the scale alone — the influence of the alloy substrate and its mechanical properties must also be considered so that a more holistic approach to predicting scale failure is adopted.

When the scale mechanically fails the damaged area can be healed by the formation of new alumina scale when the alloy is re-exposed to the oxidising environment, but only as long as there is a sufficient reservoir of available aluminium in the alloy to do so.

'Death' — Once the level of aluminium in the alloy is reduced to below a critical level, $[Al]_{crit}$, as a result of oxidation then mechanical damage to the protective alumina scale can no longer be healed. Non-protective chromium and iron oxides then form that grow catastrophically at these temperatures in what is termed 'breakaway' oxidation. The formation of these non-protective oxides ends the useful life-time of the component.

Clearly, the more mechanical damage that occurs to the scale during the 'life' of the component the quicker the aluminium reservoir will be depleted and consequently, the earlier will be the onset of breakaway oxidation, i.e. chemical failure of the component. One approach to extend the 'lifetime' of the component therefore is to increase the reservoir of available aluminium in the alloy. This approach has been adopted by Tortorelli and *et al.* [9–13] who have been examining the oxidation lifetime of iron aluminide alloys. With up to three to four times the amount of aluminium in the alloy than in FeCrAl-type alloys the iron aluminide intermetallics potentially offer significantly longer lifetimes in oxidising environments. The oxidation performance of the iron aluminide intermetallics developed so far, however, is little better than some of the existing FeCrAl-type alloys [15]. Further, the high aluminium content of the aluminide intermetallics has a detrimental effect on the mechanical properties of the alloy, making them brittle at all but high temperatures, so that the manufacture of components from such alloys is often difficult.

From the above comments it is clear that to predict the life-time of FeCrAl-type alloy components it is necessary to understand fully all the critical parameters determining the mechanical failure of their protective scales. This paper will concentrate on the effect of cycle frequency, which is dictated by the dwell time at temperature, and, with the aid of a unique thermal cycling facility at Cranfield, the effect of cooling rate.

2. Experimental

2.1. Materials

The four commercial FeCrAl-RE alloys tested were fabricated by three different routes and, as a consequence, exhibited a range of creep properties. MA956 and PM2000 are oxide dispersion strengthened alloys designed for creep resistance at high temperatures, and are therefore termed in this paper 'strong' alloys. Kanthal APM is a powder metallurgical alloy whose creep resistance is less than that of the ODS alloys but higher than that of the wrought Fecralloy steel JA13, the fourth alloy used in this study. The Kanthal APM alloy is therefore designated in this paper as an 'intermediate' strength alloy, while the Fecralloy steel JA13 is designated as a 'weak' alloy. The chemical compositions of these four alloys are given in Table 1.

2.2. Sample Preparation

For the oxidation exposures, coupons nominally 20 × 10 × 2 mm thick were polished using 1200 grit SiC paper and cleaned using a series of ultrasonic bathings in Volasil

Table 1. *Alloy compositions (mass%)*

Element	PM 2000	MA956	Kanthal	Fecralloy steel JA13
Fe	Bal.	Bal.	Bal.	Bal.
Al	5.83	4.62	5.86	5.0
Cr	18.8	19.7	21.0	16.3
Y (Y_2O_3)	(0.5)	(0.5)	<0.02	0.32
Zr	—	—	0.10	<0.05
Ti	0.49	0.41	0.21	0.01
C	0.0057	0.01	0.028	0.030
S	<0.0005	0.002	<0.001	0.001
Si	0.021	0.09	0.29	0.33
Mn	0.0495	0.12	0.11	0.12
P	—	0.016	0.018	0.013
Others	0.22 Ni 0.013 Cu 0.013 N	0.017 N 0.07 Ni 0.20 O	0.02 Co 0.03 Cu	—

344 (a degreasing solvent, supplied by Abinghurst Ltd, UK) and *iso*propyl alcohol. The bathings were followed by a final vapour degrease in *iso*propyl alcohol.

2.3. Oxidation Exposures — Effect of Cycle Frequency

All samples were contained in alumina crucibles and exposed in air at 1300°C. The samples were heated to 1300°C and cooled to room temperature at 300°C h^{-1}. The dwell time at 1300°C dictated the cycle frequency. The exposures were either isothermal, i.e. single cycle, or multi-cycle exposures with dwell times at temperature ranging from 20 h to over 300 h. At the end of each heating cycle the samples and their crucibles were weighed to determine both the total (i.e. 'gross') mass gain of the sample, which included the amount of spalled oxide, and the 'net' mass gain, which excluded the amount of spalled oxide. Traditionally the latter values have been the favoured method of presenting gravimetric data. By comparing the net mass gain values with the gross mass gain values this paper will, as a subsidiary comment, illustrate the advantages and disadvantages of each approach.

2.4. Oxidation Exposures — Effect of Cooling Rate

To assess the effect of cooling rate on the mechanical failure of the scales a unique thermal cycling facility at Cranfield was used. The automated thermal cycling rig (Fig. 1a) consists of a furnace, capable of heating up to 1400°C, held above a cooling chamber. The samples are assembled on a carousel (Fig. 1b) that is located on top of a pneumatically operated shaft. This shaft raises the carousel from the cooling

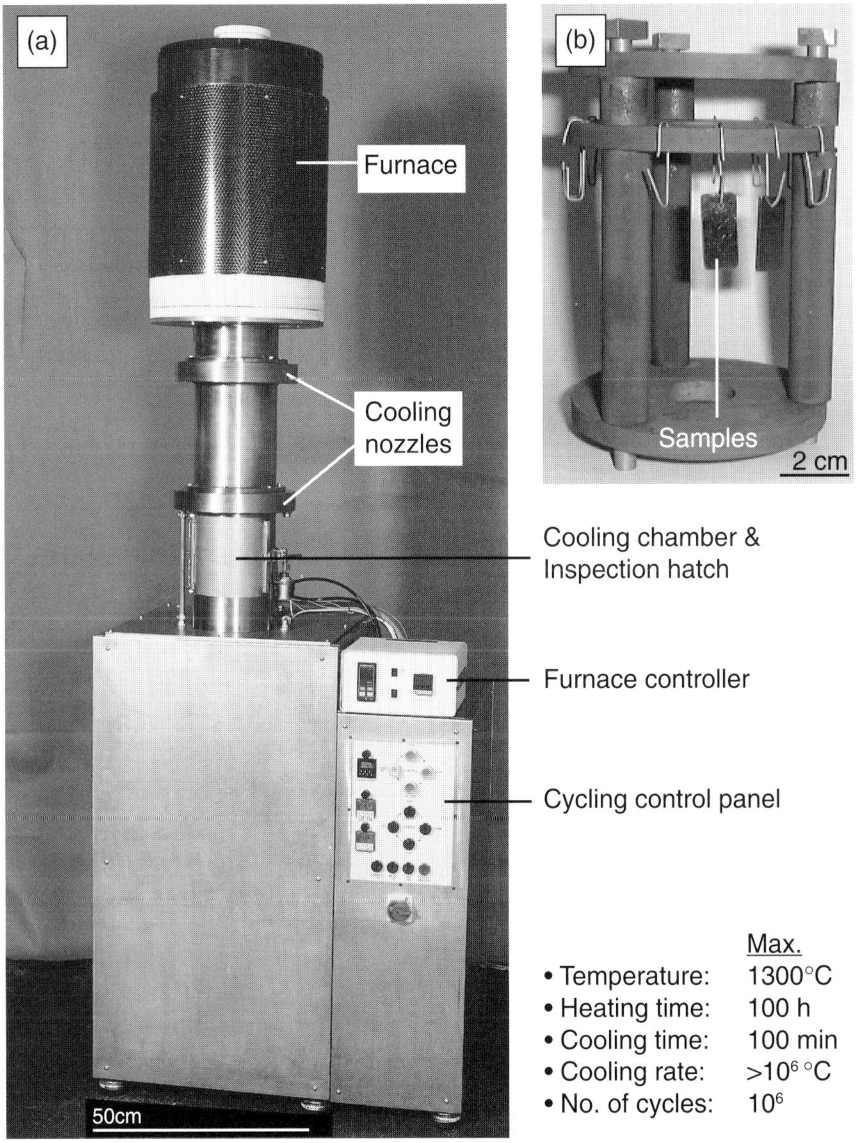

Fig. 1 *(a) Rapid thermal cycling facility and (b) sample carousel.*

chamber into the furnace and withdraws them back into the cooling chamber at the end of the dwell time. Around the periphery of the cooling chamber are positioned eight nozzles through which either compressed air or water may be fed to cool the samples. By varying the air flow rate, or by using water instead of air as the cooling medium, the cooling rate of the samples can be adjusted from ~10^4 to >10^6 °C h^{-1} (see Fig. 2). The high temperature dwell time, the cooling dwell time and the number of thermal cycles are set using the control panel. As a consequence, the thermal cycling

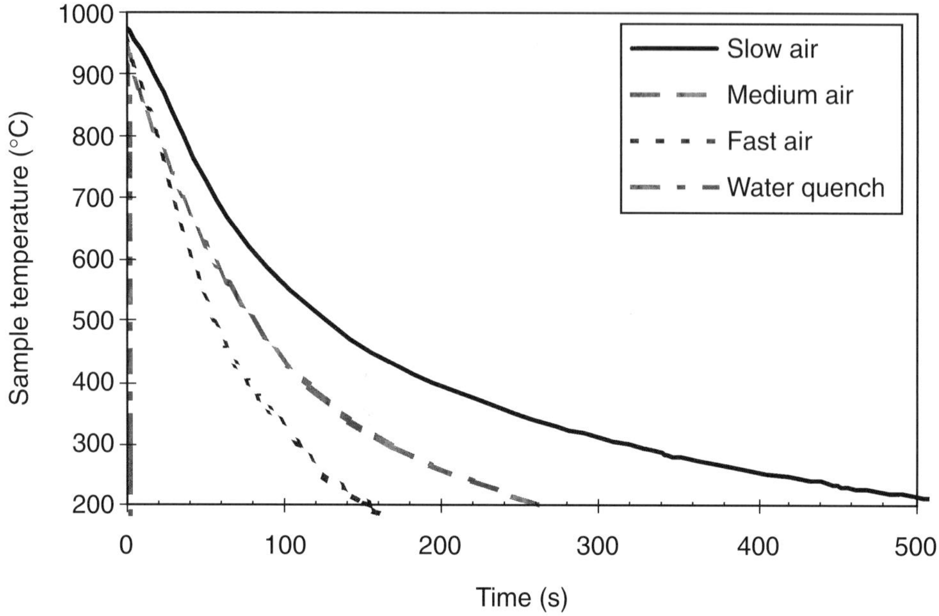

Fig. 2 Cooling rates achievable with the rapid thermal cycling facility.

exposures are executed automatically. The samples are periodically examined and weighed. For this purpose the samples are removed through the inspection hatch of the cooling chamber.

In the current study samples of MA956, Kanthal APM and Fecralloy steel JA13 were exposed using rapid thermal cycles consisting of 50 minute dwells at 1300°C followed by forced air cooling for 10 min at a rate of ~10^4 °C h^{-1}, which brought the samples to below 100°C. At regular intervals the samples were removed for measurement of their net mass gain.

3. Results and Discussion

3.1. Gravimetric Results: Net vs Gross Mass Gain Measurements

The presentation in the literature of previous gravimetric data for FeCrAl-RE alloys in oxidising environments leading, eventually, to alumina scale failure has often been in the form of net, rather than gross mass gain data. Gross mass gains are a measure of the increase in mass due to the extent of attack, i.e. of oxygen uptake, while net mass gains represent the combined effect of mass gain by oxygen uptake and mass loss by spallation of metal oxide. The advantages of using net mass changes and not measuring the amount of spalled oxide are primarily two-fold. Without the need to collect the spall debris the experimental design can be simpler and has greater flexibility. Using net mass gain measurements, significant spallation of the oxide from the alloy coupons will be evident as decreases in the mass of the coupons. This is illustrated by the net mass gain measurements of the PM2000 samples during the long term discontinuous tests with furnace cooling (series (ii)) for which large amounts of spall debris were observed after each cool to room temperature (see Fig. 3). A disadvantage of relying on net mass gain measurements is demonstrated by the corresponding data for Kanthal APM and the Fecralloy steel JA13 coupons as the net mass gain data suggest that no significant spalling occurred from these alloys during these series (ii) exposures. This is confirmed in Fig. 4 by the gross mass gain data for the Fecralloy steel JA13 but not for Kanthal APM. For the latter, the gross mass gain data indicate an increase in the oxidation rate after *ca.* 7 mg cm^{-2} mass gain, coinciding with the onset of significant spallation. The net mass gain data (Fig. 3), therefore, suggest little change in the oxidation rate and no significant spallation. Thus, it should be noted that although the net mass change data are easier to obtain and allow for greater flexibility in experimental design they can also be misleading.

3.2. Gravimetric Results — Mechanical Failure of the Scales

In order to illustrate unambiguously the extent of mechanical failure, in the form of spallation, of the scales as they grow, the gross mass gain data are best plotted against the mass of spall debris data. This has been done in Fig. 5 for the data generated during the discontinuous long term exposure (furnace cooling) tests. This illustrated both the critical characteristics of spallation and the behavioural differences between the three types of FeCrAl-RE alloys. For all alloys significant spallation occurred above a critical mass gain, the value of which was alloy specific. Above the critical mass gain the extent by which the amount of spallation increased also varied from alloy to alloy. For the creep resistant ODS alloy PM2000 the extent of spallation rapidly

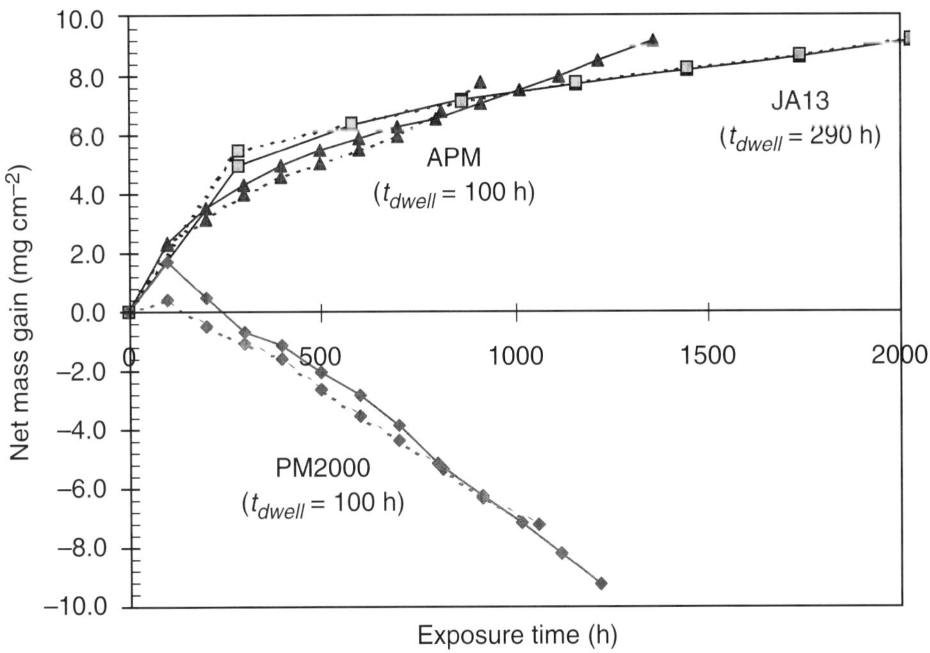

Fig. 3 *Net mass gain data for PM2000, Kanthal APM and Fecralloy steel JA13 during exposure at 1300°C using dwell times of 100-h cycles.*

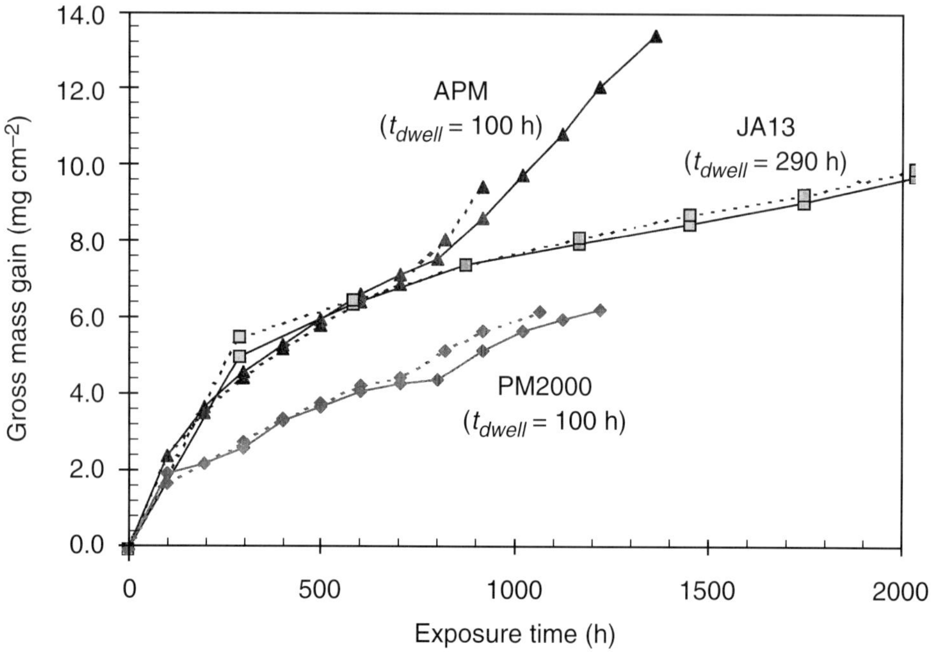

Fig. 4 *Gross mass gain data for PM2000, Kanthal APM and Fecralloy steel JA13 during exposure in air at 1300°C using dwell times of 100-h cycles.*

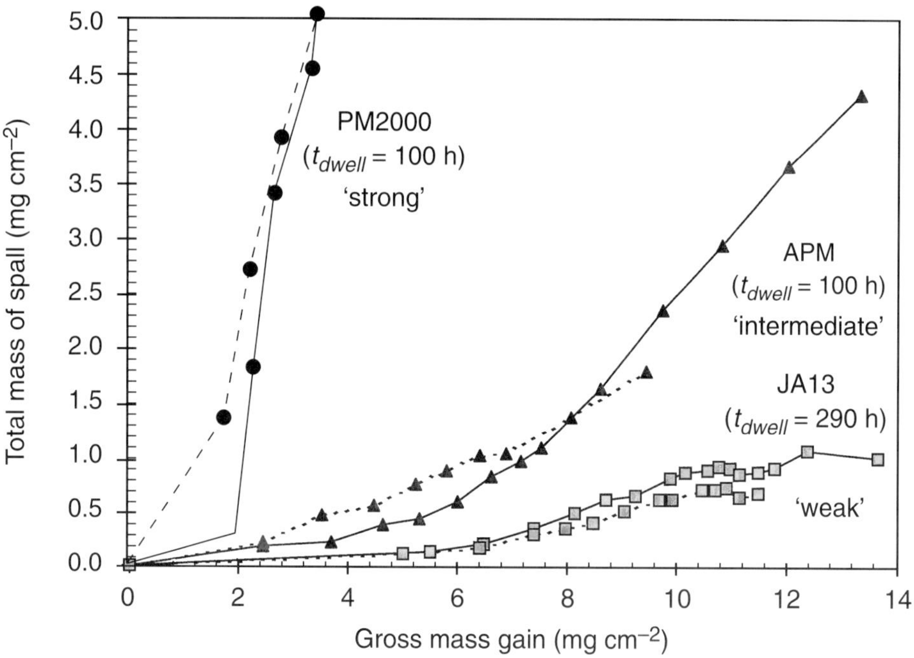

Fig. 5 *Extent of spallation from PM2000, Kanthal APM and Fecralloy steel JA13 during exposure in air at 1300°C using dwell times of 100-h cycles.*

increased after a critical mass gain of circa 2 mg cm^{-2}. Indeed, visual inspection revealed that after cooling to room temperature as much as 80% of the scale had spalled from the surface of these coupons*. For the 'intermediate strength' Kanthal APM alloy the onset of spallation occurred after about 5–7 mg cm^{-2} gross mass gain. Further, for the 'weak' Fecralloy steel JA13 there was little spalling, even after 14 mg cm^{-2} mass gain, which is approximately equivalent to a 70 μm thick scale (assuming 1 mg cm^{-2} mass gain corresponds to a 5 μm thick alumina scale). Thus, the onset and extents of spallation due to thermal transients depend, at least in part, on the extent of stress relief afforded by the plastic deformation of the alloy substrate. For the creep-resistant ODS alloys there was little stress relief and so spallation occurred early on in the 'life' of the scale once the defects in the scale had developed to what had previously been described [3] as the mode II regime. With decreasing creep-resistance of the alloy stress relief by alloy plastic deformation increased so that thicker, more defective scales developed before the scale mechanically failed due to the stresses generated in it by thermal cycling.

For all FeCrAlRE alloys examined, once the critical thickness for the onset of significant spall had been attained the cumulative mass of spall debris increased, effectively linearly, with further mass gain (see Fig. 5). The rate, with respect to mass gain, at which the extent of spallation increased was greatest for PM2000 and least for Fecralloy steel JA13. This again illustrates the effect of alloy creep resistance on the mechanical failure of the alumina scales — greater alloy plastic deformation, as in the case of the 'weaker' Fecralloy steel JA13, will result in greater relaxation of the stresses in the oxide and thus less spallation.

3.3. Gravimetric Results — Effect of Thermal Cycle Frequency on the Mechanical Failure of Scales

The influence of thermal cycle frequency on the mechanical failure of the scales was assessed by comparing the extents of spallation from Kanthal APM with those from PM2000 during the discontinuous exposure, furnace cooling tests using a range of dwell times at temperature. For Kanthal APM (Figs 6 and 7) cycle frequency had little effect on either the oxidation kinetics or the extent of spallation, at least until the scales had grown above a 'critical' mass gain (*ca.* 5 mg cm^{-2}), during thermal cycles comprising dwell times at temperatures of 100 or 290 h, or during isothermal (single cycle) exposures. Above this critical mass gain there was an initial deviation in the spallation behaviour between those samples cycled using 100 h dwells and those using 290 h dwells, possibly due to alloy batch-to-batch variations. The long term behaviour of these samples were similar, though.

For the creep-resistant ODS alloy PM2000 (Figs 8 and 9), again, no significant influence of cycle frequency on the oxidation kinetics was observed between the samples exposed using 100 h dwells at temperature and those exposed using >300 h dwells. However, there was a clear difference between the onset of spallation from these two sets of samples. The reason for the delay in the onset of spallation for the

*Note that the mass of the spall debris may exceed the gross mass gain by up to *ca.* 2.5 times. This is because, while the gross mass gain data measures the amount of oxygen uptake by the alloy, the mass of alumina spall debris includes the loss from the alloy coupon of both oxygen and aluminium.

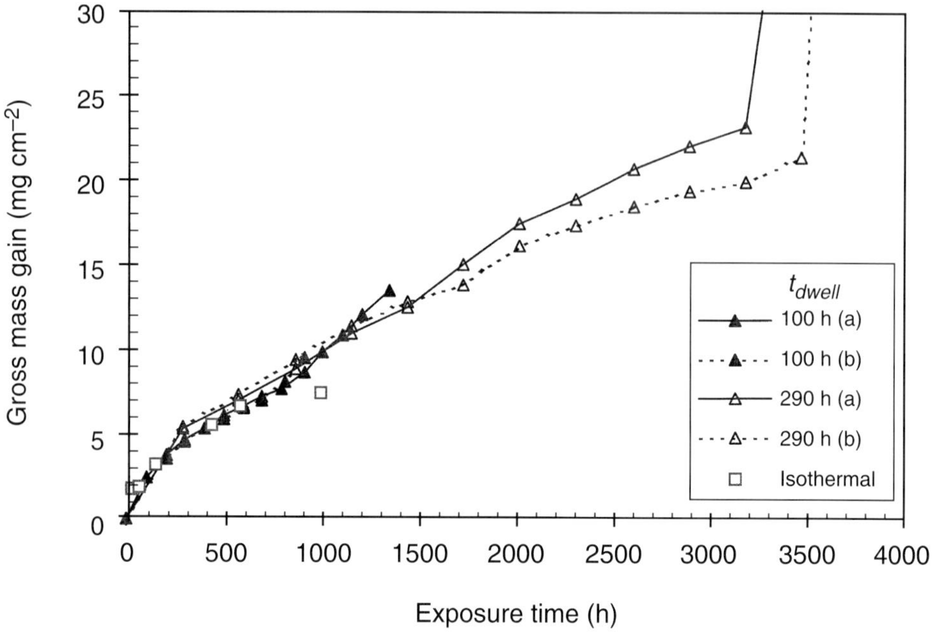

Fig. 6 The effect of cycle frequency on the oxidation kinetics of Kanthal APM at 1300°C.

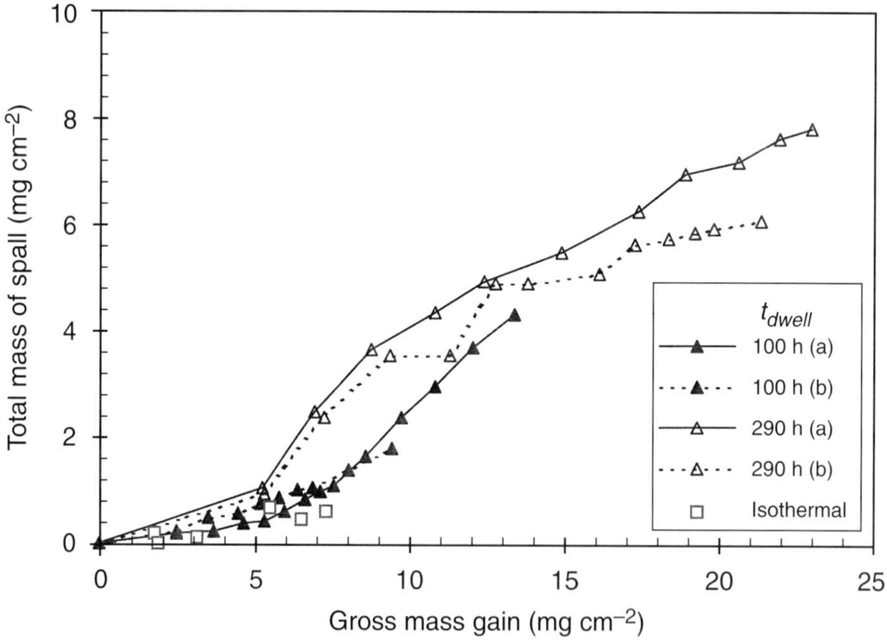

Fig. 7 The effect of cycle frequency on oxide spallation from Kanthal APM during oxidation in air at 1300°C.

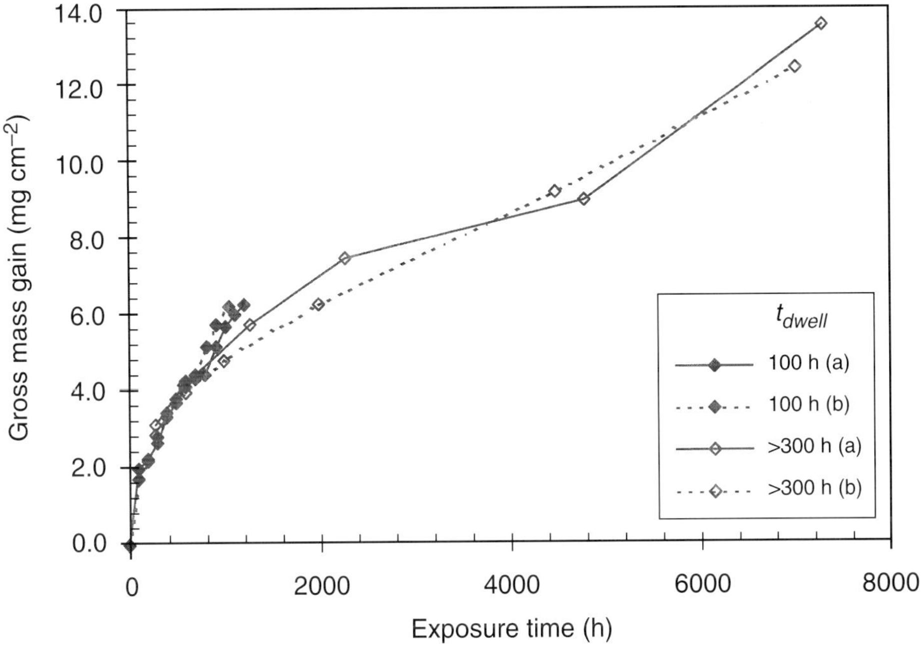

Fig. 8 *The effect of cycle frequency on the oxidation kinetics of PM2000 in air at 1300°C.*

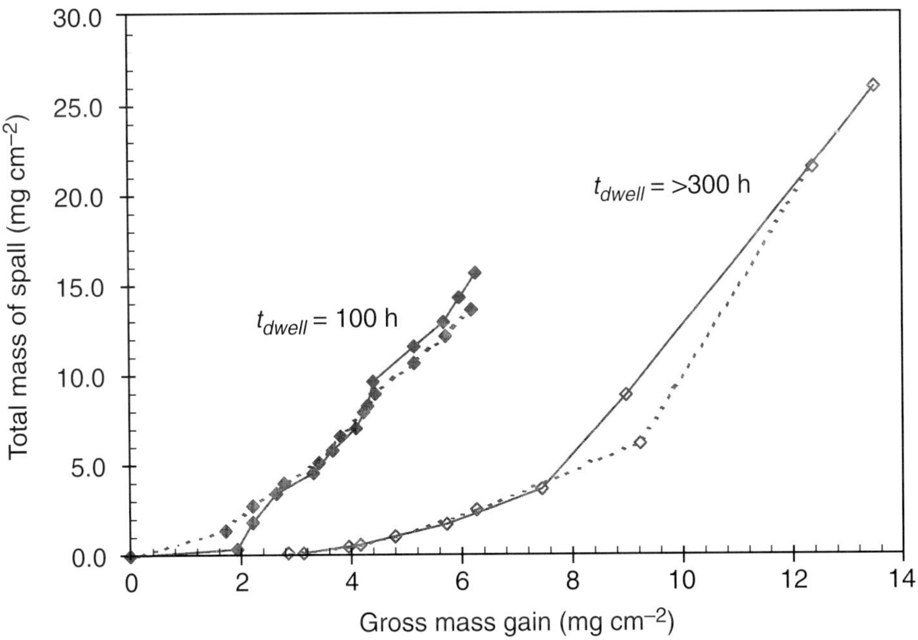

Fig. 9 *The effect of cycle frequency on oxide spallation from PM2000 during oxidation in air at 1300°C.*

>300 h cycles is not clear but may originate, in part, from the batch-to-batch variability of these commercial ODS alloys — the two sets of exposures were performed using different batches of PM2000.

The gravimetric data for Kanthal APM and PM2000 illustrate that the effect of thermal cycling frequency, as dictated by the dwell time at temperature, has little effect on the oxidation kinetics at least until the scales have grown above a 'critical' mass gain, the value of which depends on the alloy and its mechanical properties. Above the critical mass gain the effect of cycle frequency on the mechanical failure of the scale depends on the alloy upon which the scales have grown and may vary from batch-to-batch for a given alloy.

3.4. Gravimetric Results — Effect of Cooling Rate on the Mechanical Failure of Scales

For all three types of alloys examined there was no major effect of cooling rate during the early stages* of the exposure, over the range of cooling rates investigated (i.e. 10^2–10^4 °C h^{-1}) (see Figs 10–12). Kanthal APM, Fecralloy steel JA13 and MA956 were exposed in the rapid thermal cycling rig for 202, 82 and 66 1h cycles, respectively, such that the cumulative times at temperature were 168, 68 and 55 h. For all alloys these exposure times were too short for the critical mass gain for the onset of

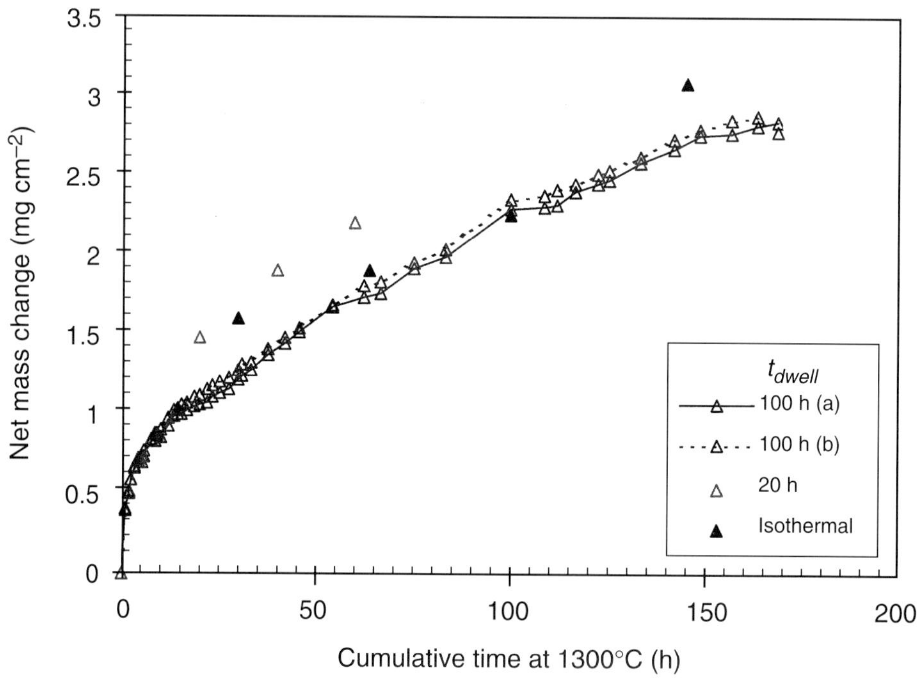

Fig. 10 Effect of cooling rate on oxidation of Kanthal APM in air at 1300°C.

*'Early stages' implies mass gains less than the critical gains for the onset of spallation as identified using the discontinuous, furnace cooling tests.

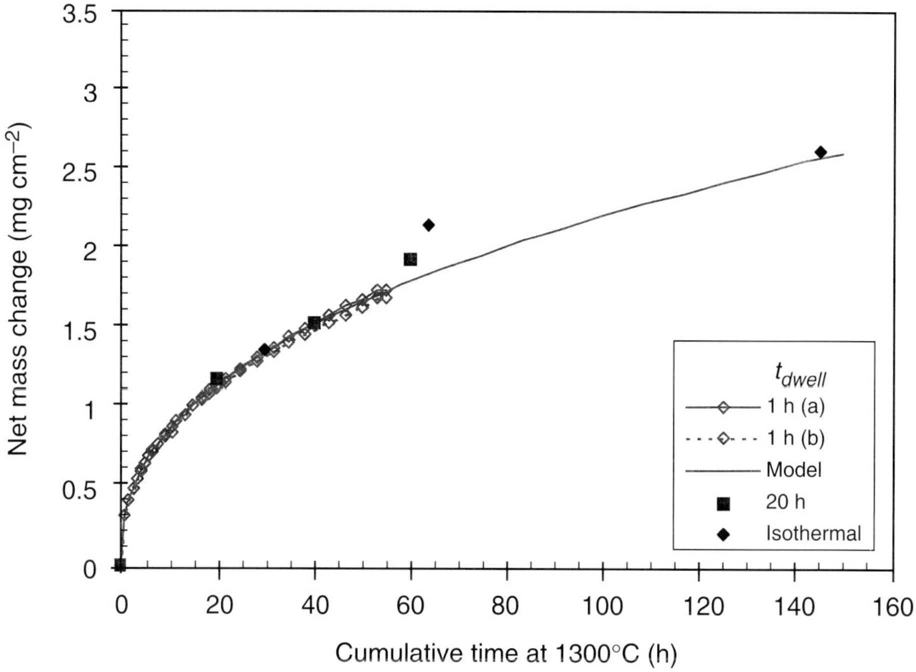

Fig. 11 *Effect of cooling rate on oxidation of MA956 in air at 1300°C.*

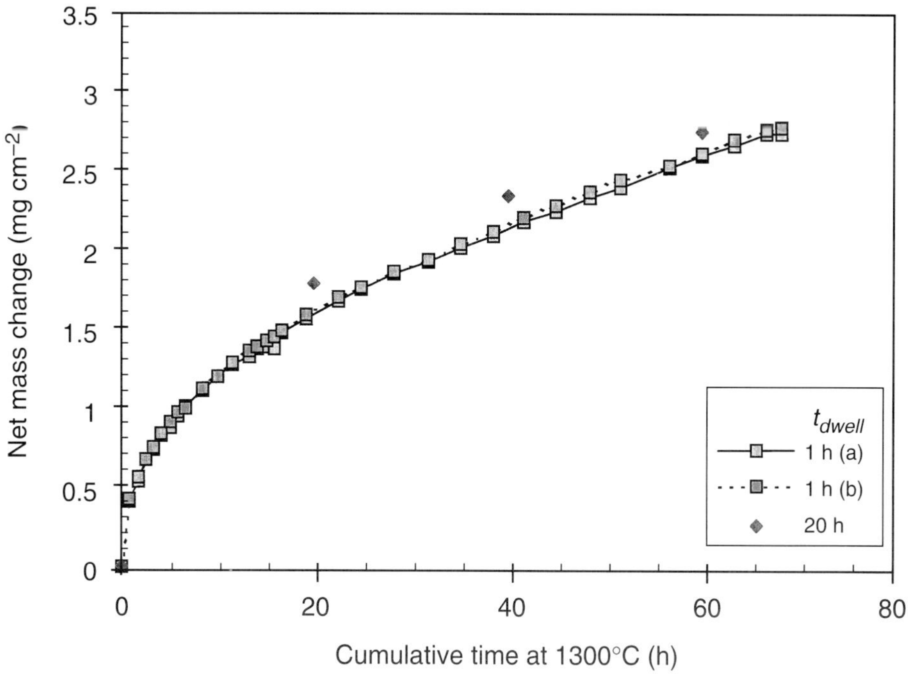

Fig. 12 *Effect of cooling rate on oxidation of Fecralloy steel JA13 in air at 1300°C.*

significant spall to be exceeded. Figures 10–12 compare the gravimetric data obtained from the rapid thermal cycling tests (1h cycles) with those for the longer term tests using slow furnace cooling of 10^2 Ch^{-1}. Note that the MA956 rapid cycling gravimetric data (Fig. 11) were modelled using a numerical best fit to the data in order to extrapolate the data out to similar exposure times as the isothermal exposures. It is expected that this extrapolation may not hold for the rapid thermal cycles once the critical mass gain (circa 2 mg cm^{-2} for MA956) has been exceeded and significant spallation occurs. For all three alloys the short cycle time/rapid cooling exposures generated similar net mass gains to those measured using the longer cycle time/ slower cooling exposures. This would suggest that cooling rate, within the range investigated, had no significant influence on either scale development or mechanical failure during scale growth prior to attainment of the critical mass gain for the onset of spallation.

The effect of cooling rate on the mechanical failure of the scales, once the critical mass gains had been exceeded, can be predicted by considering the effect of the cooling rate on the extent of stress relief, in the form of the plastic deformation of the substrate, during cooling. The creep-resistant ODS alloys exhibited little plastic deformation during either fast or slow cooling and so the cooling rate should not have any significant influence on the mechanical failure of the scales on these alloys. The extent of plastic deformation during cooling of less creep resistant alloys, such as Kanthal APM, however, will be determined by the time at elevated temperatures, which will be dictated by the cooling rate — the faster the cooling rate, the less time the substrate will have to plastically deform so that less stress relief will occur. Thus, the mechanical failure of the scales, once the critical mass gain has been exceeded, can be expected to be more severe than during slower cooling. For 'weak' alloys, such as the Fecralloy steel JA13, there may be sufficient time to relieve the stresses by plastic deformation of the substrate even during the rapid cools used in this work. Consequently, a more rapid cooling rate (i.e. $>10^4$ °C h^{-1}) may be required to inhibit substrate deformation and reveal the effect of cooling rate on the mechanical failure of the scales formed on such alloys.

4. Conclusions

- Thermal cycling plays a critical role in governing the mechanical failure of the scales formed on commercial FeCrAl-RE alloys.

- Significant spallation of alumina scales occurred above a critical mass gain (scale thickness) that was alloy dependent and associated with the extent of stress relief by plastic deformation of the individual alloy substrate during the early, high temperature part of the cooling cycle. (The greater the alloy creep resistance, the lower the critical scale thickness.)

- On all alloys examined, the extents of spallation during the thermal cycling exposures increased with mass gain for mass gains above some critical value. The rate at which the extent of spallation increased depended on the alloy creep properties, and was greatest for the creep resistant alloys, which afforded less relaxation of the stresses in the oxide scales.

- Thermal cycle frequency and cooling rate had little effect on the mechanical failure of the scales until the critical mass gain for that alloy sample has been exceeded.

- The influence of cycle frequency and cooling rate on scale failure depended on the interplay between stress generation and stress relief within the oxide/alloy system only for mass gains greater than the critical mass gain for the particular alloy on which the scale had formed.

- The cooling rate dictated the time-scale during which the alloy may plastically deform, and, as a consequence, the extent of stress relief.

5. Acknowledgements

The authors are grateful to the European Community for funding the BRITE/EURAM project 'Improve Al_2O_3' (BRE2-CT94-0605), and the BRITE/EURAM project 'LEAFA' (BE-97-4491); and to the respective partners of these projects for their assistance.

References

1. W. J. Quadakkers, K. Bongartz and F. Schubert, in *Materials for Advanced Power Engineering* (D. Coursouradis *et al.*, eds). Kluwer Academic Publishers, 1994, **2**, pp.1533–1542.
2. M. J. Bennett, R Perkins, J. B. Price and F. Starr, in *Materials for Advanced Power Engineering* (D. Coursouradis *et al.*, eds). Kluwer Academic Publishers, 1994, **2**, pp.1553–1562.
3. W. J. Quadakkers and K. Bongartz, *Werkst. Korros*, 1994, **45**, 232–241.
4. J. P. Wilber, M. J. Bennett and J. R. Nicholls, in *Materials for Advanced Power Engineering* (J. Lecomte-Beckers, *et al.*, eds). 6th Liège Conference, Sept. 1998, Liège, Belgium. Forchungszentrum Jülich GmbH, Germany, Vol. II, pp.835–846.
5. J. P. Wilber, J. R. Nicholls and M. J. Bennett, in *Microscopy of Oxidation – 3* (S. B. Newcomb and J. A. Little, eds). The Institute of Materials, London 1997, pp.207–220.
6. J. R. Nicholls, H. E. Evans and S. R. J. Saunders, *Mater. High Temp.*, 1997, **14** (1), 5–14.
7. C. A. Barrett and C. E. Lowell *Oxid. Met.*, 1975, **9**, (4), 307–355.
8. C. A. Barrett and C.E. Lowell *Oxid. Met.*, 1978, **12**, (4), 293–311.
9. B. A. Pint, P. F. Tortorelli and I. G. Wright, *Mater. Corros.*, 1996, **47**, 663–674.
10. B. A. Pint, *Mat. Sci. Forum*, 1997, **251–254**, 397–404.
11. J. H. De Van, P. F. Tortorelli and M. J. Bennett, in *Proc. 8th Ann. Conf. on Fossil Energy Materials* (N. C. Cole and R. R. Judkins, eds). Oakridge, USA 1994 US Dept. of Energy, pp.309–320.
12. M. J. Bennett, J. H. De Van and P. F. Tortorelli, in *Microscopy of Oxidation – 3* (S. B. Newcomb and J. A. Little, eds). The Institute of Materials, London 1997, pp.233–245.
13. K. B. Alexander, K Prüssner, P. Y. Hou and P. F. Tortorelli, in *Microscopy of Oxidation – 3* (S. B. Newcomb and J. A. Little, eds). The Institute of Materials, London 1997, pp.246–255.
14. H. E. Evans, *Int. Mater. Rev.*, 1995, **40**, 1.
15. J. Klöwer, *Mater. Corros.*, 1996, **47**, 685–694.

9
Cyclic Oxidation Testing and Modelling: a NASA Lewis Perspective

J. L. SMIALEK, J. A. NESBITT, C. A. BARRETT and C. E. LOWELL
NASA Lewis Research Center, Cleveland, OH, USA

ABSTRACT

The Materials Division of the NASA Lewis Research Center has been heavily involved in the cyclic oxidation of high temperature materials for 30 years. Cyclic furnace and burner rig apparatuses have been developed, refined, and replicated to provide a large scale facility capable of evaluating many materials by a standard technique. Material behaviour is characterised by mass change data obtained throughout the test, which has been modelled in a step-wise process of scale growth and spallation. This model and a coupled diffusion model have successfully described cyclic behaviour for a number of systems and have provided insights regarding life prediction and variations in the spalling process.

Performance ranking and mechanistic studies are discussed primarily for superalloys and coating alloys. Similar cyclic oxidation studies have been performed on steels, intermetallic compounds, thermal barrier coatings, ceramics, and ceramic composites. The most common oxidation test was performed in air at temperatures ranging from 800° to 1600°C, for times up to 10 000 h, and for cycle durations of 0.1 to 1000 h. Less controlled, but important, test parameters are the cooling temperature and humidity level. Heating and cooling rates are not likely to affect scale spallation. Broad experience has usually allowed for considerable focus and simplification of these test parameters, while still revealing the principal aspects of material behavior and performance. Extensive testing has been performed to model statistically the compositional effects of experimental alloys and to construct a comprehensive database of complex commercial alloys.

1. Background

Many applications of high temperature materials require repeated heating and cooling cycles, thus demanding engineering materials that form slow growing oxides resistant to mechanical damage on cooling down. Large compressive stresses arise in scales during cooling because the thermal expansion coefficients of the substrate metals are generally 30–50% greater than those of the oxides. This in turn results in scale cracking, buckling, or delamination, and some degree of spallation, either within the scale thickness or at the scale–metal interface. The subsequent growth rate of the scale in the spalled region is higher because there is less oxide thickness through which the oxidant or metal must diffuse for continued reaction. The net result is that repeated cycling generally results in a greater material consumption rate. Consequently, cycle frequency, in addition to oxidation temperature, is an important degradation factor in applied oxidation cases.

Initial attempts at simulating the effects of cycling (*ca.* 1970) were accomplished by manual loading and unloading samples in an oxidation furnace, usually every 20 h, to obtain a daily sample mass and visual observation. However, the primary NASA Lewis interests in oxidation were turbine airfoil, combustor, or exhaust nozzle applications in aircraft engines. A more realistic cycle duration would be the 1–4 h typical trip of a commercial aircraft. Consequently, manual furnace oxidation cycling evolved to once every 2 h, but this required significant attention and was only practical for total test durations of 100 h or less.

2. Cyclic Furnace Testing

Convenience and versatility dictated a more automated system, and the apparatus that evolved over the next ten years did away with the constant need of attention [1,2]. It employs a resistance furnace positioned vertically with six 3.5 cm dia. alumina tubes arranged in a hexagonal array. The specimen insertion device is a vertically oriented pneumatic cylinder above the furnace, which actuates a plate from which six platinum hanger wires and specimens are suspended. The piston travel and wire length are such that upon actuation the six specimens are lowered into the six tubes until reaching the centre of the hot zone. Typical sample dimensions are 1–2 mm thick, 1–1.5 cm wide, and 2–3 cm long, resulting in nominally 2–5 g samples for most first period transition metals. The NASA Lewis Cyclic Oxidation Facility contains ten FeCrAl Kanthal furnaces capable of 1200°C operation and four $MoSi_2$ Kanthal Super Heater furnaces capable of 1600°C operation.

One tube contains a thermocouple, which continuously monitors the temperature at the specimen position. This has been found to agree well with a thermocouple spot welded to a sample. Thermal profiles revealed less than a 10°C variation over 10 cm of the hot zone. Heating rates were generally such that the test temperature was achieved in about 2 min. Cooling was accomplished by retraction of the hanger assembly as the specimens are pulled up into a chamber to a final position *ca.*10 cm above the furnace. A baffle plate is then automatically inserted between the samples and the chamber by another pneumatic mechanism. Cooling down to 50°C takes about 5 min for typical samples. All heating and cooling rates are exponential and unassisted by any flows other than natural convection. The heating and cooling cycle duration is electronically controlled by timers, and the number of cycles is controlled by a counter.

The following discussion highlights some of the important lessons learned over the years and incorporates the main findings of two cyclic oxidation spalling and diffusional loss computer models. Many of these points were initially raised in a review paper on cyclic oxidation [3]. This paper updates that review with new supporting data and raises some additional issues.

2.1. Cooling Rate Effects

While it appears intuitive and had been widely assumed that a higher cooling rate should induce greater thermal stress from a thermal shock component, in actuality it was found that a higher cooling rate actually decreased the mass loss rate and degree of spalling [3,4]. Rather than inducing a thermal gradient in thin scales, fast cooling by water quenching was found to induce large thermal gradients in the metal substrate. This resulted in the contraction of the outer layer of substrate, opposed by the still-hot inner core. The surface is thus subjected to a shock-induced *tensile* component, and tensile deformation of the metal skin is actually observed. The scale is deformed along with the metal surface and may actually form cracks normal to the surface. This situation is much more resistant to spalling than intact scales subject to high compressive stresses. When the inner core of the metal is eventually cooled, its compressive effect on the scale is ameliorated by the intervening rigid, elongated skin of cool metal. The net effect is a significant reduction or reversal of the predicted compressive strain on the scale that would normally result from the higher thermal expansion of the substrate.

In the relevant experiments, samples of six superalloys either (i) were manually removed from a vertical tube furnace at 1100°C and immersed in room temperature water, or (ii) remained hanging as the furnace was slowly lowered away from the specimens by a motorised drive mechanism [4]. The respective cooling times were about 5 s or 2400 s (40 min), as compared to the usual 120–180 s (2–3 min) air cooling in the conventional automated cyclic test. Mass change results are shown in Fig. 1. (Compositions for these and other alloys referred to later are listed in Table 1 and refs [28–31].) B-1900 and IN 601 showed considerably more severe degradation in slow cooling than in water quenching. IN 702 was somewhat worse, while Hoskins 875 and TD–Ni were unaffected by cooling rate. TD–NiCrAl was the only alloy tested that showed more mass loss in fast cool tests than in slow cool. But this was because the hard and brittle alloy cracked itself, rather than exhibit plastic surface deformation.

The significance of tensile straining the scale (from metal deformation by thermal shock) as opposed to compressive deformation (from the thermal expansion mismatch with the metal) was further illustrated by bending an oxidised 0.3 mm thick nickel strip to a 2.5 cm radius [4]. The strip had been oxidised at 1200°C for 100 h and slow cooled, but did not spall because of minimal thermal expansion mismatch. After

Table 1. *Alloy compositions (mass%)*

Alloy	Ni	Co	Fe	Cr	Al	Nb	Ta	Mo	W	Ti	Zr	Hf	Other
B-1900 + Hf	Bal.	10.5		8	6		4.2	6		1		1.6	0.1C
IN-601	Bal.		13.4	23	1.4								0.5Si, 0.3Mn, 0.04C
TD-NiCrAl	Bal.			16.2	4.6								$2.0ThO_2$
Hoskins 875			Bal.	22.2	5.4						0.2		
TD-Ni	Bal.												$2.2ThO_2$
IN-702	Bal.			15.6	3.4					0.7			0.2Si, 0.05C
Alloy 718	Bal.		18.5	18.6	0.4	5		3.1		0.9			0.3Si, 0.2Mn, 0.04C
René N5	Bal.	8		7	6.2		7	2	5			0.2	48 ppm Y
NiAl + Zr	Bal.				30						0.2		
Ni–30Cr	Bal.			30									
MA-956			Bal.	20	4.5					0.5			$0.5Y_2O_3$

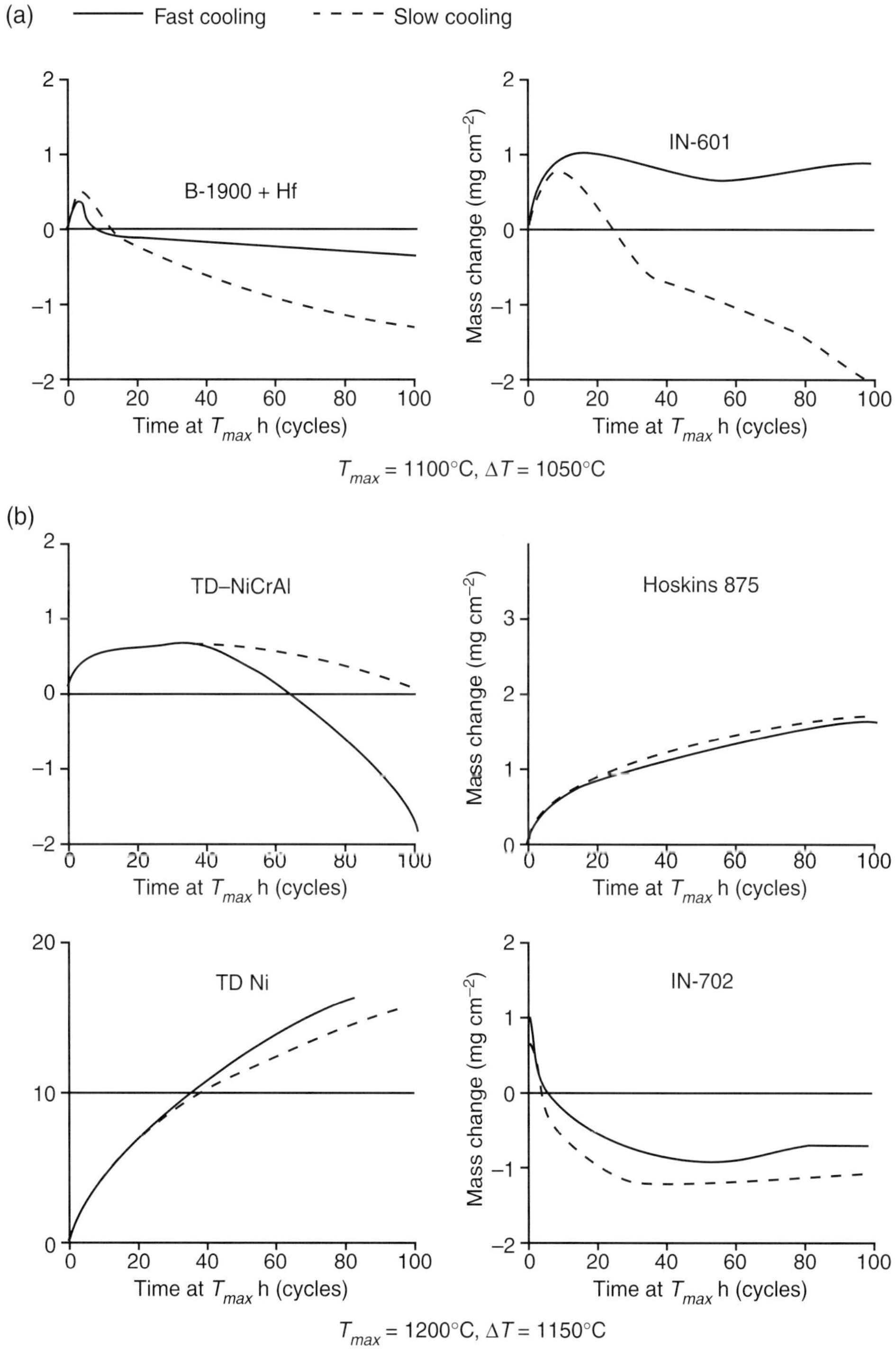

***Fig. 1** The effect of water quenching on (a) the 1100° or (b) 1200°C, 100 h cyclic oxidation behaviour of six high temperature alloys, (Lowell and Deadmore [4]).*

bending, massive spallation was found for the compressive (concave) side, while an array of tensile cracks with no spalling was found on the tensile (convex) side. The cracks were similar to tensile cracks formed on the quenched IN 702 sample with improved behaviour.

2.2. Cycle Frequency Effects

The vast majority of our tests have used a 1 h heating and 20 min cooling cycle. In special cases the cycle frequency may be increased, to provide a greater fatigue element, or decreased, to bias the test toward an isothermal exposure. It should be noted, however, that in some cases a relatively long cycle duration may provide the most severe material degradation. One explanation is that more frequent cycling may disrupt the scale by introducing a fine network of small cracks or spalled segments and so alleviate the massive strain energy built up by a single plate of oxide.

For example, in a study of undoped Ni–40Al, the most severe mass loss rates were experienced for 20-h cycle tests, followed by 50-h, then 1-h cycle tests, for the same amount of total time at temperature [5]. The measured amount of spalled area (to bare metal) was also greater for an isothermal test compared to the cyclic tests. However, in this particular case, the greater instance of interfacial void formation for long term exposures exaggerated the degree of spallation. Cycle frequency issues will be discussed further in the section on modelling.

2.3. Effect of Cooling Temperature

As discussed above, there is much data to indicate the overriding influence of thermal expansion mismatch in the generation of compressive stresses which lead to some degree of spallation. As a rule of thumb, the stress in the oxide will vary according to $E_{ox}\Delta\alpha\Delta T$. While E_{ox} and $\Delta\alpha$ are both material parameters, ΔT is fixed by the heating and cooling temperatures. The importance of the heating temperature is of course very well recognised, since it not only determines both the oxidation rate and thickness, but it also defines the upper limit to the span of the temperature change that induces the thermal stress. However, the cooling temperature is not always intentionally controlled. For example, the cooling chambers above our cyclic furnaces usually receive some degree of heating by convection. This will increase with furnace temperature, will vary with the ambient room temperature, and will also be subject to the seasonal trends in northeast Ohio.

To demonstrate whether the final cooling temperature is indeed important, Deadmore and Lowell varied ΔT by controlling the final temperatures to which the samples were cooled [3,6]. For samples oxidised at 1100° or 1200°C for 195 h, but cooled only part way to room temperature, the amount of spallation was considerably reduced as one might expect. Furthermore, samples that were cycled below room temperature by immersion into liquid nitrogen (–196°C) incurred proportionally more severe spallation. For an 1100°C test of B-1900 and IN-601, an extra 0.75 mg cm^{-2} was lost for every additional 100°C in cooling (i.e. about an extra 20% compared to the loss of about 4 mg cm^{-2} for baseline cooling to room temperature). This effect was even greater for IN-718, even with an aluminide coating. For similar tests of IN 702

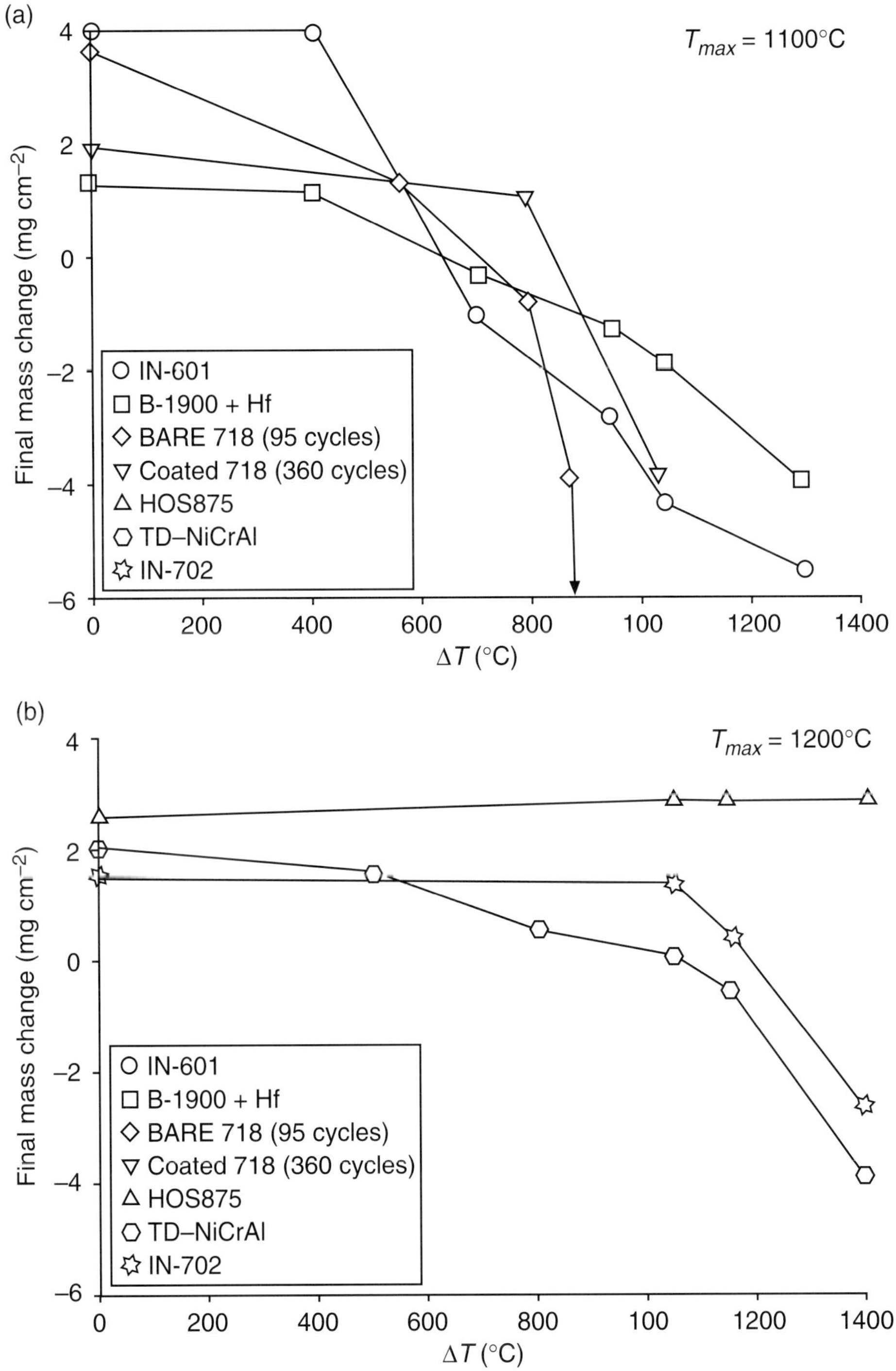

***Fig. 2** The effect of* T_{min} *on the 1100° or 1200°C, 200 h cyclic oxidation behaviour of seven high temperature alloys.* T_{min} *= −196°C achieved by liquid nitrogen (Deadmore and Lowell [6]).*

and TD–NiCrAlY at 1200°C, the increased mass loss was 1.5 mg cm^{-2} for every additional 100°C in cooling (i.e. 50% worse compared to a baseline loss of about 3 mg cm^{-2}). Only FeCrAl + Zr (Hoskins 875) was inert to this ΔT effect, having never spalled because of excellent scale adherence, the extreme ductility of the alloy, and a better match in thermal expansion.

Implicit in these results is the prediction that measurable variations in mass loss will occur with naturally occurring random variations in the cooling temperature. This uncontrolled scatter may complicate very precise scientific investigations or preclude rigorous comparisons of cyclic data obtained from various laboratories. In most cases, however, this level of precision is not required, and overall alloy performance is reasonably duplicated by the simple test where cooling takes place above the furnace. While not the most elegant experimentally, the simplicity of this process has allowed for the comparisons of nearly a thousand alloys tested in the same laboratory under similar conditions for a number of temperatures [7].

2.4. Effect of Humidity

In discussing random variations associated with the cooling temperature, ambient humidity effects must also be raised. A common observation among experimenters is that after initially cooling and weighing the sample, additional spallation may be both visually and gravimetrically observed by simply allowing the sample to sit away from the furnace area for a number of hours or days. This is related to the phenomenon referred to by the thermal barrier coating community as 'cold spalling' 'desktop spallation' or 'the weekend effect.'

In some cases this may be directly related to the decrease in interfacial bonding of alumina scales from exposure to moisture. We have made a number of observations on NiAl (1978) [5], NiCrAl (1986) [8], PWA 1480 (1989) [9], René N5 (1994) [10], and PWA 1484 (1998) [11] stating that exposure to additional moisture after cooling resulted in a measurable increase in the amount of scale spallation, especially at the oxide–metal interface. These exposures may involve water immersion, suspension over a hot beaker of water, or simply breathing on the samples. A dramatic exhibition of this phenomenon occurred when René N5 (+48 ppm mass Y) was cycled for 500 1-h cycles at 1150°C and immersed in water. Oxidation produced a net mass gain (oxygen) of 0.6 mg cm^{-2}, i.e. 1.2 mg cm^{-2} (oxide), or about 3 mm of scale. After immersion, an additional 0.7 mg cm^{-2} of oxide spalled, or about 60%. The amount of spalling to bare metal was visibly increased (Fig. 3). Similar tests have catalogued the amount of extra spallation due to moisture exposure over the duration of the test [10].

While water immersion is clearly an extreme, the fact that additional spallation occurs on extended exposure to humid conditions means that random effects in the test results may arise. The frequency with which specimens are removed from the furnace location for weighing, the length of time that samples are left out, and the humidity in both the cooling chamber or the weighing area can all have an effect on the degree of spallation and performance in a cyclic test. Since these are not normally controlled, they pose an additional degree of uncertainty that will be difficult to eliminate for a large scale test programme. For these, a common schedule for measuring mass should be followed and the ambient humidity level should be noted.

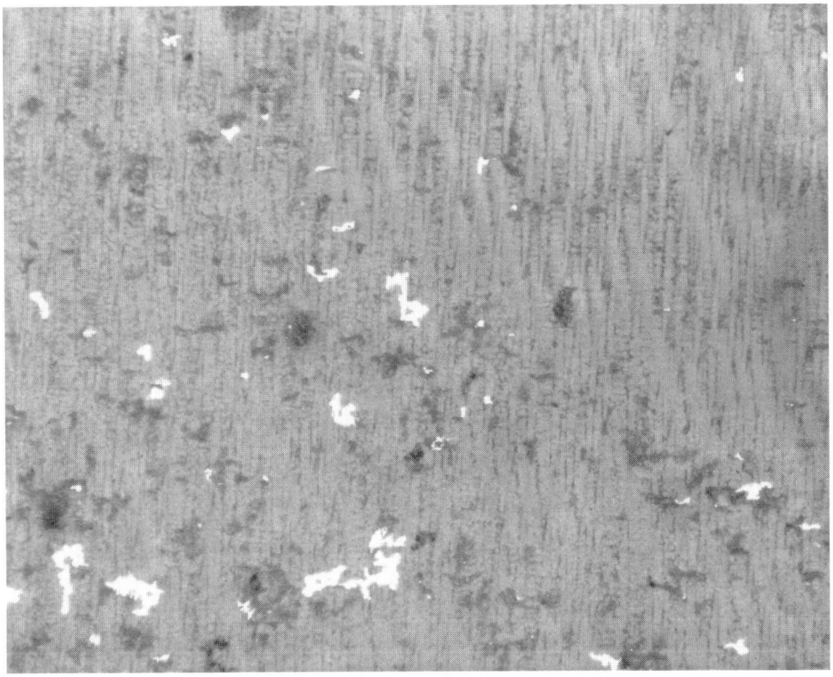

After cooling

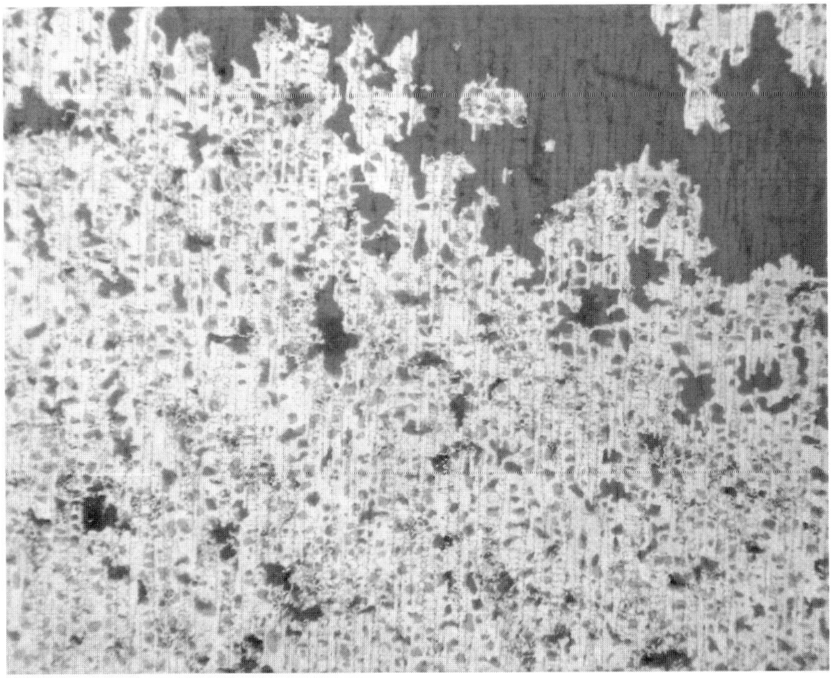

After 1 h water immersion

Fig. 3 *The effect of water immersion on the interfacial spalling for René N5 (48 ppm^{mass} Y) after oxidation for 500 1-h cycles at 1150°C.*

3. Modelling Cyclic Oxidation and Life Prediction

3.1. Spalling Model

Generally cyclic oxidation mass change curves exhibit a basic shape consisting of an initial mass gain to a maximum value, a decrease followed by crossover to negative mass changes, and finally a nearly linear rate of mass loss. To further understand this in detail, the elements of cyclic oxidation have been computer modelled in a number of our studies [1,5,12,13]. The process entails step-wise growing of a scale, spalling off a uniform portion of its thickness (or other specified spall distributions), and re-growth of the scale (at both spalled and intact regions). The intent of these efforts was to construct a reasonably faithful description of the process so that the functional dependencies of mass change curves on growth constant, cycle frequency, and spall constant could be shown. If agreement is obtained, the amount of metal consumed, very long-term behaviour, or behaviour for other cycle frequencies may all be calculated. It is not the purpose here to go into any detail on the development of the model, but rather to highlight the key findings and major implications for cyclic testing.

The model was referred to as COSP (Cyclic Oxidation Spalling Program), developed by Lowell *et al.*, and described in a number of publications [1,3,12–14]. Another version called CYCLOX is available for PC use [13].

Some of the pertinent findings obtained from model studies are now described. If F is the fraction of oxide spalled per cycle, and W'_r is the mass of the retained oxide just before cooling, then the spalling constant Q_o is defined as:

$$F = Q_o \cdot W'^{\,\alpha}_r$$

where α is an experimental constant, usually found to be approximately 1.0. (Alternatively, the mass of oxide spalled, $W_s = Q_o W_r^2$.) Then, for a parabolic growth constant of k_p, a stoichiometric constant of a (ratio of the mass of oxide to the mass of oxygen in the scale) and a cycle duration of Δt:

- the number of cycles to reach the maximum mass varies as: $(k_p \Delta t Q_o^2)^{-1/3} a^{-2}$
- the time to cross zero mass, t_o, is ~ 3.3 × the time to maximum mass, t_{max}.
- the maximum in the mass change curve $(\Delta W/A)_{max}$ varies as: $(k_p \Delta t/Q_o)^{1/3} a^{-3/4}$
- the final mass loss rate per hour, $(dW/dt)_{final}$, varies as: $(k_p\, Q_o /\Delta t)^{1/3}\, a^2$.

By simple inspection it can be seen that for a constant oxide phase (stoichiometric factor), increases in k_p *or* Δt will produce a systematic decrease in the number of cycles to reach maximum and zero mass change, as well as raise the level of the maximum mass change. An increase in the spall constant Q_o will also decrease the number of cycles to reach maximum and cross zero, decrease the level of the maximum mass gained, and increase the final loss rate. The outputs of the model are the net mass change of the sample, the amount of scale present, the cumulative amount of oxide spalled, and the cumulative amount of metal consumed. An alternative procedure is to input actual oxidation data and determine the best fit to the experimental curve. This fit will define both a k_p and Q_0 and allow re-calculation of the mass change curve, the oxide thickness, and the amount of metal consumed for any cycle frequency or total oxidation time. Thus, long-term implications may be obtained from less extensive cyclic tests.

To illustrate the parametric effects graphically, a number of cases were run by CYCLOX to generate families of curves where only one input parameter was varied. The results are shown in Fig. 4. Here some generic results for NiAl (+Zr) alloys have been modelled. In Fig. 4(a), the effect of varying k_p of NiAl from 1000 to 1400°C is shown. Here a Q_0 of about 0.0001 cm^2mg^{-1} was chosen from previous model fits of actual 1200°C, 1-h cycle data (shown in Fig. 5). The compression of the curves to higher $\Delta W/A_{max}$, shorter t_{max} and t_0, and larger dW/dt_{final} mass loss slopes is apparent. In Fig. 4(b), increases in cycle duration from 0.1 to 10 h show the steady increase in $\Delta W/A_{max}$, t_{max} and t_0 and the decrease in dW/dt_{final}. Finally, in Fig. 4(c), for a 1200°C k_p and 1-h cycles, increases in the spall constant Q_0 bring the curves from nearly isothermal behaviour, through decreasing $\Delta W/A_{max}$, t_{max} and t_0, to higher rates of dW/dt_{final}, ending with nearly immediate mass losses at $Q_0 = 0.1$ and 1.0.

Some comparisons with experimental curves are shown in the next few Figures. First, the model fit to an experimental curve is shown for Ni–48Al–0.1Zr (at.%) oxidised at 1200°C for 3000 h [14] (Fig. 5). A k_p was assumed to be 0.02 $mg^2cm^{-4}h^{-1}$ from other work. Thus Q_0 was then determined to be 0.000084 for this model fit. The agreement is very good.

Secondly, a well-behaved family of curves is shown in Fig. 6 for Ni–30Cr oxidised at 1050°C with varying cycle frequencies [1,2]. Note that as the cycle duration is increased, the number of hours to cross zero is also increased. The predicted

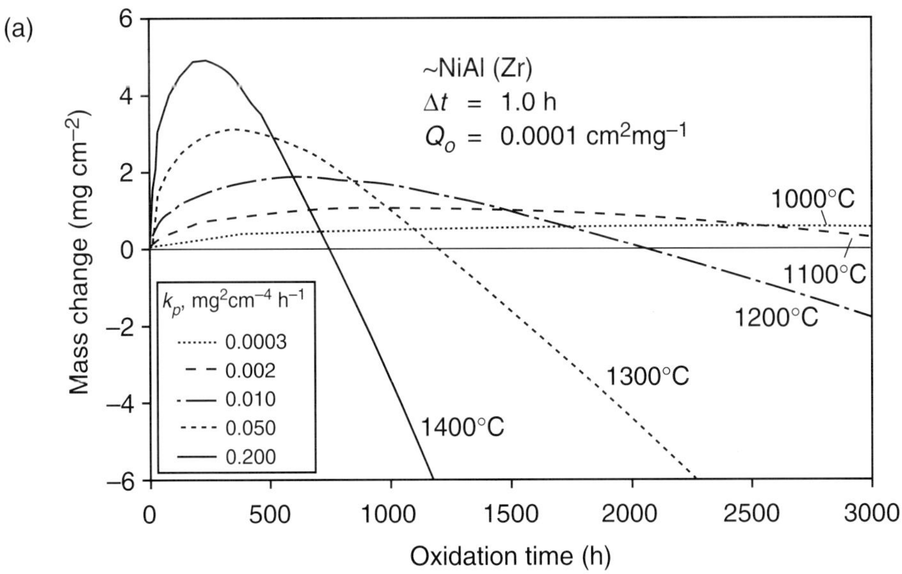

***Fig. 4 (a)** The effect of growth constant* k_p *on families of model mass change curves generated by the CYCLOX cyclic oxidation model computer program.*

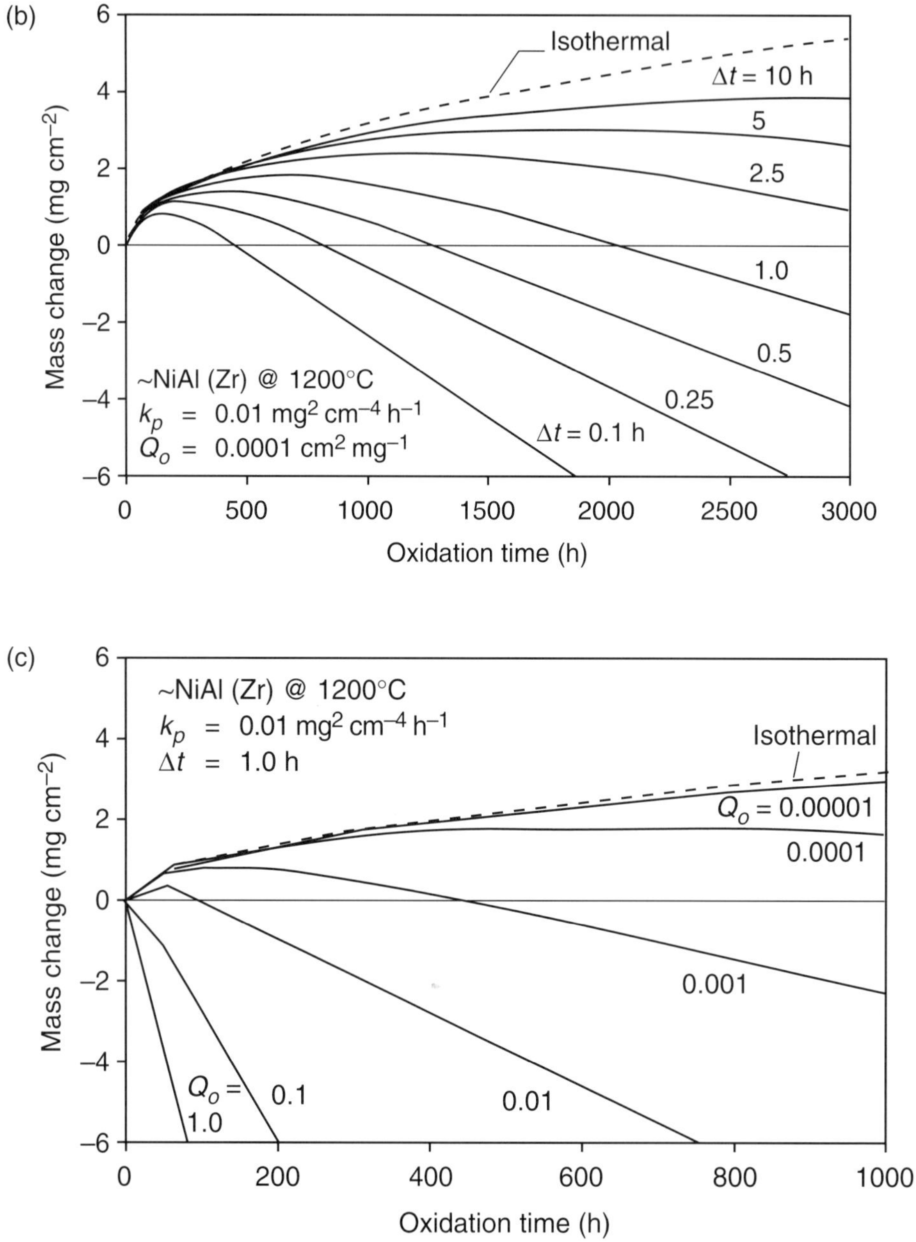

Fig. 4 *(Continued) The effect of (b) cycle duration Δt, and (c) spall constant* Q_o *on families of model mass change curves generated by the CYCLOX cyclic oxidation model computer program.*

dependency should be as $\Delta t^{2/3}$ for the number of *hours* to cross zero (i.e. as $\Delta t^{-1/3}$ for the number of *cycles* to cross zero). A comparison of the predicted and actual dependence is shown in Fig. 7 for the data obtained for cycle durations of 10 min, 30 min, 1 h (A and B) and 5 h, where the predicted slope on a log–log plot should be 2/3.

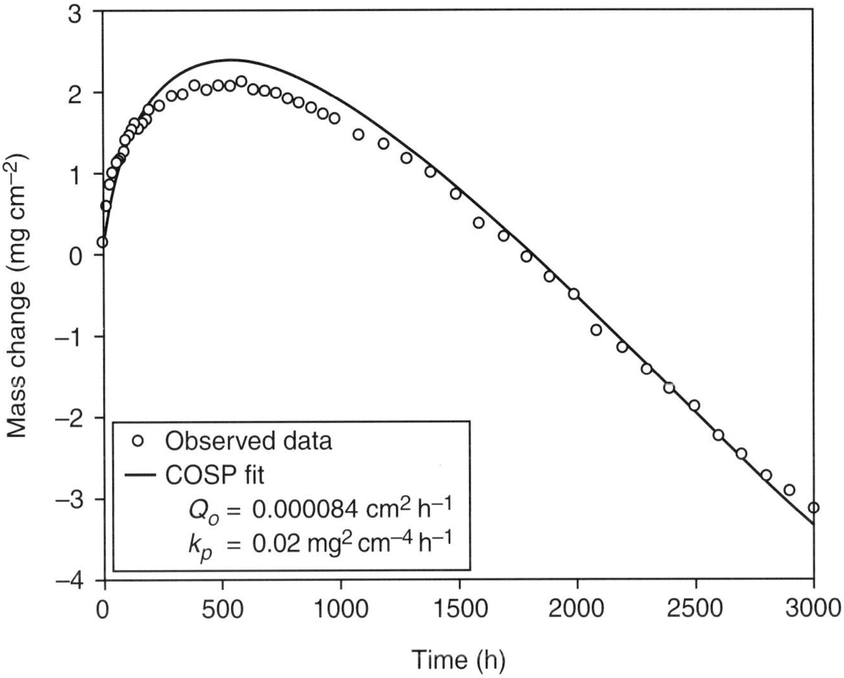

Fig. 5 *Computer fit (COSP) agreement with actual data for Ni48Al–0.1Zr (at.%) oxidised at 1200°C for 3000 1-h cycles;* k_p = *0.02 mg²cm⁻⁴h⁻¹,* Q_o = *0.000084 (Barrett [14]).*

While the basic shape of the cyclic oxidation curve can usually be fitted by this model and long term performance be approximated from a shorter term test, it is not always true that the performance measured with one cycle duration can be predicted from that measured with another. Another family of curves was obtained for MA 956 (Y_2O_3 dispersed FeCrAlY) oxidised at 1200°C using 15 min, 1-, 5- and 50-h cycles [15]. As shown in Fig. 8, the overall trend was to cross zero at longer times with a longer cycle duration, however, the 5 h Δt test actually crossed sooner than the 1-h Δt test.

Finally, some ostensibly peculiar behaviour with cycle duration was observed for pure NiAl oxidised at 1100°C and Zr-doped NiAl oxidised at 1200°, 1250° and 1300°C. Here only slight variations or even decreases in the number of hours to cross zero occurred with substantial increases in cycle duration (from 1 to 100 h), as shown in Fig. 9 [5,14–16]. This phenomenon was discussed in relation to interface void formation and scale adhesion for undoped NiAl [5]. However for the Zr-doped NiAl, spalling occurred by transgranular cleavage within the scale and appeared similar for each cycle duration. Thus there was no apparent explanation of this behaviour based on interface morphology.

In conclusion, it is important to point out that there is not necessarily a monotonic negative effect with increasing cycle frequency. Thus, each industrial need tends to evolve an approximate test standard with regard to cycle duration, such as 1–2 h cycles for aircraft turbines compared to thousands of hours for boiler applications. While some variations in cycle duration may be informative, extensive divergence from the intended use cycle may be misleading.

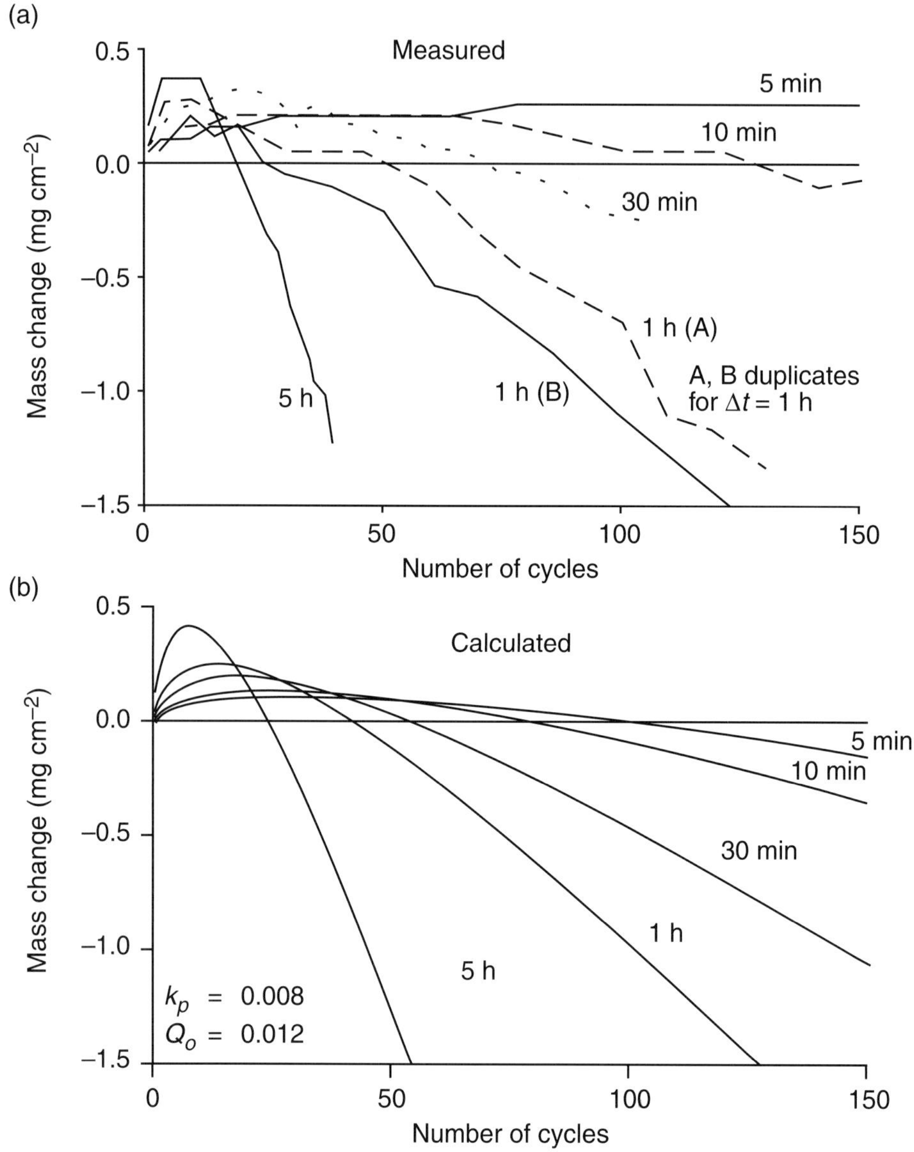

Fig. 6 *The effect of cycle duration, Δt, on the cyclic mass change curves for Ni–30Cr oxidised at 1050°C for 150 h;* $k_p = 0.008\ mg^2cm^{-4}h^{-1}$, $Q_o = 0.012\ cm^2mg^{-1}$ *(Lowell* et al.*[12]).*

3.2. Mechanical Aspects of the Spalling Model

This cyclic oxidation model was broadened to include effects arising from an increase in stress on the scale and normalised values of scale properties and fracture damage by Chan [17]. Trends in spalling behaviour were derived as a function of ΔT, the crack distribution in the scale, and the position of failure. Thus, different mass change behaviour could be accounted for by different test or material parameters. Since

NiAl scales exhibit variances in the normalised scale properties, it is not improbable that this is associated with the deviations from the expected cycle frequency effect.

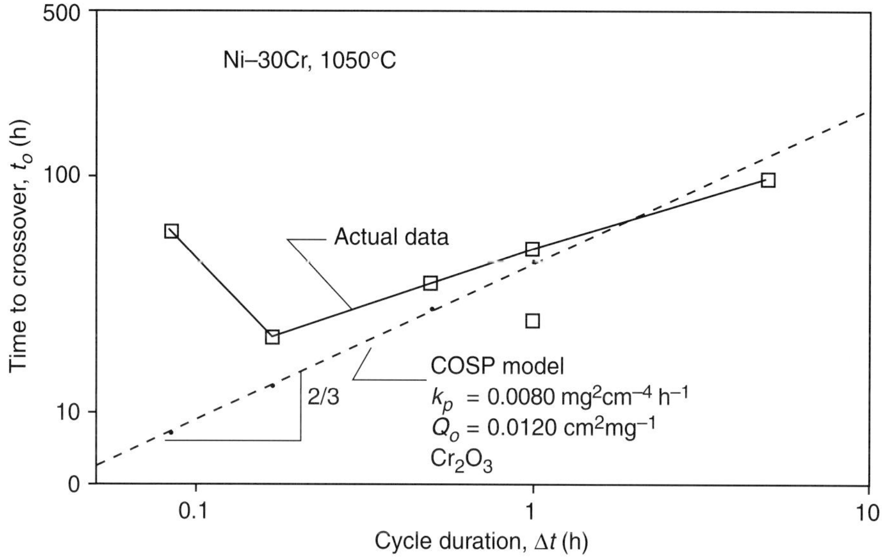

Fig. 7 Agreement of calculated and experimental effects of cycle duration, Δt, on the time to cross zero mass change for Ni–30Cr (from Fig. 5).

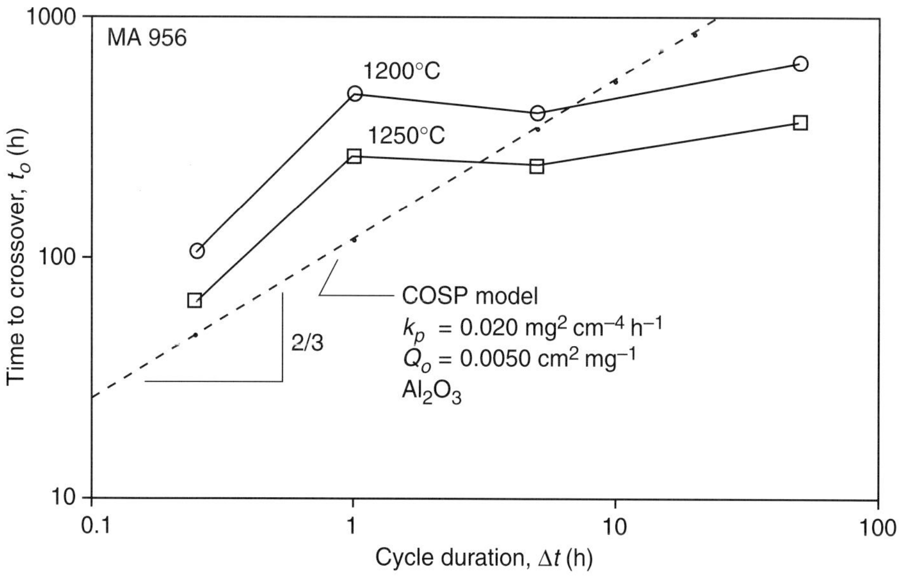

Fig. 8 The non-ideal effects of cycle duration, Δt, on the time to cross zero mass change for MA 956 oxidised at 1200°C and 1250°C (Lowell and Barrett [15]).

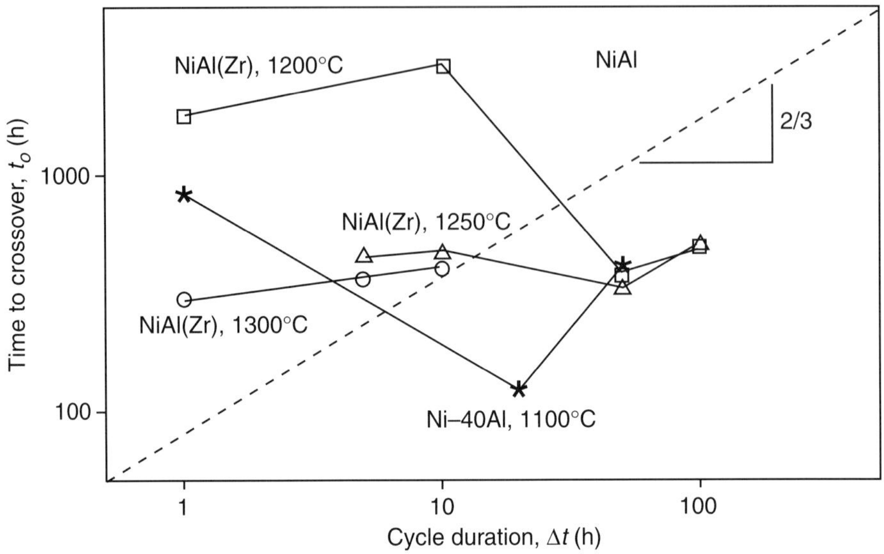

Fig. 9 *The anomalous effects of cycle duration, Δt, on the time to cross zero mass change for NiAl (±Zr) oxidised at 1100–1300°C (Smialek [5]; Nesbitt and Barrett [16]; Barrett and Lowell [15]).*

3.3. Diffusion Modelling and Life Prediction

If there are no mechanism changes, the total amount of oxidised element (i.e. Al or Cr) is eventually consumed, implying oxidation of a new, less protective component. In reality, this transition (or breakaway oxidation) will occur before total consumption. By modelling the diffusional behaviour of the oxidised element in conjunction with its consumption by cyclic oxidation, the time to arrive at a critical Al or Cr content (i.e. failure or oxidative life) can be determined. The key parameters supplied by the cyclic model are the amount of metal consumed, W_m and its rate of comsumption, dW_m/dt. Failure criteria may be based on total surface recession[18], depletion to a critial level of solute [19], or depletion to form a non-protective surface layer phase [20].

For the latter two criteria, the diffusion equation must be solved to predict concentration profiles. The rate of solute consumption dW_m/dt is used as a boundary condition at the oxide/metal interface and is equated to the solute flux in the alloy. During isothermal oxidation, the rate of metal consumption may be easily calculated from the parabolic rate constant. It is also inversely related to the oxide thickness. The time dependence of dW_m/dt can be used to yield a unique solution to the diffusion equation which, for a 'semi-infinite' sample, incorporates a time independent solute concentration at the oxide/metal interface.

During cyclic oxidation, the oxide thickness does not increase in a monotonic fashion but decreases whenever spalling occurs on cooling. Thus, the rate of metal consumption discontinuously increases following each spalling event and is higher than that during isothermal oxidation. This higher rate of metal consumption also causes the solute concentration at the oxide/metal interface to decrease with time, resulting in shorter oxidative lifetimes, especially for 'finite' thickness samples (coatings or thin sections) [21].

Because of the complex time dependence of this boundary condition, finite-difference diffusion models have been developed to predict concentration profiles for cyclic oxidation, coupled with the COSP spalling model. These models were used to predict the protective lifetime of an alumina-forming NiCrAlZr overlay coating on a simple Ni–Cr–Al [18] or Ni-base superalloy [22], as determined by a critical Al interface concentration. Figure 10 shows an example of calculated and observed compositional profiles for a coated alloy after oxidation and diffusion at 1150°C for 100 h. The coating aluminum content has decreased from 25 to 6 at.% and diffused more than 200 μm into the substrate.

Figure 11 illustrates that the spalling model and spalling parameter, Q_o, can be used to predict the cyclic oxidation lifetime effect on coated substrates. Here a NiCrAl coating was modelled for 1200°C cyclic oxidation, assuming $k_p \approx 0.01$ mg^{2}cm^{-4}h. For $Q_o = 0.02$, representing data obtained from a similar composition of bulk NiCrAl(Zr), lifetimes of about 1500 h are projected. For $Q_o = 0.4$, more representative of undoped MAl alloys, the lifetimes were only about 300 h. The model was also used to predict the oxidative lifetime in monolithic β-NiAl samples [19]. In this case, a decrease in Al content to approximately 38 at.%, resulted in the transformation from β-NiAl to less resistant γ′-Ni_3Al and was therefore used as the failure criterion.

4. Comprehensive Data Sets (Compositional Effects and Performance Ranking)

4.1. Statistically Designed Test Programme

A number of extensive cyclic oxidation studies involved statistical design of experiments. For example, a mapping study was used to define performance contours and to define an optimum composition in the Ni–Cr–Al system [23]. Also, partial factorial alloying designs were used to quantify element effects and develop an equation relating performance to composition for complex Ni–Co–Cr–Al–Ti–Ta–Nb–W–Mo superalloys [24,25]. In these and a number of other related studies, an approach was developed to combine the growth and spalling behaviour into one figure of merit, k_a, called the attack parameter, where lower is better. This value was correlated with actual metal recession for a broad selection of alloys and was approximated as $k_1^{1/2} + 10k_2$, (where these parameters correspond to the growth and spalling constants in a 'paralinear' curve fit to the cyclic oxidation mass change curve, i.e. $\Delta W/A = (k_1 t)^{1/2} - k_2 t$ [1,23,26]. Here k_1 and k_2 loosely correspond to k_p and $(k_p Q_o)^{1/3}$, which represent the growth constant and the final slope in the cyclic oxidation computer model.

Some generalisations drawn from these studies were summarised in other reviews of high temperature oxidation [27,28]. In these it was illustrated that the cyclic behaviour of both model and commercial superalloys is most improved by higher levels of aluminium and, to a lesser extent tantalum. Chromium is only somewhat beneficial for the alloy sets studied, in part due to the fact that it is usually reduced when aluminium is increased in commercial alloys. Titanium most notably decreases oxidation resistance in both the model alloy data set and in the characterisation of commercial alloys. The effect of cobalt content was not significant.

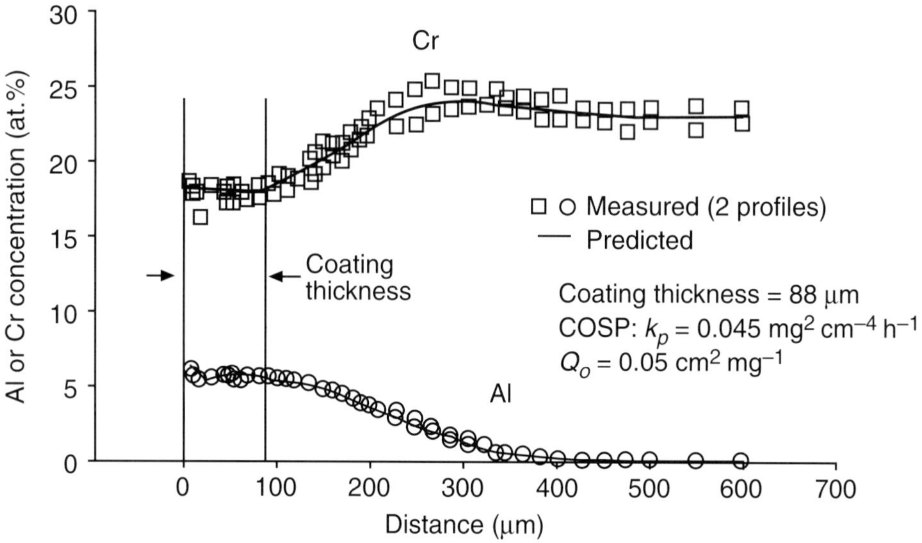

***Fig. 10** A comparison of the actual and modelled (COSIM/COSP) diffusion profiles for a Ni–16Cr–25Al coating on a Ni–22Cr substrate (at.%) after cyclic oxidation at 1150°C for 100 1-h cycles (Nesbitt and Heckel [19]).*

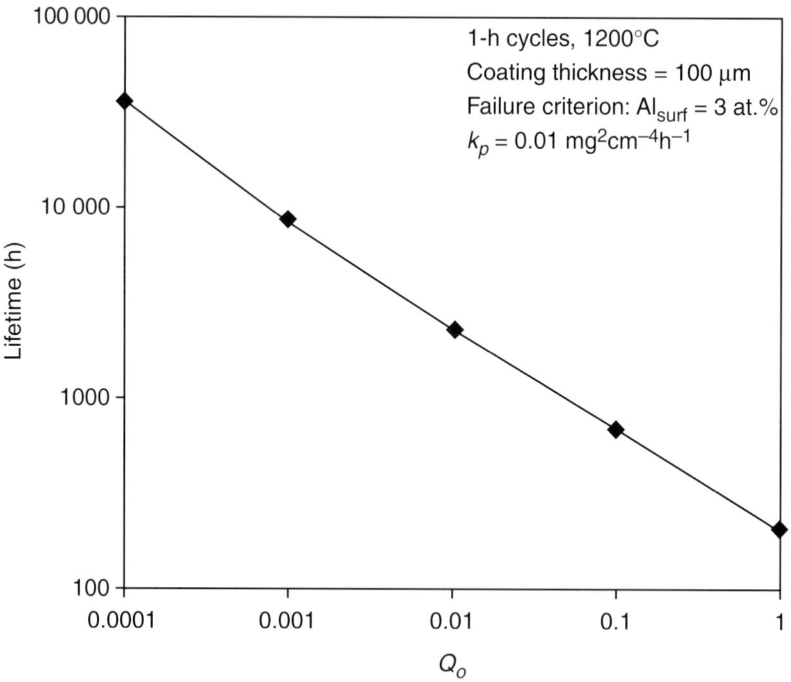

***Fig. 11** Projected lifetimes for various spall constants Q_o (from COSIM/COSP models) for a Ni–20Cr–15Al coating on a Ni–10Cr–5Al substrate (at.%) after cyclic oxidation at 1200°C using 1-h cycles (Nesbitt [22]).*

Idealised trends have been quantified as illustrated in Fig. 12. In this figure, an attack parameter was calculated from a multilinear regression equation developed from the 1100°C cyclic oxidation data of 34 commercial Ni-base superalloys [26]. A base alloy of Ni–8Cr–6Al–6Ta–4Mo–4W–1Nb (mass%) was chosen as a simplified reference composition, having nearly the optimum k_a (≈ 0.1) displayed by real alloys. The relative effects produced by varying one element at a time can be seen by

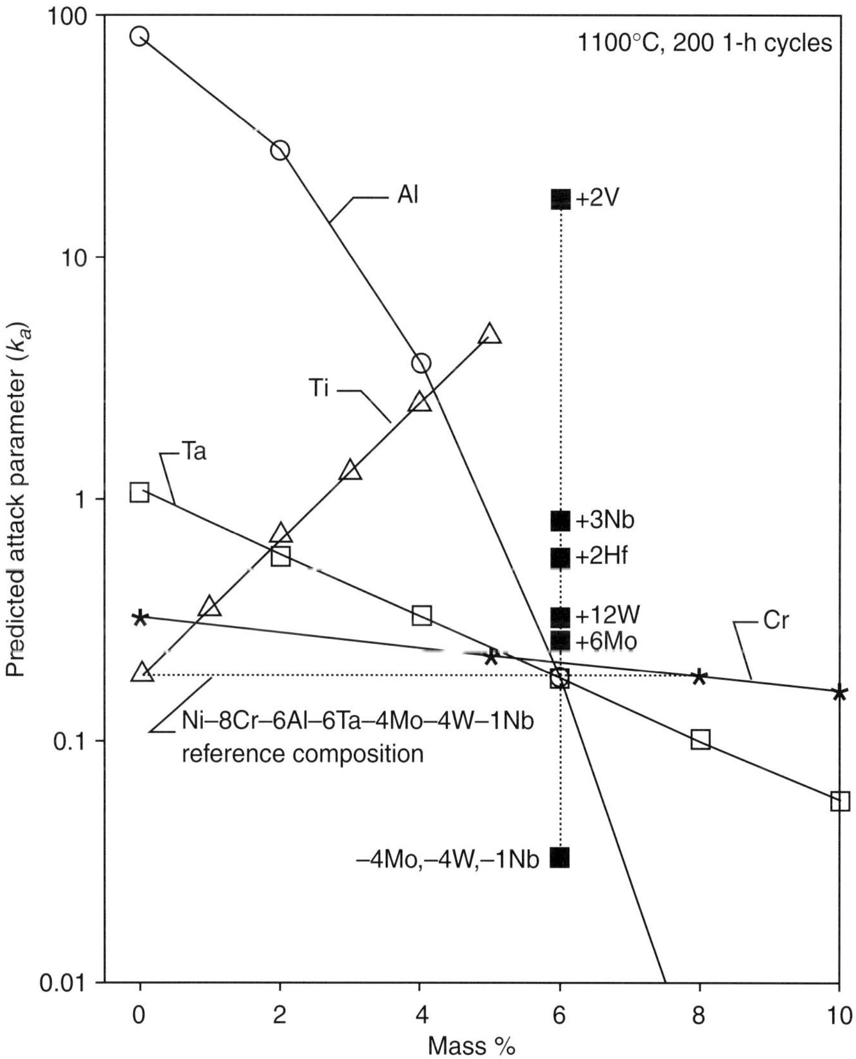

***Fig. 12** The effect of individual element variations on the 1100°C cyclic oxidation resistance attack parameter* k_a *for superalloys near the base composition of Ni–10Co–8Cr–6Al–6Ta–4W–4Mo–1Nb.* k_a *is an empirically obtained figure of merit proportional to material recession as determined from a combination of the growth rate and final mass loss rate. The curves were calculated from a 15 term multilinear regression fit to oxidation data obtained on 34 commercial superalloys (Barrett [26]; Smialek, Barrett and Schaeffer [28]).*

comparing the individual element lines. Data points for some individual alloys (with different molybdenum, tungsten, niobium, hafnium, or vanadium contents) are also shown in comparison to the reference composition (dashed line).

There are a number of other related large-scale cyclic oxidation studies [29–31], some of which had been summarised and discussed in refs [27] and [28]. Much of the mass change data from all of these test programs is presently being consolidated in one single format by means of a CD-ROM [7] to allow for quick, convenient data retrieval and alloy comparisons.

5. Summary

Various issues have been raised regarding the NASA Lewis experience with cyclic oxidation testing. It was shown that cooling rate does not necessarily increase the amount of spallation. In fact, for a number of alloys, water quenching was seen to *reduce* spalling in comparison to slow furnace cooling because of thermal shock and tensile deformation of the underlying metal surface. The *amount* of cooling, however, typically led to higher compressive stresses and spallation, even when achieved by cooling below room temperature.

Increasing the cycle frequency is generally expected to increase the amount of spalling for a fixed total test duration. However, a number of exceptions have been noted. This emphasises the need to simulate the intended application to avoid misleading conclusions. The frequency of removing the sample from the furnace area for weighing and inspection may also give rise to unexpected results. Spalling may be induced here because of further temperature reductions or exposure to room moisture. This moisture effect was most noted for alumina scales that were only partially adherent because of an intermediate sulfur content or relatively low yttrium content.

The generic mass change behavior observed in cyclic oxidation tests was modelled by an iterative, step-wise computer calculation that employed the oxidation rate, the cycle duration, a spall constant, and the scale stoichiometry factor (k_p, Δt, Q_o, and $\boldsymbol{a}$). Systematic trends were observed in the model mass change curves as a function of these input parameters. Descriptive features ($\Delta W/A_{max}$, t_{max}, t_o, *and* dW/dt_{final}), can be defined by mathematical functions of the input parameters. Actual mass change curves may be described nicely by a unique model fit (meaning k_p *and* Q_o) for a given cycle duration and scale phase (Δt *and* $\boldsymbol{a}$). However, it appears that Q_o may sometimes display rather unexpected variations with oxidation temperature or cycle duration.

The cyclic model outputs were also coupled with a finite difference diffusion model to calculate diffusion profiles and solute loss in monolithic or coated articles. This allowed life prediction to be performed, assuming a critical solute content, depletion zone formation, or total material recession as criteria for oxidative failure.

Finally, large assemblies of cyclic data have been summarised through the use of a single attack parameter, k_a, that had been correlated with metal recession. This allowed for first level ranking of alloys by oxidative performance in a simple common test. This parameter was also modelled in terms of alloy composition for both statistically designed and commercially available alloys. Both approaches identified the primary benefits of aluminium content, followed by tantalum and chromium, over the range of compositions studied. The negative effects of titanium and vanadium were clear; molybdenum, tungsten, and niobium additions were less detrimental. No significant effect of cobalt content could be determined.

References

1. C. A. Barrett and C. E. Lowell, High temperature cyclic oxidation furnace testing at NASA Lewis Research Center, *J. Testing and Evaluation*, JTEVA, 1982, **10**, 273–278. (Also NASA TM 81773, 1981.)
2. C. E. Lowell, *Oxid. Met.*, 1973, **7**, (2), 95–116.
3. C. E. Lowell, J. L. Smialek, and C. A. Barrett, Cyclic oxidation of superalloys, in *High Temperature Corrosion*, NACE 6 (1981, San Diego, CA), pp.219–226 (R. A. Rapp, ed.). NACE, Houston, TX, 1983.
4. C. E. Lowell and D. L. Deadmore, *Oxid. Met.*, 1980, **14**, (4), 325–336.
5. J. L. Smialek, *Metall. Trans.*, 1978, **9A**, 309–320.
6. D. L. Deadmore and C. E. Lowell, *Oxid. Met.*, 1977, **11**, (2), 91–106.
7. C. A. Barrett, A CD-ROM Database of High Temperature Cyclic Furnace Oxidation of Alloys (unpublished research, 1999).
8. J. L. Smialek, Adherent Al_2O_3 scales produced on undoped NiCrAl alloys, in *N.L. Peterson Memorial Symposium, Oxidation of Metals and Associated Mass Transport* (1986, Orlando, FL), pp.297–313 (M. A. Dayanada, S. J. Rothman and W. E. King, eds). TMS-AIME, Warrendale, PA, 1987.
9. B. K. Tubbs and J. L. Smialek, Effect of sulfur removal on scale adhesion to PWA 1480, in *Corrosion and Particle Erosion at High Temperatures* (1989, Las Vegas, NV), pp.459–486 (V. Srinivasan and K. Vedula, eds). TMS-AIME, Warrendale, PA, 1989.
10. J. L. Smialek, D.T. Jayne, J. C. Schaeffer and W. H. Murphy, *Thin Solid Films*, 1994, **253**, 285–292.
11. J. L. Smialek, unpublished research, 1998.
12. C. E. Lowell, C. A. Barrett, R. W. Palmer, J. V. Auping and H. B. Probst, *Oxid. Met.*, 1991, **36**, (1/2), 81–112.
13. H. B. Probst and C. E. Lowell, *J. Metals*, 1988, **40**, 18–21.
14. C. A. Barrett, The effect of 0.1 atomic percent zirconium on the cyclic oxidation behavior of β-NiAl for 3000 h at 1200°C, in *Oxidation of High-Temperature Intermetallics* (1988, Cleveland, OH), pp.67–83 (T. Grobstein and J. Doychak, eds). TMS-AIME, Warrendale, PA, USA, 1993.
15. C. A. Barrett and C. E. Lowell, unpublished research, 1998.
16. J. A. Nesbitt and C. A. Barrett, Predicting the oxidation-limited lifetime of β-NiAl, in *Structural Intermetallics* (1993, Seven Springs Resort, Champion, PA), pp.601–609 (R. Darolia *et al.*, eds). TMS-AIME, Warrendale, PA, 1993.
17. K. S. Chan, *Metall. Mat. Trans.*, 1997, **28A**, 411–422.
18. J. A. Nesbitt and C. E. Lowell, *Mat. Res. Soc. Symp. Proc.*, 1993, **288**, 107–118.
19. J. A. Nesbitt and R. W. Heckel, *Thin Solid Films*, 1984, **119**, 281–290.
20. J. A. Nesbitt and E. J. Vinarcik, Predicting the oxidative lifetime of β–NiAl–Zr alloys, in *Damage and Oxidation Protection*, **25-1**, Book No. H0692A, pp. 9–22 (K. Haritos and O. O. Ochos, eds.), ASME, 1991.
21. J. A. Nesbitt, *J. Electrochem. Soc.*, 1989, **136**, 1518–1527.
22. J. A. Nesbitt, Diffusional aspects of the high temperature oxidation of protective coatings, in *Diffusion Analysis and Applications*, pp. 307–324 (A. D. Romig and M. A. Dayanada, eds). TMS-AIME, Warrendale, PA, 1989.
23. C. A. Barrett and C. E. Lowell, *Oxid. Met.*, 1977, **11**, (4), 199–223.

24. C. A. Barrett, R. V. Miner and D. R. Hull, *Oxid. Met.*, 1983, **20**, 255–278.
25. C. A. Barrett, NASA TM 83784, Washington, D.C., 1984.
26. C. A. Barrett, NASA TM 105934, Washington, DC, 1992.
27. J. L. Smialek and G. H. Meier, High-temperature oxidation, in *Superalloys II,* Ch. 11, pp.293–326 (C. T. Sims, N. S. Stoloff and W. C. Hagel, eds). John Wiley, New York, NY, 1987.
28. J. L. Smialek, C. A. Barrett, and J. C. Schaeffer, Design for oxidation, in *ASM Handbook Vol. 20, Materials Selection and Design*, pp. 589–602, ASM International, Materials Park, OH, 1997.
29. C. A. Barrett, 10,000 hour cyclic oxidation behavior at 815°C (1500°F) of 33 high-temperature alloys, in *Environmental Degradation of Materials*, pp.319–327. Virginia Tech., Blacksburg, VA, 1977.
30. C. A. Barrett, 10,000 h cyclic oxidation behavior at 982°C (1800°F) of 68 High-Temperature Co-, Fe-, and Ni-Base Alloys, NASA TM 107394, 1997.
31. C. A. Barrett, R. G. Garlick, and C. E. Lowell, *High-Temperature Cyclic Oxidation Data, Parts I and II*, NASA TM 83665 and TM 101468, Washington, DC, 1989.

10

Significance of Experimental Procedures on the Hot Corrosion Behaviour of Nickel-Base Alloys under Cyclic Conditions

C. LEYENS[*], I. G. WRIGHT[†], B. A. PINT[†] and P. F. TORTORELLI[†]

[†]Oak Ridge National Laboratory, Corrosion Science and Technology Group, Oak Ridge, TN, USA

DLR-German Aerospace Center, Institute of Materials Research, Cologne, Germany

ABSTRACT

A simplified test procedure was established to assess the hot corrosion behaviour of MCrAlY-type nickel-base alloys under the influence of molten sodium sulfate as well as sodium sulfate/potassium sulfate salt mixtures. Salt coated specimens were exposed to 1h thermal cycles at 950°C in flowing oxygen for up to 500 cycles. Mass change data of the specimens revealed a significant dependence of the corrosion not only, as expected, on the average contaminant flux rate, but also on the initial amount of salt deposited during each recoating cycle. Furthermore, deposit removal before salt recoating was found to markedly influence the corrosion of the alloys. This was related to changes in salt chemistry by the dissolution of elements such as Cr from the alloy which shifted the basicity of the salt and thus affected the extent of attack. Substituting Na for K in sodium sulfate/potassium sulfate salt mixtures generally resulted in decreased attack. Although the high K-containing salts still caused significant attack typical of Type-I hot corrosion, the overall degradation was much less than for sodium sulfate alone.

1. Introduction

Although new materials for high-temperature applications are mainly developed to provide high mechanical strength and creep resistance, it is mandatory to consider their hot corrosion behaviour in the anticipated environment from the beginning of their development. Testing of the corrosion resistance of materials is also of great importance when well-established materials (or material systems including coatings) are to be operated in environments where the material behaviour is not well known. As the operating conditions for a specific material or material system can be quite complex, the optimum test would be to run the actual component in an existing facility or engine. However, as such a test is generally extremely costly, and as failure of the component may cause significant damage to the whole operating device, the component test is almost always the last step for a material to demonstrate its suitability for a specific application. Therefore, simplified corrosion tests are extremely desirable as a cost-effective materials evaluation for simulating industrial environments as well as a controlled experimental technique for developing an

---

*At the time this work was performed the author was on sabbatical at Oak Ridge National Laboratory.

improved understanding of corrosion mechanisms. Unfortunately, no standardised test procedures have yet been established, and most research laboratories, materials manufacturers and also their customers have developed their own tests.

Recently, the European Federation of Corrosion has started an initiative to develop guidelines and standards for high-temperature corrosion research and testing [1]. Although simplified tests cannot (and are not intended to) simulate all aspects of actual operating conditions, standardised testing methods would be a great help in generating comparable data for materials evaluation. As for non-standardised tests, the question still to answer is related to how well the results of simplified tests correlate with the more complex service conditions of the 'real world'. The degree of correlation will, in many cases, depend on the sophistication of the test and on the proper control of the relevant test parameters [2]. In some cases reasonable correlation between laboratory tests and service experience has been achieved as reported for example by Saunders [3].

Among laboratory tests, the discontinuous measurement of the high-temperature corrosion of materials is widely used (as reported for example, by Steinmetz [4] and Nicholls [5]). The most common methods are:

(a) crucible tests, where the specimens are totally or partly immersed in a crucible containing a molten salt;

(b) salt coat tests, where the specimens are coated with salt prior to exposure to high temperatures; and

(c) burner rig tests, where the specimens are exposed to a burner into which flame contaminants can be injected.

Due to considerable differences in the test methods it is not surprising that the single results obtained from one of the above test methods may differ significantly, and sometimes even do not reflect the same ranking for a variety of materials. However, even when one particular method is chosen, the results are strongly dependent on the exact testing procedure. The aim of the present paper was to assess the importance of test parameters in the salt test. Effects of salt deposit present on the specimen surface, the recoating frequency, removing the deposits before salt recoating, as well as the effect of the salt composition were investigated for the hot corrosion behaviour of cast, model MCrAlY-type bond coat compositions under cyclic conditions. The specimens were tested in 1h cycles at 950°C up to 500 h and subsequently characterised by optical microscopy and scanning electron microscopy.

2. Experimental

The alloys included in this study included cast versions of NiCoCrAlY (Ni–22Co–18Cr–12.5A1–0.06Y, mass%) and NiCoCrAlYHfSi (Ni–20Co–18Cr–12.5Al–0.55Y–0.26Hf–0.37Si, mass%) coatings. The alloys were vacuum induction melted and cast in a water-chilled copper mould. The castings were then annealed at 1250°C for 4 h in quartz ampoules. Specimens (1.5 cm dia. × 0.1–0.15 cm) were cut from the castings and then polished with SiC paper to 600 grit, avoiding preferred orientation of the grinding marks. The 1-h thermal cycles at 950°C were performed in an automated test rig in which the specimens were held in a vertical tube furnace by Pt–Rh wire hooks for 1 h and cooled to room temperature for 10 min between cycles. Dry oxygen flowed continuously into the bottom of the tube during the exposures. SO_x (of the order of 0.1 vol.%) usually used in such tests was omitted to simplify operation. Since the hot corrosion scenario under consideration was that described by Bornstein [6] in which salt is delivered to the components by periodic shedding of compressor deposits, rather than by condensation of vapour species, this was considered a reasonable approach.

Prior to testing, all specimens were ultrasonically cleaned in acetone and methanol. For hot corrosion tests, the specimens were coated with water solutions of different salt compositions (Table 1). The salt compositions were chosen to represent pure Na_2SO_4 (salt 1) which is widely used for hot corrosion studies, as well as to represent such deposits which are likely to form during combustion of biomass-derived fuels (salts 9 and 10) [7]. For the biomass-fuel related deposits, the higher K content was taken into account by three different K/Na ratios (K/Na = 0, 2, and 15). The salt solutions were deposited onto the specimens using a dropper pipette, and the water was evaporated by heating the specimens on a hot plate up to about 200°C. In order to provide a uniform salt coating even at the beginning of the tests, the specimens were pre-oxidised for 1 h at 950°C prior to coating, rather than by depositing salt on the bare metal surface.

One set of NiCoCrAlY specimens was recoated with pure Na_2SO_4 after 20 and 100, 1-h cycle intervals ('recoating frequency'), respectively, applying different 'contaminant flux rates' (CFRs) (Table 2). In some cases the specimens were washed carefully in distilled water prior to recoating in order to remove the deposits. A second set of NiCoCrAlYHfSi specimens was coated with salts of different K/Na ratios (Table 1) at an average CFR of 1.070mg $cm^{-2}h^{-1}$. These specimens were recoated every 100 cycles without deposit removal before recoating. All specimens were weighed before and after coating with salt and after washing, respectively, as well as before and after exposure, using a Mettler model AG245 analytical balance. Mass change of the specimens was calculated taking into account the continuous mass loss of the Pt–Rh hooks. Corrosion products were characterised using light microscopy (LM) and a field emission gun scanning electron microscope (FEG-SEM) equipped with energy dispersive spectroscopy (EDS).

Table 1. *K/Na and 2S/(Na + K) ratios for different salt compositions*

Designation	K/Na	2S/(Na + K)
Salt 1	0	1
Salt 9	2	1
Salt 10	15	1

Table 2. *Amount of salt, recoating frequency, contaminant flux rate (CFR) and washing procedure of* Na_2SO_4*-coated cast NiCoCrAlY*

Designation	Amount of salt (mg cm^{-2})	Recoating every	CFR (mg cm^{-2} h^{-1})	Washing before recoating
Sample A	1.016 ± 0.062	100 h	(1.016 ± 0.062) 10^{-2}	Yes
Sample B	1.061 ± 0.072	100 h	(1.061 ± 0.008) 10^{-2}	No
Sample C	0.262 ± 0.050	20 h	(1.310 ± 0.001) 10^{-2}	Yes
Sample D	0.2614 ± 0.043	20 h	(1.320 ± 0.002) 10^{-2}	No
Sample E	1.059 ± 0.101	20 h	(5.295 ± 0.005) 10^{-2}	Yes
Sample F	1.051 ± 0.156	20 h	(5.255 ± 0.008) 10^{-2}	No
Sample G	4.746 ± 0.383	20 h	(23.730 ± 1.915) 10^{-2}	Yes
Sample H	0.283 ± 0.085	100 h	(0.283 ± 0.001) 10^{-2}	No

3. Results and Discussion

3.1. Effect of Salt Deposition Procedure

An overview of the mass gain vs number of 1h cycles data for Na_2SO_4-coated NiCoCrAlY specimens tested at 950°C is given in Fig. 1. The CFR was varied by almost two orders of magnitude between 0.283×10^{-2} and 23.730×10^{-2} mgcm^{-2}h^{-1} and accordingly the extent of attack also differed widely. As expected, the highest mass loss was found for the highest CFR. Whereas at a first glance mass change data were in a scatter band for most specimens up to 350, 1-h cycles, one specimen exposed to the highest CFR (sample G) and one specimen exposed to a fresh salt coating every 20 h at an intermediate CFR (sample D) started to lose mass rapidly after 200–250, 1-h cycles indicating an accelerated attack.

In Figs 2–5 the mass change data for each specimen are given in more detail. The open circles represent the overall specimen mass change after exposure including the deposits, whereas the filled circles represent mass change data obtained from the specimens after salt removal and drying, thus excluding the mass of deposits. The mass change vs time curves for a 100 h recoating sequence revealed a strong dependency of single data points on the washing procedure (Fig. 2). The data for sample A (washed, recoated every 100 h, Fig. 2a) followed a zigzag course for the overall specimen mass change, which was characterised by mass gain after salt deposition and exposure, subsequent mass loss, and again mass gain. After 100 cycles no measurable amount of deposit was present on the specimen surface as indicated by specimen mass after washing; however, after 200, 300, and 350 cycles, the specimen had lost mass after removing the deposits. As will be discussed in more detail later, the increasing mass loss with increasing numbers of cycles after washing was attributed to partial loss of the oxide scale during washing. Although the same CFR was applied to sample B (no washing, recoating every 100 h) the mass change data were significantly different from that for sample A (Fig. 2b). Identical mass gain and

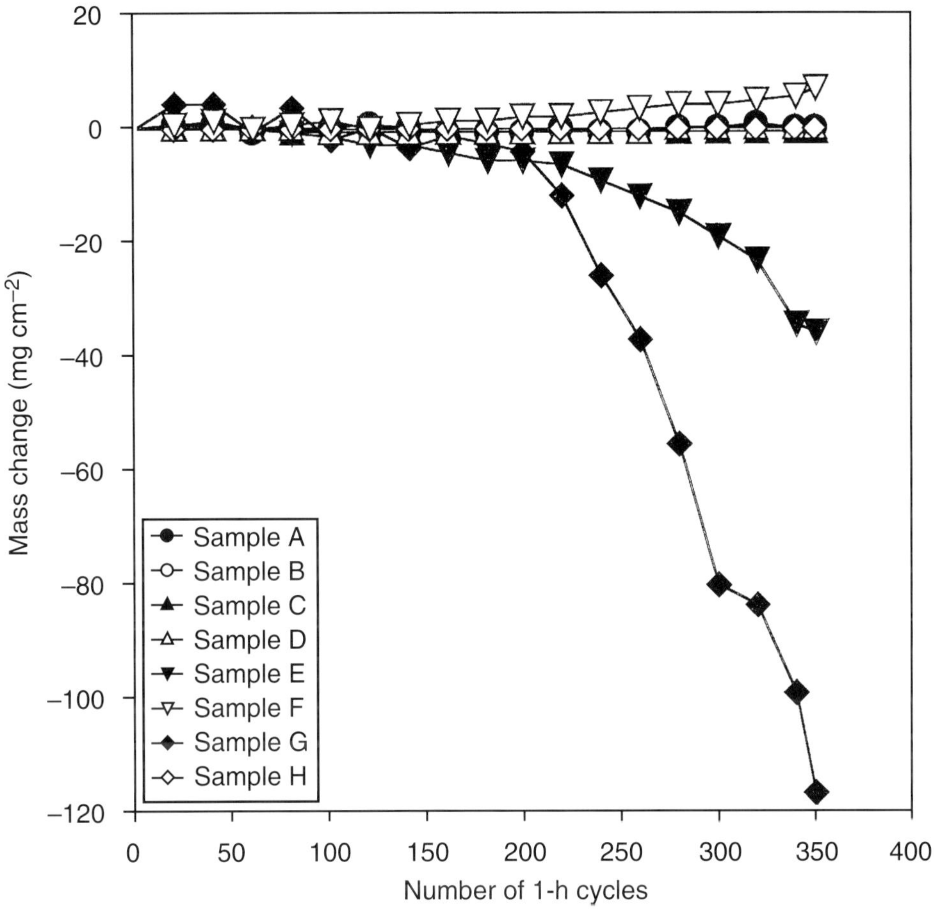

Fig. 1 *Mass change vs number of 1-h cycles curves for Na_2SO_4-coated NiCoCrAlY (overview, for details refer to Figs 2–5). Specimens were coated with different CFRs at different time intervals. Some specimens were washed to remove salt deposits before recoating with salt. Specimen designation is according to Table 2.*

mass loss behaviour were found for samples A and B after 100 cycles, indicating good reproducibility of the testing procedures in general. However, after redepositing salt onto the already exposed surface of sample B the specimen gained mass continuously rather than discontinuously, as found for sample A. Furthermore, mass loss was evident from sample B after washing at the end of the test after 350 cycles, whereas sample A exhibited a slight overall mass gain at the conclusion of the test. These results clearly indicate that, for the specific test parameters chosen in this study, deposit removal before salt recoating not only affected the mass gain data but also altered the extent of corrosion of the specimens.

Macrographs of both specimens revealed considerable attack of the initially formed alumina scale (light gray phase) and the formation of spinel phases (darker phases) on each of the specimen surfaces (Figs 6a and b). The non-uniform attack of the alumina scale was attributed to local differences of the cast alloy chemistries which, in turn, affected the local scaling behaviour. Furthermore, accelerated corrosion at

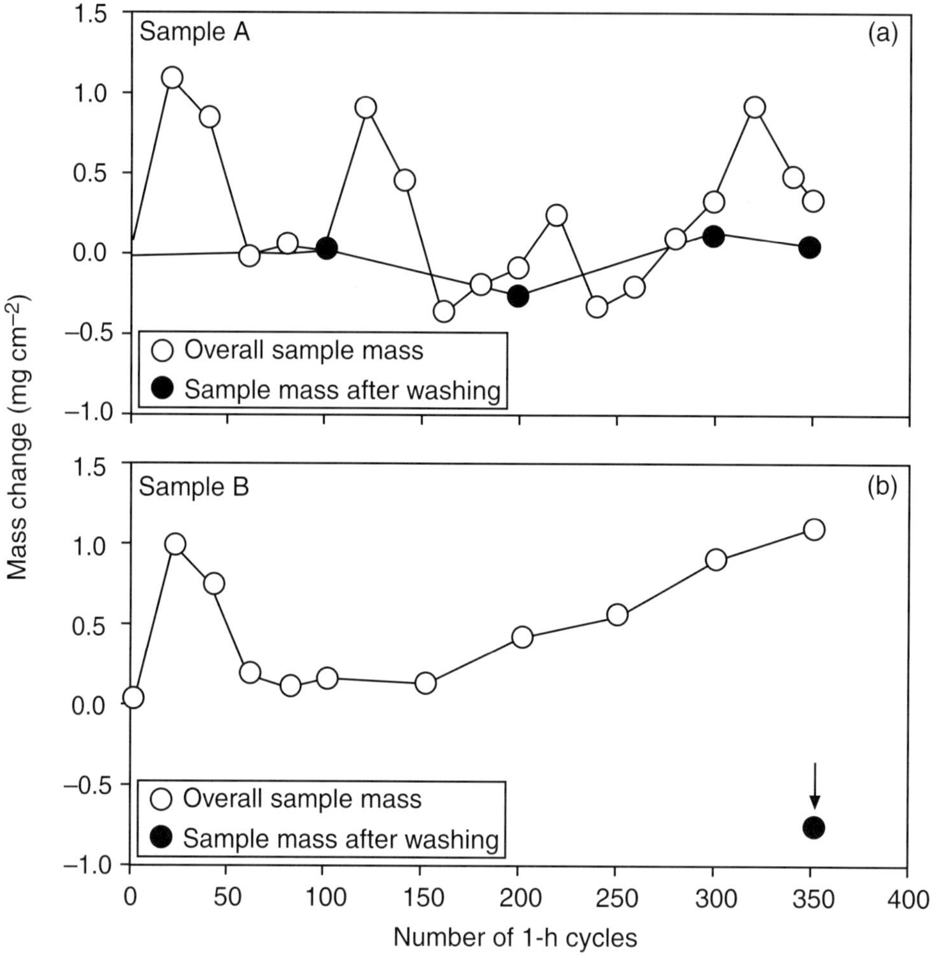

Fig. 2 *Mass gain vs number of 1-h cycles curves for Na_2SO_4-coated NiCoCrAlY. The CFR (a) was 1.016×10^{-2} mg cm^{-2}h^{-1} and (b) 1.061×10^{-2} mg cm^{-2}h^{-1}, respectively. Both specimens were recoated every 100 h and deposits were removed before recoating on* **Sample A. Sample B** *was washed only at the end of the test.*

the specimen edges was found, which was caused by salt accumulation at the edges relative to the flat specimen surfaces. Specimen edges also are susceptible to corrosion since the increased surface area leads to more rapid depletion of protective scale-forming elements, compared to the flat sides. It should be noted that salt accumulation at the edges (for the high CFRs) was not only a result of droplet formation during testing caused by gravitation, but was also caused by non-uniform wetting of the molten salt. Surface tension caused salt attraction to the edges over the entire circumference of the specimens, rather than only at the lowest point of the specimen suspended vertically in the tube furnace (see Fig. 6).

Mass change data were significantly affected by the initial amount of salt present on the surface, which depended for a given CFR on the recoating frequency, i.e. the number of cycles between salt redeposition cycles. Although the average CFRs were

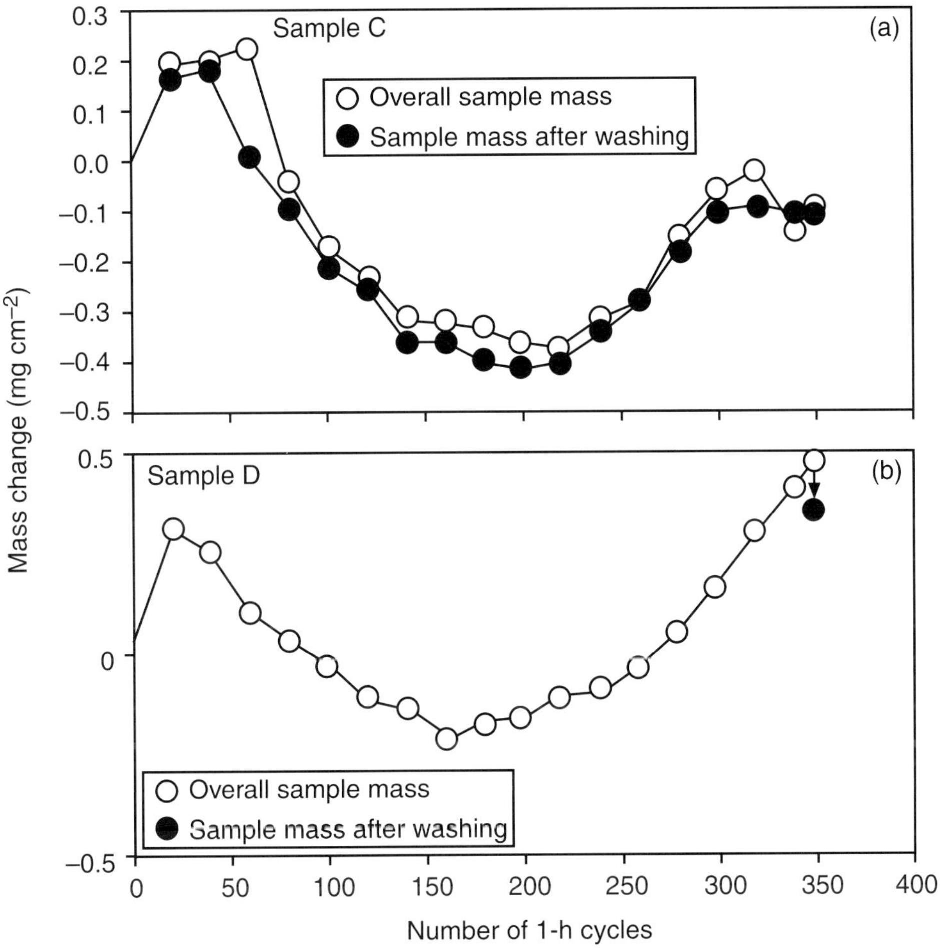

***Fig.* 3** *Mass gain vs number of 1-h cycles curves for Na_2SO_4-coated NiCoCrAlY. The CFR was (a) 1.310×10^{-2} mg $cm^{-2}h^{-1}$ and (b) 1.320×10^{-2} mg $cm^{-2}h^{-1}$, respectively. Both specimens were recoated every 20 h and deposits were removed before recoating on* **Sample C. Sample D** *was washed only at the end of the test. Note that the ordinates have different scales.*

almost identical, mass change for sample C (washed, recoated every 20 h) was apparently different from that of sample A (washed, recoated every 100 h) (Figs 2a and 3a). It should be noted that, due to the 5 times higher recoating frequency, sample C was coated with about 1/5 of the amount of salt compared to sample A on each recoating cycle. For samples A and C a similar mass gain peak at the beginning of the test was found, however, at longer times of exposure specimen C lost mass, and then started to gain mass again after about 220 cycles. Unlike the specimens recoated every 100 cycles, the overall specimen mass essentially mirrored the 'real' specimen mass curve of sample C (including the mass of deposits). Even when the deposits were not removed from the surface before recoating (sample D, Fig. 3b) the shape of the mass change curve was almost identical to that obtained for the washed specimens. However, sample D gained mass after testing, whereas sample C lost

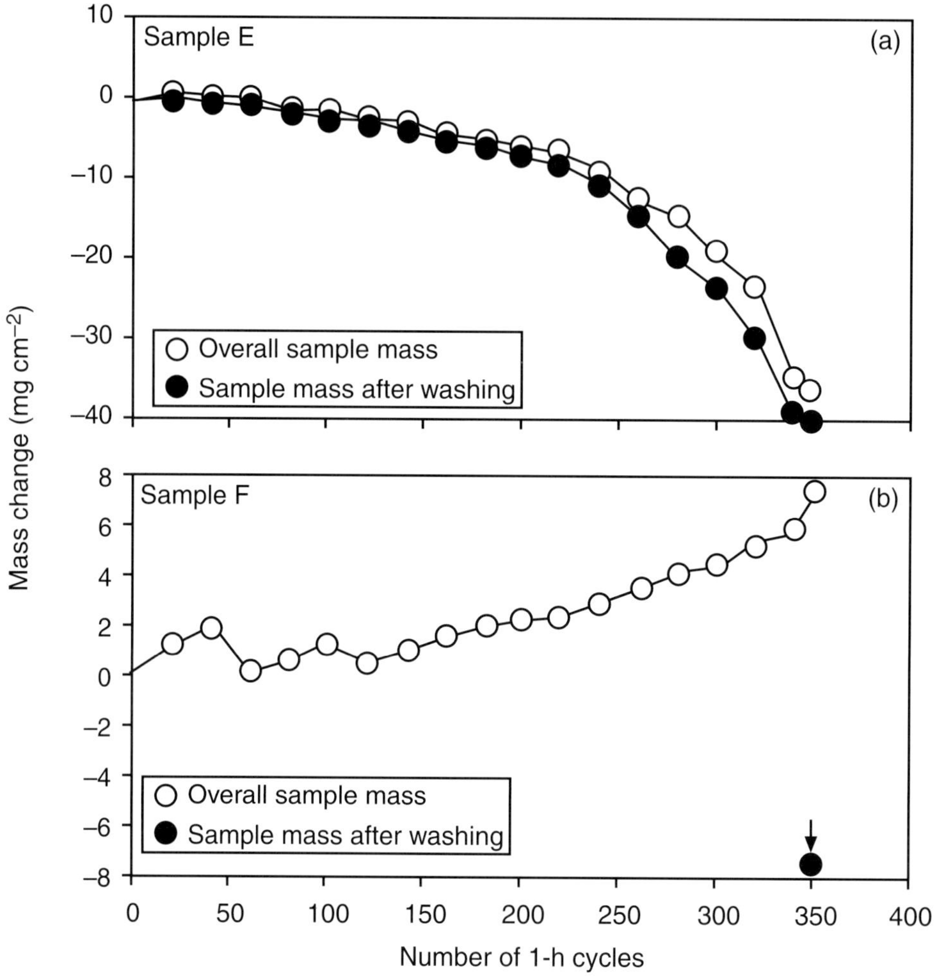

***Fig. 4** Mass gain vs number of 1-h cycles curves for Na_2SO_4-coated NiCoCrAlY. The CFR was (a) 5.295×10^{-2} mg $cm^{-2}h^{-1}$ and (b) 5.255×10^{-2} mg $cm^{-2}h^{-1}$, respectively. Both specimens were recoated every 20 h and deposits were removed before recoating on* **Sample E. Sample F** *was washed only at the end of the test. Note that the ordinates have different scales.*

mass. Comparison of Figs 2 and 3 led to the conclusions that (i) there was a significant effect on mass change not only of the average CFR but also of the actual amount of 'fresh' salt present on the surface during exposure and that therefore (ii) the recoating frequency considerably influenced the attack. The macrographs of the specimens after 350, 1-h cycles also revealed that the extent of attack was similar and independent of deposit removal before recoating when less salt was coated more frequently (Fig. 6, C and D). The amount of alumina scale present on the surface of these specimens was much higher than for those specimens coated with a higher amount of salt at a lower recoating frequency (Fig. 6, A and B). Metallographic cross-sections are in preparation to verify the different extents of corrosion.

As expected, increasing the overall amount of salt on the specimen surfaces resulted

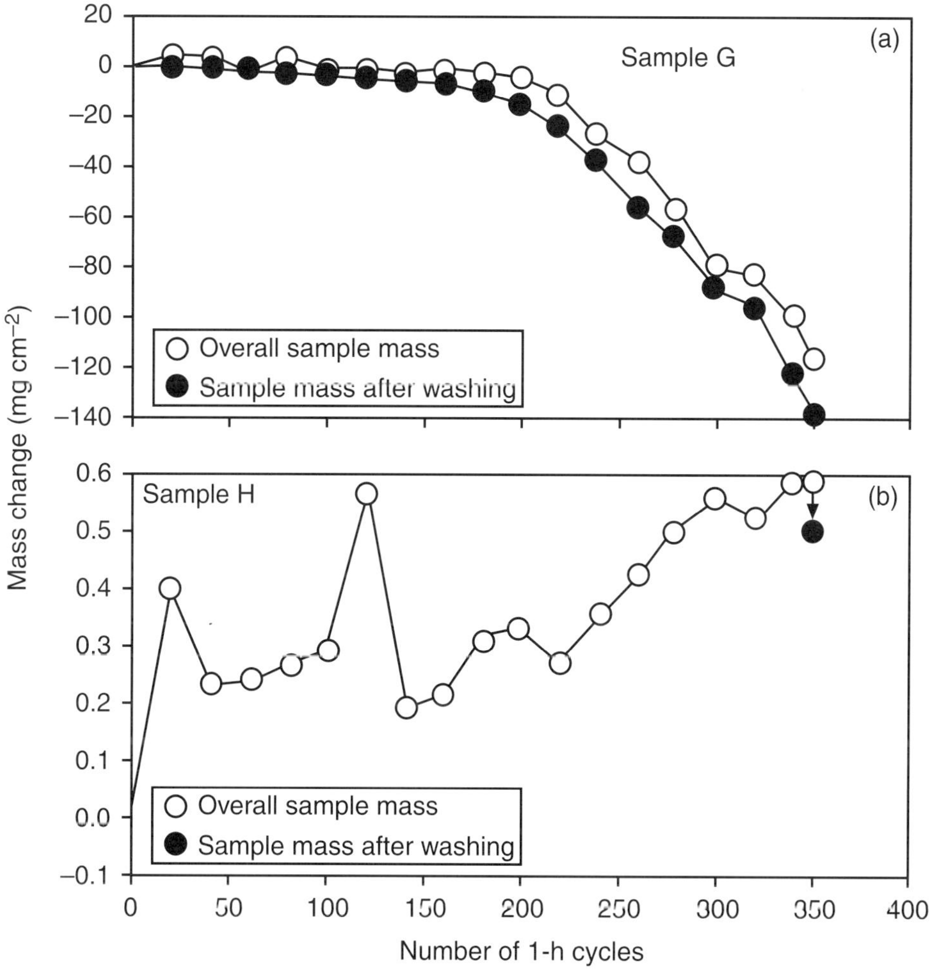

***Fig. 5** Mass gain vs number of 1 h cycles curves for Na_2SO_4-coated NiCoCrAlY. The CFR was (a) 23.730 × 10^{-2} $mgcm^{-2}h^{-1}$ and (b) 0.283 × 10^{-2} $mgcm^{-2}h^{-1}$, respectively.* ***Sample G** was recoated every 20 h and deposits were removed before recoating.* ***Sample H** was recoated every 100 h and only washed at the end of the test. Note that the ordinates have different scales.*

generally in increased corrosion (Fig. 4). Coated with about 5 times the amount of salt of sample C, sample E lost mass continuously (Fig. 4a), again except for the first few cycles where slight mass gain was observed. The overall specimen mass mirrored the 'real' specimen behaviour but, of course, the specimen mass was always lower after washing as a result of loss of deposits and, especially after longer times of exposure, of loss of parts of the oxide scale, as will be discussed later. In contrast to sample E, where the deposits were removed before recoating every 20 h, sample F (no washing before recoating every 20 h) essentially gained mass with an increasing number of cycles (Fig. 4b). Even when the deposits were removed from sample F after the test, the 'real' mass loss was not as great (about 7.5 mg cm^{-2}) as that for sample E (about 40 mg cm^{-2}). Therefore, Fig. 4 provides further evidence of the

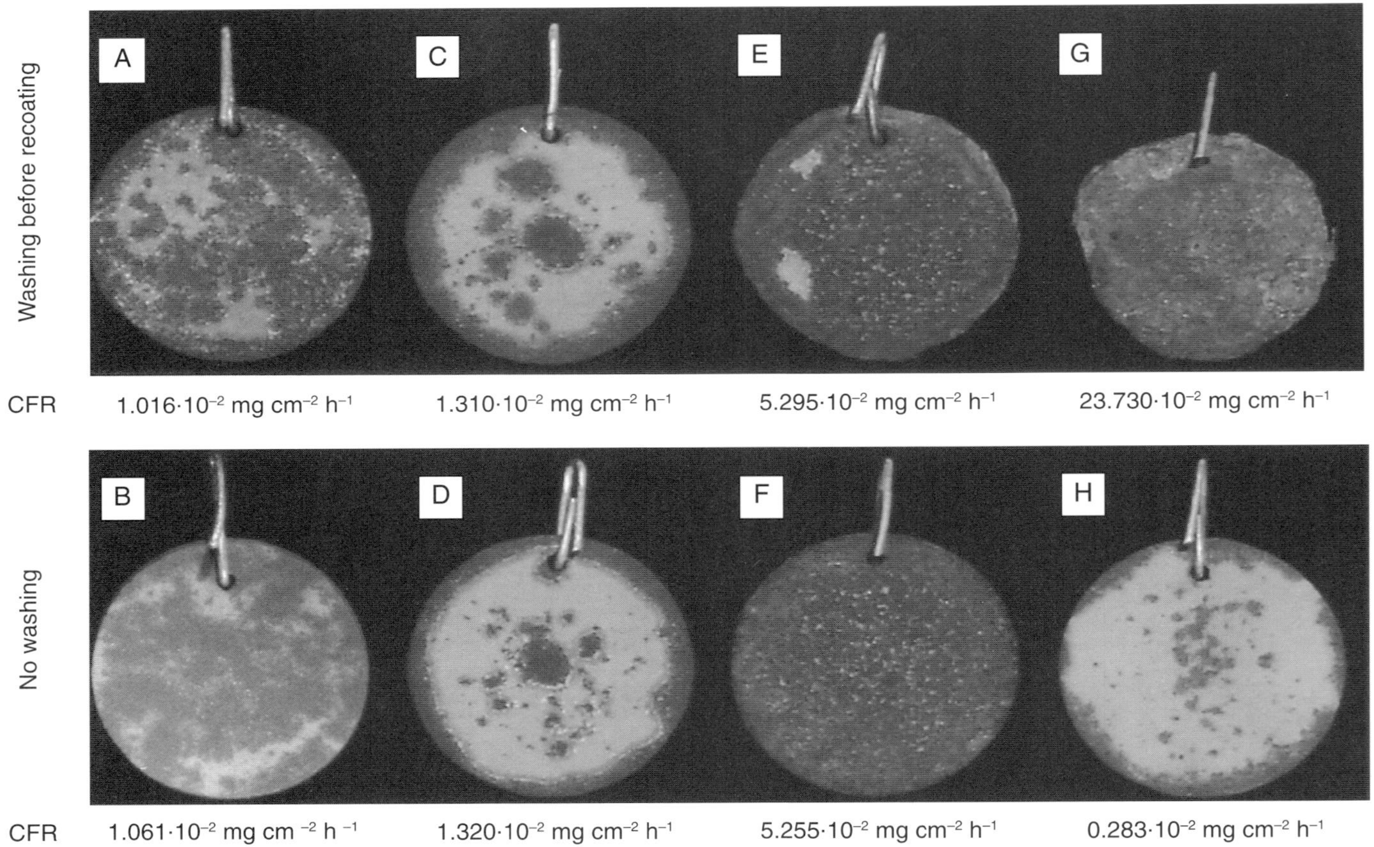

Fig. 6 *Macroscopic images of Na_2SO_4-coated NiCoCrAlY specimens after 350, 1-h cycles. Light grey oxide scales represent alumina whereas darker parts of the specimens surfaces mainly consist of spinel phases. Letters designating the specimens correspond to Table 2. Note that the inital diameter of the specimens was 15 mm.*

significance of deposit removal before recoating. Even though the amount of deposits present on the surface of sample F was steadily increased, the severity of the corrosion was less than for sample E which was confirmed by the macroscopic appearance of the specimens (Fig. 6, E and F), showing that sample E was much more degraded than sample F.

Comparison of mass data in Figs 3 and 4 suggested a general trend of corrosion associated with deposit removal. For both the lower CFR applied to samples C and D (Fig. 3) and the higher CFR applied to samples E and F (Fig. 4) deposit accumulation apparently decreased the attack. Although the amount of salt was steadily increased on samples C and E the presence of the earlier deposit seemed to decrease the intensity of the attack. Based on the current understanding of the mechanisms of hot corrosion under salt deposits, fluxing of the alumina scale in the presence of a molten salt is strongly dependent on the basicity of the salt [8]. Both basic and acidic fluxing are associated with a high and low oxide ion activity, respectively, and thus cause degradation of the alumina scale [9]. Although Cr_2O_3 is also soluble in fused Na_2SO_4 [10] Cr was reported to be effective in retarding accelerated attack of alumina scales. Goebel *et al.* [9] showed that for various Ni-base alloys Cr_2O_3 established an oxide ion activity in Na_2SO_4 that was not sufficient to maintain basic fluxing and is not low enough to cause acidic fluxing. Rapp *et al.* [11] subsequently showed that Cr_2O_3 dissolution buffers the oxide ion activity at a value near that where Al_2O_3 has minimum solubility. Chromate fuel additives are known to effectively combat hot corrosion.

In the present study the yellow colour of the deposits removed from the NiCoCrAlY specimens indicated the presence of sodium chromate (Na_2CrO_4), i.e. Cr was leached out of the substrates by the molten salt. In the case of salt deposition without deposit removal before recoating, the fresh salt was mixed with the salt that was already present on the surface for several cycles which already contained a certain amount of Cr. With the addition of fresh salt the basicity of this salt was presumably less changed due to the presence of the 'Cr buffer' compared to samples exposed to a fresh salt without any Cr-buffered deposit on the surface. In the latter case, the alumina scale was attacked more significantly.

The strongest attack of cast NiCoCrAlY was experienced with a CFR of 23.73×10^{-2} mg cm^{-2} h^{-1} (sample G, recoated every 20 h, Fig. 5a). The specimen started losing considerable mass after about 200, 1-h cycles, and was heavily corroded after the test (Fig. 6, G). It should be noted that, for the chosen geometry of the specimens and their vertical suspension in the furnace, approximately 5 mg cm^{-2} was the maximum amount of salt that could be coated without losing salt due to droplet formation and subsequent dripping during heating. In contrast to sample F, where a similar amount of salt deposit was accumulated after 5 recoating cycles, deposition of 'fresh' salt every 20 h on sample G caused a very aggressive form of attack. Again, the general shape of the overall mass change curve for sample G corresponded well with the 'real' specimen mass data after deposit removal.

Surprisingly, the lowest CFR (0.283×10^{-2} mg cm^{-2}h^{-1}) applied in the tests to sample H did not result in a significantly lower mass change (Fig. 5b), although the specimen contained the highest amount of alumina on the surface after testing (Fig. 6, H). As for sample A, a zigzag mass change curve was obtained for sample H due to salt recoating after 100 cycles, which was somewhat in contrast to sample B (Fig. 2b) which continuously gained mass after the initial mass gain and mass loss sequence. Mass loss after washing of sample H at the end of the test revealed a slight salt accumulation during the test; however, it should be noted that some debris associated with some loss of oxide scale was found in the container after deposit removal.

3.2. Effect of Salt Evaporation

Comparison of the mass change data associated with recoating and deposit removal revealed that, for low amounts of salt coated onto the specimen, the mass loss obtained with washing before recoating was fairly low (Fig. 7a). It should be noted that mass loss in general included not only salt deposits but also loss of oxide during washing which, however, was found to be marginal in the case of CFR = 1.310×10^{-2}mg $cm^{-2}h^{-1}$. For higher amounts of salt deposited on the surface (Fig. 7b) the absolute amount of salt present on the surface before washing was much higher indicating that the salt had not completely evaporated within a period of 20, 1-h cycles. However, after about 200 overall cycles the mass loss after deposit removal started to increase. This disproportionally high mass loss of the specimen was associated with partial loss of the oxide scale in addition to the mass loss resulting from deposit removal during washing. The onset of an increasing amount of oxide loss during washing was correlated with the onset of accelerated specimen mass loss (compare Fig. 7b with Fig. 4a). As a result of rapid scale growth and continuing attack by fresh salt the mechanical integrity of the oxide scale was apparently significantly affected after longer times of exposure. Once part of the oxide scale spalled, the next salt coating encountered either a thinner oxide scale or bare metal (depending on the type of spallation), thus further accelerating the attack.

A more detailed analysis of the evaporation behaviour of the salt deposits is given in Fig. 8. The relative mass change of the specimens denotes the mass of the specimen during exposure in relation to its starting mass, including the fresh salt coating. It should be noted, that the specimen mass was affected by superposition of mass gain as a result of corrosion and mass loss caused by salt evaporation and possibly oxide spallation. A more accurate determination of the salt evaporation rate would demand a very corrosion resistant alloy forming a slow-growing well-adherent oxide scale. Because of the significant attack of the NiCoCrAlY alloy after extended exposure times associated with scale spallation, the evaporation rate was calculated from only a few data points. With these data the evaporation rate for pure sodium sulfate (salt 1) at 950°C in flowing oxygen was calculated to be roughly of the order of 0.01 mg $cm^{-2}h^{-1}$. However, it should be noted that the salt evaporation rate was found to depend significantly on the alloy chemistry [9,12] as this in turn affected the deposit chemistry and, therefore, its vapour pressure.

It is apparent from Fig. 8 that the sodium sulfate was evaporated from the NiCoCrAlY specimens at an almost linear rate, as observed for both CFR = 1.320×10^{-2} mg $cm^{-2}h^{-1}$ and CFR = 5.255×10^{-2} mg $cm^{-2}h^{-1}$. The data were determined at different times between different overall numbers of cycles. Deviation from linear behaviour was observed with higher numbers of cycles and, to a somewhat greater extent, for the higher CFRs. For higher numbers of cycles, side effects like oxide spallation and accelerated scale re-growth during the 20 cycle periods accounted for the deviation from the linear behaviour. For CFR = 5.255×10^{-2} mg $cm^{-2}h^{-1}$, and

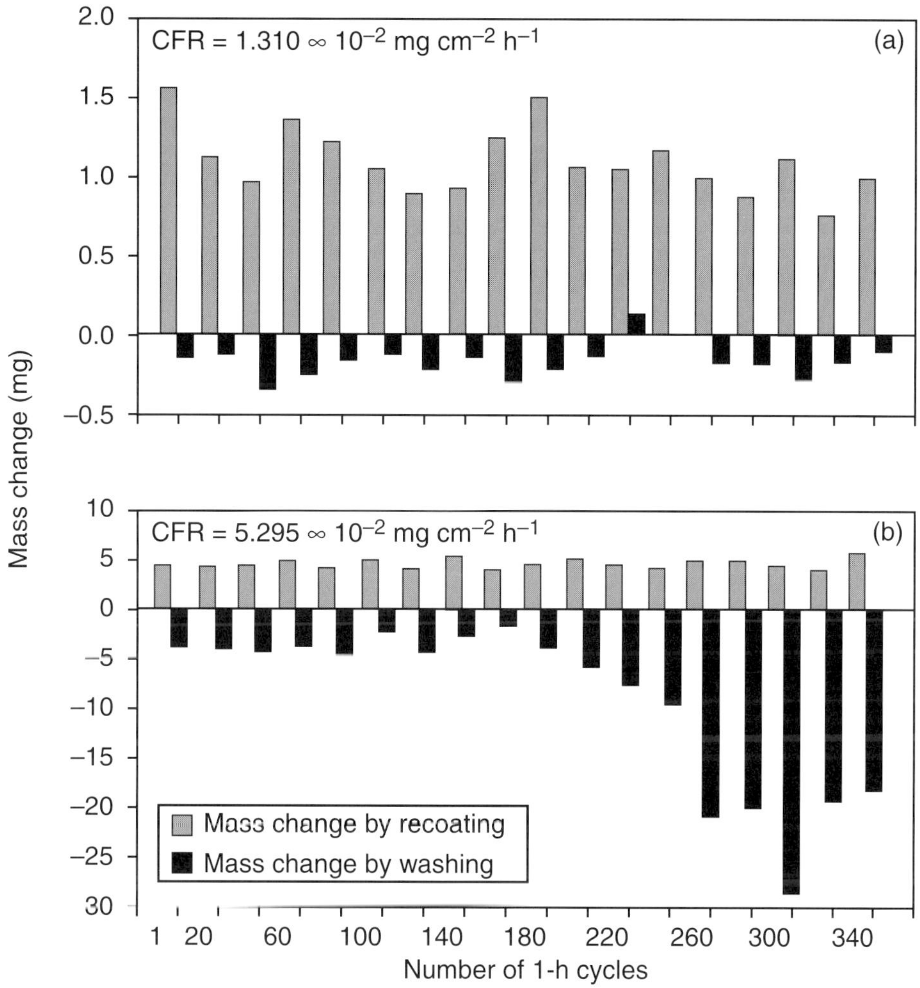

Fig. 7 *Relative mass change vs number of 1-h cycles for two NiCoCrAlY specimens Na_2SO_4-coated at different CFRs. Different sets of data points were taken after different numbers of cycles, e.g. between 280 and 300 cycles data were recorded after 281, 283, 285, 290, 295, and 300 h. Mass loss of the specimens is partly a result of salt evaporation during exposure (refer to text for details).*

more clearly for CFR = 23.730 × 10^{-2} mg $cm^{-2}h^{-1}$ (not shown here), droplet formation and subsequent loss by dripping caused disproportionally high mass loss after the first few cycles (Fig. 8b) but, once the 'excess' salt was lost, a linear evaporation rate was again established (Fig. 8b, data points for 280–300 h). It should be noted again that these data did not accurately reflect the salt evaporation behaviour, because they included side effects related to oxide spallation. This was clearly indicated by the significant scatter of the data after 20 cycles, ranging from about zero to 40% relative mass change (Fig. 8).

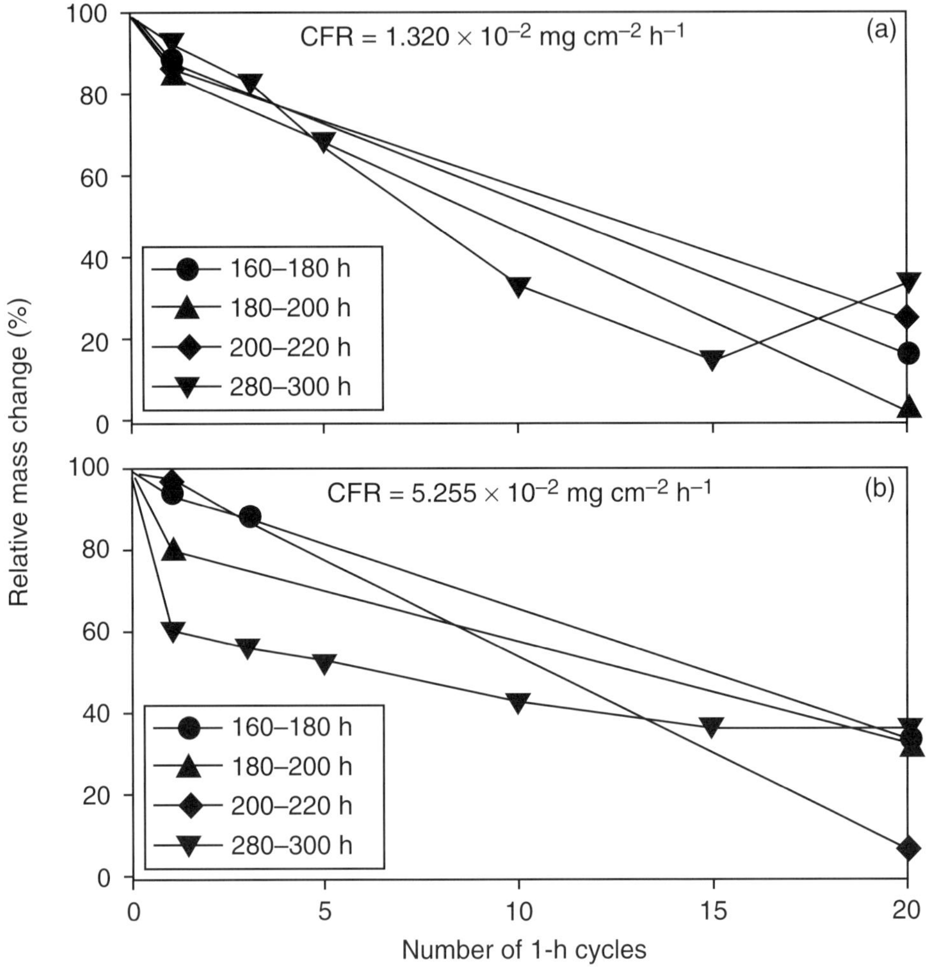

Fig. 8 *Mass change vs number of 1-h cycles for two NiCoCrAlY specimens Na_2SO_4-coated at different CFRs. Mass gain is attributed to the amount of salt added whereas mass loss is mainly due to salt evaporation and loss of oxide scale during cleaning (refer to text for details).*

3.3. Effect of Salt Composition

To assess the extent of corrosion likely to occur with the presence of contaminants typical of combustion products of biomass-derived fuels, NiCoCrAlYHfSi was exposed to different salt compositions representing K/Na ratios of 0, 2, and 15. A stoichiometric sulfur content was established by using Na_2SO_4 and K_2SO_4 for preparation of the salt solutions. The average CFR was 1.0×10^{-2} mg cm^{-2}h^{-1}, and specimens were coated every 100 h without deposit removal before recoating. The mass change vs number of 1-h cycles curves showed a significant effect of the K/Na ratio on the corrosion behaviour. A decreased level of attack was associated with an increased K/Na ratio (Fig. 9). An alumina scale formed on all specimens in the initial stages of oxidation, However, degradation occurred after about 20 cycles for K/Na = 0 (pure Na_2SO_4) and after about 100, 1 h cycles for K/Na = 2 and 15. In the case of K/Na = 0, the molten salt attack led to a rapid mass loss by oxide scale spallation, whereas K/Na = 2 caused only moderate mass loss that started after recoating after 100, 1-h cycles. Mass loss was caused by the formation of spinel phases at the expense of protective alumina as a result of salt attack, and subsequent oxide spallation during thermal cycling. For K/Na = 15, a mass gain was measured up to 500, 1-h cycles, indicating reasonable adherence of the oxide scale.

After 500, 1-h cycles at 950°C, FEG-SEM/EDS analysis of the specimen surfaces revealed spinel phases as the main oxide phases (Fig. 10). Arrows in Fig. 10 indicate that the oxide scales spalled locally during cycling on all samples. The effect of (sulfate) salt composition can be summarised by the fact that the corrosion was significantly

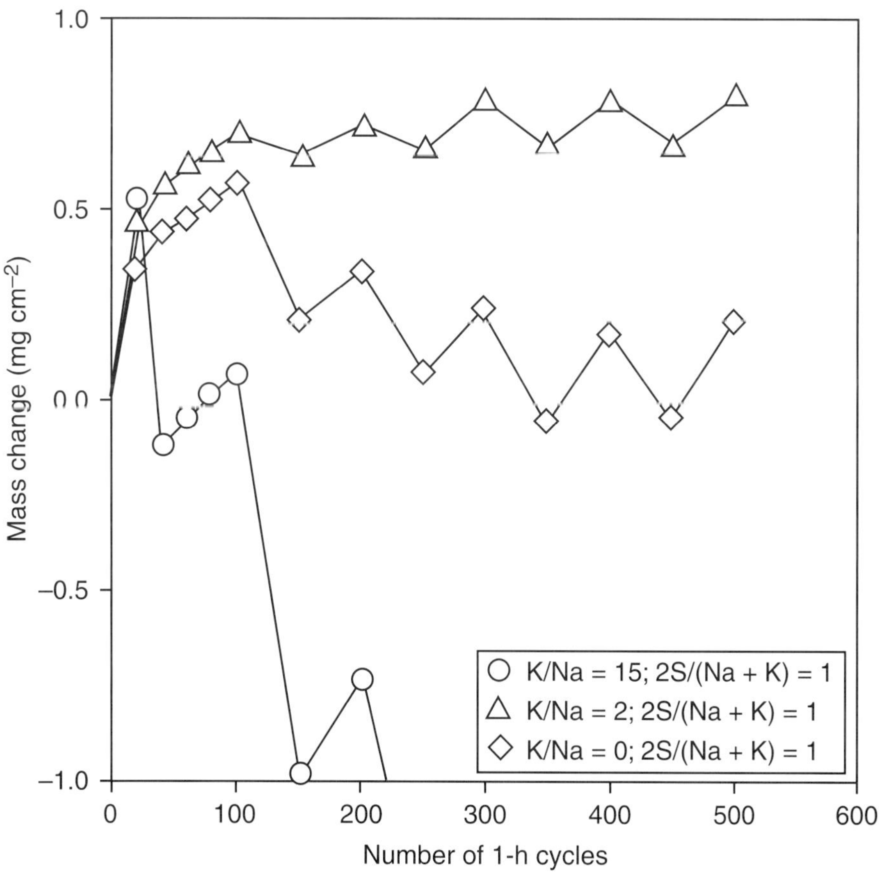

Fig. 9 *Mass change vs number of 1-h cycles for NiCoCrAlHfSi alloy salt coated with three different K/Na ratios and 2S/(Na + K) = 1 the latter representing a stoichiometric S content. The average CFR was 1.070 mg* $cm^{-2}h^{-1}$*. Mass loss for K/Na = 0 was as high as 5.8 mg* cm^{-2} *after 500 1-h cycles.*

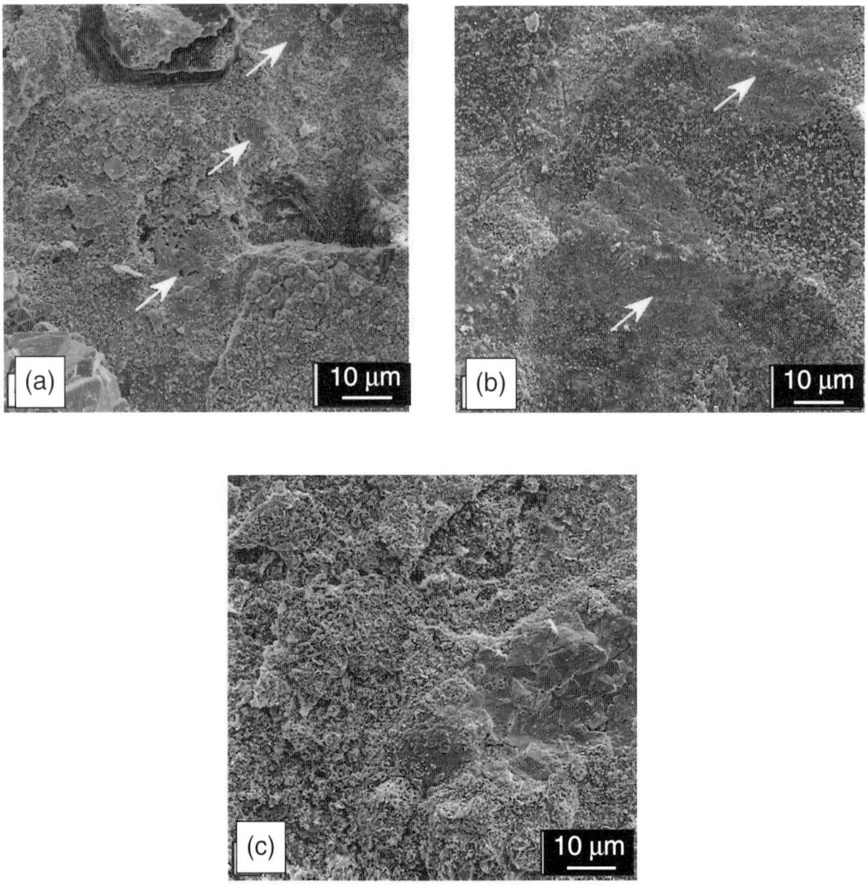

Fig. 10 SEM micrographs of NiCoCrAlYHfSi specimens after 500 1-h cycles. (a) K/Na = 0, (b) K/Na = 2, (c) K/Na = 15. Arrows indicate surface areas where parts of the oxide scale spalled during exposure at 950°C or on cooling down to room temperature between the cycles. Note that spinel phases were the main oxide phases present on the surfaces on all specimens.

decreased with increasing K content in the molten salt. This effect was found earlier also for a NiCoCrAlY alloy [13]. A more detailed study of the effect of the K/Na and 2S/(Na+K) ratio intended to simulate excess alkali conditions typical of some biomass fuels on the hot corrosion of NiCoCrAlYHfSi is discussed elsewhere [14].

Based on the present results, a 20 h salt recoating frequency and an average contaminant flux rate on the order of 1.0×10^{-2} mg cm^{-2}h^{-1} with deposit removal before recoating was considered to provide the most consistent results regarding correlation between mass change data and actual corrosion of the specimens. However, the 20 h salt recoating procedure was very labour-intensive. Therefore, for the sake of a reasonable specimen throughput (100, 1-h cycles per week) in the laboratory, the regular salt deposition procedure was chosen to be 1.0 mg cm^{-2} salt recoated every 100 h without deposit removal before recoating.

4. Summary and Conclusions

The hot corrosion behaviour of NiCoCrAlY-type alloys under thermal cycling conditions was found to depend significantly on the testing procedures. The use of the contaminant flux rate alone as an average measure of the amount of deposits applied onto the specimens cannot be considered as a sufficient measure of the corrosion when the contaminants are not applied continuously during exposure unlike the case, for example, with burner rig tests. For intermittent deposition of contaminants onto the specimens, a significant dependency of the corrosion on the recoating frequency was evident. Even when the same average contaminant flux rate was applied, mass change data revealed that the corrosion was less severe for those specimens coated more frequently but with a lower amount of salt on each recoating step. Therefore, it is suggested that for hot corrosion tests with intermittent salt deposition, the 'recoating frequency' must be always considered along with the average 'contaminant flux rate'. Furthermore, the effect of deposit removal should be taken into account when data from different corrosion tests are compared. However, the effect of recoating frequency is minor compared to that of the contaminant flux rate and deposit removal or retention. Deposit removal (washing) before recoating may result in an accelerated attack, due to a change of the basicity of the molten salt compared to deposits which where present on the surface for longer exposure times.

Whenever material behaviour is to be assessed by simplified corrosion tests it is necessary to simulate the anticipated environment as closely as possible. For different K/Na ratios which were considered to be relevant to combustion product obtained from biomass-derived fuels, significantly different corrosion was observed, showing that seemingly minor changes in the deposit chemistry may cause a different extent of corrosion.

5. Acknowledgements

Research sponsored by the US Department of Energy, Assistant Secretary for Energy Efficiency and Renewable Energy, Office of Industrial Technologies, as part of the Advanced Turbine Systems Program under contract DE-AC05-96OR22464 with Lockheed Martin Energy Research Corporation, and the German Aerospace Center. M. P. Brady, DF. Wilson, and JR. DiStefano at ORNL provided comments on the manuscript.

References

1. *Guidelines for Methods of Testing and Research in High Temperature Corrosion* (H. J. Grabke and D. B. Meadowcroft, eds). Publication No. 14 in European Federation of Corrosion Series, Published by The Institute of Materials, London, 1995.
2. P. Hancock, *Corros. Sci.*, 1982, **22** (1), 51–65.
3. S. R. J. Saunders, *Mater. Sci. Technol.*, 1986, **2,** 282–289.
4. P. Steinmetz, C. Duret and R. Morbioli, *Mater. Sci. Technol.*, 1986, **2**, 262–271.

5. J. R. Nicholls, in *Guidelines for Methods of Testing and Research in High Temperature Corrosion* (H. J. Grabke and D. B. Meadowcroft, eds). Publication No. 14 in European Federation of Corrosion Series Published by The Institute of Materials, London, 1995, pp.11–36.
6. N. S. Bornstein and W. P. Allen, *Mater. Sci. Forum*, 1997, **251–254**, 127–134.
7. I. G. Wright, C. Leyens and B. A. Pint, "An Analysis of the Potential Corrosion Threat to Gas Turbines Fueled by the Products of Biomass Gasification" submitted to Turbo Expo 2000, to be held in Munich, Germany May 2000.
8. R. A. Rapp and Y. S. Zhang, *JOM*, 1994, **12**, 47–55.
9. J. A. Goebel, F. S. Pettit and G. W. Goward, *Metall. Trans.*, 1973, **4**, 261–278.
10. Y. S. Zhang, *J. Electrochem. Soc.*, 1986, **133**, 655–657.
11. R. A. Rapp, *Mater. Sci. Eng.*, 1987, **87**, 319–327.
12. C. Leyens, unpublished research, 1999.
13. C. Leyens, I . G. Wright and B. A. Pint, in *Elevated Temperature Coatings: Science and Technology III* (J. Hampikian and N. B. Dahotre, eds).The Minerals, Metals and Materials Society, Warrendale, 1999, pp.79–90.
14. C. Leyens, I. G. Wright and B. A. Pint, *Oxid. Met.*, submitted 1999.

11
Advantages of Ferritic Stainless Steels in Cyclic High Temperature Conditions

L. ANTONI and J. M. HERBELIN*
Ugine Research Center, 73403 Ugine Cedex, France
*Ugine Cold Rolling Department, BP 15, 62330 Isbergues, France

ABSTRACT

The resistance to thermal cycling is an important consideration when developing exhaust system materials, since they are subjected to severe temperature fluctuations (from room temperature up to 1273 K), particularly during starting and stopping of the vehicle. The upstream part of the exhaust line is an excellent example of a high temperature application for stainless steels based on their ability to form durable protective oxide scales. The effect of such cycling on the materials is emphasised by the fact that the physical properties of the metal and its protective oxide scale are not the same, the system behaving as a composite.

Several austenitic and ferritic stainless steels have been investigated. Cycle conditions (about 30 min) have been chosen in order to be representative of an actual vehicle use. The cyclic oxidation results, after 1200 thermal cycles, demonstrate the importance of spalling phenomena.

The austenitic 18-9D, 18-10T and R20-12 grades, with higher thermal expansion coefficients than ferritic grades, subject the scale to higher stresses during each thermal cycle. This leads to marked spalling and to an increase in oxidation kinetics and loss of metal.

In the case of the ferritic stabilised F17TNb, F17MNb and F18MT grades, continuous protective Cr_2O_3 scales are formed and spalling remains sufficiently small so that parabolic oxidation kinetics (as measured by mass loss) are retained so proving ferritic stainless steels to be more appropriate grades for such cyclic high temperature conditions. Thus, ferritic stabilised stainless steels guarantee a long life for exhaust systems in these more severe conditions.

1. Introduction

Since the introduction of catalytic converters in automotive exhaust lines in order to meet increasingly strict environmental standards for car gas emission, the temperature of the gases at the inlet of the exhaust system has increased to improve conversion efficiency. The high temperature resistance of the exhaust system materials has thus taken on a much greater importance, particularly since the temperature in the upstream region can reach as high as 1273 K.

Moreover, greater time warranties are being offered on those components that now have to be considered as durable parts of a vehicle. This implies the need for complex and new designs for exhaust systems. It generates growing demands from car manufacturers for durable, heat-resistant and corrosion resistant materials with high temperature mechanical strength, such as stainless steels.

The resistance to thermal cycling of these materials is an important consideration since they are subject to severe temperature fluctuations [1] (from room temperature up to 1273 K), particularly during starting and stopping the vehicle. The use of stainless steels is based on their ability to form protective oxide scales. The effect of such cycling on these materials is emphasised by the fact that the physical properties of the metal and its protective oxide scale are not the same, the system behaving as a composite.

Initially, austenitic stainless steels R20-12 (1.4828), 18-10T (1.4541) were used on up-market vehicles, but their price restricts them to a limited market. Stabilised ferritic grades, particularly F12T (1.4512), which are less costly than austenitic alloys, give rise to an increased market share of stainless steel. In severe conditions, other stabilised ferritic grades are also used such as the 17% chromium-containing stainless steels: F17MNb (1.4526), F17TNb (1.4509) and F18MT (1.4521).

This paper aims at providing an overview of the oxidation properties of these austenitic and ferritic stainless steels under cyclic test conditions. In the first part, the testing technique that has been developed and adopted for meaningful simulation of upstream service conditions is presented. The data processing used in order to take into account oxide spallation is described.

In the second part, the results obtained after cyclic oxidation at 1223K up to 1200 cycles are presented. The comparative performance of ferritic and austenitic stainless steel grades is discussed.

2. Experimental

2.1. Sample Preparation

Cyclic oxidation tests were performed on different industrial ferritic and austenitic stainless steel grades. Their nominal chemical compositions are listed in Table 1. These ferritic grades are stabilised by addition of titanium and/or niobium which bond residual carbon and sulfur in the alloy. They improve the high temperature ductility and promote the stability of the ferritic phase even at high temperature. Niobium additions lead to the formation of Fe_2Nb the intermetallic precipitates of which improve the high temperature mechanical properties particularly the creep strength. Molybdenum addition in the F17MNb grade improves further the corrosion resistance of this stainless steel.

Rectangular samples $20 \times 30 \times 1.5\ mm^3$ were punched out of industrial coils. Before oxidation, the coupons were ground under water using abrasive paper through

Table 1. Nominal chemical compositions (mass%) of stainless steels investigated

<table>
<tr><th>Structure</th><th>Ugine</th><th>WNr.</th><th>AISI</th><th>C</th><th>Si</th><th>Mn</th><th>Cr</th><th>Mo</th><th>Ni</th><th>Ti</th><th>Nb</th></tr>
<tr><td>Ferritic</td><td>F17TNb</td><td>1.4509</td><td>441</td><td>0.002</td><td>0.5</td><td>0.5</td><td>17.8</td><td>–</td><td>–</td><td colspan="2">0.7</td></tr>
<tr><td></td><td>F17MNb</td><td>1.4526</td><td>436</td><td>0.04</td><td>0.4</td><td>0.5</td><td>17.5</td><td>1.25</td><td>–</td><td>–</td><td>0.6</td></tr>
<tr><td></td><td>F18MT</td><td>1.4521</td><td>444</td><td>0.02</td><td>0.4</td><td>0.4</td><td>17.7</td><td>2.0</td><td>–</td><td colspan="2">0.45</td></tr>
<tr><td>Austenitic</td><td>18-9D</td><td>1.4301</td><td>304</td><td>0.04</td><td>0.5</td><td>1.5</td><td>18.2</td><td>–</td><td>8.7</td><td>–</td><td>–</td></tr>
<tr><td></td><td>18-10T</td><td>1.4541</td><td>321</td><td>0.03</td><td>0.5</td><td>1.3</td><td>17.5</td><td>–</td><td>9.2</td><td>0.3</td><td>–</td></tr>
<tr><td></td><td>R20-12</td><td>1.4828</td><td>–</td><td>0.04</td><td>2.0</td><td>1.7</td><td>19.8</td><td>–</td><td>12.7</td><td>–</td><td>–</td></tr>
</table>

1200 grit. Samples were cleaned ultrasonically in an equi-volume alcohol–acetone bath. A minimum of three samples of each selected grade was then located in a high temperature resistant metallic sample holder and placed into the furnace.

2.2. Cyclic Oxidation

The intent of the cyclic test was to reproduce exhaust line service conditions, especially during starting and stopping of the vehicle, with short alternating heating and cooling of the engine. A home-made set-up was designed (Fig. 1).

A classical vertical furnace was employed. Temperature was regulated directly with a thermocouple placed in the middle of the sample holder. Before each experiment, the temperature homogeneity was carefully controlled in the central zone to make sure that all samples went through the same heating cycle. A special automated apparatus was added to introduce successively the samples into the furnace and take them out for intermediate cooling cycles. The emersion and immersion periods and the number of cycles are chosen before the experiment begins.

The heating / cooling cycle comprised a twenty minutes heating stage (time count started only when the sample reached a temperature 20°C below the testing temperature), and a five minutes cooling stage under air jets. The heating time from room to test temperature was about 13 min while cooling to room temperature (303K) lasted about 2 minutes.

Only the holding time was considered in calculating the total experiment time. The cyclic oxidation tests were performed in laboratory air. Measurements were made after 50, 100, 200, 400 h, i.e. 150, 300, 600 and 1200 cycles using new samples for each time.

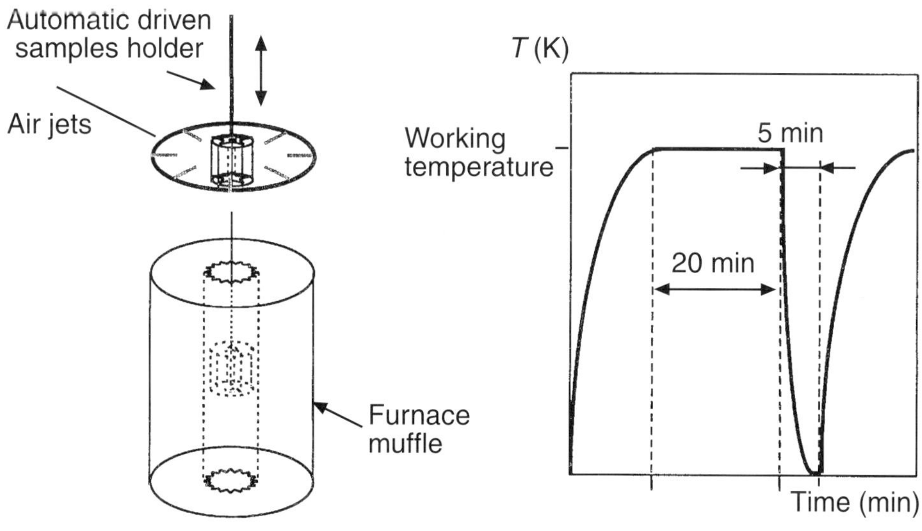

Fig. 1 Schematic illustration of the experimental cyclic oxidation set-up and adopted heat cycle.

2.3. Data Processing

The conventional assessment of oxidation kinetics is to follow the mass change ΔW as a function of the exposure time t at testing temperature. The oxidation rate k_p can then be determined using the relation $\Delta W^n = k_p t$, with n the mass change exponent. A value of about 2 for n indicating a near-parabolic rate law and a diffusion controlled rate is generally reported for iron-chromium alloys under continuous oxidation [2]. These oxidation rates allow comparisons to be made of the oxidation resistance of different materials. However, this calculation makes the assumption that no scale spallation occurs during oxidation tests.

Several models have been reported which take the spallation phenomena into account [3,4]. Another method is to remove the oxide scale completely and then measure the mass loss ΔW_{loss} of the samples. To do this, samples were placed in a NaOH–$KMnO_4$ bath for several hours. Chromium oxides are then transformed into potassium chromates which are soluble in water or acidic media. The mass loss ΔW_{loss} could then be converted into a mass gain assuming a constant oxygen/metal ratio in the scale, or into a material thickness loss Δe using the following relation:

$$\Delta e = \frac{\Delta W_{loss}}{\rho S} \tag{1}$$

with S the surface of the sample, ρ material density (7.7 g cm^{-3} for ferritic stainless steels and 7.9 g cm^{-3} for austenitic stainless steels). Internal oxidation is not taken into account by this approach but on the grades investigated it remains small enough in front of the extra formed scales to be considered as negligible. This point has been verified on grades that do not spall, like ferritic grades under isothermal conditions.

Oxide spallation information was obtained using the calculation recommended by Rundell [4]. The tendency to spalling could be characterised by an adherence factor Ad defined by:

$$Ad(\%) = \frac{W_{oxide\ on\ sample}}{W_{total\ oxide}} = \frac{W_{after\ heat\ treatment} - W_{after\ oxide\ stripping}}{W_{total\ oxide}} \tag{2}$$

Total oxide mass was calculated using mass loss values assuming a constant oxygen/metal ratio (0.42) in the scale. Low Ad values indicate significant scale spalling whereas Ad values near to 100% indicate no spalling phenomena.

3. Results

3.1. Oxidation Kinetics

To compare the high temperature properties of different stainless steels, cyclic oxidation tests were performed in air at 1223 K. These conditions reproduced severe conditions of vehicle use. Manufacturers' measurements on vehicles indicated that 1223 K can actually be considered as the highest temperature expected in the exhaust line.

Results are presented in Fig. 2. Austenitic grades exhibited lower oxidation resistance

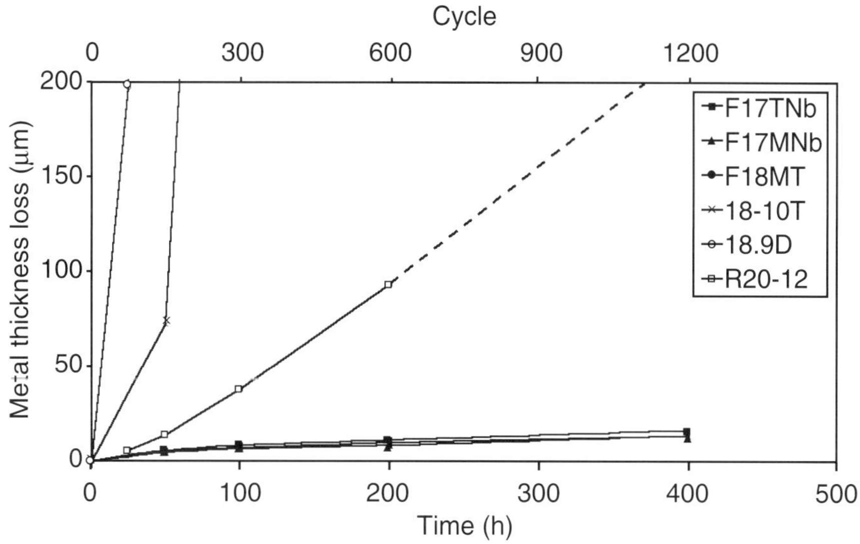

Fig. 2 *Metal thickness loss vs exposure time under cyclic oxidation in air at 1223K.*

than ferritic grades. In particular, metal thickness loss was very rapid for 18-9D and 18-10T grades indicating a non-protective oxide layer; 1223K was above their temperature application range. Refractory austenitic R20-12 grade showed the best cyclic oxidation resistance of the austenitic grades but also remained not suitable for such high temperature conditions.

The three stabilised ferritic F17TNb, F17MNb and F18MT grades exhibited for their part a good cyclic oxidation behaviour. Metal thickness loss was small even after 1200 cycles (or 400 h exposure time at 1223K). As mass gain is proportional to metal thickness loss, one can note that ferritic grades obeyed a near-parabolic rate law typical of a protective oxide layer.

Continuous oxidation tests at 1223K in air were also investigated in order to illustrate the importance of cyclic conditions to high temperature resistance. Results are presented in Fig. 3. In this case, the austenitic refractory R20-12 grade exhibited good oxidation resistance and of the same order as ferritic grades. The oxidation kinetics remained nearly parabolic indicating ionic diffusion controlled build-up of the protective scales. For ferritic grades, results for isothermal exposures fell into the same range as those for cycling conditions.

These results pointed out the importance of cyclic conditions in the high temperature resistance of austenitic stainless steels, whereas little differences were observed on the investigated stabilised ferritic grades. Such simulations were thus important when designing new exhaust line materials.

Scale adherence is illustrated in Fig. 4. The adherence factor of ferritic grades remained above 80% even after 1200 cycles thus indicating little scale spallation, whereas the austenitic 18-9D and 18-10T exhibited a rapid decrease of the adherence factor thus indicating important spalling phenomena from the first heating cycles.

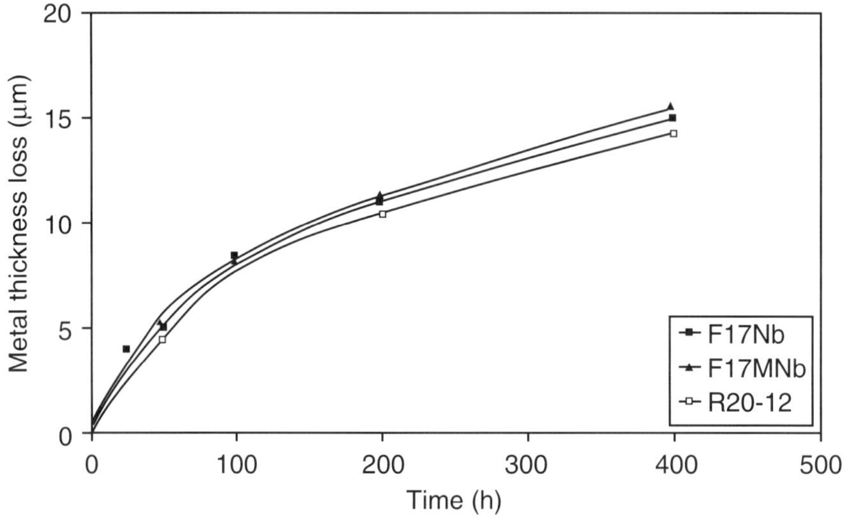

Fig. 3 *Metal thickness loss vs exposure time under continuous oxidation in air at 1223K.*

3.2. Scale Examination

Investigations on scale formation were carried out mainly for 18-9D and F17TNb samples after 150 cycles (and also 1200 cycles for F17TNb). Scanning Electron Microscopy (SEM) surface micrographs are shown in Fig. 5. Alloy 18-9D presented a facetted surface covered by some scattered chips as detailed in Fig. 5(c). Energy

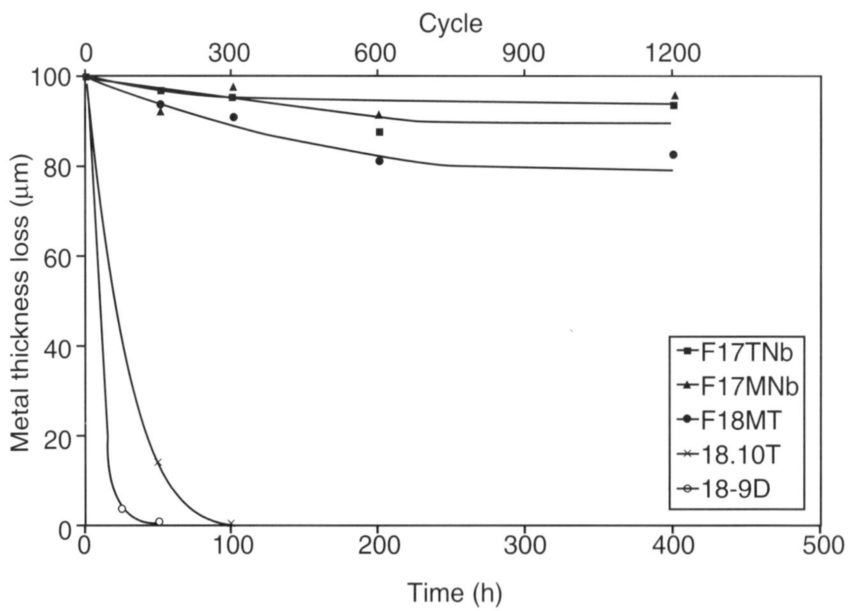

Fig. 4 *Scale adherence factor vs exposure time under cyclic oxidation in air at 1223 K.*

dispersive X-ray analysis (EDX) indicated that these chips are iron-chromium oxide pieces, whereas matrix components were found around these.

These features confirmed the importance of spallation on this austenitic grade and this was confirmed by cross-section observations (Fig. 6a). There was no continuous oxide scale on the sample and only a few compact oxides composed of iron–chromium oxide could be seen. Analysis by EDX and EPMA indicated that the outer part of the compact oxide sections was iron-rich and the inner part contained more chromium oxide. Iron and chromium internal oxidation was observed up to 5 to 10 μm under the surface. No silicon or manganese enrichment could be observed through the scale.

Figures 5(b) and (d) show the surface morphology and cross-section of the F17TNb grade. The general view of this specimen indicated a smoother surface uniformly covered by small nodules. Cross-section observation of this sample is presented in Fig. 6(b). It can be seen that the oxide scale is thin (about 5 μm), uniform, dense and covered by more porous nodules. EDX and EPMA analysis indicated that this

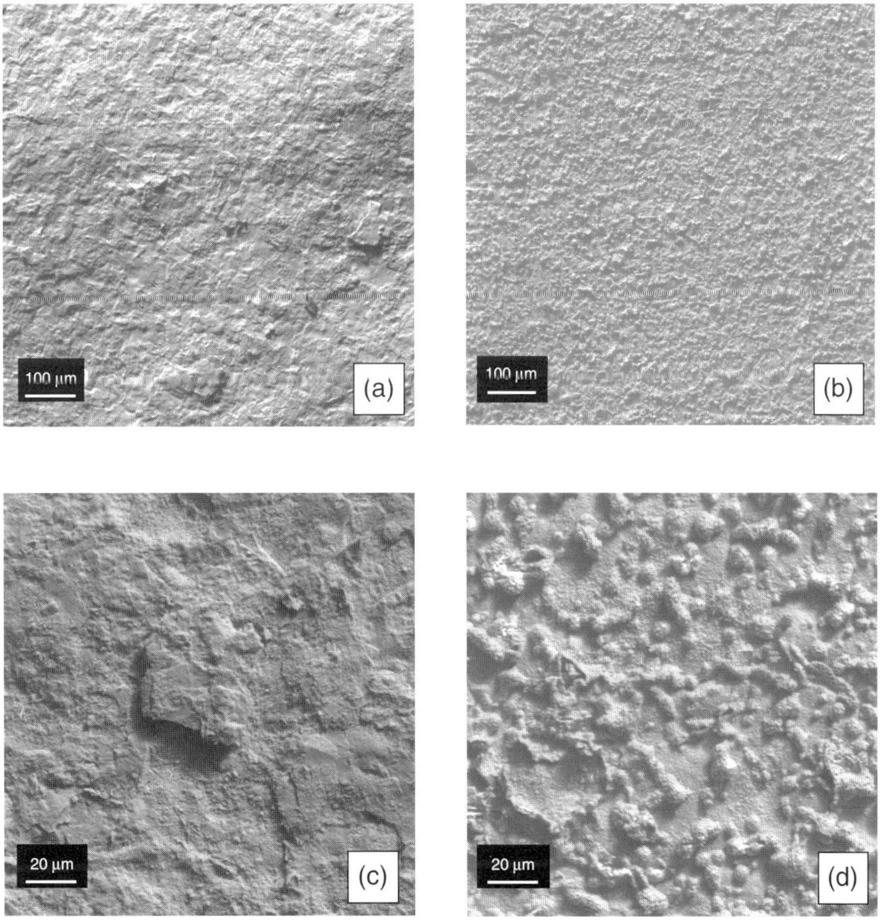

Fig. 5 *SEM surface micrographs of 18-9D (a, c) and F17TNb (b, d) after 150 cycles at 1223 K.*

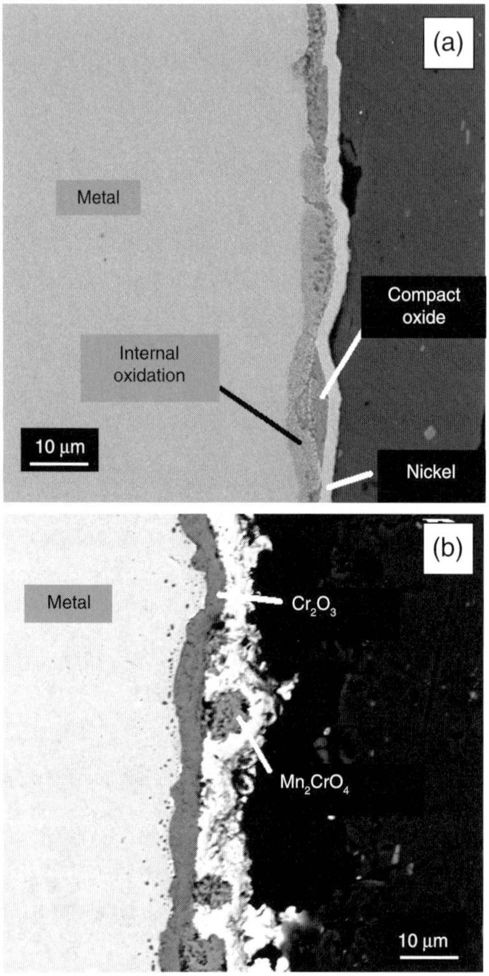

Fig. 6 SEM cross-section micrographs on (a) 18-9D and (b) F17TNb after 150 cycles at 1223 K.

compact layer was composed of Cr_2O_3 with a thin SiO_2 layer at the metal/oxide interface. The nodules were composed of $MnCr_2O_4$. Some titanium internal oxidation could be observed.

The surface morphology and cross-section of a F17TNb grade after 1200 cycles is shown in Fig. 7. The same shape as before can be noted with a more significant covering of $MnCr_2O_4$ and a thickening of the already uniform and dense Cr_2O_3 scale. Similar observations were made on F18MT and F17MNb grades after 150 and 1200 cycles.

4. Discussion

The resistance to thermal cycling is an important consideration when developing exhaust system materials, since they are subjected to severe temperature fluctuations. The effect of such cycling on the materials is accentuated by the fact that the physical

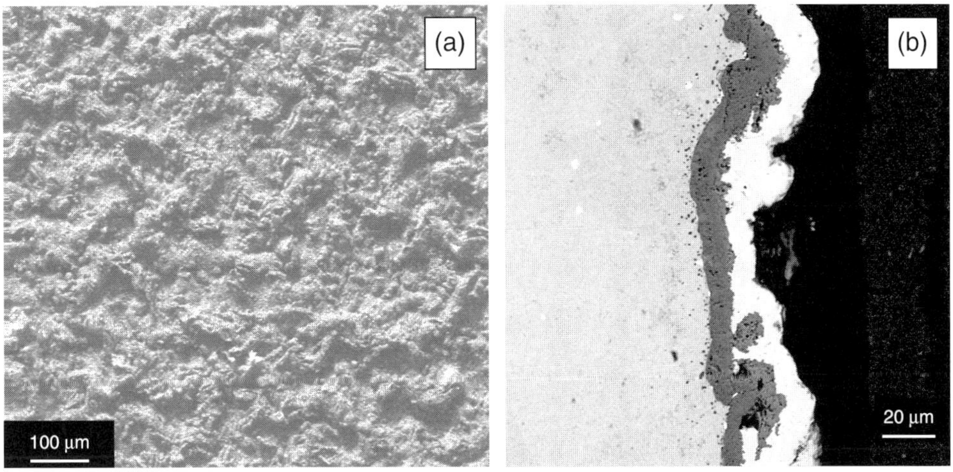

Fig. 7 (a) SEM surface and (b) cross-section micrographs on F17TNb after 1200 cycles at 1223 K.

properties of the metal and its protective oxide scale are not the same, the system behaving as a composite.

During isothermal oxidation, the main generated stresses are those resulting from scale growth. They are generally compressive in nature, with fairly low values, and can usually be relaxed by creep of the metal or oxidation at the scale surface [5]. No difference in metal thickness loss could be seen between the austenitic R20-12 grade and ferritic stabilised F17TNb, F17MNb and F18MT grades even after long exposure times. Thus, isothermal oxidation is not able to discriminate the oxidation resistance of these stainless steels. Moreover, isothermal exposures are not very representative of most service applications, in particular in exhaust systems.

In contrast, the stresses generated during cyclic oxidation (heating or cooling) are caused by the thermal expansion difference between the metal and the scale. These thermally generated stresses can be very large (from about 100 MPa to several GPa). Figure 8 illustrates the different phases of thermal cycling. During heating, the base metal expands more than the scale because of its higher thermal expansion coefficient (CTE). The scale is subjected to tensile stresses which can only be relieved by cracking. During isothermal hold, damaged areas are protected by a newly oxidised layer on the exposed metal. During subsequent cooling, the metal contracts more than the oxide leading to high compressive thermal stresses in the scale, which may be relaxed by spalling [6]. The damage induced by thermal cycles includes also creep in the metal substrate. This occurs in a narrow zone a few microns deep beneath the scale, and plastic strain occurs throughout the thickness of the oxide, causing cracking and spalling.

The stresses are higher in austenitic stainless steels than in ferritic grades, since CTEs of the former are typically of about 20×10^{-6} K^{-1} between 293 and 1173K, compared to about 12×10^{-6} K^{-1} for ferritic grades. Values of CTE for oxides are much lower than for metal (1×10^{-6} K^{-1} for SiO_2, 8×10^{-6} K^{-1} for Cr_2O_3). The difference in CTEs subjects the scale to high thermal stresses during each thermal cycle. When the scale reaches a certain thickness, it is no longer able to withstand the imposed

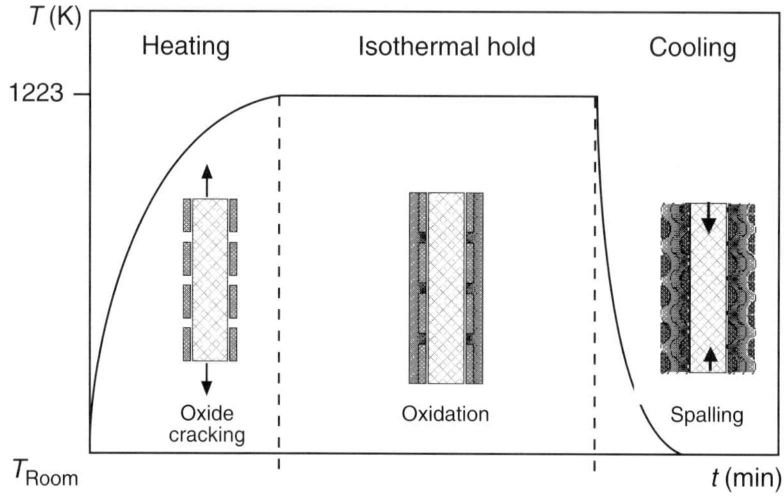

Fig. 8 Behaviour of a metal/scale composite during the different stages of a thermal cycle.

stresses. Partial spalling occurs during each cycle and results in a facetted surface morphology. No durable continuous protective oxide scale, like a Cr_2O_3 layer, can be formed and internal oxidation becomes important. It leads to a marked increase in oxidation kinetics or loss of metal.

The CTEs of ferritic stabilised grades are closer to those of oxides, leading to lower stresses. Furthermore, the presence of stabilising elements also stabilises the ferritic structure at high temperatures, preventing phase transformation in the metal which could cause changes in volume and affect the mechanical stability of the scale [7].

Moreover, the body-centered cubic structure of ferritic stainless steels leads to high diffusion coefficients for iron and chromium ($4–8 \times 10^{-10}$ cm^2s^{-1} at 1273K compared to $1–4 \times 10^{-11}$ cm^2s^{-1} for austenitic grade). This facilitates homogenisation of chromium concentration gradients, which reduces excessive chromium depletion at the metal/ scale interface. This property, combined with a lower oxygen solubility compared to the fcc structure, reduces the tendency for internal oxidation and promotes the formation of a protective Cr_2O_3 layer as could be observed on the investigated ferritic grades.

5. Conclusions

Various stainless steels were oxidised in air at 1223K for periods up to 400 h. A cyclic oxidation test was used for meaningful simulation of the severe service conditions of exhaust upstream components. The results pointed to differences in behaviour between continuous and cyclic oxidation for austenitic grades, whereas there were only a few differences with ferritic stabilised grades.

The cyclic oxidation results and surface examinations on different stainless steels investigated in this work demonstrate the importance of spalling phenomena, in particular for austenitic 18-9D, 18-10T and R20-12 grades. Due to the higher CTE of the austenitic grades compared to the ferritic grades the scale is subjected to higher thermal stresses than the scales on the ferritic grades during each thermal cycle. Partial spalling occurs during each cycle, leading to a marked increase in oxidation kinetics or loss of metal and to a facetted surface morphology. Protective scale may be formed at temperature but this spalls on cooling and internal oxidation takes place.

In the case of ferritic stabilised F17TNb, F17MNb and F18MT grades, spalling remains small enough to preserve parabolic oxidation kinetics thus indicating an ion diffusion controlled build-up of a protective oxide scale. This scale is composed of a uniform and dense Cr_2O_3 oxide with a thin SiO_2 layer at the metal/oxide interface and covered by $MnCr_2O_4$ nodules.

Thus, the ferritic stabilised stainless steels examined exhibit better cyclic high temperature oxidation resistance and consequently offer longer life exhaust system, even in severe high temperature conditions.

References

1. B. Baroux, M. Mantel and P. de Veyrac, Luxembourg, 2–6 May 1998, Le Courrier de l'OTUA, No. 3 — Fascicule 4.
2. F. Armanet and J. H. Davidson, *Les aciers inoxydables*, Les Editions de Physique, 1990, p.449 489.
3. C. A. Barrett and C. E. Lowell, *Oxid. Met.*, 1978, **12**, 4, 293–311.
4. G. R. Rundell, *Met. Prog.*, 1985, 51–58.
5. M. Schutze, *Oxid. Met.*, 1995, **44** (1/2), 29–61.
6. D. L. Deadmore and C. E. Lowell, *Oxid. Met.* 1977, **11**, (2), 91–106.
7. M. Mantel, B. Baroux, J. Ragot and P. Chemelle, *Memoires et Etudes Scientifiques Revue de Metallurgie*, 1989, **(10)**, 637.

12
Investigation of a New Methodology in High Temperature Oxidation

P. VANGELI
Avesta Sheffield AB, R&D, 774 80 Avesta, Sweden

ABSTRACT
In Sweden, since 1975, Avesta Sheffield R&D has been evaluating the scaling temperature for heat resistant steels from gravimetric measurements of cyclic exposures. This scaling temperature is supposed to rank alloys and to determine the maximum temperature at which they can be used. But this is not fully reliable and no extrapolation for long time exposures can be made from it. Since this scaling temperature does not give a real picture of the steel behaviour, another method of testing should be used. In this paper, the old method is fully described. A new methodology, based on the results of isothermal and cyclic tests, is proposed and put forward for discussion.

Oxidation of Avesta Sheffield 253MA (Fe21Cr11Ni1.5SiCe) steel has been studied under isothermal and cyclic conditions in air in the temperature range 800–1200°C. A typical cyclic test is 2 h of oxidation at the temperature of the test followed by 10 min at room temperature. This method has been tested with the higher alloyed steel 353MA (Fe25Cr35Ni1.5SiCe).

For the 253MA steel, oxidation rates obey a parabolic law below 1150°C. The activation energy was calculated. At 1150°C, after 45 h of cyclic exposure, spallation starts. A breakdown in the adherence of the oxide scale is observed. A critical mass gain is linked to the limit of the oxide scale adherence. This critical mass change, responsible for spallation, is about 2.2 mg cm^{-2} for 253MA, corresponding to an oxide thickness of 12–16 µm. Steel 353MA has the same type of behaviour i.e. parabolic up to 1150°C. Under cyclic conditions, a critical mass gain is also observed in this case: 3.8 mg cm^{-2}.

1. Introduction

Heat resistant alloys based on the Fe–Cr–Ni system have useful oxidation resistance at high temperature. They are used in the cement, gasification, steel and metal industries, in power plants, for heat exchangers, furnaces, combustion chambers, etc. Although oxidation resistance is an important factor, this property has not been determined by any standardised test. It is difficult to find data obtained with the same method for more than a few alloys. The need for a universally accepted method to extrapolate from quick laboratory tests into industrial situations has pushed people to start investigations in this direction. A first guideline [1] was published in 1995 on recommendations for high temperature corrosion testing which is a first step to progress in standardisation. A questionnaire was circulated and laboratories asked about their test methods [2]. Most laboratories have their own test methods but these vary widely in exposure time, amount and duration of thermal cycles, and criteria to evaluate oxidation resistance. The European Commission's Standards Measurements and Testing Committee is supporting a programme concerned with corrosion testing, called Testcorr [3]. ASTM has a standard practice for simple static oxidation testing [4] but it does not include cyclic testing. Predictive modelling for cyclic oxidation has been investigated [5,6,7] by a number of workers.

Alloys 253MA and 353MA are austenitic heat resisting stainless steels designed for service in the temperature range 850–1100°C for applications where high creep strength and good corrosion resistance are important. They have good resistance against carbon and nitrogen pick-up and very good rupture and creep strength. In addition to the alloying elements chromium, nickel and silicon, these grades contain small amounts of rare earth metals (REM), which significantly improve the resistance to oxidation.

In order to have a good picture of alloy behaviour, a combination of different tests is used at Avesta Sheffield:

(a) Long time exposures in box furnaces in which 3000 h tests are performed in the temperature range 850–1150°C in air. A cycle consists of a week in the hot furnace followed by 2 h at room temperature. The specimens are rapidly cooled in air by extracting them directly from the furnace and letting them cool naturally in the laboratory air. They are returned to the hot furnace after weighing.

(b) Tests in gas furnace (still in progress).

(c) Field tests.

(d) Isothermal tests in thermogravimetric analysis (TGA).

(e) Cyclic tests in TGA to test the adherence of oxides.

The investigated method described in this paper is based on the kinetic behaviour under isothermal and cyclic conditions (d) and (e). The tests have to be reproducible, easy to perform and must give, in a minimum time, a good knowledge of the steel oxidation resistance. Combination of tests should give us a base to predict lifetime in service conditions and oxide behaviour in term of adherence and protectiveness. This method should give the following information:

- Oxidation law (parabolic, linear or other) at the respective temperatures.
- If the law is parabolic: k_p (parabolic constant) for different temperatures.
- Arrhenius diagram giving the critical temperatures and activation energies.
- From the cyclic tests: limit of adherence of oxide scale, is there a critical mass gain for which significant spallation occurs?

2. Experimental Conditions

The study was carried out using 253MA and 353MA alloys (chemical composition in Table 1). These alloys are austenitic and contain 1.5% Si and Ce added as mischmetal. Coupons of 25 mm × 15 mm × 6 mm are cut from plates. Before each corrosion test, the specimens are ground on #180 SiC papers and then cleaned with alcohol and dried.

Corrosion tests conducted in a TGA Setaram TG 96, in flowing synthetic air O_2/N_2, in the temperature range 800–1200°C under isothermal and cyclic conditions, for 24 up to 120 h. A computer monitors the balance and the furnace. The furnace is heated to the required temperature, while the specimen is kept in the cold zone at room temperature. The gas is introduced in the thermobalance. A typical test starts when the specimen is lowered into the reaction chamber. A cyclic test has a cycle consisting of 2 h in the hot furnace and 10 min in the cold zone (room temperature in the laboratory air). Time and mass gain are recorded continuously, including time in the cold zone.

3. Results from Isothermal Tests

The two grades were oxidised in flowing synthetic air, under isothermal conditions in the temperature range 800–1200°C for 253MA and 850–1200°C for 353MA. The mass change was monitored continuously using a Setaram TG 96 microbalance.

Figure 1(a) shows the result for 253MA presented as normalised mass gain vs time. The curves appear to follow a parabolic oxidation rate for the temperature range 900–1150°C. For the temperatures 1130 and 1150°C, the parabolic rate is reached after about 25 h and less than 5 h is needed for the lower temperatures. At 800°C, the mass gain is too small to be able to distinguish if the curve is parabolic since after 150 h, the mass gain was only 0.026 mg cm^{-2}. The curve obtained at 1200°C shows an accelerated oxidation rate corresponding to the formation of a non-protective oxide scale. A transformed curve $(\Delta m/S)^2$ vs time is calculated and plotted in Fig. 1(b) and 1(c). The slope of linear curves gives the parabolic rate constant k_p. Oxidation rate is related to temperature by the Arrhenius equation:

$$k_p = Ce^{-E_a/RT} \tag{1}$$

where k_p is the rate constant ($g^2\,cm^{-4}s^{-1}$), C a constant, R the universal gas constant (8.3 $JK^{-1}mol^{-1}$), T the absolute temperature (in K) and E_a the activation energy in $Jmol^{-1}$. An Arrhenius diagram (lnk_p vs $10^4/T$ with T in Kelvin and K_p in $g^2\,cm^{-4}s^{-1}$) is shown in Fig. 2. Two domains are observed, below and above a critical temperature

Table 1. *Composition in mass% of 253MA and 353MA used in this study*

	C	Si	Mn	Cr	Ni	Mo	N	Ce
253MA	0.086	1.63	0.55	20.84	10.95	0.21	0.155	0.04
353MA	0.052	1.58	1.42	24.79	34.83	0.09	0.150	0.059

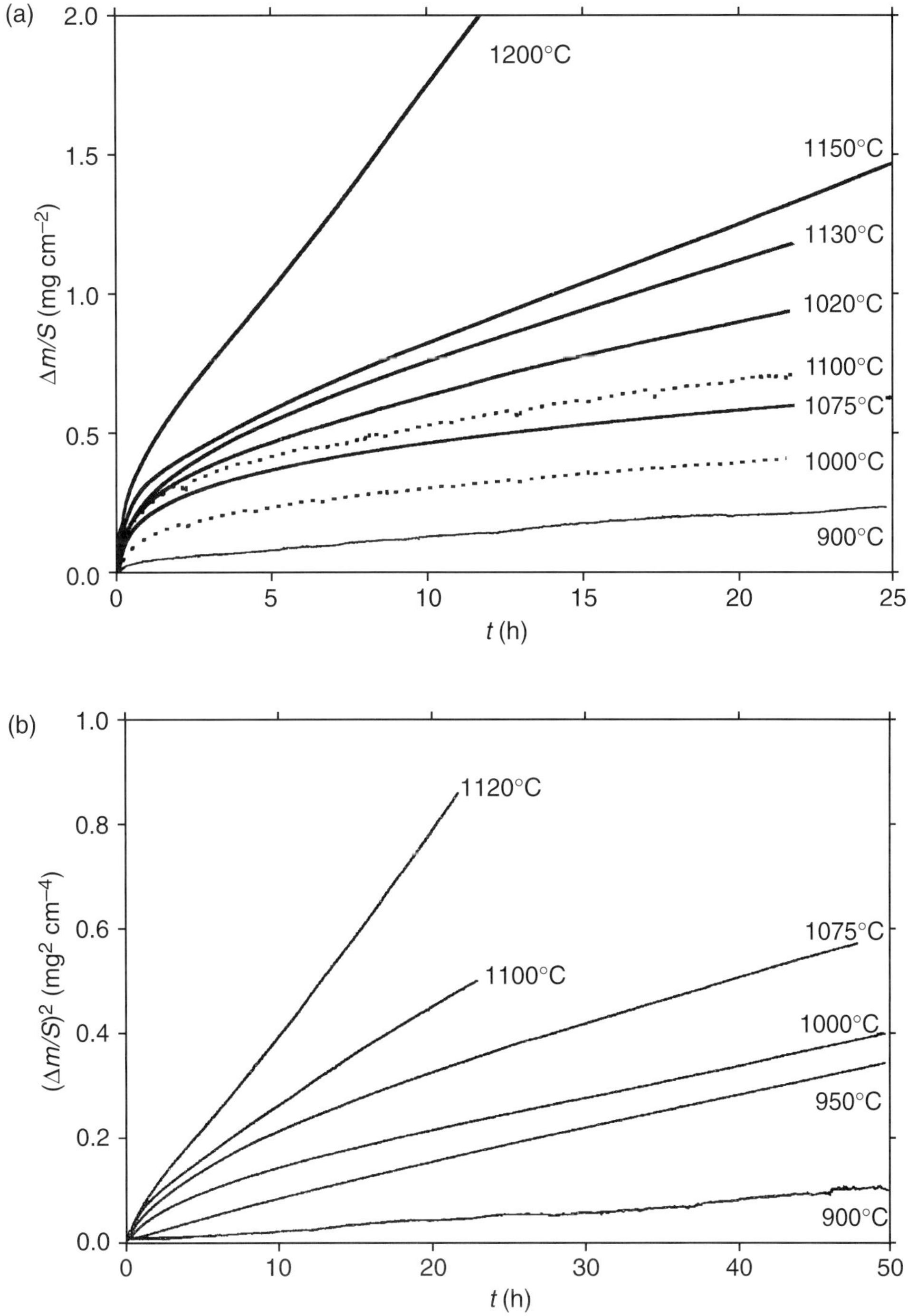

Fig. 1 *(a) Mass gain during isothermal oxidation of 253MA in air at the indicated temperatures, and (b) transformed curves (mass gain squared vs time) at lower temperatures.*

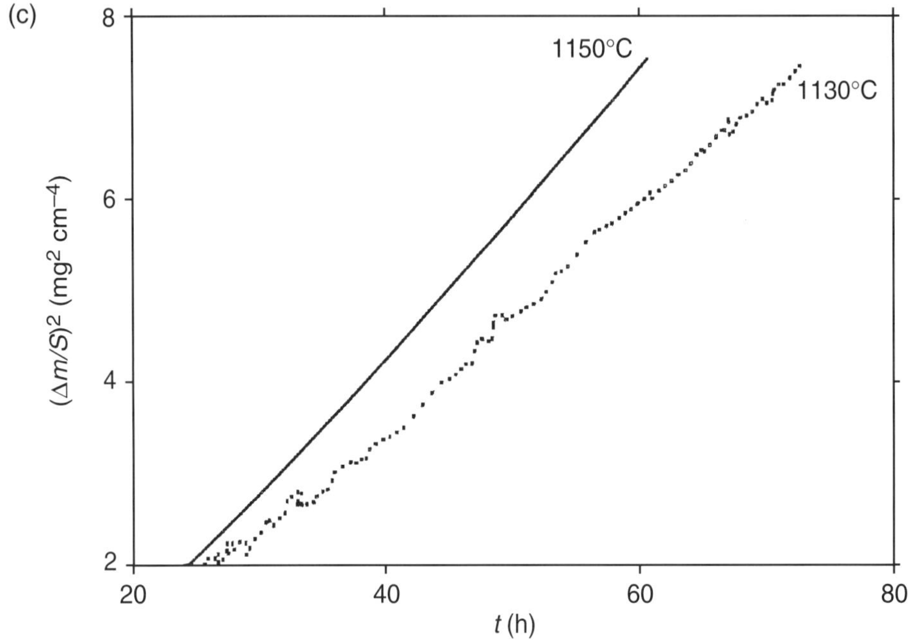

Fig. 1 *(Continued) (c) transformed curves (mass gain squared vs time) at higher temperatures.*

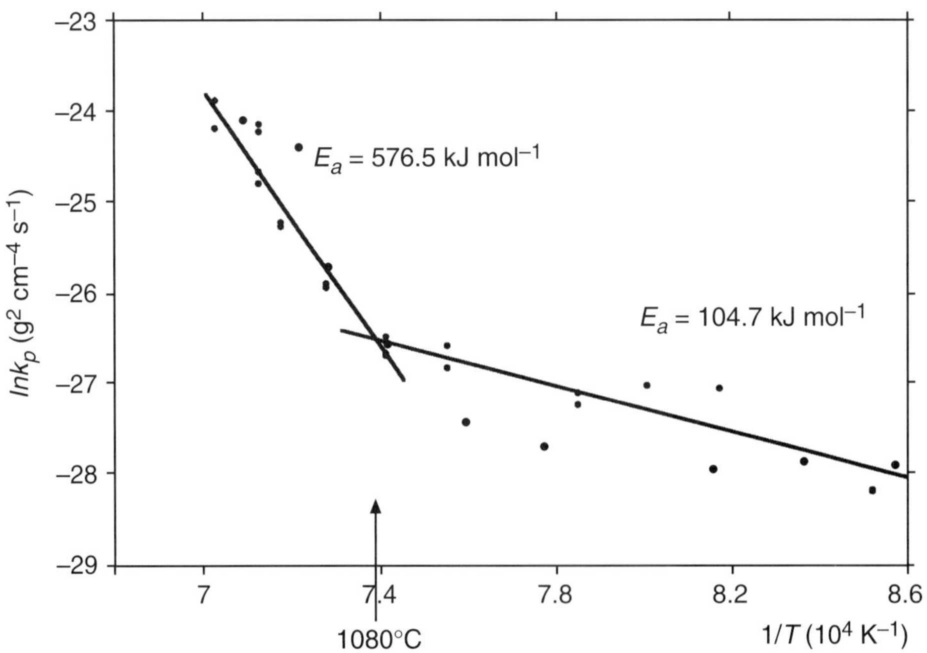

Fig. 2 *Arrhenius diagram and activation energies for parabolic oxidation of 253MA in air.*

of 1080°C. Below 1080°C, the oxidation rate is parabolic, slow and the activation energy low (104.7 kJmol^{-1}). The alloy 253MA is not very sensitive to temperature change up to 1080°C. From 1080 to 1150°C, E_a is much higher, 576.5 kJmol^{-1}, corresponding to a quick but still parabolic oxidation rate. Rundell [8] found 130 kJmol^{-1} for the temperature range 980–1095°C and 612 kJmol^{-1} for 1095–1200°C.

Alloy 353MA shows the same type of behaviour. The results are presented in Figs 3(a) and 3(b). The oxidation rate follows a parabolic law up to 1150°C, which is reached almost from the beginning of the experiments. Curiously, the mass gain at the same temperature after the same time of oxidation is sensibly higher than for 253MA. The parabolic constants are higher for the temperatures up to about 1120°C. From the Arrhenius diagram (Fig. 4), a critical temperature of 1075°C is observed, sensibly the same as for 253MA. The activation energy below 1075°C (107 kJmol^{-1}) is much the same as for 253MA, whereas its value above 1075°C is lower (298 kJmol^{-1}).

4. Results from Cyclic Tests

The study carried out under cyclic conditions gives us information on the adherence of the oxide scales formed in the temperature range 1000–1200°C. The experimental conditions are: 1 cycle of 2 h at the temperature *T* of the experiment followed by 10 min at room temperature. The cyclic oxidation curves are represented for the entire cycle (time of exposure and time of cooling to room temperature). The mass losses during cooling are represented in the curves by a vertical step down. The plots obtained under cyclic conditions for the tested temperatures are reported in Fig. 5 for 253MA and Fig. 6 for 353MA.

Oxidation rates obey a parabolic law for temperatures up to 1100°C. No spallation is observed, even after 120 h exposure for 253MA and 310 h for 353MA. For 253MA, at 1130°C, when the mass gain reaches the critical value of 2.1 mg cm^{-2} (breakaway point in the curve), the sample loses more mass during cooling than it gains during oxidising. At 1150 and 1175°C, this critical mass gain is about 2.3 mg cm^{-2} and at 1200°C is 2.5 mg cm^{-2}. For 353MA, the same observations can be made and the critical mass gain can be approximated to about 3.8 mg cm^{-2}. Under cyclic conditions, 353MA is much more resistant to oxidation than 253MA. For all the temperatures and all the grades, the curves before the breakaway point are the same as in an isothermal oxidation test.

5. Discussion

Previously, oxidation experiments at Avesta Sheffield were carried out in order to determine the scaling temperature of alloys, according to an internal standard procedure developed in 1975 [9]. In this the specimen was exposed to the test temperature for 45 h. During this time, the specimen was cooled to room temperature 5 times: after 4, 20, 23, 26 and 28 h. The rate of oxidation was calculated as the total normalised oxygen pick-up divided by 45 h. Spalled oxide was collected in the specimen holder and the sample and the spalled oxide weighed together. The

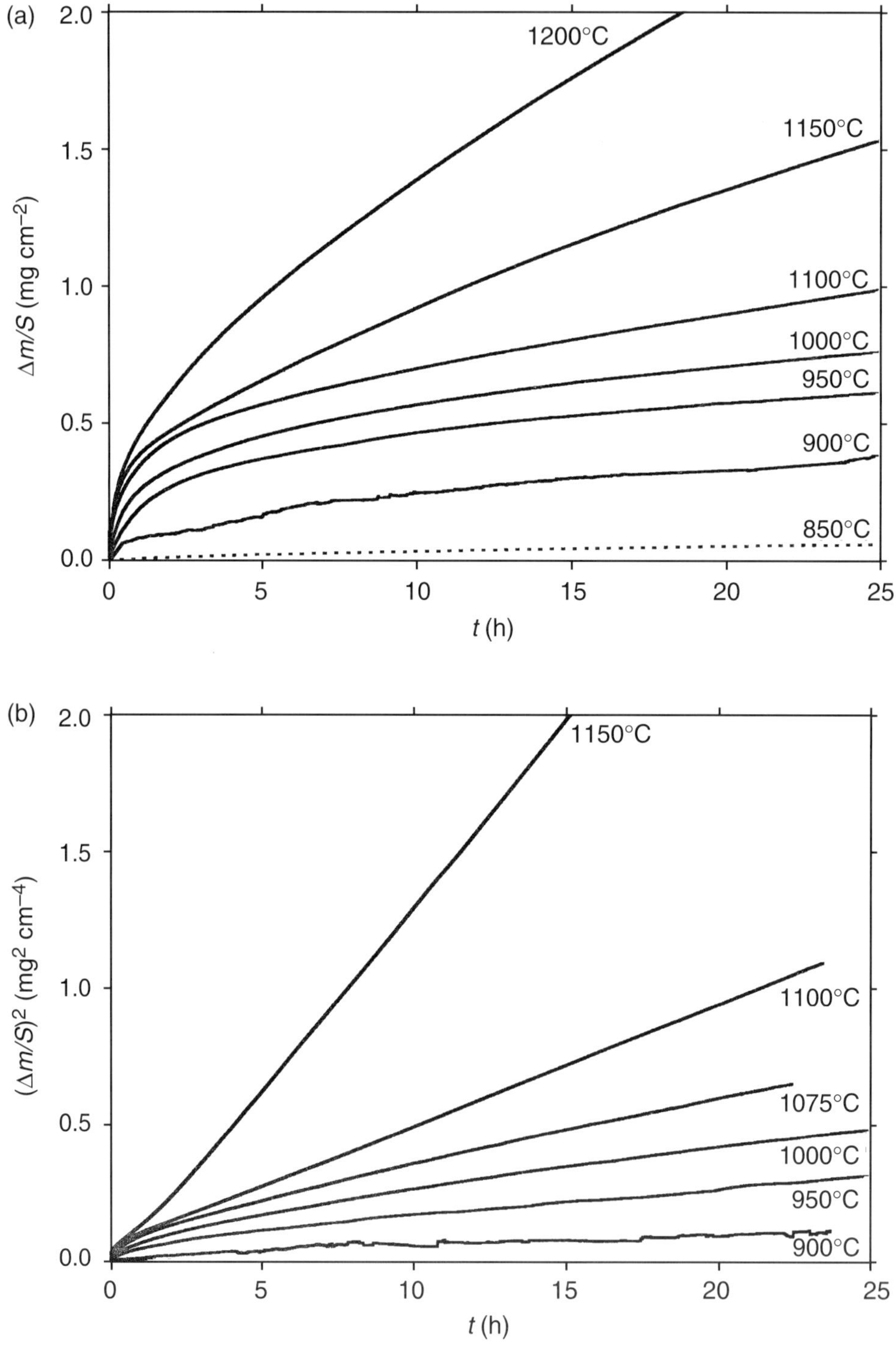

Fig. 3 *(a) Mass gain during isothermal oxidation of 353MA in air at the indicated temperatures; (b) transformed curves (mass gain squared vs time) of 353MA at the indicated temperatures.*

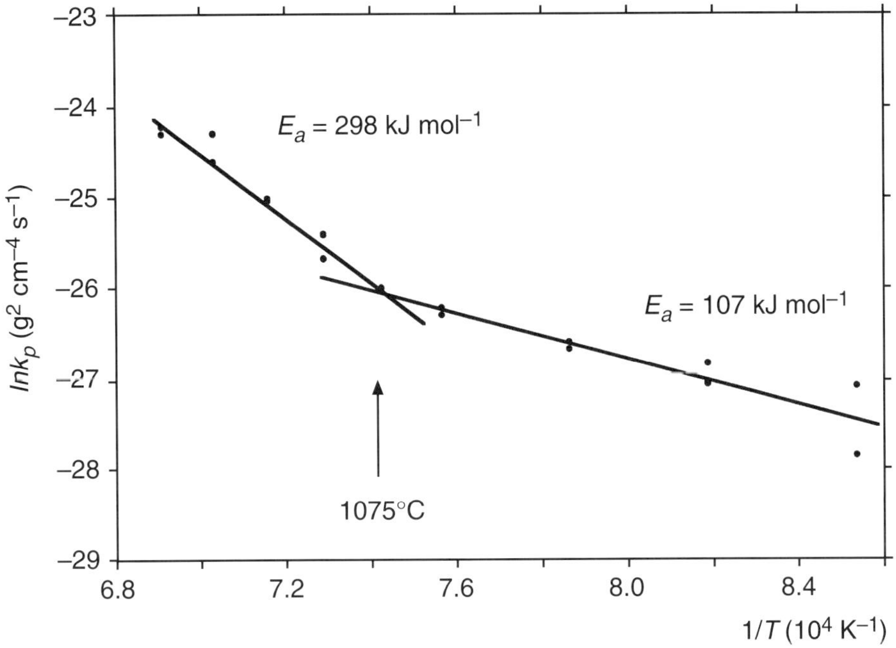

Fig. 4 *Arrhenius diagram and activation energies for parabolic oxidation of 353MA in air.*

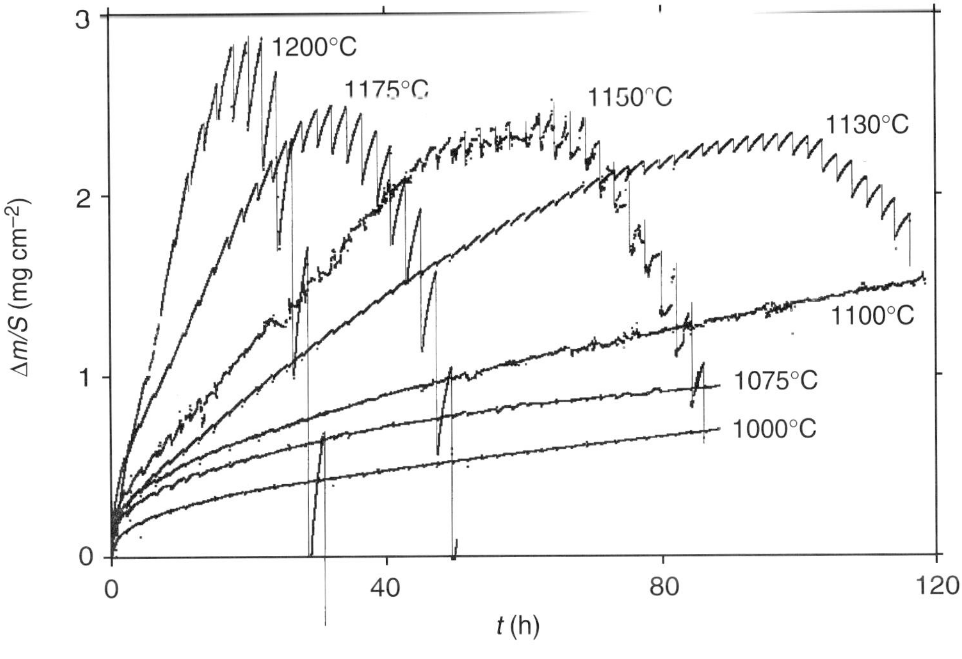

Fig. 5 *Mass gain during cyclic oxidation of 253MA in air at the indicated temperatures.*

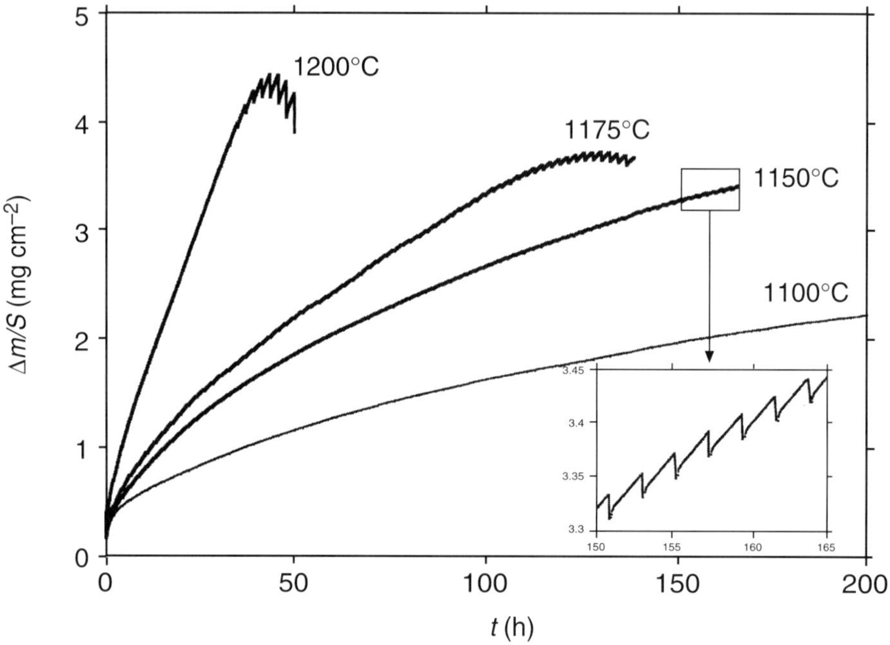

Fig. 6 *Mass gain during cyclic oxidation of 353MA in air at the indicated temperatures.*

evaluation of the scaling temperature is made from a graph of rate of oxidation vs temperature. The scaling temperature is defined as the lowest temperature of the following:

- temperature for which the oxidation rate is 1 g $m^{-2}h^{-1}$ or
- 50°C below the temperature for which the oxidation rate is 2 g $m^{-2}h^{-1}$.

The scaling temperature was then used to rank alloys and determine the maximum temperature at which they can be used. Alloy 253MA was found to have a scaling temperature of 1150°C and 353MA of 1175°C. The difference between these two values was not big enough to be a good sales argument, as it was well known that 353MA can be used for much longer and at higher temperatures than 253MA. With this method, the difference in oxidation behaviour was not well represented. According to all users and experts, this scaling temperature is not fully satisfactory and reliable. Moreover, it does not give us information on the oxidation mechanism or the oxide scale adherence.

With the new method, we are able to have a better picture of the steel behaviour. The results obtained in the microthermobalance are reproducible and they agree with the results obtained by Rundell [8]. Under isothermal conditions, 253MA and 353MA have a similar behaviour which is parabolic below 1150°C. From these tests, it is difficult to draw conclusions on the oxidation resistance of these two alloys. The cyclic tests are needed to be able to show that alloy 353MA develops an oxide scale that is more resistant to thermal shocks than 253MA.

This method allows us to have access to the following information:

- Oxidation law.
- Temperature range for which the oxidation is parabolic.
- k_p (parabolic constant) for different temperatures.
- Critical temperature where a change occurs in the Arrhenius diagram.
- E_a (activation energy).
- Spallation under cyclic conditions.
- Critical mass gain for which a breakdown in the adherence of the oxide scale is observed.

With this method, we can extrapolate to 3000 h and use the cyclic tests in models predicting lifetime. With a set of tests, the influence of the surface, of additional elements (Ce, Si), and of heat treatment can be tested. But to get a complete description of the oxidation behaviour of an alloy, other studies should be carried out on morphology, scale and crack observation, the depletion zone, the adherence of oxide using a descaling method [10], and so on.

6. Conclusions

The alloys 253MA and 353MA have been investigated under isothermal and cyclic conditions in air in the temperature range 800 to 1200°C. The steels have a very similar oxidation rate under isothermal conditions but a very different resistance under cyclic conditions. The values obtained for the two alloys are shown in Table 2.

There is an absolute necessity of cyclic tests to rank these two alloys. It is of great importance that different people in different laboratories use a common method to compare different alloys.

Table 2. *Values obtained after isothermal and cyclic tests in air for alloys 253MA and 353MA*

	253MA	**353MA**
Parabolic oxidation rate up to	1150°C	1150°C
Temperature corresponding to a change in the slope in the Arrhenius diagram	1080°C	1075°C
E_a below 1070°C	104.7 kJ mol^{-1}	107 kJ mol^{-1}
E_a above 1070°C	576.5 kJ mol^{-1}	298 kJ mol^{-1}
Critical mass gain in cyclic tests	2.2 mg cm^{-2}	3.8 mg cm^{-2}

7. Acknowledgements

The author wishes to acknowledge B. Ivarsson and O. Axelsson for their contributions and helpful discussions.

References

1. EFC Working Party Report on Guidelines for Methods of Testing and Research in High Temperature Corrosion (H. J. Grabke and D. B. Meadowcroft, eds). Publication EFC No. 14, in European Federation of Corrosion Series. Published by The Institute of Materials, London, 1995.
2. M. F. Stroosnijder and F. Starr *Corrosion '98,* Paper No. 205, NACE International, Houston, Tx, 1998.
3. 'Testcorr', European programme to develop a Code of Practice for Discontinuous Testing in High Temperature Gaseous Atmospheres, started in 1996.
4. ASTM G 54-84, Standards for Simple Static Oxidation Testing, 1984, re-approved 1996.
5. H. E. Evans and R. C. Lobb, *Corros. Sci.,* 1993, **35** (5/8), 999–1005.
6. C. O. Moon and S. B. Lee, *Oxid. Met.,* 1993, **39** (1/2), 1–13.
7. C. E. Lowell *et al, Oxid. Met.,* 1991, **36** (1/2), 81–112.
8. G. Rundell, ACOM, Avesta Sheffield Corrosion. Management and Application Engineering, 1985, No. 2.
9. Avesta Sheffield, internal standard MFK 710, 1975.
10. G. Rundell and J. McConnell, *Oxid. Met.,* 1991, **36** (3/4), 253–263.

13
Effect of Implanted Silicon on the Cyclic Oxidation Behaviour of AISI 304 and AISI 430 Stainless Steels

F. J. PÉREZ, M. J. CRISTÓBAL*, P. HIERRO and F. PEDRAZA
Dpto de Ciencia de los Materiales, Facultad de Química, Universidad Complutense, 28040-Madrid, Spain
*Dpto de Ciencia de los Materiales, E.T.S.I.I. y M., Universidad de Vigo, Lagoas-Marcosende, 36210-Vigo, Spain

ABSTRACT
Cyclic oxidation tests were carried out to study the influence of implanted silicon on the oxidation resistance of two different stainless steels, austenitic AISI 304 and ferritic AISI 430. The effect of silicon, as a protection against cyclic oxidation, was tested in air at 1173K. Characterisation of the oxidised specimens was carried out by means of X-ray diffraction (XRD), scanning electron microscopy (SEM) and energy dispersive X-ray analysis (EDS). The experimental results confirm that silicon implantation, at the nominal dose of 10^{16} Si ions cm^{-2}, does not have significant effect on the cyclic oxidation behaviour of the austenitic AISI 304 steel in air at 1173K. However, it appears to enhance the oxidation resistance of the ferritic AISI 430 at longer oxidation times. It is suggested that further work should be carried out using higher doses. This study provides evidence that ferritic stainless steel has better resistance than the austenitic stainless steel to cyclic oxidation in air at 1173K.

1. Introduction

An essential requirement for the application of a metallic material at high temperature is its resistance to high-temperature oxidation or corrosion. This is normally achieved when protective oxide scales are formed on its surface. The protective action of these oxide scales depends on their ability to reduce the rate of the corrosion reaction by acting as a diffusion barrier between the metallic substrate and the corrosive environment. Generally, these protective oxide scales consist of Cr_2O_3, Al_2O_3 or SiO_2 [1].

Most high-temperature materials systems are used in applications where thermally cycling is the major factor in determining their useful life. The stresses induced by the different thermal expansion coefficients between oxide scale and alloy, are added to the residual growth stresses that accompany oxide formation. Moreover, with temperature changes, phase transformation can occur in both the oxide and in the substrate, leading to the development of new stresses in the scale. Christl *et al.* [2] found that after about 30 cycles the oxide scale developed on an Fe–Cr–Mo alloy became multi-layered. A magnetite layer was formed beneath haematite multilayers.

These stresses can cause cracking and spallation of the scale. Therefore, to reveal the failure conditions thermal cycling experiments have today become of major importance.

At the moment, no general standardisation of testing methods has been made. Nevertheless, it should be mentioned that several process parameters of the cyclic oxidation test could affect the behaviour of the tested materials. The most significant process parameters appear to be: heating and cooling rates, high temperature (T_{high}) and low temperature (T_{low}) of the thermal cycling test, the isothermal holding time at high temperature, surface roughness, duration of the experiments and static or flowing gas atmosphere. For example, Deadmore and Lowell [3] investigated several high alloy steels with the high-temperature of the cycle between 1100°C and 1200°C. Not only the spallation behaviour but also the oxide composition and structure change depended on the maximum temperature of the cyclic oxidation. Murris *et al.* [4] found that PM (powder metallurgy) chromium (Ducropur™) showed a significant spallation rate of the oxide during thermal cycling up to 1000°C. The spallation rate was reduced by a factor of about 50 at 900°C. Mervel [5] reported that in general the thermal stresses in the oxide scale are proportional to the temperature difference, and are independent of the heating and cooling rates provided that the temperature is uniform in the oxide and the alloy. Barnes *et al.* [6], on the other hand, supposed a relaxation effect of the oxide stresses at low cooling rates. Therefore, care should be given to the test conditions, before a correct interpretation and comparison of the different test results can be achieved.

The beneficial role of reactive elements, such as yttrium, cerium and other rare earth elements, on the oxidation behaviour of high temperature alloys containing chromium or aluminium has received considerable attention in recent years [7–9]. Small amounts (less than 1%) of these elements added to the bulk alloy or to the surface, may decrease the oxidation rate and dramatically improve the scale adherence in the case of chromia-formers. Several mechanisms have been proposed [10–12] to explain the beneficial influences of rare-earth additions to high-temperature alloys. However, no conclusive mechanisms have yet emerged to explain those effects.

The possible use of stainless steels at higher temperatures of operation than those of normal use, has not been studied extensively [13–15]. However, in recent years a few works related to the effect of the rare earth elements on those steels have appeared in the literature. Seal *et al.* [16] studied the influence of superficial applied CeO_2 coatings on the isothermal as well as cyclic oxidation behaviour of three grades of austenitic steel (AISI 321, 316 and 304) in dry air at 1273K. The results clearly showed that coatings not only reduced the rates of scale growth for all three varieties of steel, but also improved the scale adhesion significantly. Stroosnijder *et al.* [17] investigated the influence of implanted cerium on the corrosion resistance of the Alloy 800 H (Fe–19.6Cr–32.2Ni–0.7Mn–0.45Si–0.46Ti–0.38Al–0.09C (mass%)) in a simulated coal gasification atmosphere at 973K. It was shown that the corrosion in the H_2S-containing gas mixture could be reduced for the cerium implanted material if the implantation dose is sufficiently high. Noli *et al.* [18] investigated the oxidation behaviour of Mg-implanted AISI 321 austenitic stainless steel. Two different mechanisms of Mg ion diffusion and oxidation of the implanted steel were observed. At low temperatures an improvement in the oxidation resistance was found to occur. However, at higher temperatures, the implanted magnesium reduced the scale adherence and therefore enhanced the oxidation rate.

Nevertheless, little attention has been given to the implantation of silicon ions although Hou and Stringer recently proposed that this element could act as a reactive element in high temperature oxidation [19]. Amano [20–22] studied the high temperature oxidation behaviour of Ni–20Cr alloys containing 1 and 3% of silicon. According to the results obtained the additions of silicon reduced the oxidation rate, but showed little or no effect on the adherence of the surface oxide.

One approach to introduce the reactive element into the surface of an alloy is ion implantation, and this has received a great attention in recent years. Ion implantation offers several advantages: a wide range of elements can be implanted, the number of implanted ions are independently controlled, penetration depths are typically of the order of 0.01–1 μm, the process temperatures are low, no internal interfaces are introduced and the implantation process has good reproducibility.

Ion implantation and oxide coatings are also the most practical procedures for ensuring locally high active element concentration within the scale. For example, 10^{17} Ce^{+} ions cm^{-2} reduced scale growth and spallation on Fe–20Cr–25Ni stainless steel at 900–950°C, whereas the addition of 0.13% cerium to the alloy surface only improved spallation resistance [23]. The radiation damage generated during irradiation might have an important additional effect at low temperature. Furthermore, new and metastable phases could be formed. However, currently, such damage may be removed by subsequent annealing at moderate temperature [24].

It is clear from this review, that the effect of the rare earth elements on the corrosion resistance of stainless steels still requires further investigation in order to analyse the possible surface modification for enhancing the life time at higher temperatures of operation. Therefore, this paper makes a contribution to these further investigations.

2. Experimental Procedures

Specimens of AISI 304 steel (Fe–18Cr–8Ni–1.42Mn–0.35Si–0.056mass%C) and AISI 430 steel (Fe–16.5Cr–0.21Ni–0.30Mn–0.32Si–0.064mass%C) in the form of 10 mm × 7 mm × 2 mm coupons were cut from cold-rolled plates. Before implantation all the samples were ground on silicon carbide paper up to 600 grit and then ultrasonically degreased in acetone and cleaned in ethanol. The uniform implantation of silicon, at a nominal dose of 10^{16} Si^{+} ions/cm^2, into one of the principal coupon faces was undertaken using a 150 keV acceleration potential.

The experiments were performed in static laboratory air at atmospheric pressure, up to 320 cycles, each consisting of a period of 60 min at the oxidation temperature (1173 K) and 15 min. at room temperature — this appeared to be long enough to cool the specimens below 60°C. The specimens were inserted into and removed from the heat zone in a few seconds to guarantee rapid heating and cooling.

The techniques used to characterise the structure and composition of the oxidation products formed, included optical metallography (OM), X-ray diffraction (XRD), scanning electron microscopy (SEM) and energy dispersive X-ray analysis (EDS).

3. Results and Discussion

3.1. Oxidation Kinetics

Figure 1 shows the mass change measurements after 320 cycles for the implanted and non-implanted steels. The mass changes excluded the amount of oxide scale that spalled off during each cooling. It should be remembered that these data were not corrected for non-implanted surface area of specimens, which was equal to approximately 65% of the total surface. Thermogravimetric results show that silicon implanted AISI 304 steel exhibits better oxidation behaviour up to approximately 200 cycles. However, for longer oxidation times the spallation is greater.

Figure 2 shows in detail the mass gain for the implanted and non-implanted AISI 430 steel. The growth of the oxide present, after a transient period of a faster corrosion rate, showed a para-linear kinetic growth (linear zone from 25 h). The linear rate constants for non-implanted and silicon implanted AISI 430 steel approached a constant value of $k_p = 6.6 \times 10^{-10}$ g cm^{-2}s^{-1}, $k_p = 5.8 \times 10^{-10}$ g cm^{-2}s^{-1}, respectively. The mass gain data show a small improvement in the oxidation behaviour for the implanted AISI 430 steel after 200 cycles.

3.2. Scale Morphology and Structure

For the nucleation and growth studies, the behaviour of implanted and non-implanted sides on the same specimen were compared directly, thus excluding the possibility of scattering between various tests.

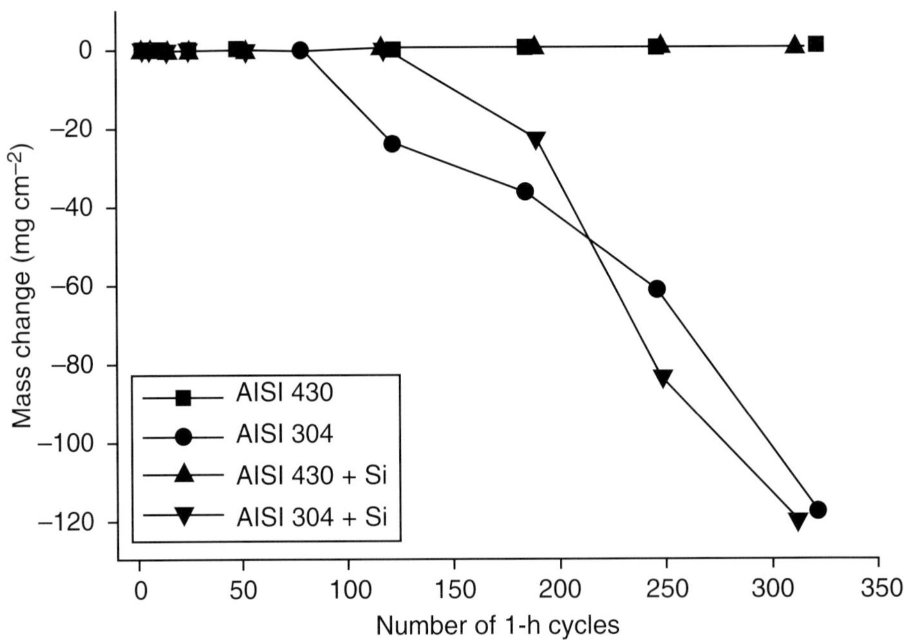

***Fig. 1** Mass change vs number of 1-h cycles for implanted and non-implanted AISI 304 steel, and for implanted and non-implanted AISI 430 steel, in air at 1173K.*

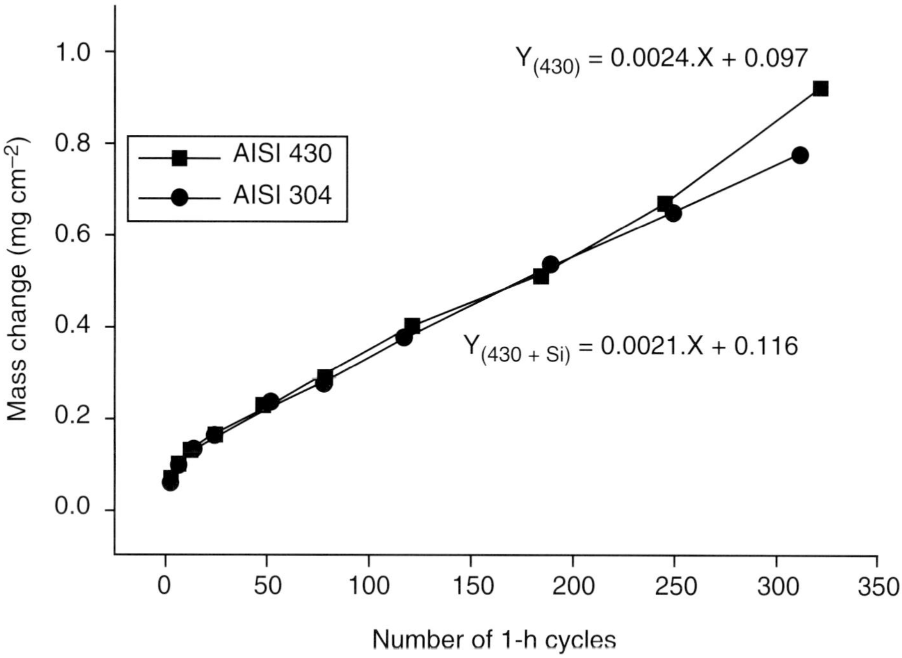

***Fig.* 2** *Mass gain vs number of 1-h cycles for non-implanted and silicon implanted AISI 430 steel, in air at 1173K.*

Figure 3 shows the XRD spectra of the silicon implanted AISI 430 steel oxidised for different cyclic oxidation times. X-ray analysis showed the same corrosion products, on both the non-implanted and the implanted AISI 430 steel. The scale formed on the AISI 430 under cyclic conditions was mainly composed of Cr_2O_3 together with small amounts of a spinel of chromium and manganese. Peaks corresponding to both phases grew simultaneously with the oxidation cycles, but the ratio $Cr_2O_3/Mn_{1.5}Cr_{1.5}O_4$ is smaller at higher oxidation times. The EDS analysis of the top surface oxide scale formed after 2, 48, 80 and 320 cycles corroborate these results in both implanted and non-implanted AISI 430 steel. The EDS spectra show that the oxide scales are rich in chromium and manganese, but that the ratio Mn/Cr increases with the oxidation time.

X-ray diffraction spectra obtained in both implanted and non-implanted AISI 304 steel showed the same results. Figure 4 shows the XRD spectrum of the implanted steel oxide scale after 2 and 80 cycles; the XRD diffractograms showed that the oxide spalled after 180 and 320 cycles. At shorter oxidation times, the peaks correspond to the base alloy together with one hexagonal oxide close to Cr_2O_3 and one cubic oxide close to Fe_3O_4 according to the powder diffraction files. At longer oxidation times, the peaks corresponding to the initially protective chromia changed to less protective $(Cr,Fe)_2O_3$. The XRD spectra also show the presence of an α' (bcc) phase in addition to the oxides and austenitic patterns. The lattice parameters of this phase are almost identical to that of the α'-martensite (bcc) obtained after polishing. This layer of α'-martensite (bcc) vanishes entirely after a short time of heat treatment, as has been observed by other researchers [25]. This phase could be explained in terms of the

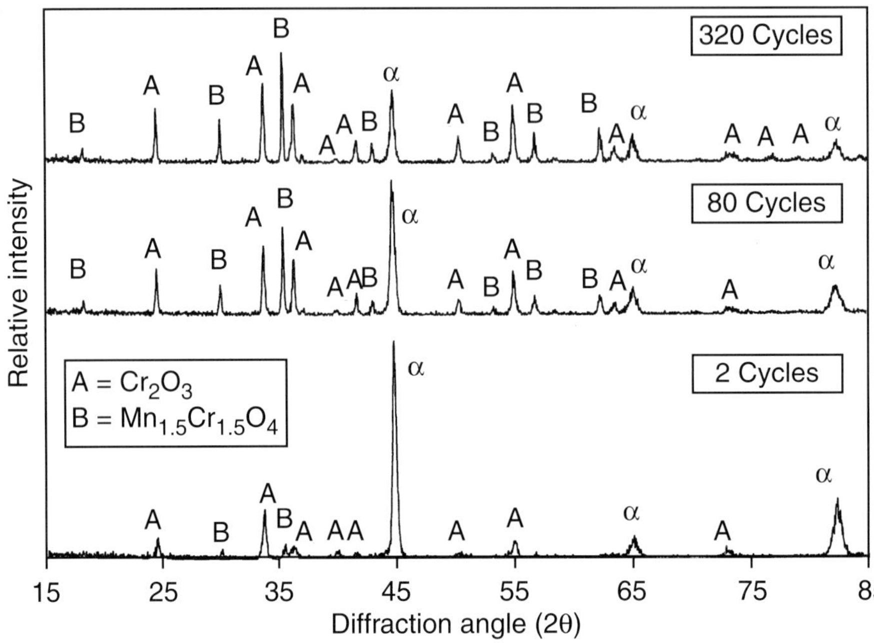

***Fig.* 3** *XRD spectra of the 10^{16} Si^+ ions cm^{-2} implanted AISI 430 steel, after exposure for different cycles in air at 1173K.*

chemical composition of a thin layer substrate adjacent to the oxide since chromium enrichment induces a phase transformation γ (fcc) $\rightarrow \alpha'$ (bcc).

In all the specimens studied the scale morphology of the implanted and the non-implanted AISI 304 steel was the same. The oxide surface morphology consists of a compact base layer of polygonal crystals, many of them with triangular faces with grain size increasing with the oxidation time. Figure 5 shows the surface morphology of the oxide scale developed after 4 cycles. After a relatively small number of cycles, about 24, some nodules began to appear over the compact oxide scale (Fig. 6a). The EDS analysis of the nodules indicated that chromium, iron and oxygen were the dominating elements in them. The EDS analysis of the flat surface shows that the amount of iron was significantly high on the nodules (see diagram opposite). The formation of micro-cracks (Fig. 6b) in well-adherent scale indicated the initial degradation of the compact oxide scale. Micro-crack formation is essentially the consequence of shear stresses or secondary tensile stresses in compressively loaded scales [1]. After 80 cycles, the surface shows significant blistering, with an extensive cracking on the top of the blistered scale (Fig. 7a). The number and size of the nodules has been shown to increase with increasing exposure time. After longer exposures (120–180 cycles) the surface presents larger spalled areas of a circular shape. Although there is no absolute evidence, the nodules rich in iron are probably the source of the spalled areas, as indicated by the circular shape on both morphological structures. After approximately 200 cycles, the oxide surface spalled completely. Inspection of the spalled surface shows that vertical oblique platelets appeared, in the newly developed oxide scale and they seem to have grown along preferred crystallographic directions (Fig. 8a). The EDS analysis showed that iron and oxygen were the major

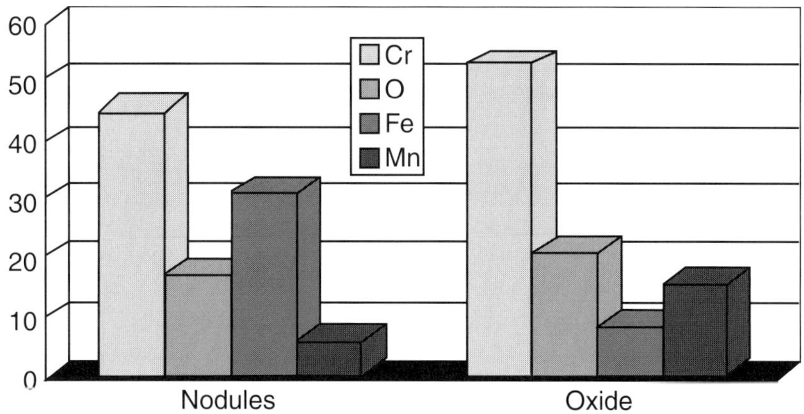

Diagram 1 *EDS analysis of the oxide scale developed on the AISI 304 steel after 24 cycles exposure in air at 1173K*

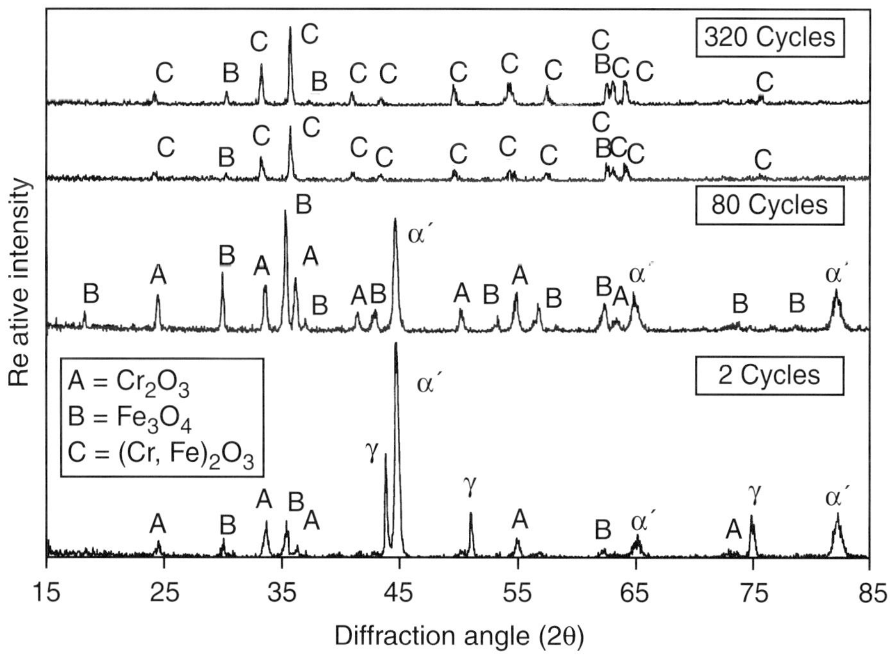

Fig. 4 *XRD spectra of the 10^{16} Si^+ ions cm^{-2} implanted AISI 304 steel, after exposure for different cycles in air at 1173K.*

constituents of these platelets (probably Fe_2O_3, with hexagonal crystallographic structure). It is clear from these results, and the XRD results, that silicon does not inhibit the growth of non-protective and non-adherent iron-rich oxides such as mainly Fe_3O_4 and $(Cr,Fe)_2O_3$.

Figure 9(a) shows the cross-section for the implanted AISI 304 after 80 cycles. The oxide scale presents a uniform thickness together with noticeable tails of internal

Fig. 5 SEM surface morphology of the oxide scale developed on the implanted AISI 304 steel, after 4 cycles of exposure in air at 1173K.

oxidation. The EDS analysis shows the enrichment of silicon in these zones of internal oxidation (Fig. 10a). (The Au peaks correspond to the surface layer that has been deposited to avoid charging problems.) The height of the iron peak could be explained by the matrix effect in this zone with a small volume of oxide. The tails of the internal oxidation could be explained by inward diffusion of oxygen through the grain

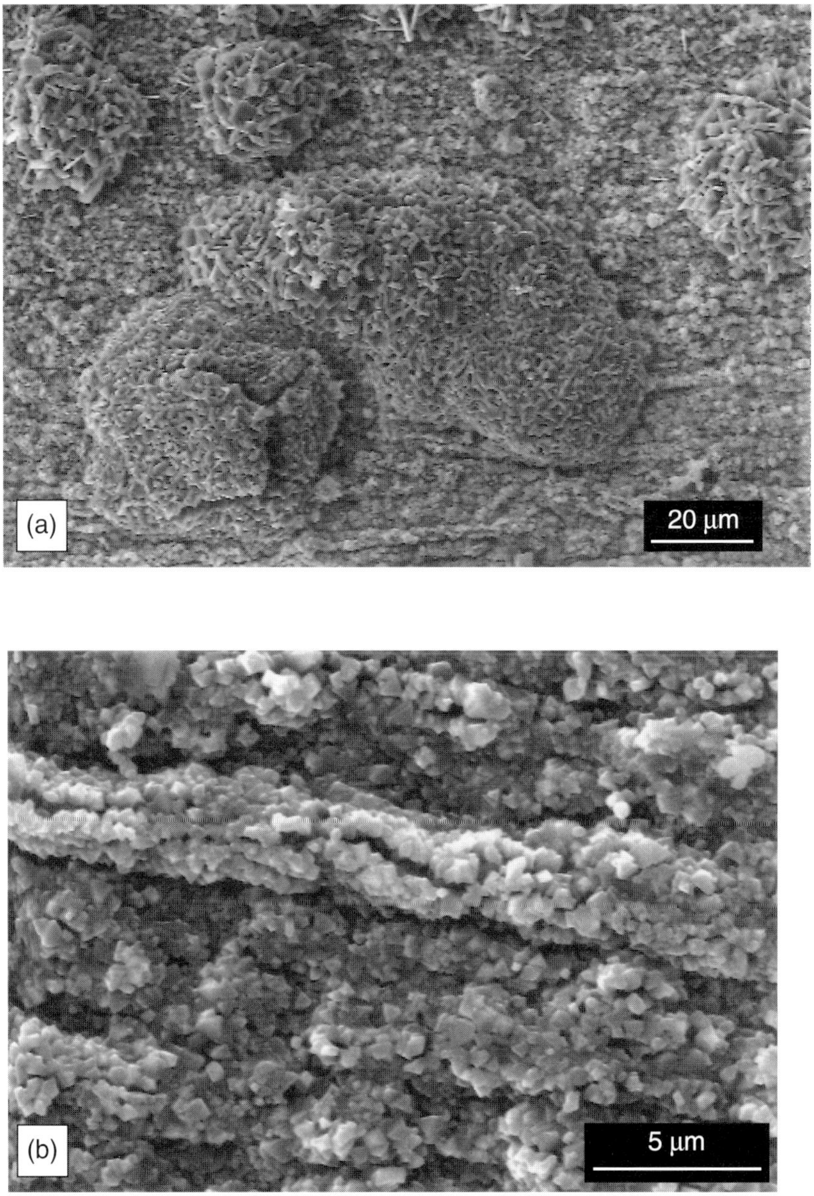

Fig. 6 *SEM surface morphology of the oxide scale developed on the implanted AISI 304 steel, after 24 cycles of oxidation in air at 1173K; (a), showing appearance of nodules over the compact oxide scale; (b), formation of micro-cracks in well adherent scale.*

boundaries that are widened by outward diffusion of metallic cations and condensation of vacancies from the surrounding lattice [18].

Figure 11(a) shows the scale morphology for the implanted AISI 430 steel after 4 cycles of oxidation. Again, the morphology is the same in the non-implanted steel for all oxidation cycles. The surface of the substrate exhibits a compact layer of

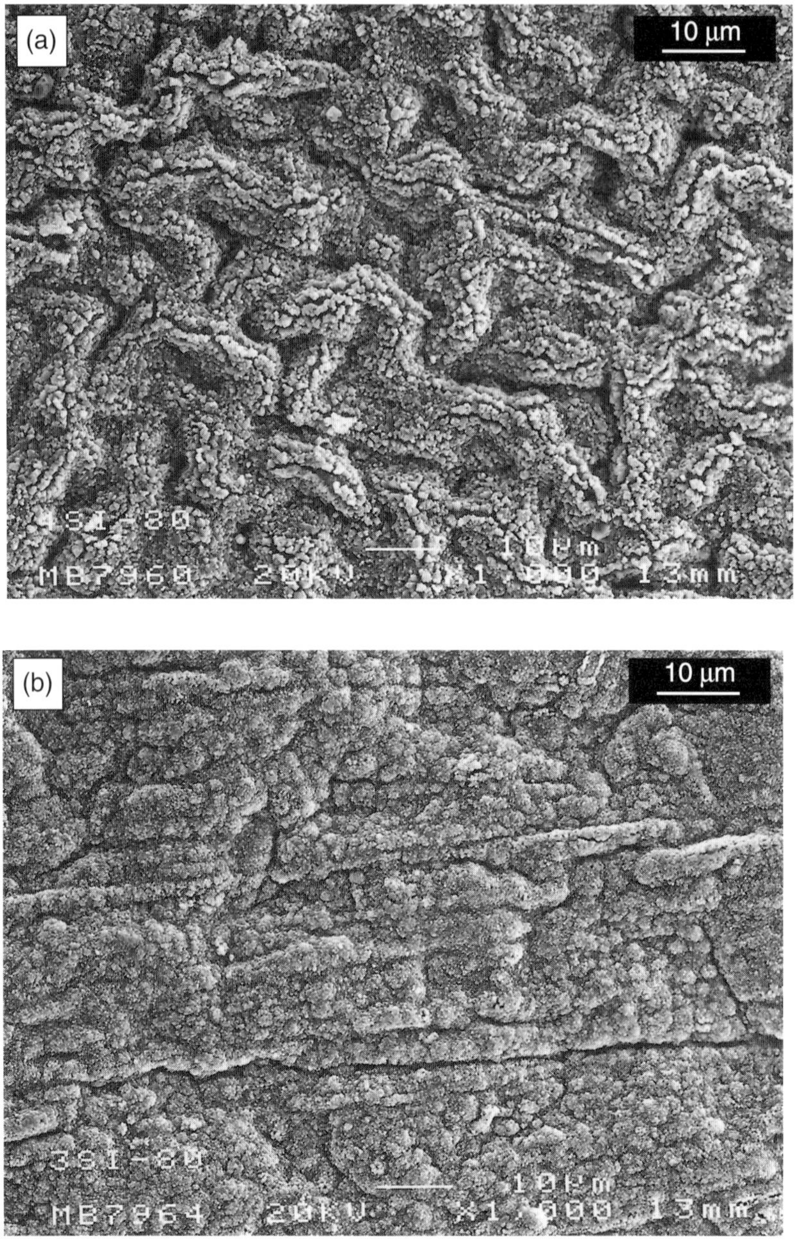

Fig. 7 *SEM surface morphology of the oxide scale developed after 80 cycles of exposure in air at 1173K; (a) on the implanted AISI 304 steel, and (b), on the implanted AISI 430 steel.*

polygonal crystals, similar to those developed on AISI 304 at the same oxidation times (see Fig. 5), but with smaller dimensions. On the samples oxidised during 24 cycles a few isolated groups of vertical oblique platelets appear which seem to have grown along preferred crystallographic directions (Fig. 11b). EDS analysis show that those platelets are rich in chromium and oxygen (probably Cr_2O_3). Further, it

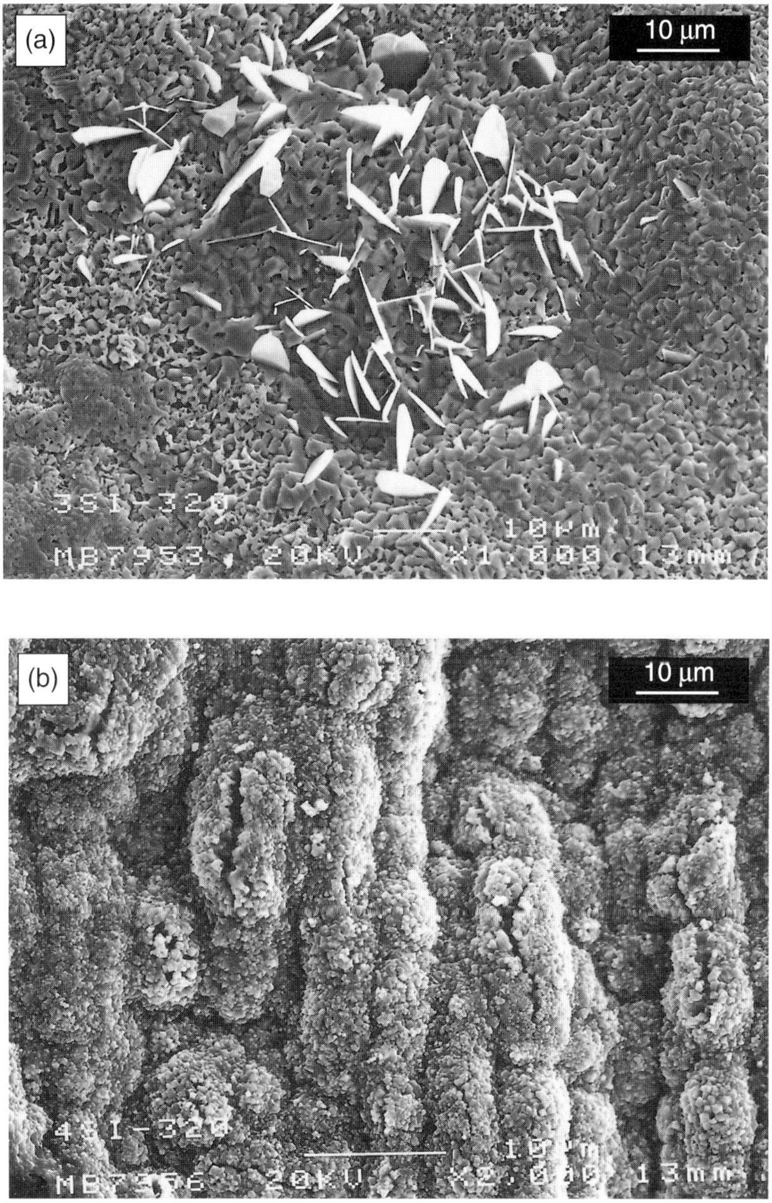

Fig. 8 *SEM surface morphology of the oxide scale developed after 320 cycles of exposure in air at 1173K; (a) on the implanted AISI 304 steel, and (b), on the implanted AISI 430 steel.*

can be seen that the compact layer begins to crack (Fig. 7b). At longer oxidation times, SEM micrographs provide evidence for the blistering of the scale together with significant cracking on the top of the blister morphology (Fig. 8b). Another important feature concerns the adhesion of the scale to the substrate; both implanted and non-implanted AISI 430 showed no appreciable spalling on cooling up to 320 oxidation cycles.

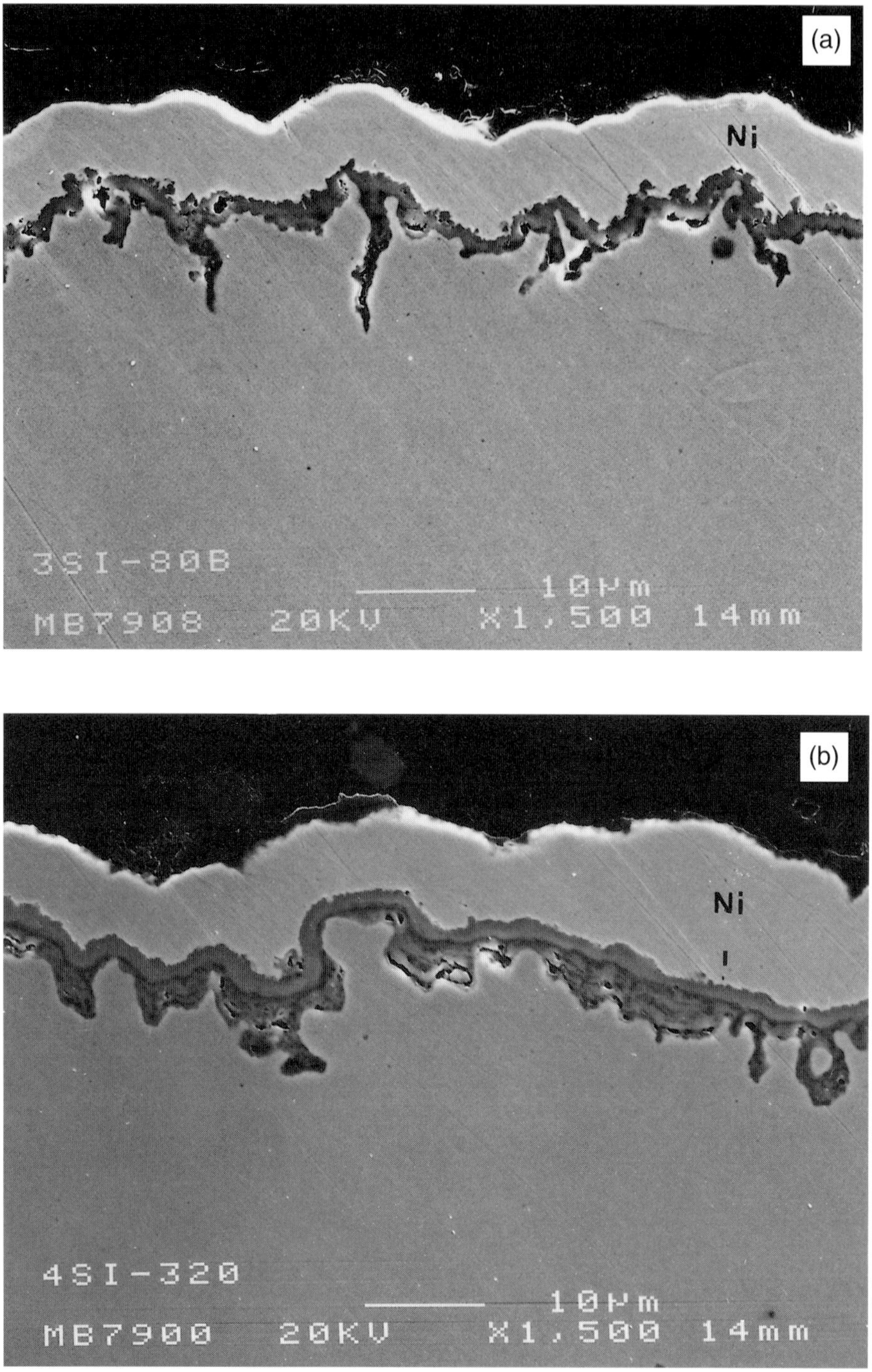

Fig. 9 *SEM cross-sections of the oxide scales (a) implanted AISI 304 steel, after 80 cycles of oxidation in air at 1173K, and (b) implanted AISI 430 steel, after 320 cycles of oxidation in air at 1173K.*

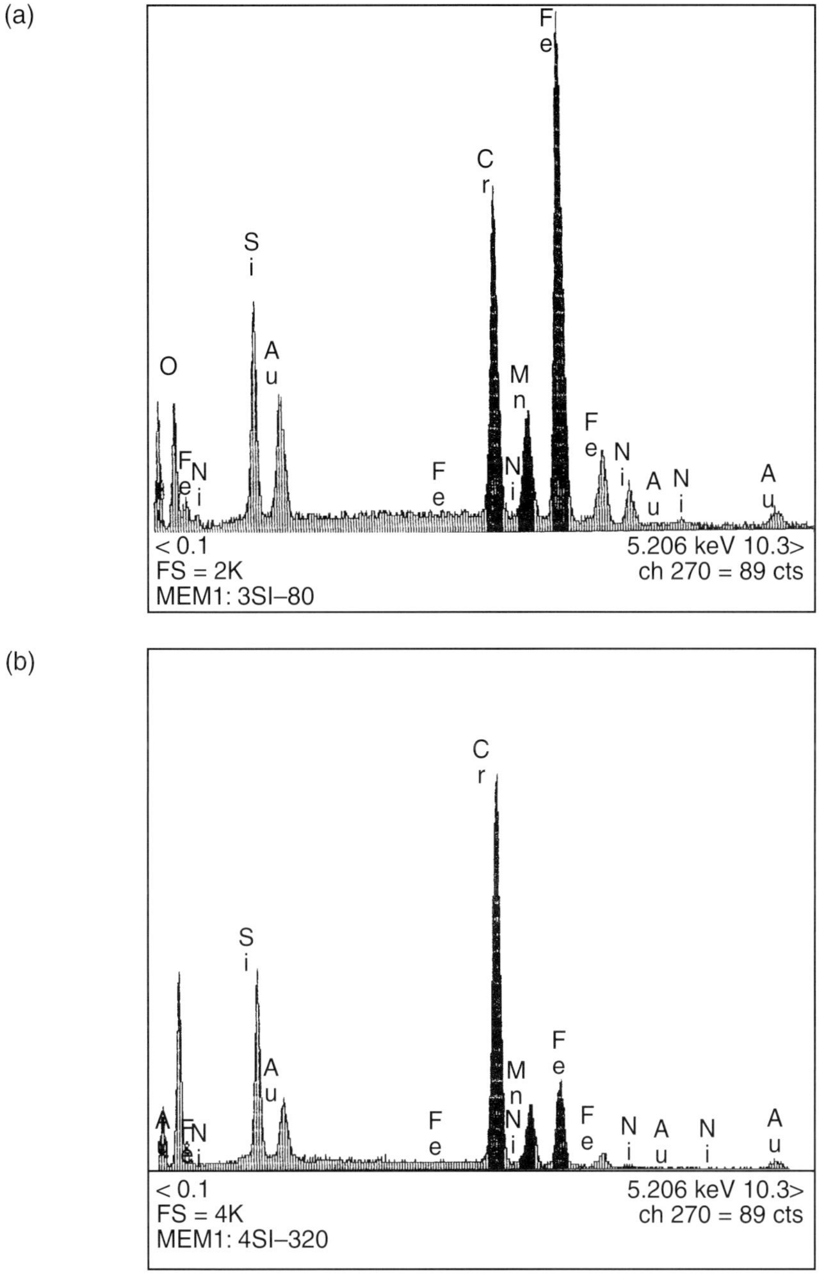

***Fig. 10** EDS analysis on the cross-section for the (a) Si-implanted AISI 304 steel after 80 cycles of exposure in air at 1173K; (b) Si-implanted AISI 430 steel after 320 cycles of exposure in air at 1173K.*

Figure 9(b) presents the cross-section for the implanted AISI 430 steel after 320 cycles which has the same structure as that of the non-implanted steel. Cross-sectional micrographs revealed that the oxide scale has a non-uniform thickness, with many

***Fig. 11** SEM surface morphology of the oxide scales developed on implanted AISI 430 steel, after exposure in air at 1173K; (a) 4 cycles, and (b) 24 cycles.*

intrusions at the scale/metal interface. Some of these intrusions are very big, with a thickness between 4 and 8 μm. It is clear from the SEM micrographs that the scale is divided into two zones. The outer zone has a uniform thickness of about 1.5–3.5 μm, and compact morphology. The inner zone forms intrusions with an irregular shape and thickness, and a very porous aspect. EDS analysis (Fig. 10b) shows that silicon is incorporated in the inner zone in the implanted 430 steel. Another observation of the EDS analysis is that chromium is the major constituent of those inner zones.

It should be noted that ferritic stainless steel exhibits better resistance to the cyclic oxidation than the austenitic stainless steel. This is probably due to a combination of factors. One is that the ferritic steel has a lower coefficient of thermal expansion than that of the austenitic steel, which leads to lower thermal stresses [27]. Another factor is that the oxide scale formed on the AISI 304 is mainly composed of less protective and adherent oxides such as Fe_3O_4 and $(Cr,Fe)_2O_3$. A further explanation, related to the AISI 304 stainless steel, could be that tensile growth stresses in the oxidised surface scale could result from a volume increase in the metal sub-surface zone arising from internal oxidation. On the other hand, Cr diffusion in ferrite is higher than in austenite, so that the reconstruction of the protective layer is much faster in ferritic stainless steels. Moreover, the possibility of σ phase formation for longer oxidation times must be considered in the mechanical design.

It follows from these results that silicon implantation at a nominal dose of 10^{16} Si ions cm^{-2} at 150 keV does not play a significant role in the kinetics or in the composition and morphology of the oxide scale on the AISI 304 steel. However, any implanted silicon on the AISI 430 steel slightly reduces the growth of the scale at longer oxidation times, without any change in the composition and morphology of the oxide scale. This is not in agreement with similar observations on the effects of reactive elements on the oxidation behaviour of chromia forming materials (see e.g. [8,9,28,29]). Furthermore, other work has been recently published by the present authors [30] in which it has been demonstrated that ion implantation of 1×10^{15} ions cm^{-2} of silicon at 150 keV (compared with 1×10^{16} in the present paper) improved the oxidation resistance of austenitic AISI 304 steel during isothermal oxidation in air, at 1173 K.

These considerations point to the need to continue this work using different doses of silicon, to check if there is a minimum dose to impart a beneficial effect on the cyclic oxidation for these stainless steels.

4. Conclusions

These results lead to the following conclusions:

- The presence of implanted silicon at a nominal dose of 10^{16} Si^+ ions cm^{-2} on the surface layer of the AISI 430 seems to enhance the oxidation resistance at long oxidation times. However, silicon does not play any effect in the composition and morphology of the oxide scale.

- Thermogravimetric mass change shows that silicon implanted 304 steel presents better oxidation behaviour up to approximately 200 cycles, but at longer oxidation times the effect disappears. Silicon implantation does not change the structure and morphology of the oxide scale and in consequently, does not improve the spallation resistance.

- The AISI 430 steel (ferritic) has significant better resistance to the cyclic oxidation than the AISI 304 stainless steel (austenitic).

5. Acknowledgements

The authors want to express their gratitude to the Universidad Complutense de Madrid for the financial support of this work: Project No PR156/97-7169.

References

1. M. Schütze, in *Protective Oxide Scales and their Breakdown* (D. R. Holmes, ed.). J. Wiley & Sons, Chichester, UK. 1997, p.1.
2. W. Christl, A. Rahmel and M. Schütze, *Mater. Sci. Eng.*, 1987, **87**, 289.
3. D. L. Deadmore and C. E. Lowell, *Oxid. Met.*, 1977, **11**, 91.
4. M. I. Murris, M. J. Cristóbal and M. F. Stroosnijder, 'The influence of yttrium addition on the cyclic oxidation resistance of PM chromium', *4th Europ. Conf. on Advanced Materials and Processes*, Venice, Italy, 25–28 September, 1995, pp.179.
5. R. Mervel, *Mater. Sci. Technol.*, 1987, **3**, 531.
6. J. J. Barnes, J. G. Goedjen and D. A. Shores, *Oxid. Met.*, 1989, **32**, 449.
7. J. Stringer, *Mater. Sci. Eng.*, 1989, **A120**, 129.
8. Y. Saito, B. Onay and T. Maruyama, *J. Phys.* (France) IV, 1993, **3**, 217.
9. K. Przybylski and G. J. Yurek, *Mater. Sci. Forum*, 1989, **43**, 1.
10. J. Stringer, A. Z. Hed, G. Wallwork and B. A. Wilcox, *Corros. Sci.*, 1972, **12**, 625.
11. H. Pfeiffer, *Werks. Korros.*, 1957, **8**, 574.
12. C. Giggins, B. Kear, F. S. Pettit and J. K.Tien, *Met. Trans.*, 1975, **5**, 1685.
13. N. Hussain, K. A. Shahid, I. H. Khan and S. Rahman, *Oxid. Met.*, 1994, **41**, 215.
14. Z. Kubes, J. Veselà and Z. Weiss, *J. Mater. Sci. Lett.*, 1995, **14**, 876.
15. I. Saeki, H. Konno, R. Furuichi, T. Nakamura, K. Mabuchi and M. Itoh, *Corros. Sci.*, 1998, **40**, 191.
16. S. Seal, S. K. Bose and S. K. Roy, *Oxid. Met.*, 1994, **41**, 139.
17. M. F. Stroosnijder, J. F. Norton, V. Guttmann, M. J. Bennett and J. H. W. de Wit, *Mater. Sci. Eng.*, 1989, **A116**, 110.
18. F. Noli, P. Misaelides, G. Giorginis and H. Baumann, *Oxid. Met.*, 1997, **48**, 225.
19. P. Y. Hou and J. Stringer, *Oxid. Met., 1990,* **33**, 357.
20. T. Amano, S. Yajima, T. Kimura and Y. Saito, *Corros. Eng.*, 1975, **24**, 19.
21. T. Amano, *Trans. Japan Inst. Metals.*, 1980, **21**, 721.
22. Y. Saito, T. Maruyama and T. Amano, *Mater. Sci. Eng.*, 1987, **87**, 275.
23. M. J. Bennett, J. B. Price and D. J. Savage, UKAEA Report AERE-R11228, 1984, Harwell Laboratory, UK.
24. G. K. Hubler, O. W. Holland C. R. Clayton and C. W. White (eds), 'Ion implantation and Ion beam Processing of Materials', New York, 1984.
25. R. Guillament, J. Lopitaux, B. Hannoyer and M. Lenglet, *J. Phys.*, (France), IV, C9, 1993, **3**, 349.
26. H. J. Grabke and G. H. Meier, *Oxid. Met.*, 1995, **38**, 147.
27. A. Godner, *Weld. Res. Counc. Bull.*, 1956, **31**, 3.
28. R. Guillament, J. Lopitaux, B. Hannoyer and M. Lenglet, *J. Phys.*, 1993, IV, C9, **3**, 349.
29. M. F. Stroosnijder, J. D. Sunderkotter, M. J. Cristóbal, H. Jenett, K. Isenbugel and M. A. Baker, *Surf. Coat. Technol.*, 1996, **83**, 205.
30. F. J. Pérez, *et al., Surf. Coat. Technol.*, 1998, **108–109**, 127.

14
The Importance of Relevant Experimental Parameters for High Temperature Cyclic Oxidation Experiments

An attempt for a checklist

V. A. C. HAANAPPEL and M. F. STROOSNIJDER
Joint Research Centre – European Commission, 21020 Ispra (Va), Italy

ABSTRACT

Interpretation of published cyclic oxidation data and comparison of results from different organisations is often difficult. This is due to the fact that there exists a lack of standards and guidelines for testing the high temperature oxidation behaviour of metals and alloys. Actually, the important experimental parameters are still poorly known. In this presentation a check-list for the most relevant aspects for performing cyclic oxidation tests, will be discussed ; it is divided into three major sections, i.e. specimen characteristics, test characteristics and specimen evaluation methods.

The importance of these for the cyclic oxidation behaviour of metals and alloys will be illustrated by some examples. Many organisations are performing cyclic oxidation experiments with different test parameters, because the conditions used by individual laboratories are simulating a variety of industrially relevant conditions. As a result of this, on the one hand, it is not relevant to demand all organisations to perform cyclic oxidation experiments under identical conditions and, on the other hand, a more complete documentation of all the important experimental parameters would make a correct interpretation and comparison of test results from different organisations possible.

This checklist will help to support the need for setting up a Code of Practice for cyclic oxidation testing, and to develop a more fundamental approach for preparing documents in which a detailed description of all aspects of cyclic oxidation testing is given.

1. Introduction

Cyclic oxidation is one of the most severe phenomena responsible for lifetime-limitation of materials and components exposed in many industrially relevant applications, e.g. in power generation, the aerospace industry, and the automotive industry. Generally, the formation of an oxide layer results in the protection of the underlying material against ongoing degradation processes, such as corrosion and erosion. With thermal cycling under practical conditions a high amount of stress will be incorporated in the oxide scale as a result of differences in thermal expansion coefficients between the oxide and the underlying substrate. Generally, thermal stresses are much larger in absolute value than growth stresses. This adversely affects the stability of the oxide scale by inducing cracking and/or spallation. As a consequence, the loss of the protective scale may result in an increased degradation rate of the underlying material.

Even small differences in experimental methods or material properties can significantly affect the cyclic oxidation behaviour of materials under investigation. This can cause misleading or even wrong conclusions, indicating that care should be taken with those parameters employed for cyclic oxidation testing. For more than 70 years high temperature corrosion has been an important scientific and technological subject. However, there have been, surprisingly, no significant efforts in the past to develop standards and guidelines to allow the optimum experimental methods to be agreed and proper analysis and inter-comparison of results to be made. During the last five years steps have been made to rectify this position. For example, in 1996 a questionnaire was circulated world-wide asking for details of the test methods currently used for cyclic corrosion testing. The results from this showed that the current forms of cyclic oxidation testing are not all sufficiently able to provide data of what occurs in industrial applications, or to provide a fundamental basis for the development of predictive models and new materials.

The objective of this study is to support the need for preparing a Code of Practice for cyclic oxidation testing, and to develop a more fundamental approach for preparing documents in which a detailed description of the most important aspects of cyclic oxidation testing is given. On the basis of these aspects a checklist has been developed concerning material and test characteristics. Some of these aspects will be illustrated by examples. However, it is not the aim of this study to review critically the diversification of on-line and post-exposure specimen evaluation methods and equipment for performing cyclic oxidation tests. The on-line analysis methods that are mentioned, i.e. thermogravimetry, thin layer activation, and acoustic emission, each has its own advantages and disadvantages. In order to obtain a complete and detailed understanding of the phenomena occurring during cyclic oxidation tests, a combination of these techniques will be recommended. Ideally, it should be possible to use these three techniques together whilst tests are in progress. However, from a practical point of view, this is not always possible. Consequently, repetition of the experiment(s) under equal conditions but with different analysis techniques is sometimes necessary to provide a full picture of the material under specific and well-defined thermal oxidation conditions. Recommendations for the selection and full characterisation of equipment of high temperature oxidation/corrosion testing, including furnaces and furnace tubes, flow and pressure controllers, and thermocouples, have already been discussed in associated papers [1,2].

2. Checklist of Specimen, Test and Evaluation Characteristics

The checklist shown in Table 1 gives an overview of the most important material and test parameters useful for preparation of reports. In particular, in high temperature cyclic oxidation tests, these parameters are of major importance in affecting the cracking and spallation behaviour of the oxide layer. It is also important to monitor the oxidation and spallation process continuously without influencing the degradation process itself. Several techniques are already available, each with its own specific advantages and disadvantages.

3. Specimen Characteristics

3.1. Introduction

For a correct interpretation of test data and to make comparison of results from different organisations possible, it is important to provide as fully as possible the definition and preparation of test specimens. These include **material manufacturing route**, **chemical composition**, **microstructure** (grain phase-type, size and orientation), **specimen production route** of the samples including pre-treatment processes, **surface finishing procedure**, **shape and dimensions** of the specimens, **surface cleaning procedure**, **sample-support**, and **handling procedure**.

Each of the above mentioned aspects can affect the oxidation behaviour. For example, shape and dimensions of specimens can be an important parameter in cyclic corrosion tests, particularly when small samples are used. It is well known that local specimen geometry affects the growth and thermal stress in the oxide scale, where highly curved surfaces, i.e. edges and corners, lead to high local strains in the oxide scale. From this point of view it can be expected that oxide scales from small specimens are more susceptible to scale cracking and spallation than those from larger specimens with a lower 'edge-corner/total surface area' ratio. Another aspect affecting the oxidation behaviour is the specimen handling procedure, which has already been reported elsewhere [3]. The main conclusion from that work was that some handling procedures might damage the oxide scale thus influencing the final mass change in an uncontrollable way. This can be misleading and even result in errors in the interpretation of the data.

In the following section some of the specimen characteristics (listed in Table 1) will be discussed in detail and illustrated with some examples.

3.2. Chemical Composition

Oxidation experiments were performed with a TiAl-based intermetallic alloy from two different batches (further indicated as batch A and B) of nominal composition Ti–48Al–2Cr (at.%). Both batches were produced by an identical production route and heat treatment, resulting in an identical microstructure. Sheets of this material with a thickness of about 1 mm were produced using a 'modified pack-rolling' process. A primary annealing heat treatment of 2 h at 1000°C in vacuum was applied to flatten the sheets after the rolling process. This was followed by 2 h at 1200°C to obtain the near-gamma microstructure. More information concerning the manufacturing route can be found elsewhere [4]. The chemical composition (most important elements) of the materials is given in Table 2.

The cyclic oxidation experiments were performed in static laboratory air. Each cycle consisted of a period of 1 h at 800°C and 12 min at ambient temperature (AT) sufficient to cool the specimens below 50°C. The specimens were inserted into and removed from the hot zone of the furnace within a few seconds to guarantee rapid heating and cooling. At the initial stage of the experiment, the specimens were removed and weighed after each cycle. At a later stage, the specimens were measured at intervals of more cycles.

The mass change data as a function of the number of cycles of Ti–48Al–2Cr from

Table 1. Checklist of the most relevant characteristics for performing cyclic oxidation tests; divided into three major sections

Specimen characteristics	
• Material manufacturing route	Description of the route of processing, e.g. modified rolling process with particular heat treatment procedures to obtain a specified and well-defined microstructure.
• Chemical composition	Chemical composition of the material in at.% or mass.%, whereas minor constituents can be listed in ppm.
• Microstructure	Grain size (including variation), grain orientation, and phase-type.
• Specimen production route	Description of the handling procedure to manufacture specimens from the as-received material.
• Surface finish	Description of the surface preparation route including the finishing step, e.g. grinding with 600 grade SiC-paper or polishing with 1μm diamond paste.
• Shape and dimensions	Specimen (includes irregular)geometry: Rod: diameter and length Disc: diameter and thickness Block: length, width, and thickness
• Surface cleaning procedure	Cleaning procedure of the surface in order to remove organic substances (degreasing procedure), by e.g. RBS soap, ethanol and hexane.
• Support	Method, is e.g. a small hole is necessary to suspend the sample with a Pt or quartz wire.
• Handling procedure	Description of the handling procedure in which mode the specimens are treated in order to measure the mass change, e.g. mechanical contact by using a pair of tweezers.

the two batches are shown in Fig. 1. In both cases the specimens revealed a slow transient oxidation during the first few hours of exposure followed by a more rapid oxidation after longer exposure times, indicative of a loss in protectiveness of the oxide scale. During the initial stage of oxidation no differences were found between the mode of mass change of specimens from the different batches. More details on the oxidation kinetics and mechanism, as well as the composition and structure of the oxide scale can be found elsewhere [5,6]. After longer exposure times, i.e. after about 175 cycles, a significant amount of spallation of the oxide scale occurred, in particular with the specimens derived from batch B. However, with specimens from

Table 1. *Checklist of the most relevant characteristics for performing cyclic oxidation tests; divided into three major sections*

Test characteristics	
• Test apparatus design	Detailed description of the apparatus design, or a reference to a paper in which the used test facility is described.
• High temperature	In °C or K.
• Low temperature	In °C or K.
• Total cycling time	Total number of cycles with cycle characteristics (…h at …°C, and …h at …°C).
• Heating rate	In °C or Ks^{-1} or min^{-1} or h^{-1}.
• Cooling rate	In °C or Ks^{-1} or min^{-1} or h^{-1}.

Specimen evaluation characteristics
• Thermogravimetry: – continuous – discontinuous • Thin layer activation • Acoustic emission • Post-exposure examination techniques (including visual inspection)

Table 2. *Chemical composition of the intermetallic alloy Ti–48Al–2Cr*

	Ti (at.%)	Al (at.%)	Cr (at.%)	O (mass ppm)	N (mass ppm)	C (mass ppm)	H (mass ppm)
Batch A	Bal.	48	2	750	220	55	<10
Batch B	Bal.	48	2	580	370	65	30

batch A, almost no mass gain or mass loss was measured, with increasing number of cycles indicating that the sum of the scale growth rate and the re-healing rate more or less equalled the spallation rate.

The differences in oxidation behaviour of the material from the two different batches cannot be explained by the microstructure [5,6]. A difference in alloy purity might be more obvious. Several alloying elements are known to alter the corrosion properties even if they are present in small quantities. So, small differences in composition can significantly affect the cyclic oxidation behaviour under certain specific conditions. Therefore, it is not unlikely that small differences in, for example, oxygen, nitrogen, or carbon concentration can influence the oxidation behaviour of Ti–48Al–2Cr beneficially or detrimentally.

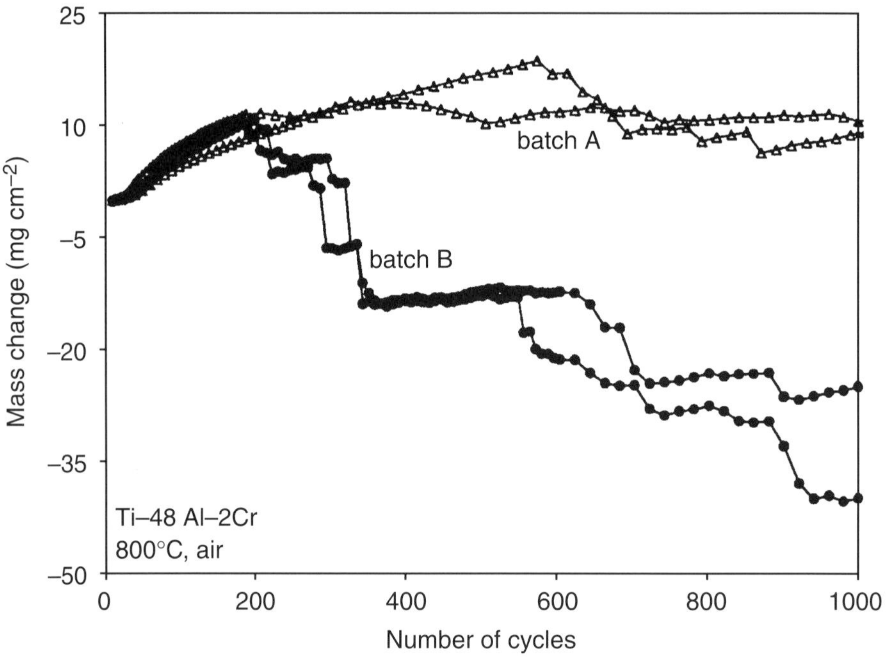

Fig. 1 Mass change vs number of cycles of near gamma Ti–48Al–2Cr in air at 800°C. Measurements were performed with material obtained from two different batches. (Each cycle corresponds to 1 h at 800°C and 12 min at ambient temperature.)

3.3. Microstructure

The influence of microstructure on the cyclic oxidation behaviour of Ti–48Al–2Cr (at.%) has already been discussed in detail [5,6]. The most important aspects of this topic will be summarised here. More information concerning the preparation of the specimens, the performance of the cyclic oxidation tests, (similar to those described in Section 3.2) and microstructure parameters can be found elsewhere [4–6]. To obtain specific microstructures, various heat treatments were performed, i.e. near gamma (1200°C/2 h), the duplex (1300°C/2 h), the nearly lamellar (1360°C/0.5 h) and the fully lamellar structure (1400°C/0.5 h).

From the cyclic oxidation experiments performed at 700 and 800°C it was observed that the growth rate of the oxide scales, as well as their composition, structure and morphology, were unaffected by the microstructure. The somewhat different mass changes observed on the fully lamellar specimen (see Fig. 2) after longer test periods at 700°C are possibly related to the large grain size of this material.

On the other hand in a study by Gil *et al.* [7,8] it was found that the oxidation resistance of TiAl-based intermetallics was strongly affected by the amount and distribution of the various phases that are present in the alloy. However, these materials showed coarse-grain microstructures, which differed significantly from the materials used by Haanappel *et al.* [5,6], which were fine grained and more homogeneous and expected to be more representative for technical applications.

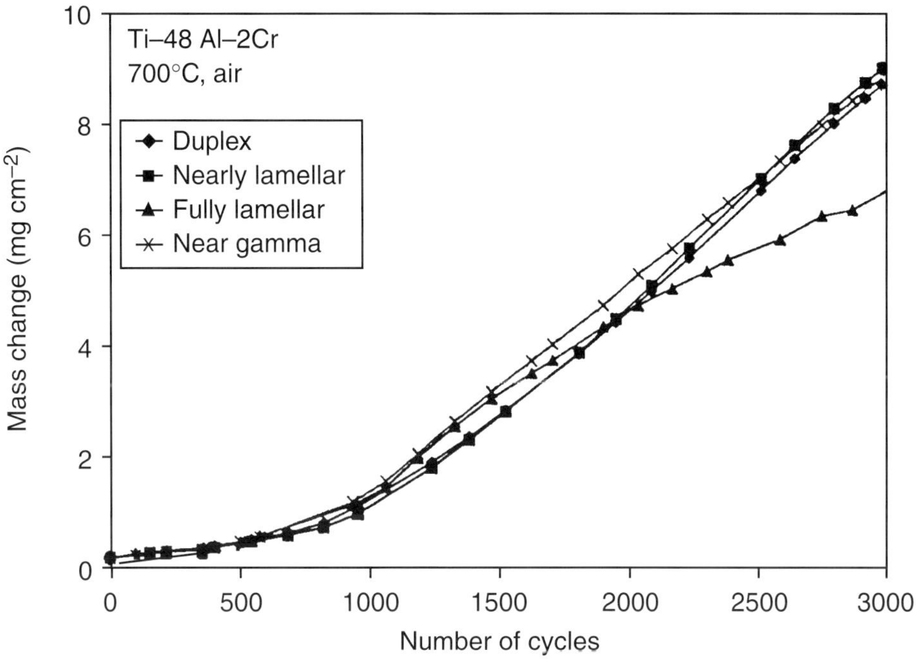

Fig. 2 *Mass change vs number of 1-h cycles of near gamma, duplex, nearly lamellar, and fully lamellar Ti–48Al–2Cr in air at 700°C [5].*

From these conclusions it is clear that under certain specific conditions the microstructure of the tested material may also influence the oxidation behaviour. As a consequence, published data, which show some scatter in cyclic oxidation data can at least partially be explained by this effect and probably might cause misleading conclusions.

3.4. Sample Surface Finishing Procedures

The main objective of surface finishing is to eliminate any differences in surface conditions such as residual cold work, surface contamination and oxidation, and/or near-surface depletion, which can be introduced during specimen manufacturing where elevated temperatures are sometimes needed. However, no standardisation has been developed yet, and as a consequence different surface treatments are used, ranging from electrochemical polishing to grinding and blasting. This can partly be explained by the fact that the use of different procedures is sometimes closely related to the different forms of industrial applications.

Due to the diversification of surface polishing techniques it is not always possible to compare results, even when the thermal cycling process conditions are chosen to be identical [5,6,9–12]. Choudhury *et al.* [12] had already observed that the oxidation behaviour of TiAl-based intermetallics, i.e. kinetics, scale morphology and constituents, was strongly dependent on the sample surface preparation procedure prior to oxidation. A more recent study on this subject was made by Dettenwanger *et al*. [9] in studying the influence of surface preparation on the oxidation behaviour of γ-TiAl. It was shown (Fig. 3) that ground specimens (600 grit SiC-paper), exposed in pure oxygen, oxidised more slowly than specimens polished with 1 μm diamond paste. These results correspond well with those found by Choudhury *et al*. [12]. For exposure in air, however, no differences were found in the oxidation behaviour.

4. Test Characteristics

4.1. Introduction

High temperature alloys are frequently used in power plants, gasification systems, the petrochemical industry, combustion processes, and in automotive and aerospace applications [13]. Depending on the application, materials are subjected to contaminated atmospheres and thermal cycling. For example, aircraft turbines represent thermal cycles under oxidising conditions with isothermal hold-times of a few hours, whereas ground power turbines need a hold-time of at least one thousand hours [14]. In order to investigate the behaviour of these materials under practical conditions, several simulated test methods have been developed, but no general standardisation of these test methods has so far been made [15]. It is already known that many organisations are performing cyclic oxidation experiments with different test parameters since the laboratory conditions that they use simulate a variety of

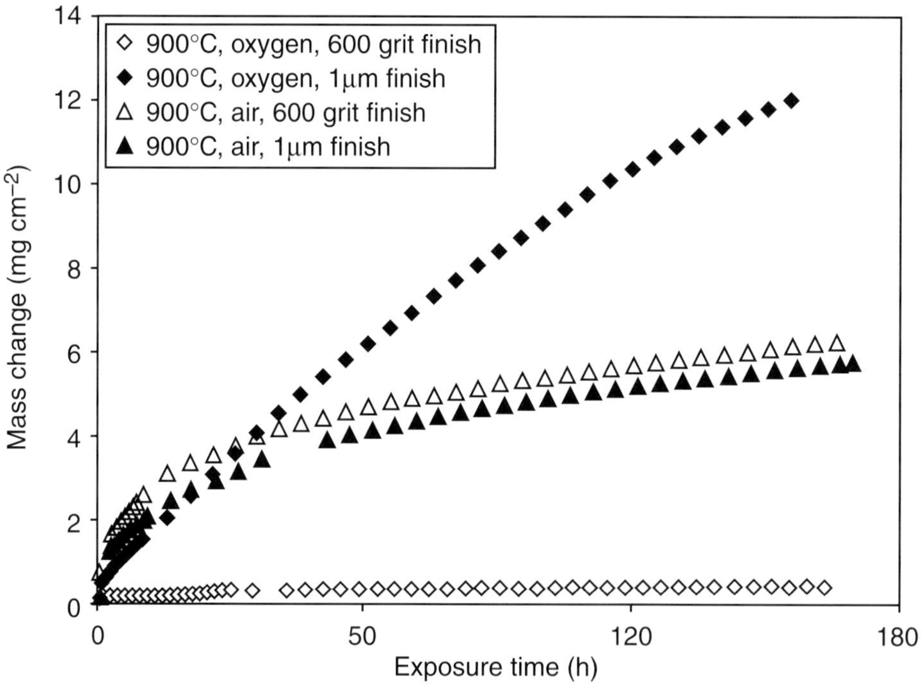

Fig. 3 *Mass change vs time for isothermal oxidation in air and oxygen at 900°C showing the influence of surface finish and atmosphere on the oxidation of γ-TiAl [9].*

industrially relevant conditions [16]. As a result of this, it is impossible to demand from all organisations that they perform cyclic oxidation tests under identical circumstances. The main process parameters related to cyclic oxidation tests are; **design of equipment, maximum test temperature, minimum test temperature, hold-time at the highest temperature, hold-time at the lowest temperature, total cycling time, heating rate, and cooling rate**.

The current forms of cyclic oxidation testing that are in use are not always sufficiently able to provide data relevant to what is occurring in industrial applications or to provide a fundamental basis for the development of predictive models and new materials. Two examples, shown in Table 3, illustrate the differences in cycle time and number of cycles between different organisations. The importance of test characteristics for the high temperature cyclic oxidation behaviour of metals and alloys will be illustrated by two examples, i.e. total cycling time, and effect of hold-time.

4.2. Total Cycling Time

The total cycling time of specimens exposed at high temperatures is generally one of the most important test parameters. During laboratory experiments this particular test parameter of experimental performance is generally well defined. The main objective in performing laboratory experiments is to expose materials during a relatively short period of exposure to atmospheres simulating high temperature processes, e.g. combustion processes in gas turbines and incinerators, coal gasification processes, and aerospace applications. The kinetics and oxidation mechanism of materials at high temperatures can be investigated and often extrapolation models will be used to predict the oxidation behaviour for longer exposure times to be used during the laboratory test. However, choosing the total exposure time incorrectly can lead in the worse case to wrong or misleading conclusions.

In the described study the importance of the total cycling time is illustrated, and it will be shown that the outcome of the results is significantly affected by changing the total duration of the experiment. The materials under investigation were near gamma Ti–48Al–2Cr (at.%) and pure Ti. Coupons with dimensions of $10 \times 10 \times 1$ mm^3 were machined from the materials. The cyclic oxidation experiments were performed

***Table** 3. Frequency analysis of the number of responding organisations using specific 'cycle times' and 'number of cycles'. (Total number of organisations: 26 involved in cyclic high temperature oxidation research.) [16]*

Cycle time (h)	No. of organisations	Number of cycles	No. of organisations
0.25–0.9	2	<10	2
1.0–1.8	12	10–90	6
2.0–4.5	6	100–450	4
5.0–9.0	3	500–900	4
10–14	2	1000–10000	7
15–30	4		
40–120	2		

in static laboratory air. Each cycle consisted of a period of one hour at 800°C and 12 min at ambient temperature (AT), the latter being sufficient to cool the specimens below 50°C. The specimens were inserted into and removed from the hot zone of the furnace within a few seconds to guarantee rapid heating and cooling.

Figure 4(a–c) shows the mass change data of Ti–48Al–2Cr vs the number of cycles. From Fig. 4(a) it can be concluded that during exposures up to 40 cycles the mass of the material increased exponentially. This was explained by a loss of protectivity of the scale formed during the initial stage (first minutes of exposure), associated with the dissolution of a protective α-Al_2O_3 barrier [5,6]. Extrapolation of these data might result in the conclusion that severe corrosion will still continue with time. However, if experiments were performed up to 160 cycles, the kinetics, particularly during the final stage, could be described by a parabolic oxide growth, mainly due to the formation of TiO_2. It is obvious that a semi-protective oxide scale is formed resulting in a reduced mass gain rate with time which will probably continue to be maintained after longer exposure times. However, experiments have shown that after longer exposure times, in particular after 175 one-hour cycles, a different pattern of the mass gain curve was obtained, which could no longer be explained by a simple oxidation model. The detailed shape of the curve depends on the rate of mass loss by spallation and on the rate of mass gain by (re-) growth of the oxide scale. This phenomenon continued even up to 3000 cycles [6]. The mass loss as a function of time (mass loss rate) during this period is an indirect measure for the degradation rate of the material. It is especially during this period that a more reliable prediction of the oxidation behaviour of this material for long-term exposures under structural applications can be made.

From the above results it can be recommended that experiments should be conducted for relatively long periods (depending on the practical application) in order to avoid misleading or wrong conclusions. However, care has also to be taken in these cases as will be illustrated by the next example. Figure 4(d–f) shows the oxidation behaviour of titanium in air at 800°C as a function of exposure time. Up to an exposure period of 10 one-hour cycles all the materials showed more or less parabolic oxidation behaviour. This indicates the formation of a rather protective oxide scale, which probably can protect the underlying substrate against severe degradation even after longer exposure times. Experiments performed up to 100 cycles showed that the mass gain initially increased exponentially, but after 40 one-hour cycles the kinetics were more or less linear. Increasing the total exposure time up to 1500 cycles suggested that the oxide scale was rather protective. No further mass gain was determined. However, cross sectional analysis showed that the constant mass change was due to the fact that all the substrate material was oxidised. No oxygen pick-up could take place anymore to provide a positive mass change.

From this point of view it is clear that conclusions based on the cyclic oxidation behaviour of materials are strongly related to the experimental performance. Different exposure times show different oxidation kinetics, which can affect the conclusion(s) based on extrapolation of the obtained data. In addition, but not discussed here, is the point that shorter tests at higher temperature can also be misleading. Therefore, it is recommended that care be taken in a well-defined choice of the total cycling time.

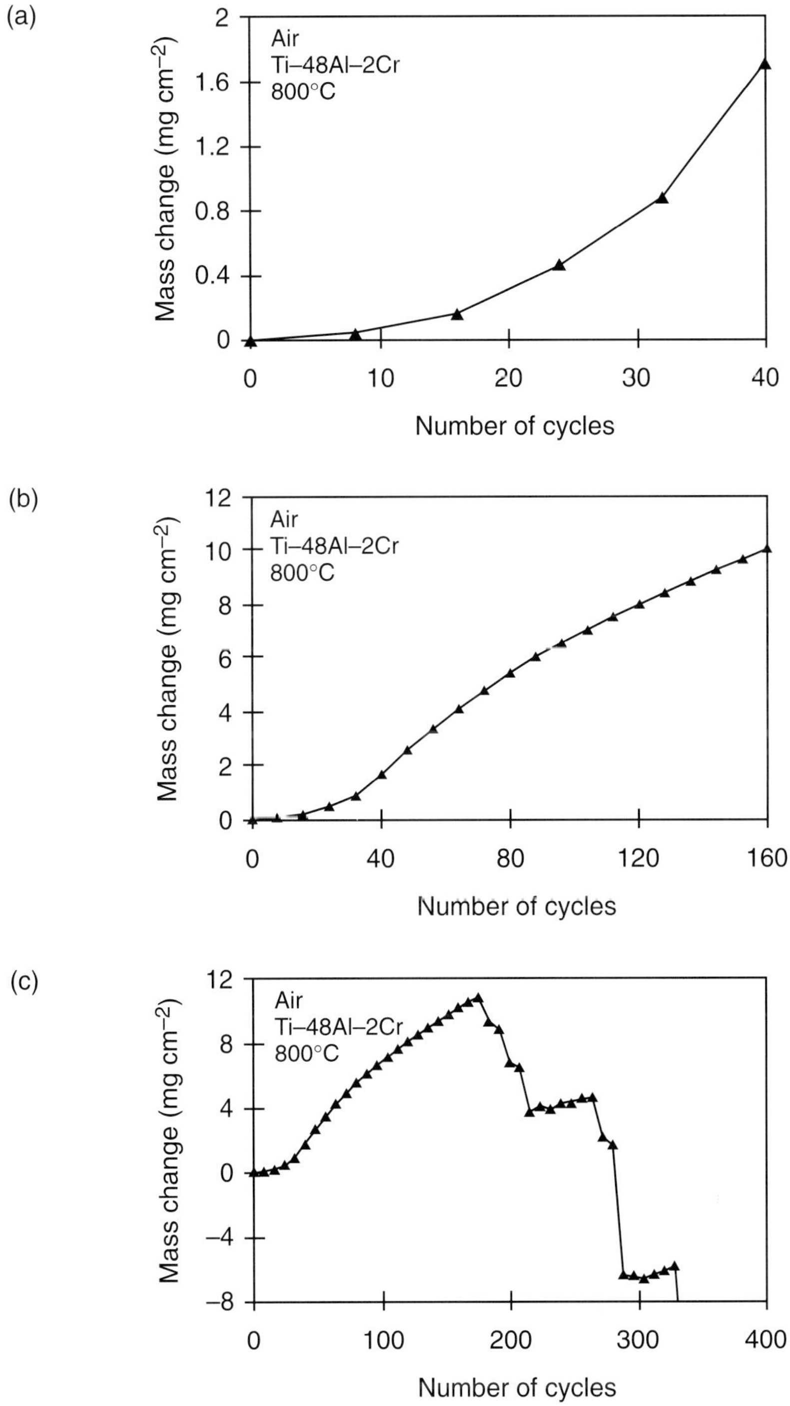

Fig. 4 *Mass change vs number of 1-h cycles (1 cycle corresponds with 1 h at 800°C in air and 12 min at ambient temperature) of (a–c) Ti–48Al–2Cr (at.%).*

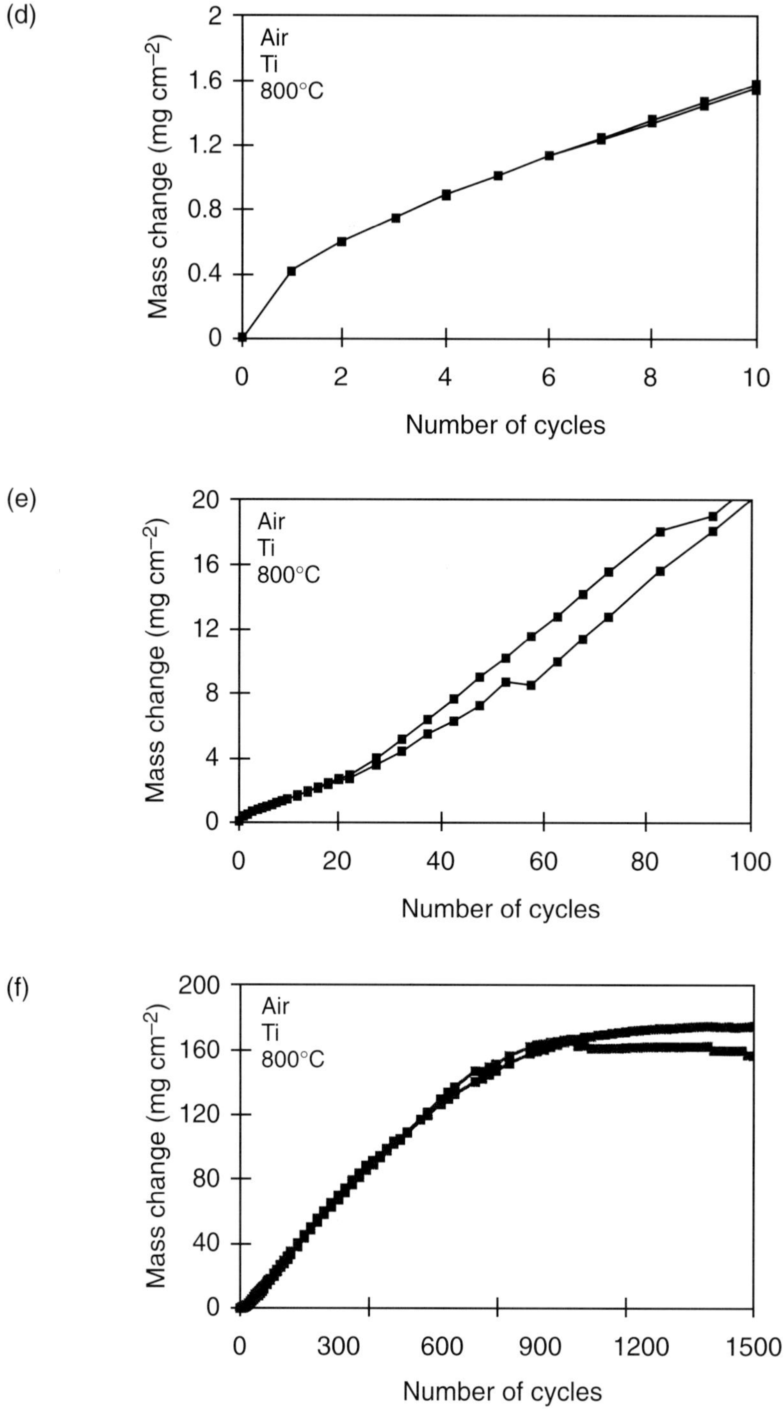

Fig. 4 *(Continued) Mass change vs number of 1-h cycles (1 cycle corresponds with 1 h at 800°C in air and 12 min at ambient temperature) of (d–f) pure Ti.*

4.3. Effect of Hold-time

The isothermal hold-time at the maximum and minimum temperature can also affect the kinetics of materials during cyclic oxidation tests. The variety of cycling periods for testing the cyclic oxidation resistance is generally based on the attempt to simulate as closely as possible the industrial situation. The hold-times at the highest and the lowest test temperature should be chosen after careful consideration bearing in mind the demands and conditions related to industrial components.

Two intermetallic alloys, i.e. Ti–48Al–2Cr–2Nb and Ti–48Al–2Mn–2Nb were exposed to two different cyclic oxidation tests; (a) 1 h at 800°C and 12 min at room temperature, and (b) 20 h at 800°C and 4 h at ambient (AT) temperature. Under these specific circumstances, the cyclic oxidation behaviour of these materials was not significantly affected by differences in hold-time (see Fig. 5), with equal exposure times at the test temperature. However, it is not always possible to predict whether the cyclic oxidation behaviour of the test specimens will be susceptible to changes in the hold-time at the highest and/or lowest temperature. In any case, these periods can have a significant influence on the course of oxidation or corrosion. In particular, chromium does have a relatively high sensitivity for oxide spallation and this is a function of the specific experimental parameters related to the cyclic oxidation testing.

Figure 6 shows the spallation behaviour as measured using Thin Layer Activation (TLA) of chromium as a function of the cycle period [17]. From these results it is clear that, with similar total exposure times, cycles of 20 h/800°C and 4 h/AT. induce

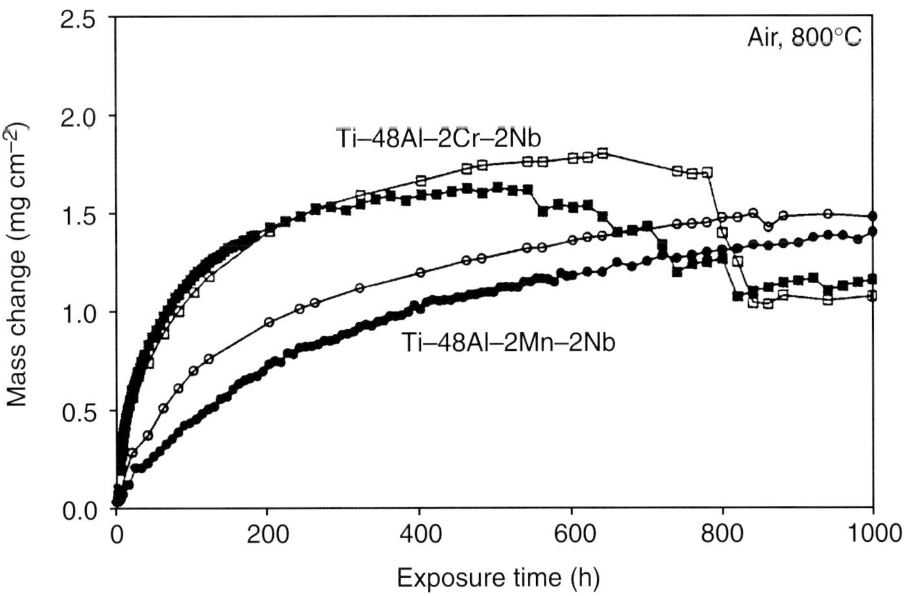

Fig. 5 *Mass change vs number of cycles of Ti–48Al–2Cr–2Nb and Ti–48Al–2Mn–2Nb in air at 800°C. Measurements were performed with two different cycle periods, 20 h at 800°C and 4 h at ambient temperature (open symbols) and 1 h at 800°C and 12 min at ambient temperature (closed symbols).*

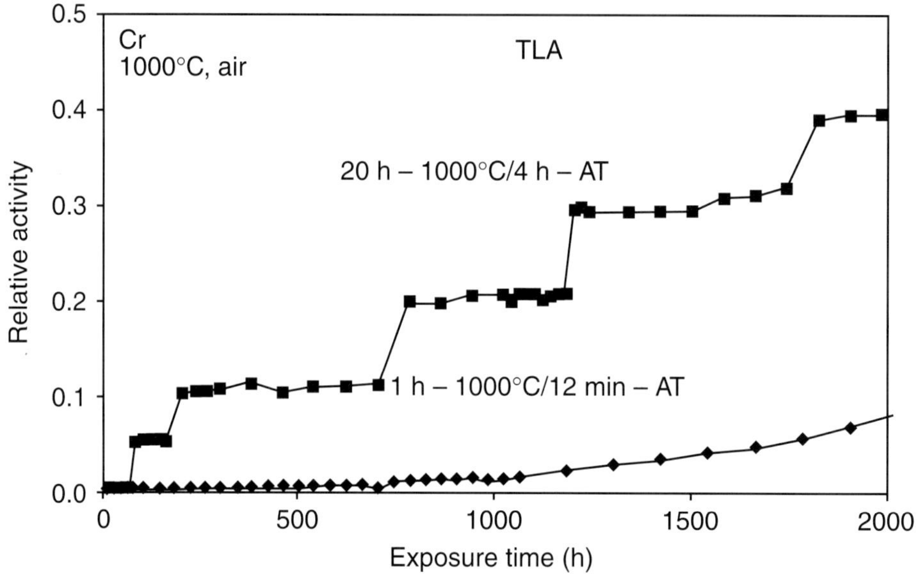

Fig. 6 *Spallation behaviour as measured using Thin Layer Activation of chromium due to thermal cycling from 1000°C to ambient temperature. Two different cycle periods were used, 20 h at maximum and 4 h at ambient temperature (AT), and 1 h at maximum and 12 min at ambient temperature.*

a significantly higher spallation rate as compared with cycles of 1 h/800°C and 12 min/AT.

The effect of isothermal hold-time during thermal cycling tests was also discussed by Schütze *et al.* [15]. It was concluded that these periods can have a significant effect on the course of corrosion, and can influence the overall kinetics of the material during cyclic oxidation tests. In order to compare data from several potential alloys, it is recommended that the isothermal periods at the lowest and highest temperature should be chosen as carefully considered values.

5. Conclusions

A checklist has been presented with the most relevant characteristics for performing cyclic oxidation tests, i.e. specimen characteristics, test characteristics, and specimen evaluation methods. This checklist will be helpful in preparing studies and in preparing documents, such as reports and papers. Moreover, a full documentation of all the important experimental parameters will make a correct interpretation and comparison of test results from different organisations possible. The checklist can be considered as a precursor for the development of standards and more extended and definitive guidelines on how to conduct cyclic oxidation tests and finally for the establishment of a Code of Practice.

References

1. H. J. Grabke *et al.*, *Werks. Korros.*, 1993, **44**, 345.
2. EFC Working Party Report on Guidelines for Methods of Testing and Research in High Temperature Corrosion (H. J. Grabke and D. B. Meadowcroft, eds). Publication No.14 in European Federation of Corrosion Series. Published by The Institute of Materials, London, 1995.
3. V. A. C. Haanappel and M. F. Stroosnijder, *Corros. Sci.*, 1997, **39**, (6), 1083.
4. H. Clemens *et al.*, in *Gamma Titanium Aluminides.* (Y. W. Kim, R. Wagner and M. Yamaguchi, eds). Published by The Minerals, Metals & Materials Society (TMS), Warrendale, Pa, 1995, pp. 717–726.
5. V. A. C. Haanappel, W. Glatz, H. Clemens and M. F. Stroosnijder, *Mater. High Temp.*, 1997, **14**, (1), 19.
6. V. A. C. Haanappel *et al.*, *Oxid. Met.*, 1997, **48**, (3/4), 263.
7. A. Gil, E. Wallura, H. Grübmeier and W. J. Quadakkers, *J. Mater. Sci.*, 1993, **28**, 5869.
8. A. Gil, H. Hoven. E. Wallura and W. J. Quadakkers, *Corros. Sci.*, 1993, **34**, 615.
9. F. Dettenwanger, E. Schumann, M. Rühle, J. Rakowski and G. H. Meier, *Oxid. Met.*, 1998, **50**, (3/4), 269.
10. M. J. Bennett, R. Perkins, J. B. Price and F. Starr, in *Materials for Advanced Power Engineering*, (D. Coutsouradis *et al.*, eds). 1994, 1553.
11. M. Walter, M. Schütze and A. Rahmel, *Oxid. Met.*, 1993, **39**, 389.
12. N. S. Choudhury, H. C. Graham and J. W. Hinze, in *Proc. Symp. of Properties of High Temperature Alloys*, 1976, (Z. A. Foroulis and F. S. Pettits, eds). Published by The Electrochemical Society, Princeton, N. J. pp. 668.
13. P. Kofstad, *High Temperature Corrosion*, Published by Elsevier Applied Science, London, 1988.
14. D. L. Deadmore and C. E. Lowell, *Oxid. Met.*, 1977, **11**, (2), 91.
15. M. Schütze, V. Guttmann and M. F. Stroosnijder, in *Guidelines for Methods of Testing and Research in High Temperature Corrosion*. (H. J. Grabke and D. B. Meadowcroft, eds). Publication No.14 in European Federation of Corrosion Series. Published by The Institute of Materials, London, 1995, pp.104.
16. F. Starr and M. F. Stroosnijder, Internal Report (1997).
17. M. F. Stroosnijder, I. Murris, V. A. C. Haanappel and Y. P. Jacob, 'The Use of Thin Layer Activation in Cyclic Oxidation Testing', to be published.

15
Corrosion of Heat Exchanger Steels in Cycled Gas Atmospheres at 600°C

M. MAIER and J. F. NORTON
Joint Research Centre, Institute for Advanced Materials, 1755 ZG Petten, The Netherlands

ABSTRACT
In heat exchanger sections of present generation power plants environmental conditions often vary between oxidising atmospheres and highly reducing conditions, i.e. with very low oxygen levels. Both strongly depend upon process conditions and fuel quality. Additionally, these atmospheres can contain sulfur in the form of SO_2 or H_2S as well as chlorine in the form of HCl. This may result in severe corrosion of the materials. The data presented in this paper are derived from an investigation conducted at a temperature of 600°C in cycled nitrogen based atmospheres containing 12% CO_2, 7% H_2O with either 1%O_2 + 0.15%SO_2 or 1% CO + 0.15% H_2S for the oxidising and the reducing conditions, respectively. The atmospheres were cycled at 25 h or 100 h intervals, starting with either an oxidising or reducing step. Additionally, experiments under simple oxidising or reducing conditions were performed to produce base-line data. The results give information concerning the corrosion kinetics of the stainless steels NF616, X20 CrMoV 12 1, Alloy 800H and AISI 304 under these conditions and present microstructural analysis data from XRD, SEM and OM investigations. Due to the large variations in Cr and Ni contents of the alloys significant differences are found for the corrosion reactions under oxidising and reducing conditions and consequently also for the cycled conditions. The investigations focus on establishing the type and quality of the scales formed during the various stages of the reaction and their influence in providing further protection to the alloy surface, crucial for the general corrosion resistance of the alloy under the given conditions.

1. Introduction

Heat exchanger tubes in many present-day power plants are frequently exposed to atmospheres which cycle between oxidising and reducing conditions at elevated temperatures. As S- and Cl-containing species are often present, severe corrosion of the materials can occur. The present study focuses on establishing the type and quality of the scale formed on selected 9–20% Cr steels under atmospheric cycling and its influence in providing continuing protection to the alloy under the given conditions.

The tests conducted in the Joint Research Centre, Institute for Advanced Materials (JRC-IAM) Petten laboratories simulate these extreme variations of conditions by cycling between two different gas atmospheres, at a given temperature, slightly increased gas pressure and for two cycling dwell times.

2. Experimental Technique

Four candidate heat exchanger alloys were selected for this study, i.e. the ferritic 9% and 12%Cr steels NF616 and X20 CrMoV 12 1 for comparison with the austenitic Alloy 800H and AISI 304 materials. The compositions of the alloys are given in Table 1.

Rectangular coupons, nominally 10 mm × 8 mm × 6 mm or 20 mm × 10 mm × 1 mm were machined from each alloy. The major flat parallel surfaces were abraded with 320 grit SiC paper on a stationary polishing table under standardised conditions to remove machining marks. All specimens were individually identified, measured and degreased in acetone in an ultrasonic bath before weighing to constant mass.

Duplicate specimens of each alloy were placed in pure Al_2O_3 crucibles so that the total mass gain (including spalled products) could be measured at each inspection. The specimens were inserted in the reaction chamber of a vertical top-loading autoclave and the test atmosphere established by evacuating the chamber and back filling with the N_2-based test gas several times. The autoclave was heated to 600°C and the gas flow rate adjusted to approximately. 25 Lh^{-1} at 1.5 bar_{abs} (1.5×10^5Pa).

The compositions of the gas mixtures used for the oxidising (OX) and the reducing (RED) conditions are listed in Table 2.

Assuming equilibration, the gas mixtures have pO_2 values of 1×10^{-2} bar (10^3Pa) and 6×10^{-22} bar (6×10^{-17}Pa) for the OX and RED gases, respectively; pS_2 values are 1×10^{-36} bar (10^{-31}Pa) for OX and 5×10^{-7} bar (5×10^{-2}Pa) for RED; carbon activity, a_c values are $< 4 \times 10^{-38}$ and 5×10^{-4}, respectively.

The gas atmosphere was changed after each cycle (25 or 100 h), during which the specimens were removed from the autoclave and the mass change of each specimen measured. After four cycles one of the duplicate specimens was discontinued and examined by surface and cross-sectional analysis techniques (XRD, SEM, OM), the other was continued for a total of 10 cycles and then examined.

Four test series were conducted for each of the alloys, i.e. (a) atmosphere cycling, starting with OX, (b) atmosphere cycling, starting with RED; (c) reference (only OX); and (d) reference (only RED).

Table 1. *Composition of the alloys (Fe = balance)*

mass%	C	Mn	Si	Ni	Cr	Nb	Mo	W	V	Ti	Al
NF616	0.12	0.46	0.03	0.07	8.85	0.07	0.42	1.85	0.2	—	—
X20	0.19	0.50	0.32	0.66	11.2	—	0.86	—	0.25	—	—
800H	0.07	0.70	0.38	30.6	20.4	—	—	—	—	0.33	0.28
304	0.03	0.83	0.55	8.66	17.6	—	—	—	—	0.01	0.02

Table 2. *Composition of the gas mixtures*

vol.%	N_2	CO_2	H_2O	O_2	CO	SO_2	H_2S
OX	bal.	12	7	1	—	0.15	—
RED	bal.	12	7	—	1	—	0.15

3. Results

3.1. Mass Gain Data

The mass change data collected at each interruption to the tests were plotted as specific mass change vs time. Figures 1 and 2 show these for the four test series for the 25-h and 100-h dwell periods, respectively.

Significant differences in corrosion behaviour are found between exposures in pure oxidising and reducing conditions and consequently also under cyclic conditions. Mass change curves for the cyclic experiments show that there is a mass reduction during the oxidising cycle, the extent of which is dependent upon alloy composition. This is most probably due to oxidation of the previously formed sulfides with simultaneous evaporation of sulfur dioxide.

Furthermore, it is obvious that during exposure to the simple oxidising gas mass gains (and thus corrosion rates) are lower by a factor of at least 2 compared with the reducing gas mixture. Differences are also seen in the cyclic experiments, where the tests that were started in an oxidising environment show similarl lower mass gains than the cyclic tests which had been started in a reducing gas atmosphere, with the exception of the AISI 304 alloy.

Overall mass gains for the cyclic experiments which had commenced in an oxidising regime were found to be similar to the mass gains during pure oxidation, especially for the NF616 material. Similarly, cycling starting with a reducing atmosphere, showed final mass gains corresponding to those following exposure in purely reducing conditions, for all alloys.

Obviously, the type of atmosphere used during the first step of the cyclic treatment strongly influences the general trend of the reaction kinetics during the following exposures. The scale formed initially appeared to remain largely intact and thus diffusion and reaction properties of this scale determine the growth of subsequent layers.

3.2. Surface SEM

3.2.1. 25 hours dwell time

The surfaces of specimens that had been exposed during tests starting in an oxidising gas and ending in the reducing mixture revealed an outer scale consisting of large grains which EDX analysis indicated was mostly an iron-rich sulfide. This morphology was most pronounced in the case of the NF616 material, with sulfide grain sizes varying between the X20, 304 and 616 alloys (Fig. 3). Only very small amounts of iron- (and, in this case, nickel)-containing sulfide were found covering the surfaces of Alloy 800H. These had a considerably smaller grain size than that observed with the lower-alloyed materials.

Cyclic experiments which had started in a reducing gas mixture with the specimens being exposed to the oxidising gas during the last cycle, exhibited a morphology consisting of fine-grained oxide crystals, which appeared very loosely packed in the case of Alloy 800H. On the NF616 material, the fine whisker crystals observed obviously covered larger sulfide grains grown during the previous cycle. Coverage,

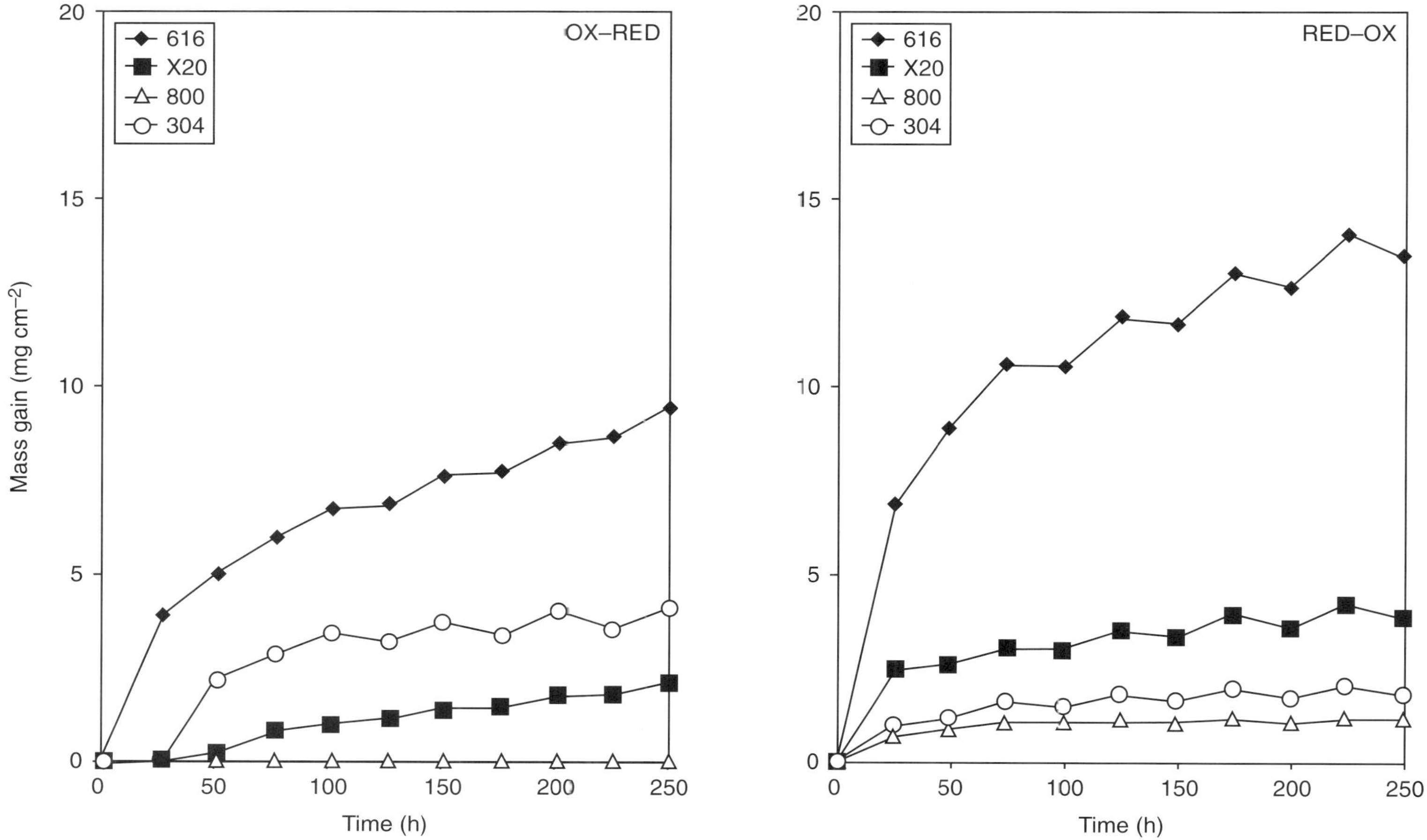

Fig. 1 Specific mass gain vs time for the cyclic experiments with 25-h dwell time.

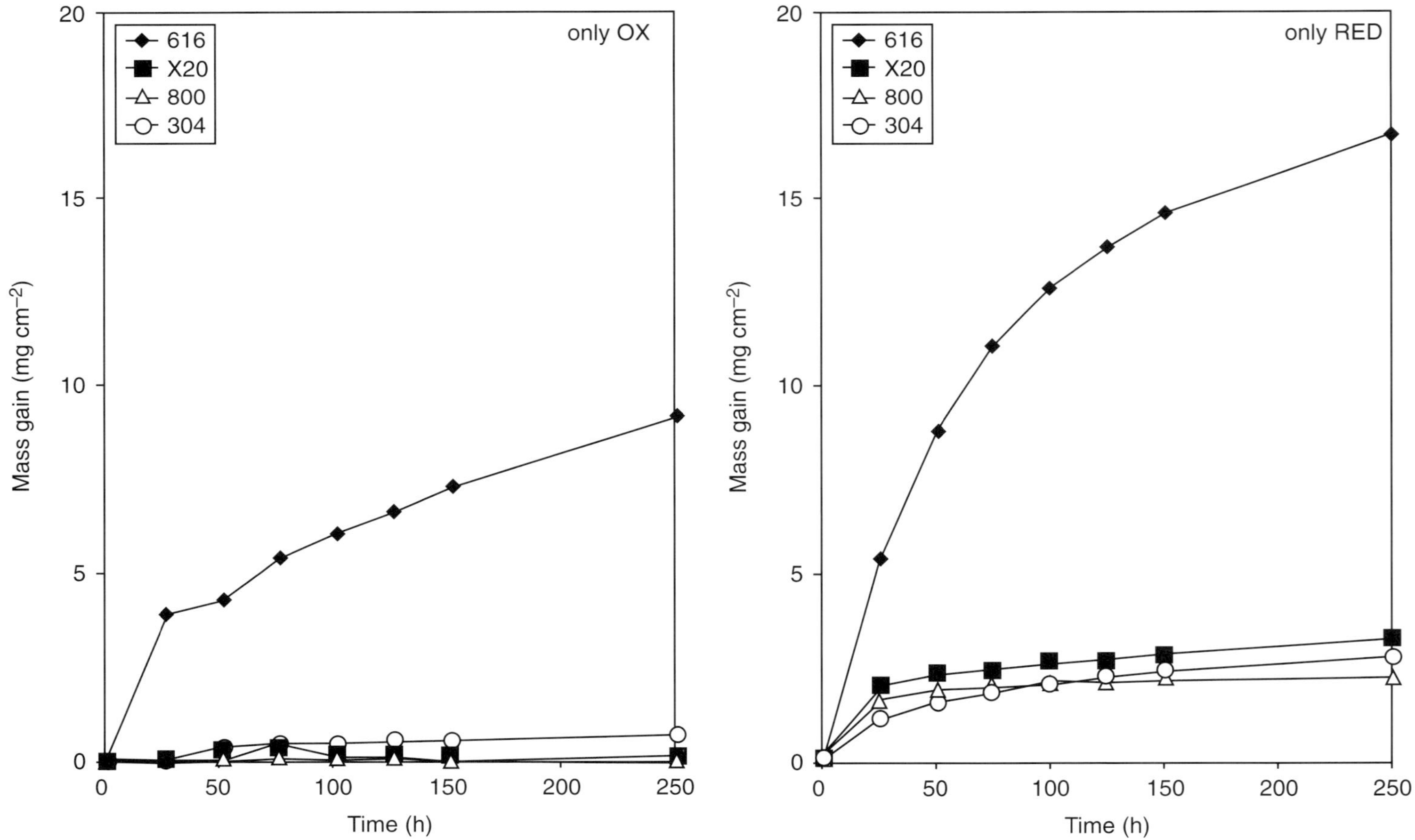

Fig. 1 *(continued) Specific mass gain vs time for the cyclic experiments, with 25-h dwell time.*

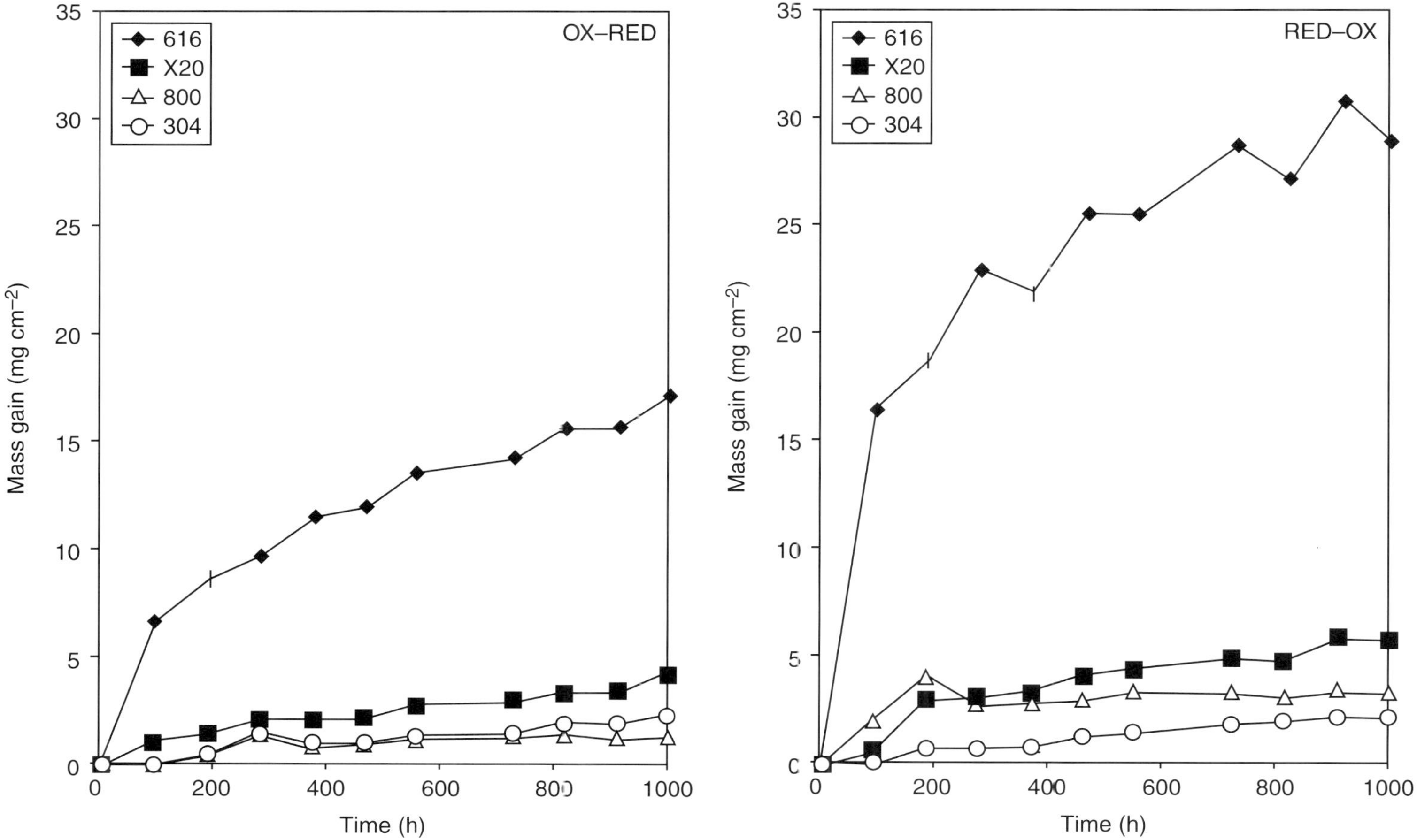

Fig. 2 Specific mass gain vs time for the cyclic experiments, with 100 h dwell time.

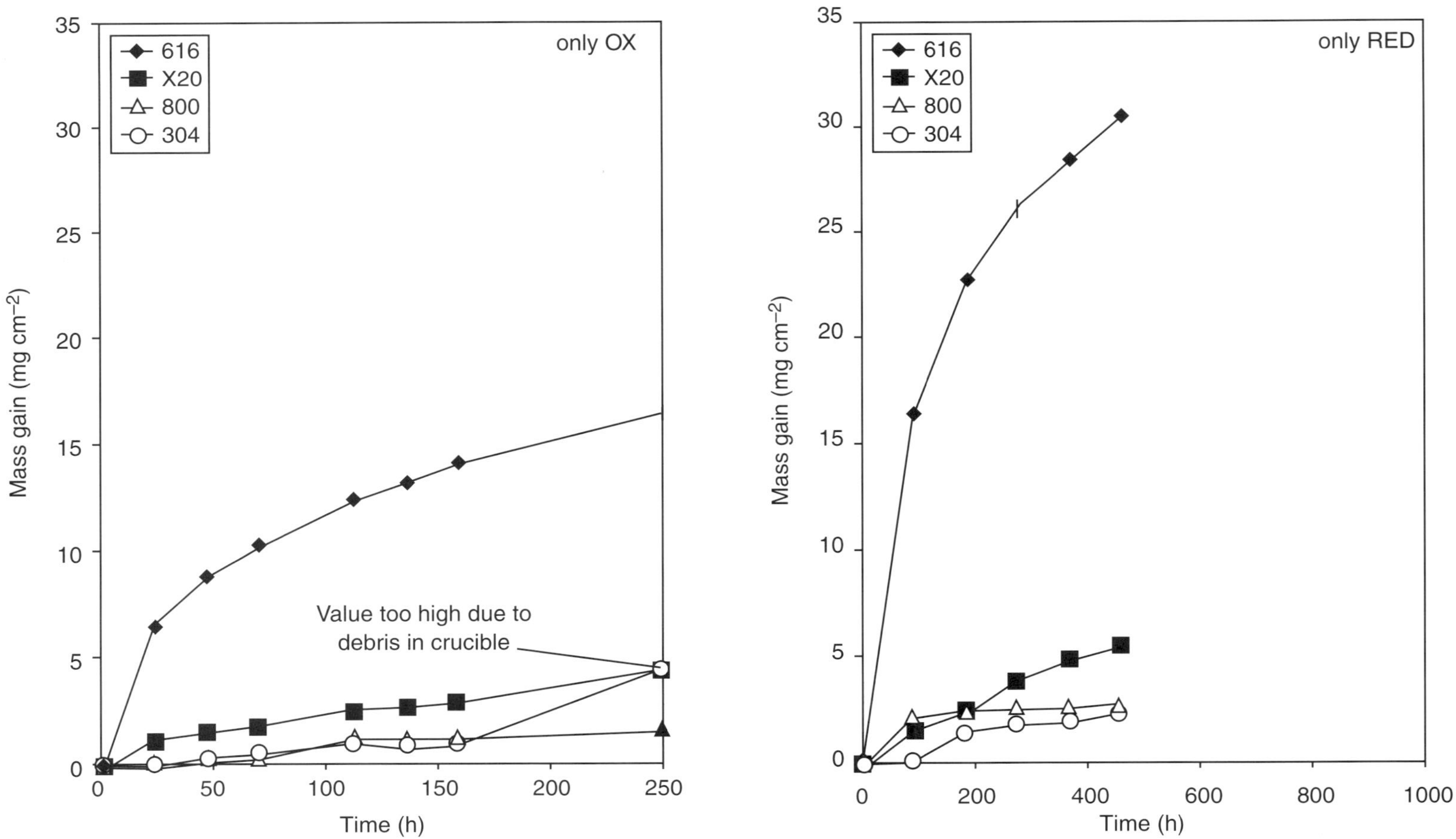

Fig. 2 *(Continued) Specific mass gain vs time for the cyclic experiments, with 100-h dwell time.*

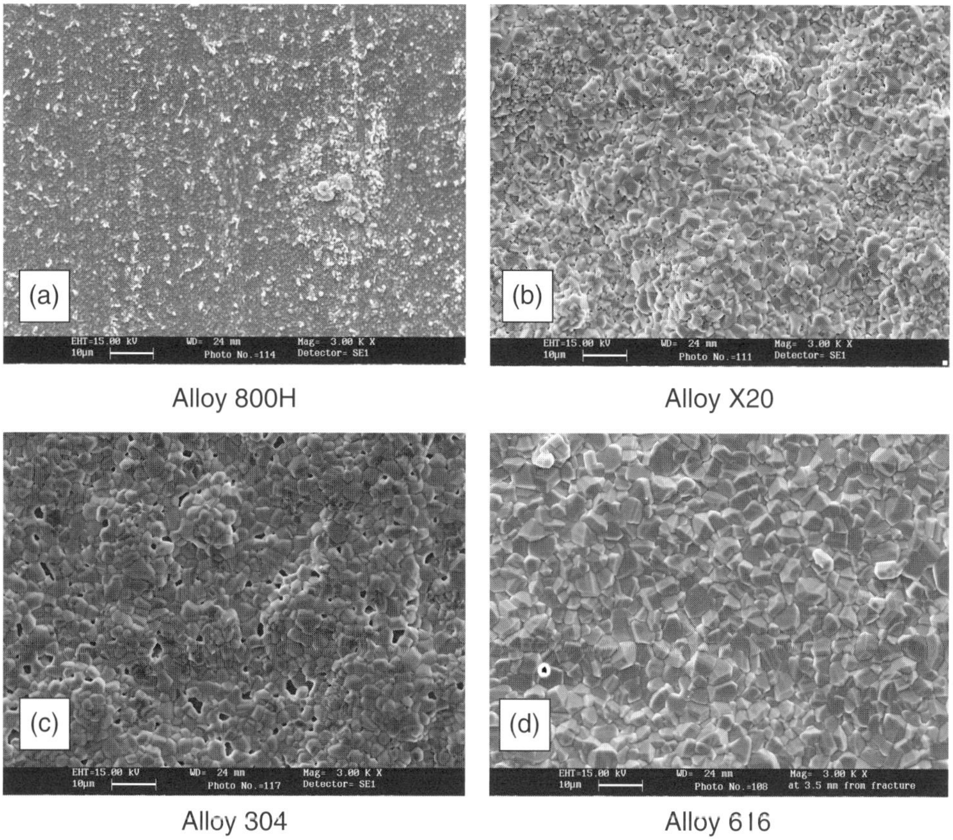

Fig. 3 *Surface SEM micrographs after 10 cycles of 25 h in OX–RED conditions.*

particularly of Alloy 800H and AISI 304 was not uniform, the reason being the non-uniform growth of the sulfide phases and, in some cases, severe spallation of the scales (Fig. 4).

Reference specimens exposed only to the oxidising gas exhibited whisker type oxide layers the roughness of which increased with decreasing chromium content. The scale on the 304 and 616 materials consisted of relatively large oxide whiskers and platelets (Fig. 5).

Surface scales formed during exposure to the reducing gas atmosphere were found to be almost pure sulfide, having large average grain sizes. Although these scales appeared relatively compact, cracking and heavy spallation was observed (Fig. 6).

In the case of Alloy 800H material, the grain structure was significantly finer, EDX also detecting traces of chromium and nickel oxides in the scale.

3.2.2. Comparison of cyclic regimes and dwell times

A general comparison of the scales formed during the four types of tests is presented in Fig. 7, taking the NF616 materials as an example.

A thick, uniform sulfide scale consisting of large grains, was observed on specimens that had been exposed to the reducing gas mixture immediately prior to examination

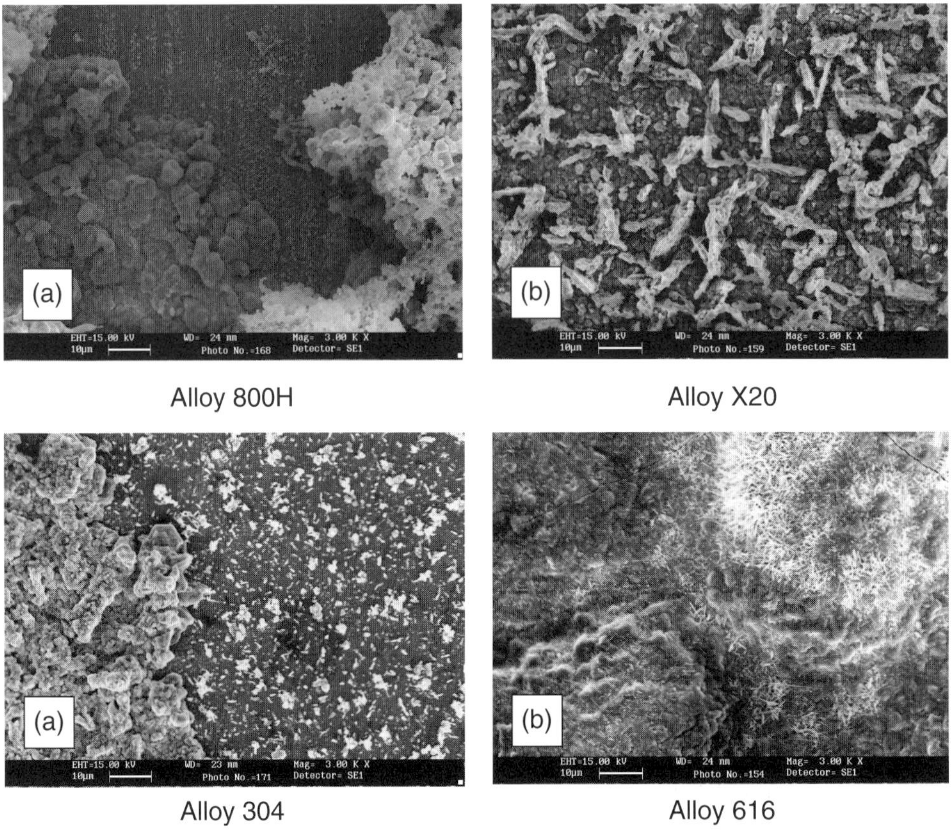

Fig. 4 *Surface SEM micrographs after 10 cycles of 25 h in RED-OX conditions.*

in the microscope. This was the case for both reference coupons exposed to the simple reducing gas, (Fig. 7d), as well as for specimens exposed under cyclic conditions (Fig. 7a).

In contrast, a typical oxide scale showed a much more porous structure of fine whiskers or platelet-shaped small crystals, often incompletely covering the surface. In the case of the specimens exposed to the simple oxidising gas, this type of oxide uniformly covered the surfaces, (Fig. 7c), whereas for specimens exposed to the cycled atmosphere, the scale still clearly showed signs of the sulfide scale protruding from beneath the oxide which had subsequently grown above it (Fig. 7b).

A similar comparison is made in Fig. 8 for the behaviour of the 4 alloys exposed under the OX–RED conditions for 10 cycles of 100 h each and for a companion series under RED–OX conditions in Fig. 9. All alloys show a clear distinction between the morphologies of the scales formed when the last exposure was in a reducing atmosphere, (Fig. 8(a–d)) or an oxidising gas, (Fig. 9(a–d)).

Comparison of the morphologies of these scales formed on the four materials during environmental cycling conditions with those observed on specimens exposed in simple oxidising or reducing atmospheres is made in Figs 10 and 11.

Energy dispersive X-ray (EDX) spectra obtained on samples of the NF616 alloy after exposure to the purely oxidising and reducing atmospheres are shown in Fig. 12.

The principal peaks are for iron oxide and iron sulfide, respectively. In the case of the higher alloy materials, Alloy 800H and AISI 304, peaks for nickel were also evident.

Similar spectra were observed on environmentally cycled specimens. After cycling, the spectra contained a combination of the two examples shown above in Fig. 12, indicating that the scale was a mixture of sulfides and oxides. Nevertheless, the major phase (oxide or sulfide) was clearly dependent upon the type of atmosphere used during the last cycle of the experiment.

3.3. Cross-Sectional SEM

A duplex scale had formed on all alloys exposed to the OX-RED cyclic regime (Fig. 13(a–d)), with the outer layer rich in iron and sulfur whilst the inner scale contained more chromium and oxygen. The outer scale appeared to be very defective, containing numerous pores and voids, together with cracks and spalled areas (on the lower chromium materials). The inner scale was more dense and in most cases well adherent to the alloy substrate.

The type of scale formed appeared to be independent of whether the initial cycle had been under OX or RED conditions, (compare Figs 13 and 14). In the case of the RED-OX type of experiment, one would expect to find the oxide layer observed in

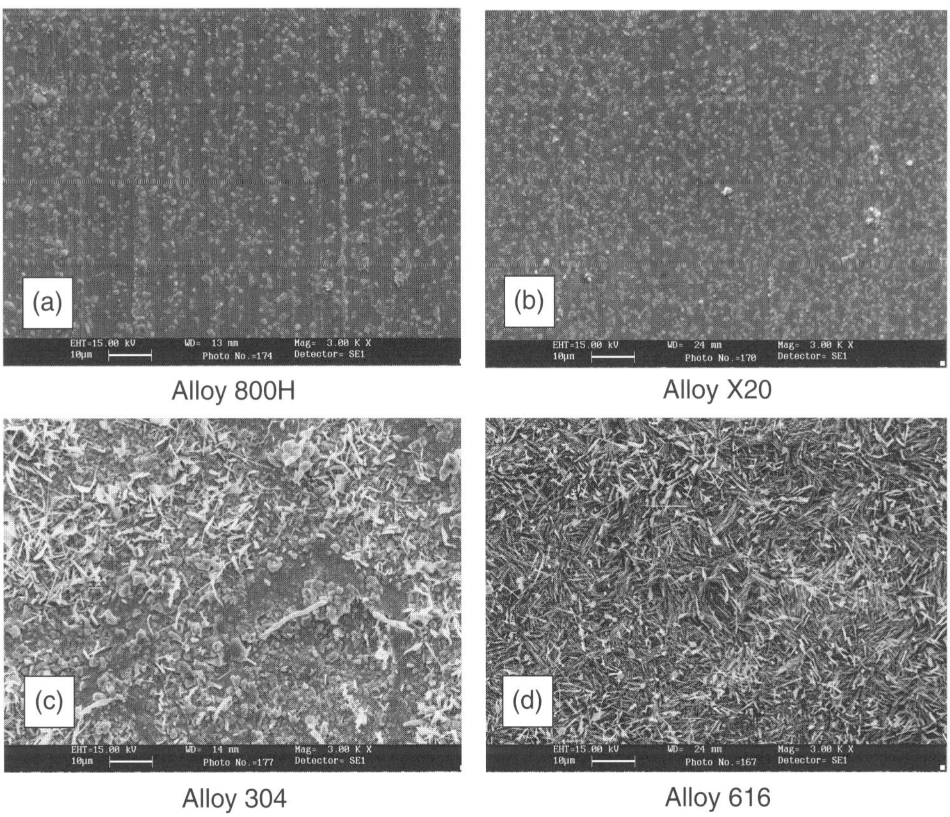

***Fig. 5** Surface SEM micrographs after 10 cycles of 25 h in OX gas.*

the surface SEM micrographs as the outermost scale but obviously this layer was too thin to be visible at these magnifications.

4. Discussion

During exposures involving cycling between an oxidising and a reducing gas atmosphere, kinetic curves of mass change vs time have been derived. These results differ considerably from what is known and expected from standard corrosion experiments conducted under constant atmospheric conditions. As seen in Figs 1 and 2, for both types of cycling tests, starting with oxidising (OX–RED) or reducing (RED–OX) gas conditions, there is no steady mass increase with time. Rather, a

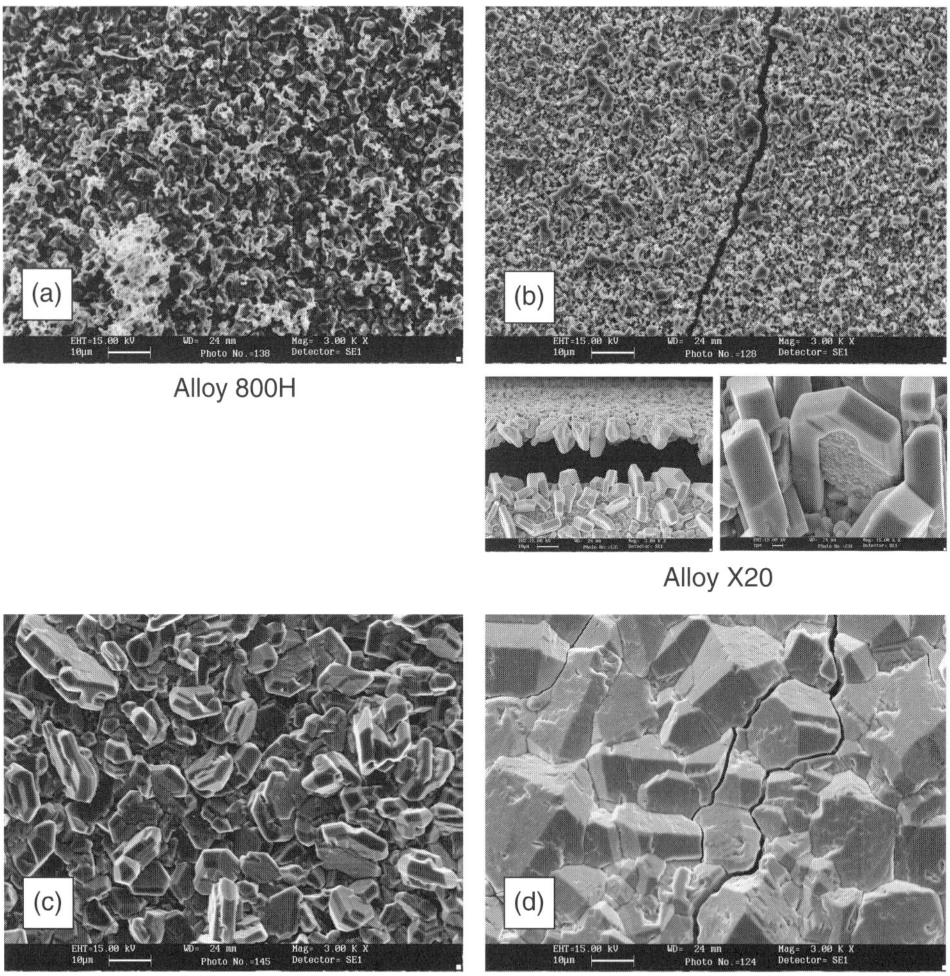

Fig. 6 Surface SEM micrographs after 10 cycles of 25 h in RED gas.

periodic reduction in mass gain rate is observed during each OX step, or even a mass loss. Levels of mass gain are lower for the OX–RED conditions, and are, in fact, generally similar to the exposures in only simple OX conditions. Similarly, the kinetics and mass gain levels under RED–OX conditions are comparable with those measured under simple RED conditions.

All mass gain results show the NF616 alloy to have by far the highest mass gain of the four candidate materials. The behaviour of the three other materials, with nominal chromium contents of 12mass%, and above, lie much closer to each other. Alloy 800H exhibits the lowest mass increase during the 25-h dwell time tests, whereas in the 100-h dwell time experiments, AISI 304 seems to be slightly superior. A general ranking (best to worst) corresponds well with the Cr and Ni content of the materials, i.e. Alloy 800H-AISI 304-X20 CrMoV 12 1-NF616. Nevertheless, all alloys show a decreasing rate of mass gain with time indicating the generally protective nature of the scales formed.

Surface SEM investigations show two general types of morphology on the surfaces of the specimens which is found to varying extents on all alloys tested. One is an iron oxide phase, with some nickel dissolved in the case of Alloy 800H and AISI 304. This surface scale consists of thin needle- or platelet-shaped crystals, not very densely

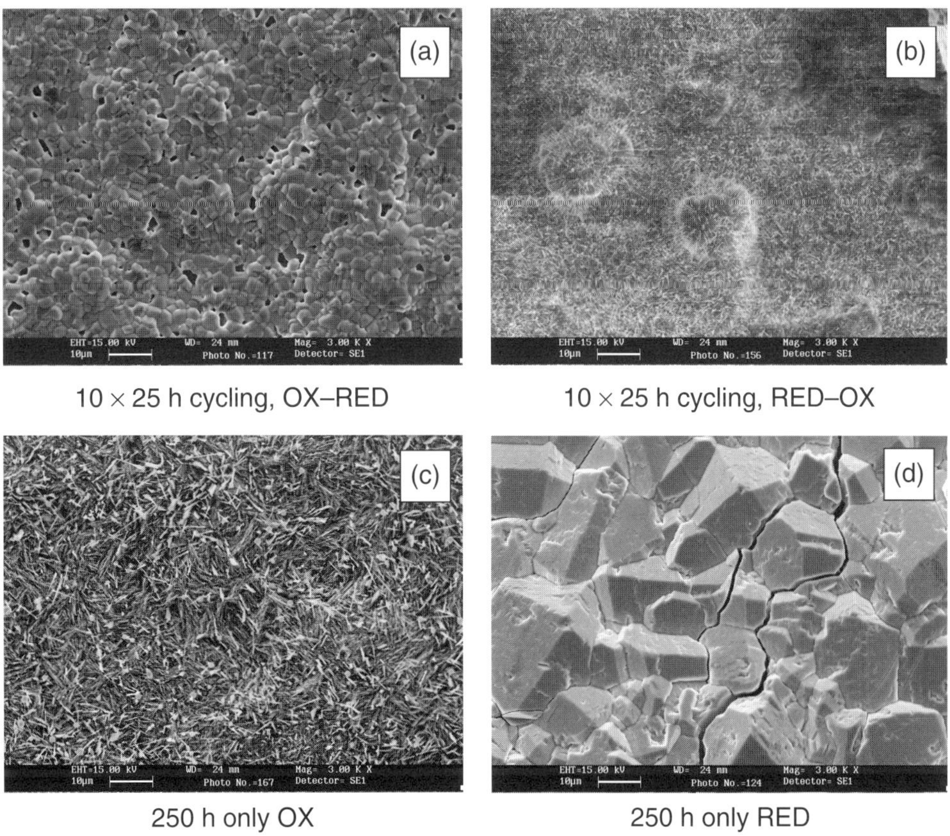

Fig. 7 SEM images for NF616 after 10 cycles of 25 h under 4 types of exposure.

packed, covering the surface of the test coupons. In the case of samples exposed in simple OX conditions, coverage is non-uniform, some areas only being covered by a thin and uniform oxide scale. The second type is a sulfide phase, present as large voluminous crystals, which appear to be much more densely packed at the surface. This scale is not well adherent, however, and cracking, with the scale eventually spalling off in large flakes, occurs.

On specimens treated in cyclic atmospheres, these two morphologies can be present adjacent to each other, on one specimen. This is most pronounced in the RED–OX case, where oxide whiskers cover the previously formed coarse sulfide grains, almost in the form of a 'fur' coating. In both cyclic conditions, the dominant phase on the surface of the material is dependent upon the type of atmosphere used during the last cycle.

Figure 12 shows two typical EDX spectra of the surface phases found on the NF616 alloy, after exposure in the simple OX and RED atmospheres. After cycling in the two atmospheres, the surface spectra show a combination of these two phases, with some nickel incorporated in the case of the Ni-containing Alloy 800H and AISI 304 materials.

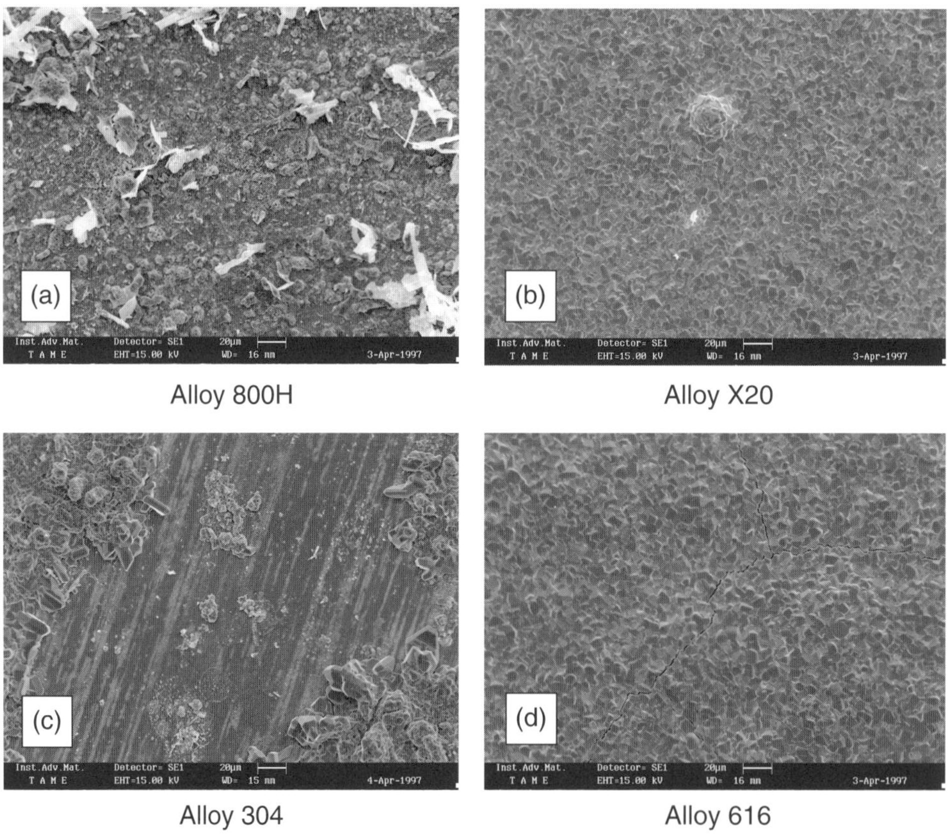

Fig. 8 *Surface SEM micrographs after 10 cycles of 100 h in OX–RED conditions.*

Cross-sectional micrographs of the scales on specimens exposed to the OX–RED and RED–OX cyclic conditions show the scales to consist of a dense inward-growing layer and a more porous outward-growing part. Obviously, in those instances where the outermost layer consists of fine oxide crystals on the surface of RED–OX treated specimens, this layer is extremely thin. The scales are dominated by the presence of mixed iron and chromium sulfide/oxide phases; in the OX–RED case the outer scale consists mainly of an iron sulfide phase, the inner layer being a mixed iron/chromium oxide/sulfide phase. Chromium is concentrated mostly in the inner part of the scale. Specimens treated in a RED–OX regime show a more oxide dominated outer scale, with only a little sulfur in the inner layers, although in some cases, internal sulfidation and chromium depletion of the underlying alloy was found.

In comparing the two series of experiments using different dwell periods of 25 and 100 h there is some inconsistency in the ranking of the corrosion resistance of the alloys, except for the NF616 material, which always shows the poorest behaviour. It is noteworthy that after an exposure time of 250 h, for both the RED-OX and the simple RED treatment, the 25-h cycles result in lower mass gains for NF616 than those measured on companion specimens exposed to 100-h cycles, whereas in OX–

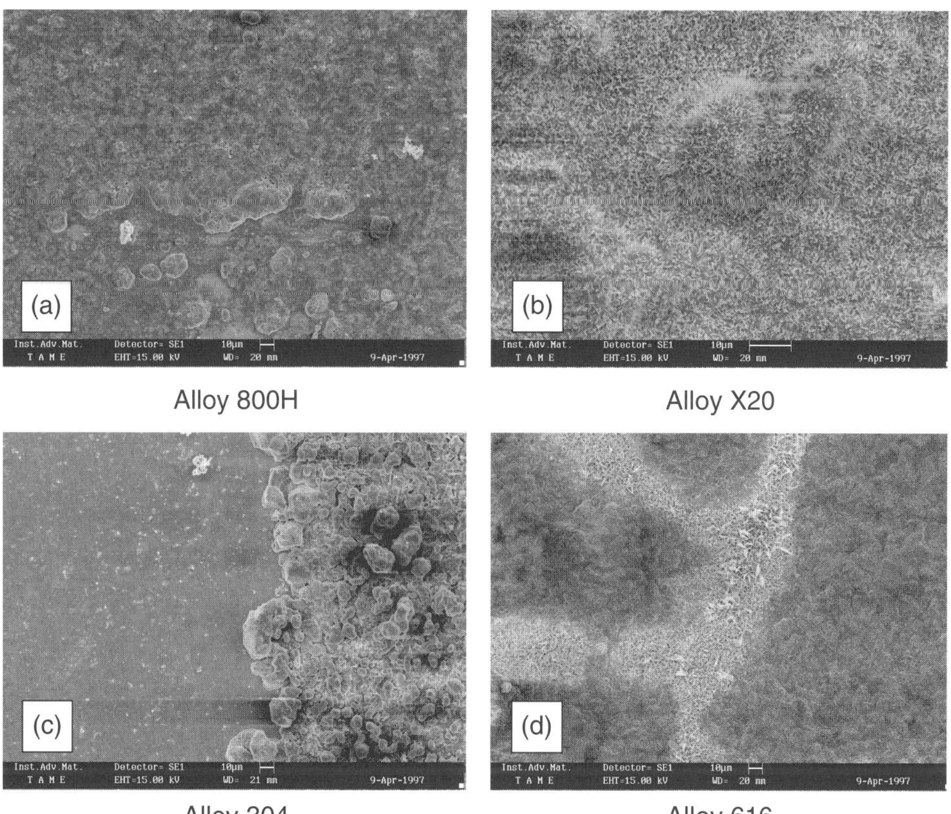

Fig.9 *Surface SEM micrographs after 10 cycles of 100 h in RED–OX conditions.*

***Fig. 10** Surface SEM micrographs after 4 cycles of 100 h, OX.*

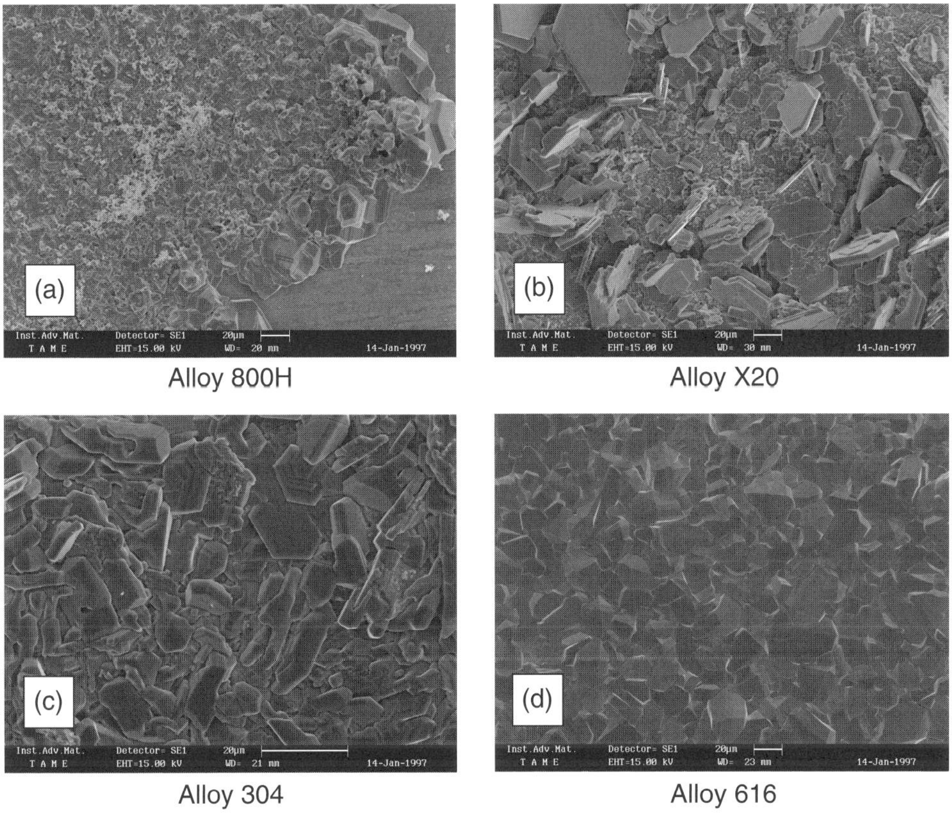

***Fig. 11** Surface SEM micrographs after 4 cycles of 100 h, RED.*

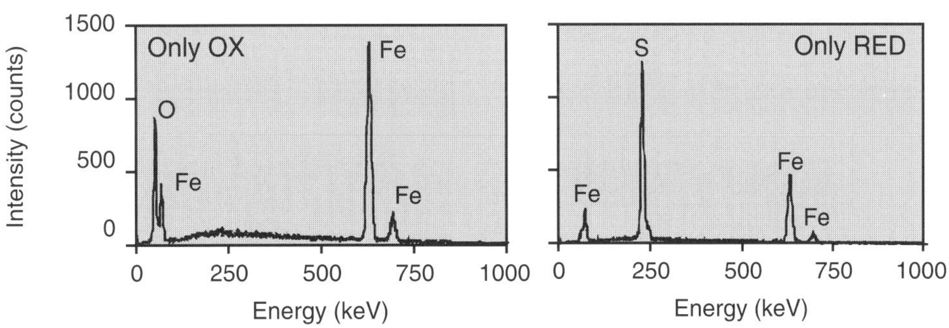

***Fig. 12** Typical EDX spectra of NF616 after 1000 h exposure; (a) non-cyclic OX and (b) non-cyclic RED.*

RED and simple OX, both series of curves correspond reasonably well. Apparently, this least corrosion resistant alloy seems to benefit from short-term cycling, even in a simple reducing atmosphere.

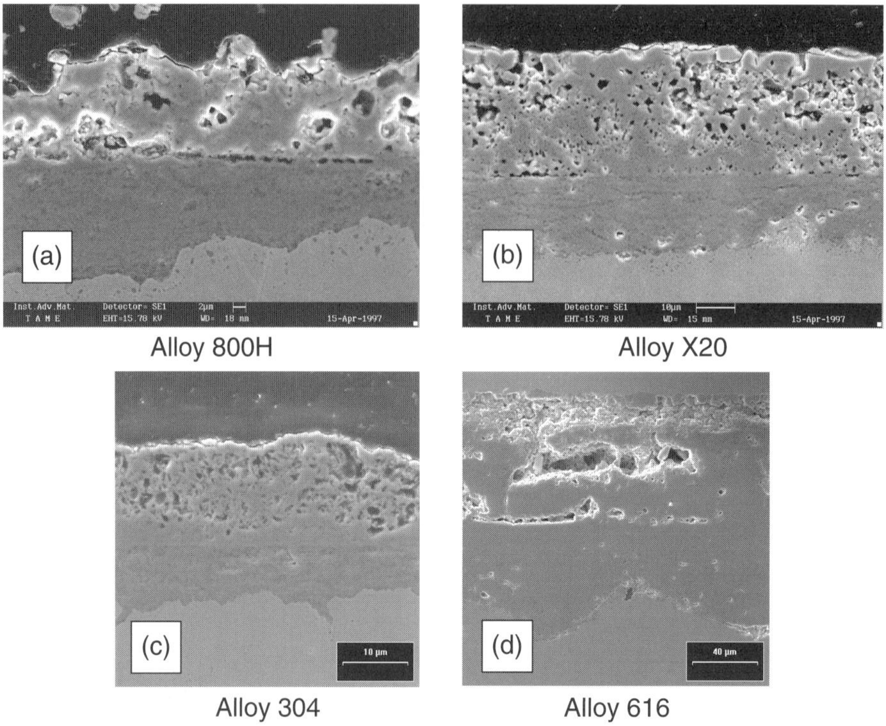

Fig. 13 *Cross-sectional SEM micrographs after 10 cycles of 100 h under OX–RED conditions.*

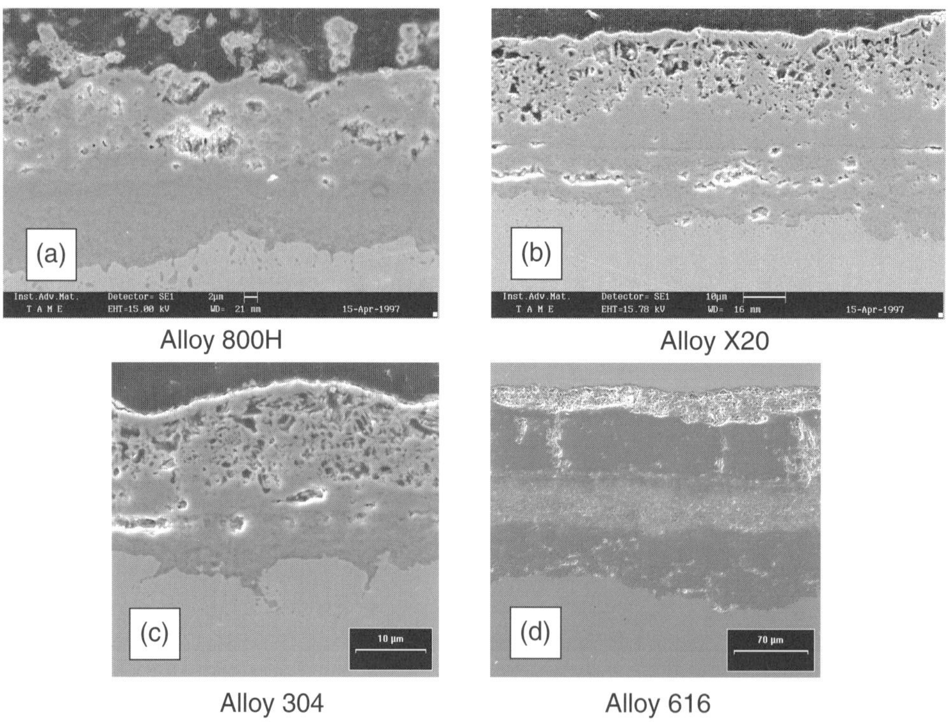

Fig. 14 *Cross-sectional SEM micrographs after 10 cycles of 100 h under RED–OX conditions.*

5. Conclusions

Corrosion studies carried out on four alloys at 600°C in cyclic atmospheres, using two different dwell periods of 25 and 100 h, for up to 10 cycles each, enabled the following conclusions to be drawn concerning the behaviour of the materials tested:

1. Formation of iron, nickel and chromium oxides and mainly iron sulfides is observed, the final morphology of the outer scale depending upon the last exposure.

2. In the cyclic tests, the materials show a lower mass gain rate in the OX step than during the RED cycle. In some cases mass losses were observed during the OX cycle.

3. In the case of the NF616 alloy, higher mass gain levels were observed for the longer dwell time of 100 h, the shorter cycle time of 25 h appearing to promote less corrosion, for a given exposure period.

4. The RED–OX type of cyclic test proved to be more corrosive than the OX–RED test.

5. The general OX–RED kinetics are similar to those derived in the simple OX regime whilst the RED–OX kinetics are in line with those of the kinetics established under RED conditions. This clearly indicates that the test conditions of the first cycle govern the behaviour of the materials in the longer term.

6. It follows that the first exposure cycle largely determines the overall behaviour of the alloy, due to the formation of either an initial protective oxide layer or a significantly less protective, defective sulfide scale.

7. The rate of mass gain decreases with time for all alloys indicating that all scales show some level of protective behaviour.

8. The general ranking of the materials tested, from greatest to least corrosion resistance, is: Alloy 800H–AISI 304–X20 CrMoV 12 1–NF616, which corresponds well with the Cr and Ni content of these alloys.

Thus, for practical purposes, the recommendation derived from this study is that in order to ensure good corrosion protection of alloys exposed to process atmospheres which cycle between oxidising and reducing conditions it is necessary to either pre-oxidise the relevant components as part of an initial treatment before service or, if that is not possible, to make sure that the first regime is an oxidising one. This promotes the formation of an oxide layer on the surface of the component with a capability of conferring protection against corrosion during future exposures to more corrosive reducing conditions.

In any case, a chromium content of at least 12%, and preferably 18 mass% or higher is preferred for these kinds of operating condition. The stainless steel AISI

304 containing 18 mass% Cr and 9 mass% Ni appears to retain sufficient corrosion resistance, whilst the higher alloyed type 800H material offers only minor additional improvements in these regimes.

6. Acknowledgements

This research study was carried out at the Institute for Advanced Materials, Petten, as part of the European Commission's Research and Development Programme, with support from the Human Capital & Mobility Programme. Special thanks go to S. Canetoli, K. Schuster and P. D. Frampton, for their support in the technical and microstructural analysis parts of the work.

16
Behaviour of Protective Coatings against Sulfidation under Cyclic Conditions

T. WEBER and M. SCHÜTZE
Karl-Winnacker-Institut, DECHEMA e. V. Theodor-Heuss-Allee 25, D-60486 Frankfurt, Germany

ABSTRACT
Various thermal spray coatings using the APS-technique as well as a commercially available dip-coating based on Al were applied to ferritic grade low alloy and stainless steels and tested in an Ar–5%H_2–1%H_2S (vol.%) atmosphere, which is typical for several processes in the chemical and petrochemical industries, energy conversion or waste treatment plants.

Experiments were run isothermally and under cycling conditions for up to 1000 h at a temperature of 700°C. The thermal spray coatings based on TiAl and $TiSi_2$ showed particularly promising results in improving corrosion resistance.

1. Introduction

Several processes e.g. in the field of oil refining, energy conversion or waste treatment are characterised by highly reducing sulfidising atmospheres. Metallic components in contact with this atmosphere at temperatures exceeding 500°C suffer heavy corrosion. An effective corrosion protection system does not exist and even high-alloy materials have only limited life as well as having specific problems of manufacturing (e.g. weldability) and high costs. Thus, conventional steels are used for components, which will need to be oversized and regularly replaced. The failure of these materials is caused by their inability to form protective oxide scales in such atmospheres which have very low oxygen partial pressure ($pO_2 \leq 10^{-20}$ atm) and high sulfur partial pressure ($pS_2 \geq 10^{-10}$ atm) [1]. Furthermore conventional alloys form sulfide scales with a much higher number of lattice defects and growth rates usually several orders of magnitude above those of the oxide scales [2–4].

Materials with a potential to form oxide scales under these conditions are aluminium, silicon and, with some limitations, also titanium. However, the amount of these that can be used as alloying elements in high-alloy steels is limited because of their effects on mechanical properties. Thus, the basic idea of this project is the use of high Al-, Si- and Ti-containing phases as coatings for common low alloy or ferritic steels [5]. These coatings can be applied using thermal spray techniques, chemical vapour deposition or dipping.

2. Experimental Procedure

The coatings investigated were:

Dip-Coating: A commercially available dip-coating based on Al (Duralco* 200) was used which can be applied by painting or spraying as well as dipping. For the present investigations the dipping method was chosen using a suitable device to withdraw the specimen from the coating fluid with constant velocity to ensure a homogeneous coating. After dipping the coatings were dried and hardened according to the specifications of the supplier.

Thermal Spray Coatings: Three different thermal spray coatings, applied by the atmospheric plasma spraying method (APS) were investigated. The spray powders were Ti–48Al–1.5Cr (at.%) with 32–115 μm particle diameter and Ti–50Al (at.%) as well as $TiSi_2$ with 20–63 μm particle diameter, respectively.

All coatings were applied to low-alloy 1.7335 steel and to stainless 1.4742 steel (German designation according to DIN17006). The compositions of the alloys are given in Table 1. The specimens had cylindrical shape (14 mm dia., 14 mm length, rounded edges) for the stainless steel and tubular shape (38 × 4 mm dia., 10 mm length) for the low-alloy steel. The low-alloy tubular shaped specimens were used only for the APS–TiAlCr coating. All other coatings were applied to the stainless steel specimens.

All specimens were tested under experimental conditions as follows:

- reducing, sulfidising atmosphere (Ar–5%H_2–1%H_2S) ($pS_2 = 10^{-6}$atm, $pO_2 = 4 \times 10^{-27}$atm — calculated from the residual H_2O content of the gas;
- temperature 700°C;
- discontinuous isothermal testing up to 1000 h, i.e. intermediate cooling for mass change measurements after 30, 100 and 300 h; and
- thermocyclic testing up to 1000 h with $\Delta t = 24$ h, minimum temperature 100°C.

The preferential formation of sulfides or oxides as corrosion products can be estimated using a stability diagram as shown in Fig. 1, where the stabilities of various oxides

Table 1. *Composition (mass%) of the materials used as substrate for the coatings*

Material	Composition	Fe	C	Si	Cr	Al	Mn	Mo
Stainless steel	(1.4742)	Bal.	0.12	0.70–1.40	17.0–19.0	0.70–1.20	—	—
Low-alloy steel	(1.7335)	Bal.	0.10–0.18	0.10–0.35	0.70–1.10	—	0.40–0.70	0.45–0.65

*Cotronics Corp., Brooklyn, NY, USA.

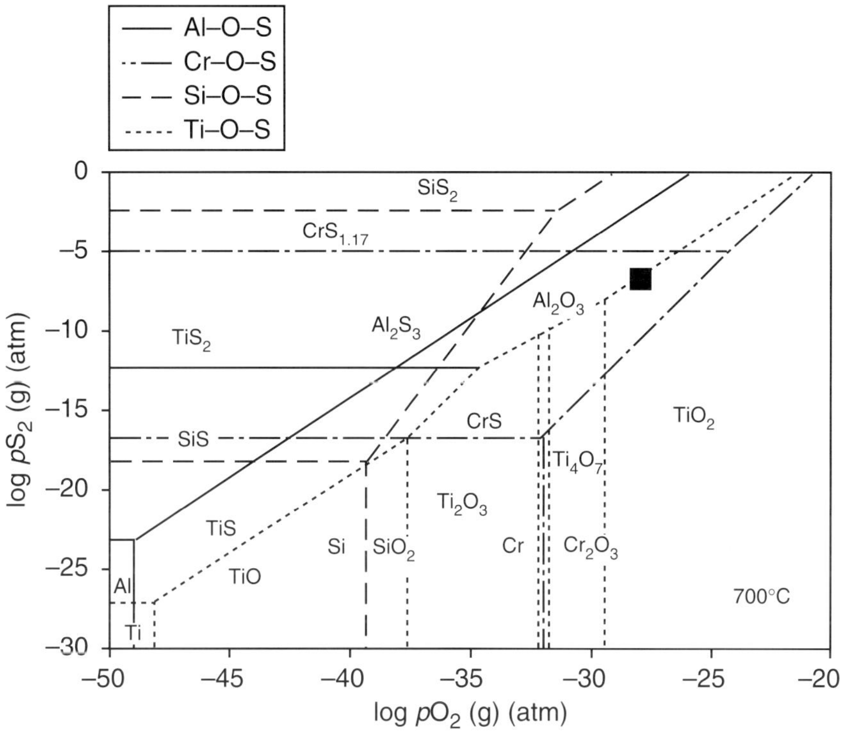

***Fig. 1** Stability diagram of oxides and sulfides of Al, Cr, Si, Ti at 700 °C. The black area represents the experimental conditions of the present investigations.*

and sulfides are given for a temperature of 700°C. The black marked area corresponds to the test conditions. According to this diagram the only oxides which can be formed in this atmosphere are Al_2O_3 and SiO_2. For Ti the test atmosphere lies right at the borderline between TiS_2 and TiO_2.

After exposure under testing conditions the specimens were prepared for metallographic investigations including EDX (energy dispersive X-ray analysis) and EPMA (electron probe micro-analysis). Before embedding in a conductive mounting resin a thin Ni-layer was electrolytically applied in order to obtain a better contrast to the black coloured mounting resin.

3. Results and Discussion

The specific mass gain curves for all coatings and uncoated samples are given in Fig. 2(a) for isothermal conditions and Fig. 2(b) for thermocyclic experiments. Compared to the uncoated specimens a significant reduction of corrosion rates was obtained with all coatings.

Dip-coating: After the dipping, hardening and drying procedure the coating exhibited a homogeneous layer of 60 to 80 μm thickness as can be seen in a cross-section prior to exposure in Fig. 3 (left). It consisted of Al-particles, size 2 to 10 μm, densely packed

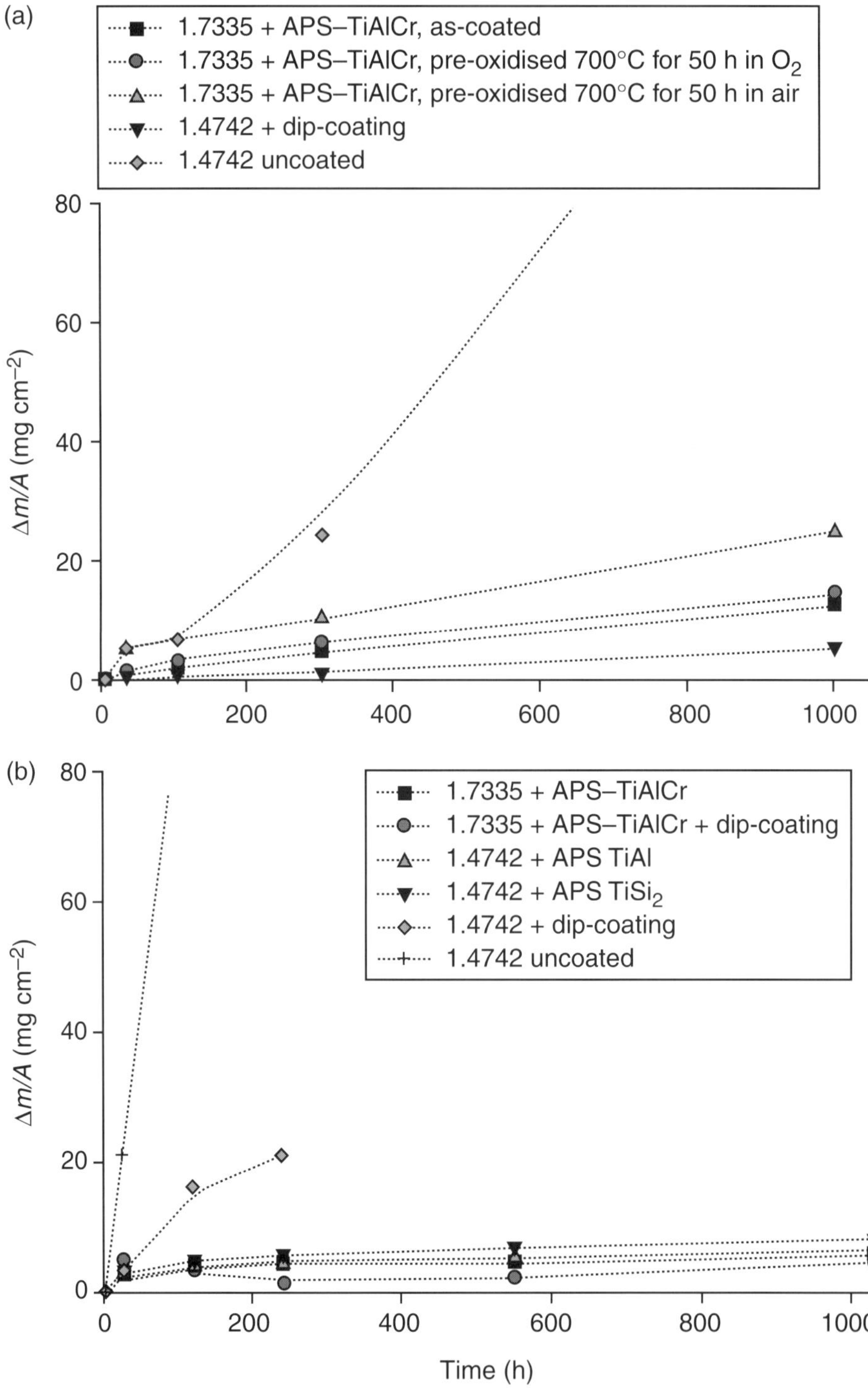

Fig. 2 *Mass gains of the investigated coatings at 700°C in Ar–5%H_2–1%H_2S; (a) isothermal and (b) thermocyclic testing.*

in an amorphous matrix consisting of chromium, oxygen and phosphorus (most probably chromium phosphate). This coating has an aluminising effect up to test durations of 300 h. A layer of Al-oxide had formed on the surface of the coating. An Al-enrichment with continuously decreasing content extending about 20 µm inward from the surface into the substrate was found by EPMA-linescan investigations. However, during further exposure in the test atmosphere the positive effect disappeared due to inward diffusion of Al. A specimen examined after an isothermal experimental run of 1000 h already showed heavy corrosion (Fig. 3, right). A thick layer of Fe- and Cr-sulfide needles had formed on the surface over a thin film of chromium phosphate which was still present. Circular cracking had occurred near the former interface between the coating and substrate. The surface of the substrate was severely attacked and along the grain boundaries internal sulfidation had taken place. This zone consisted of Fe-, Cr-sulfides, alumina and silica.

As can be seen from the mass gain curves, the behaviour under thermocyclic conditions is worse and therefore, this coating in the present form does not seem to be suitable for application under reducing sulfidising conditions.

Thermal spray coatings: The APS–TiAlCr–coating prior to exposure is shown in a cross-section in Fig. 4. This coating exhibits a lamellar structure with dispersed oxides and pores. It has a thickness between 150 µm and 230 µm and a surface which is quite rough. The interface between coating and substrate is interspersed with pores and oxides. The nominal coating composition is Ti–48Al–1.5Cr(at.%) but different phases with an enrichment of Ti or Al up to 80% were observed locally. Also some Fe-inclusions resulting from the powder-milling process and W-inclusions from the spray electrodes were present.

The TiAlCr-coatings generally showed a good sulfidation resistance. A cross-section of the inner coating on the tubular specimen after 1000 h isothermal exposure in the sulfidising test atmosphere is shown in Fig. 5. Spallation had occurred at the outside during preparation.

On the surface of the coating Fe-sulfides can be observed. At locations without sulfides either TiO_2-needles with alumina and Ti-sulfide below, or a thin seam mainly of alumina are found by EDX and EPMA investigations. Between the lamellae of the spray-particles and at the interface of the coating and the substrate oxides and sulfides

Fig. 3 *Al-based dip-coating on stainless 1.4742 steel prior to exposure (left) and after 1000 h isothermal testing in Ar–5%H_2–1%H_2S at 700°C (right).*

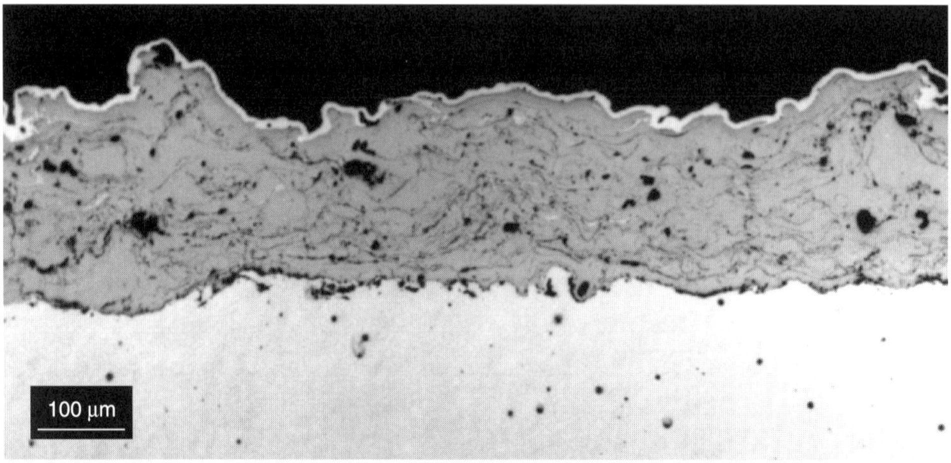

Fig. 4 *Plasma sprayed Ti–48Al–1.5Cr coating on low-alloy steel 1.7335 before exposure.*

are present. The substrate shows a zone of internal sulfidation attack along the grain boundaries with a depth of around 20 μm.

The sulfides on the coating surface are mainly present at those locations where the coating–substrate interface shows imperfections and thus where a large amount of sulfides could form. These sulfides presumably grew outward from the substrate to the coating surface between pores and crevices of the coating which were still open because of the low growth rate of Ti- and Al-oxides. Fe–sulfides grow through cation outward diffusion with a diffusion coefficient of Fe in Fe-S of an order of magnitude above the diffusion coefficient of Fe in FeO [5]. Element mappings of oxides and sulfides derived by EPMA are shown in Fig. 6.

For the isothermal runs specimens with a pre-oxidation treatment were also used. The general consideration of the pre-oxidation method is that a stable oxide scale formed during pre-oxidation could provide a better corrosion protection in highly aggressive environments. These specimens were exposed to laboratory air and oxygen, respectively, for pre-oxidation at 700°C and for 50 h before testing in the sulfidising atmosphere. However, in the present case pre-oxidation did not seem to

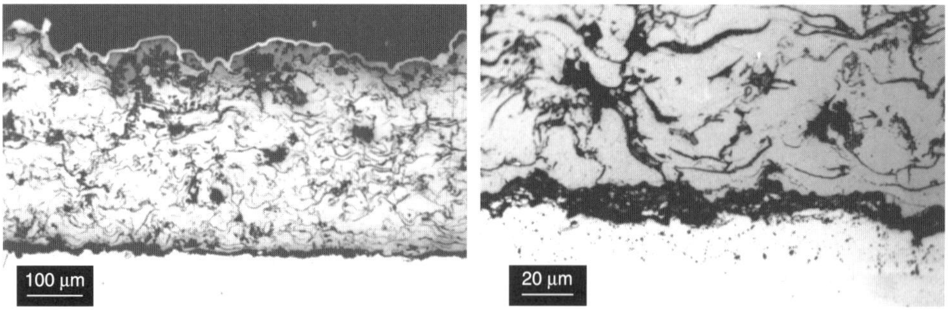

Fig. 5 *Plasma sprayed Ti–48Al–1.5Cr coating on the low-alloy steel 1.7335 after isothermal sulfidation (700°C, 1000 h), zone of internal corrosion (right).*

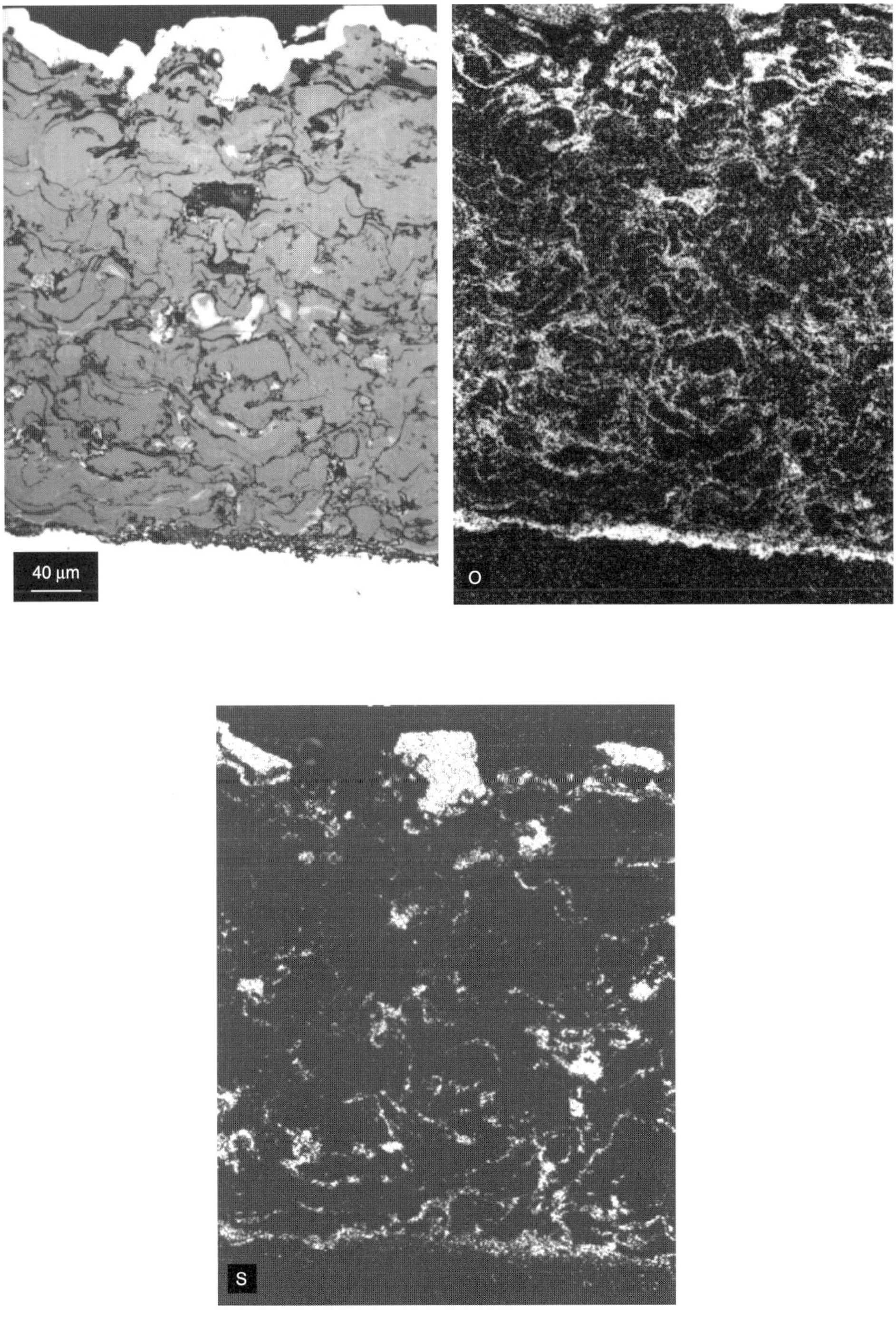

Fig. 6 *Element mappings of oxygen and sulfur for the TiAlCr-coating on 1.7335 after 1000 h isothermal sulfidation at 700°C.*

have a beneficial effect. Because of the physical defects within the coating internal oxidation of the substrate occurred with a depth of about 15 μm (Fig. 7 left). Cr- and Fe-oxides had formed, which are not stable under sulfidising conditions, so a transformation of these oxides to sulfides occurred. Thus, the internal sulfidation of the pre-oxidised specimen proceeded further compared with the non pre-oxidised specimen — as can be seen in Fig. 7 (right).

The pre-oxidised specimen showed a lower mass gain than the as-coated. The reason for this is the shape of the tubular specimens which have sharp edges. At these sharp edges coating quality is often insufficient and will vary between different specimens. Sulfidation attack occurred preferentially at these edges.

For the thermocyclic experiments cylindrical specimens made of stainless steel 1.4742 were again used. These were coated with Ti–50Al and $TiSi_2$; a cross-section of the as-coated state is shown in Fig. 8. Because of the finer spray-particles the surface is quite smooth. However, a large amount of pores and crevices is present with even larger pores in the case of the $TiSi_2$ coating (Fig. 8, right). Here the total coating thickness is around 450 μm.

All APS-coatings showed good corrosion resistance under thermocyclic conditions at 700°C for 1000 h in the sulfidising atmosphere. Spallation and crack formation at the surface of the coating due to thermally induced stresses were not observed. EDX investigations of the surface revealed mainly Ti- and Al-oxides for the TiAl-coating and Ti- and a few Si-oxides in the case of the $TiSi_2$-coating. The S-contents were 12.5 and 1.3 at.% for the TiAl- and $TiSi_2$-coatings, respectively. The cross-sections of the TiAl- and $TiSi_2$-coated specimens after exposure are shown in Fig. 9. The TiAl-coating (Fig. 9, left) is well adherent, Al-, Ti-oxides with regular distribution and some Ti-sulfides with their number decreasing in the inward direction are found along the boundaries between the spray-particles, as shown in Fig. 10. Only a little sulfidation attack is present in the substrate.

In the case of the $TiSi_2$-coating, however, circumferential cracking is observed. This cracking presumably occurred at the beginning of the experiment or during specimen manufacturing, because near the crack a sulfide concentration was found as high as in the surface zone. At the surface a two-layer structure is observed with

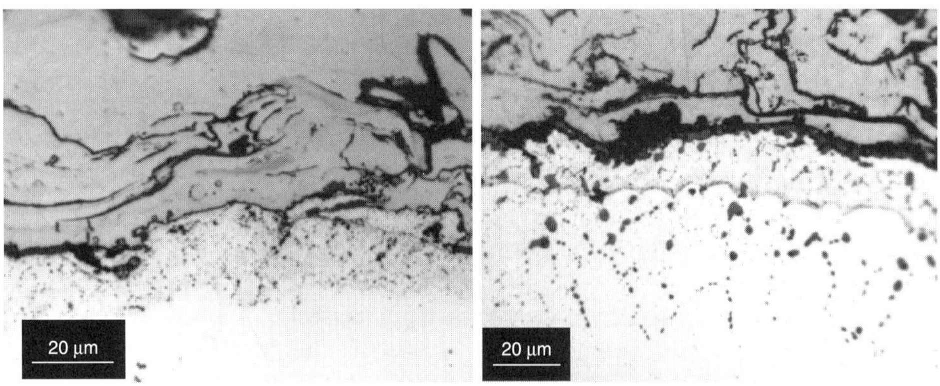

Fig. 7 *APS-TiAlCr on 1.7335, pre-oxidised in laboratory air, 50 h, 700°C, isothermal, (left), exposed to sulfidising atmosphere, 1000 h, 700°C, isothermal (right).*

Fig. 8 *APS–TiAl (left) and APS–$TiSi_2$ (right) coatings on stainless steel 1.4742.*

Ti-sulfides beneath Ti-oxides. At the interface to the substrate a Si-enrichment of about 10 μm thickness is detected by EPMA investigations. Probably due to this Si-enrichment only a small amount of sulfidation attack occurred in the substrate.

4. Conclusions

Different coatings for the protection of construction materials in reducing sulfidising atmospheres were tested at 700°C for up to 1000 h. The Al-based dip-coating is not suitable because of phases which are easily sulfidised and of spallation. Promising results were achieved especially with APS-coatings based on TiAl and $TiSi_2$ under isothermal as well as under thermocyclic conditions. The thermocycling experiments were used as a tool to assess the mechanical stability of the coatings under thermally induced stresses. Even in the case of thick coatings spallation and crack formation

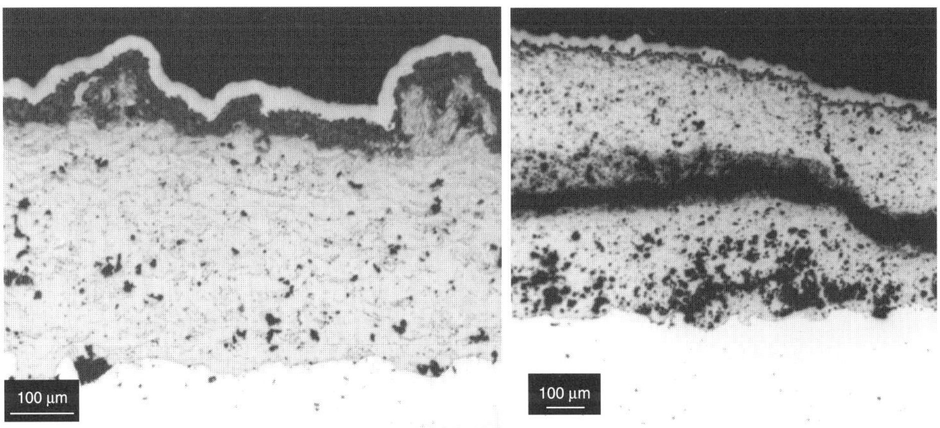

Fig. 9 *APS–TiAl (left) and APS–$TiSi_2$ (right) on 1.4742, exposed to sulfidising atmosphere, 1000 h, 700°C, cyclic test.*

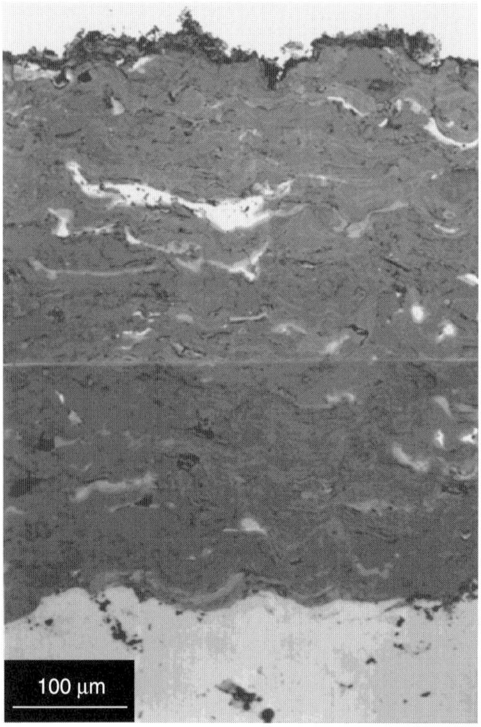

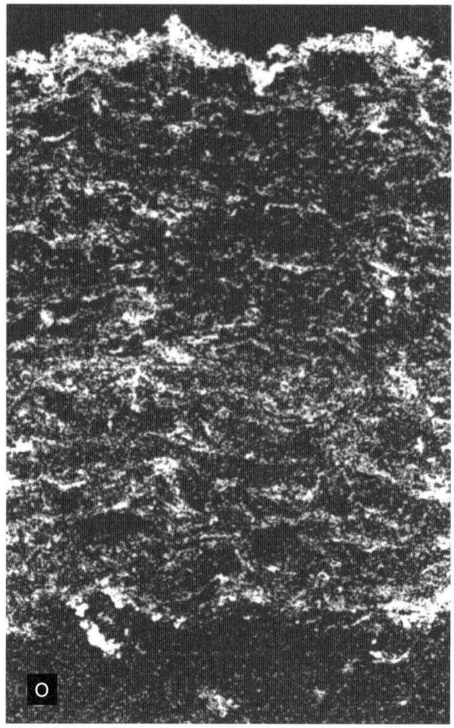

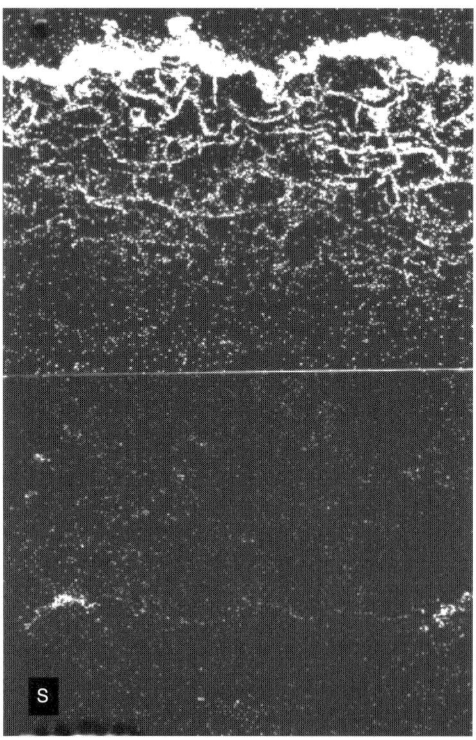

Fig. 10 *Element mappings of oxygen and sulfur for the TiAl-coating on 1.4742 after 1000 h sulfidation at 700°C, cyclic test.*

did not occur. The reason lies in the similarity of the thermal expansion coefficients. The maximum CTEs (0...1000°C) for TiAl and $TiSi_2$ have values around $\alpha = 12 \times 10^{-6}K^{-1}$ and $\alpha = 13 \times 10^{-6}K^{-1}$, respectively, while the ferritic alloys have maximum CTEs between $\alpha = 13 \times 10^{-6}K^{-1}$ for the 1.4742 and $\alpha = 14 \times 10^{-6}K^{-1}$ for the 1.7335 alloys.

The typical appearance of thermal spray coatings with a certain process-inherent defect structure results in high gas diffusivities. Thus, the corrosive atmosphere penetrates through pores and crevices directly to the substrate, where internal corrosion occurs. An increase of the coating thickness may help to a certain extent.

However, the key to an optimum protective effect lies in the density (i.e. the coatings have to be gas-tight) of the coatings. Thus, for further optimisation of the corrosion resistance the enhancement of the coating quality is most important and investigations with the SPS-method (shrouded plasma spraying) and a post-treatment using a laser technique to remelt the top layer of the coating are under way. In a second step the composition of the coating has to be optimised. In particular, ternary additions to the TiAl system will be tested.

5. Acknowledgements

This project is funded by the Bundeswirtschaftsministerium via Arbeitsgemeinschaft industrieller Forschungsvereinigungen (AiF) under contract no. 11243N, which is gratefully acknowledged. The authors also would like to thank Mrs. Berghoff-Hasselbächer for the metallographic work, Mrs de Leon for the photographs, and Mrs Schorr and Mr. Gawenda for the EPMA-, SEM/EDAX-results. The APS-coatings were applied by ATZ-EVUS, Vilseck and the Institute of Materials Engineering at Dortmund University.

References

1. D. J. Baxter and K. Natesan, *Oxid. Met.*, 1989, **31**, 305.
2. H. Habazaki *et al.*, *Corros. Sci.*, 1994, **36**, 199.
3. H. Habazaki, J. Dabek and K. Hashimoto, *Corros. Sci.*, 1993, **34**, 183.
4. K. N. Strafford and P. K. Datta, *Corros. Sci.*, 1993, **35**, 1053.
5. St. Mrowec, *Werkst. Korros.*, 1980, **31**, 371–386.
6. M. Schütze and M. Nöth, *Corrosion '98*, Paper No. 188, NACE International, Houston, Tx, USA, 1998.

Part 3

Long Term Behaviour and Near-Service Conditions

17

Long-Term, Cyclic Oxidation Behaviour of Three Chromia-Forming Alloys Exposed to Air at 982°C

M. A. HARPER and B. GLEESON*

Haynes International, 1020 West Park Avenue, Kokomo, IN 46904, USA

*Dept of Materials Science & Engineering, Iowa State University, Ames, IA 50011, USA

ABSTRACT

The cyclic oxidation behaviour of three chromia-forming commercial alloys was studied in air at 980°C for up to 720 days. Each thermal cycle consisted of 30 days at 982°C followed by air cooling to room temperature. Two of the alloys were iron base (800HT alloy: Fe–32.5 Ni–21Cr mass% and HR-120: Fe–37 Ni–25Cr mass%) and one was nickel base (230 alloy: Ni–22Cr–14W mass%). Thermogravimetric analysis, X-ray diffraction, and scanning electron microscopy confirmed that all the alloys initially formed and maintained a protective external chromium oxide (Cr_2O_3) scale. The total metal affected i.e. metal recession plus internal attack) in each of the alloys corresponded primarily to internal attack. The nickel-base 230 alloy performed the best owing to its ability to maintain a thin Cr-rich oxide scale at the alloy/scale interface during thermal cycling. The 800HT alloy performed the worst, undergoing rapid mass loss kinetics after about 150–180 days. The results clearly showed the importance of long-term testing and the need to correlate spallation behaviour with the interdependence between the subsurface changes which occur in an alloy and the type of scale which forms.

1. Introduction

Many high-temperature alloys used in industrial applications depend on the formation of a continuous and adherent Cr_2O_3 scale for prolonged service life. This requires that the alloys have a sufficient chromium content to both form and sustain the continued growth of a Cr_2O_3 scale. In certain instances, the Cr_2O_3 scale may eventually breakdown locally, leading to the formation of a less-protective scale. As discussed by Wood and Whittle [1] in the context of Fe–Cr alloys, breakdown can occur under isothermal conditions as a consequence of localised lifting and cracking of the initially-formed Cr_2O_3 scale, thus exposing the Cr-depleted alloy surface to oxygen.

Breakdown of protective Cr_2O_3 scale growth is more prone to occur under practical situations involving thermal cycling. Thermal stresses are induced during a thermal cycle (e.g. cooling from the process temperature) on account of Cr_2O_3 having a lower coefficient of thermal expansion (CTE) than the alloy substrate on which it forms [2]. The induced thermal stresses can cause cracking and eventual spallation of the scale. Repeated cracking and spallation can ultimately lead to the formation of less protective oxides as the substrate becomes depleted in chromium.

Analytical treatments to predict the changes in the subsurface alloy composition during cyclic oxidation have been developed by Whittle [3] and Wahl [4] for the case of complete scale spallation after each cycle, and by Nesbitt [5] for the case of partial spallation. The main purpose of those treatments was to arrive at a criterion to predict the minimum bulk solute concentration required to repeatedly reform a protective solute–oxide scale during cyclic oxidation. The most advanced treatment is that by Nesbitt [5], who combined a diffusion model based on finite-difference techniques with an oxide-spalling model (developed empirically by Lowell *et al.* [6,7]) to predict the rate of solute consumption associated with partial scale spallation. Nesbitt's treatment under-predicted by a factor of 2.5 the minimum aluminium concentrations necessary to reform an Al_2O_3 scale on Ni–Al alloys at 1200°C, but showed good agreement with experiment for Ni–Al–Cr alloys containing more than 20 mass% Cr. Notwithstanding the usefulness of Nesbitt's treatment, it is limited to those alloys for which the scale spalling behaviour and interdiffusion coefficients have been determined. These data are typically not available for commercial high-temperature alloys, particularly for the temperatures and cycle durations found in practice.

Harper *et al.* [8] and Gleeson and Harper [9] recently reported the results of a long-term cyclic oxidation study of Cr_2O_3-forming high-temperature alloys containing less than 2 mass% Al. The alloys were subjected to 30-day cycles for a total of up to about two years (720 days) at 982°C and one year (360 days) at 1093°C and 1149°C. It was found that alloys which exhibit adequate oxidation resistance for several thousand hours (accumulated time) can unexpectedly undergo rapid, breakaway oxidation. It was also found that many of the alloys form internal oxides and voids, which can account for 80–90% of the average metal affected by the oxidation process. Such subsurface degradation of the alloys is an aspect not considered in the analytical treatments of cyclic oxidation [3–5], and it highlights the complexity of modelling the oxidation behaviour of commercial, multicomponent alloys.

The purpose of this study was to characterise more thoroughly the cyclic oxidation behaviour of the chromia-forming alloys HR-120, 230, and 800HT. The nominal compositions of these alloys are given in Table 1. Of particular interest in this study was the correlation of oxidation behaviour with the subsurface changes which occur as a function of oxidation cycles.

Table 1. *Nominal compositions (in mass%) of the alloys studied. HR–120 and 230 are trademarks of Haynes International and 800HT is a trademark of INCO family of companies*

Alloy	Cr	Ni	Fe	Co	Mo	W	C	Others
800HT	21	32.5	Bal.	—	—	—	0.08	0.8Mn, 0.5Si, 0.4Cu, 0.4Al, 0.4Ti
HR–120	25	37	Bal.	3*	2.5*	2.5*	0.05	0.7Mn, 0.7Nb, 0.6Si, 0.2N, 0.1Al, 0.005B
230	22	Bal.	3*	—	2	14	0.1	0.5Mn, 0.4Si, 0.3Al, 0.02La, 0.015B*

*Maximum.

2. Experimental Procedures

Wrought alloy samples with approximate dimensions 1.3 cm × 2.5 cm × 2.5 cm were prepared from mill-annealed plates. Each sample was polished down to a 120-grit finish, degreased in acetone, and then weighed prior to testing. Oxidations were carried out at 982±5°C in still air using a box furnace. Samples were placed on a piece of alumina cloth which itself was placed on a tray of 214™ alloy (a trademark of Haynes International, Inc.). The 214 alloy forms an adherent Al_2O_3 scale which essentially prevents any interaction with the alumina cloth or the test samples.

At the completion of a 30-day period (720 h), the tray with all the samples was removed from the furnace and the samples were allowed to air-cool to room temperature. No attempt was made to retain any scale which might have spalled during cooling to room temperature. The samples were typically weighed within four hours of cooling and then placed back in the furnace for the next 30-day cycle. The average metal-thickness loss was approximated by dividing the final mass loss of the sample by the alloy density.

At the completion of a given test, the oxidised samples were analysed using *in situ* X-ray diffraction (XRD), optical microscopy, and a scanning electron microscope (SEM) with a thin-window energy dispersive spectrometer (EDS) attached. The EDS analysis allowed for standardless, semi-quantitative analysis of the alloys and any of the remaining scales. Cross-sections of the oxidised samples were prepared using standard metallographic techniques. The average internal attack (i.e. voids and internal precipitates) in a given sample was determined from optical metallographic examination.

3. Results

3.1. Cyclic Oxidation Kinetics

The mass-change behaviour of each alloy as a function of accumulated time during thermal cycling is shown in Fig. 1. The alloys were tested for up to about two years (720 days or 17280 h), which corresponds to 24 thermal cycles. As seen in Fig. 1, negative mass change was observed for the HR-120, 800HT and 230 alloys after 60, 150, and 390 days, respectively. After 720 days of testing, the 230 alloy showed the least amount of mass loss (1.4 mg cm^{-2}), followed by HR-120 (33.7 mg cm^{-2}), and then 800HT (417.8 mg cm^{-2}). The rates of mass loss for 230 and HR-120 were relatively constant. By contrast, 800HT underwent a rapid rate of mass loss after about 180 days. The reproducibility for the onset of this breakaway behaviour is shown in Fig. 2, where it is seen that three different 800HT alloys, oxidised for a different total number of cycles, all commenced breakaway degradation after about 150–180 days.

It is interesting to note that on the basis of mass-change kinetics, a test duration of up to 180 days, or 4320 h, would rank 800HT superior to HR-120. However, as shown in Fig. 1, this would be an incorrect ranking if service beyond 180 days were a selection criterion. The present results clearly show the importance of long-term testing and suggest that the often recommended [10] maximum exposures times of 2000–5000 h may be too low, particularly for temperatures below about 1000°C.

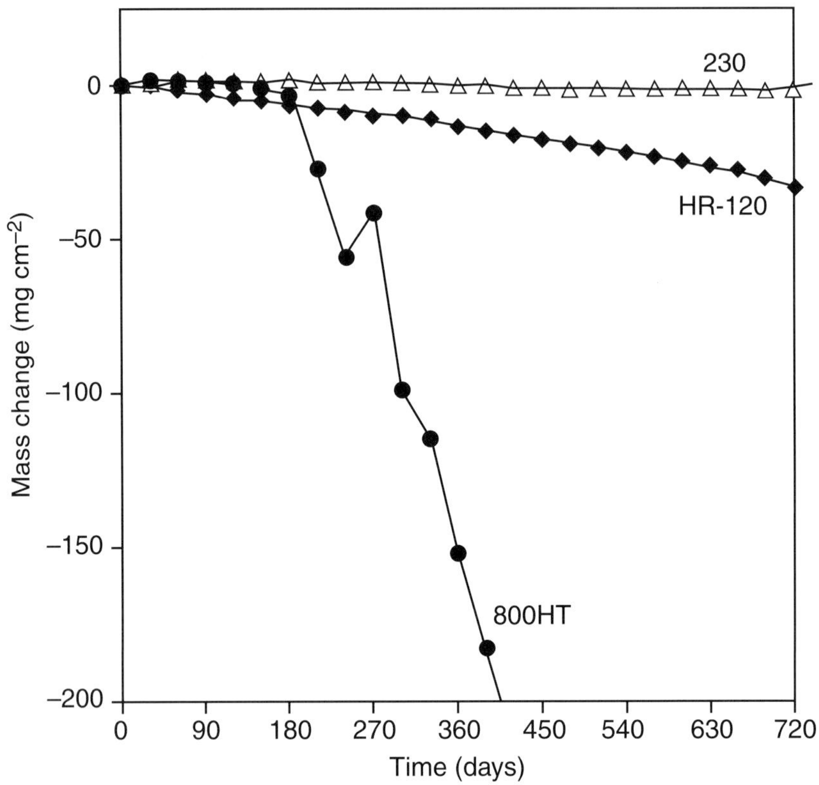

Fig. 1 *Mass change vs time plots for the cyclic oxidation in still air of alloys 230, HR-120 and 800HT at 982°C.*

3.2. Microstructural and Compositional Analyses

Figure 3 compares the surface and cross-sectional microstructures of the three alloys after three thermal cycles (i.e. 90 days). The cross-sectional microstructures are considered to be representative of unspalled regions. The surface images of the 800HT and HR-120 alloys show relatively large smooth and rough areas, indicating that gross, localised scale spallation occurred. The surface of the 230 alloy was quite different in that it was finely nodular in structure, with comparatively smaller regions appearing smooth. In agreement with the mass-change behaviour, the scale on 230 had the appearance of being more tenacious than the scales on the other two alloys. The scale that formed on each of the alloys was primarily Cr_2O_3, although, as seen in Fig. 3, the thicknesses varied. The Cr_2O_3 layer was thickest on 800HT, thinnest on 230, and of intermediate thickness on HR-120. X-ray diffraction and EDS analyses showed that the outer scale region on each alloy consisted of $MnCr_2O_4$. The presence of this $MnCr_2O_4$ layer was most predominant on 800HT.

The cross-sectional images in Fig. 3 show that 800HT underwent a considerable amount of internal oxidation which was primarily intergranular in form. Microanalysis of the internal precipitates in 800HT showed that a number of oxides were present. Near the alloy/scale interface, the precipitates were typically rich in

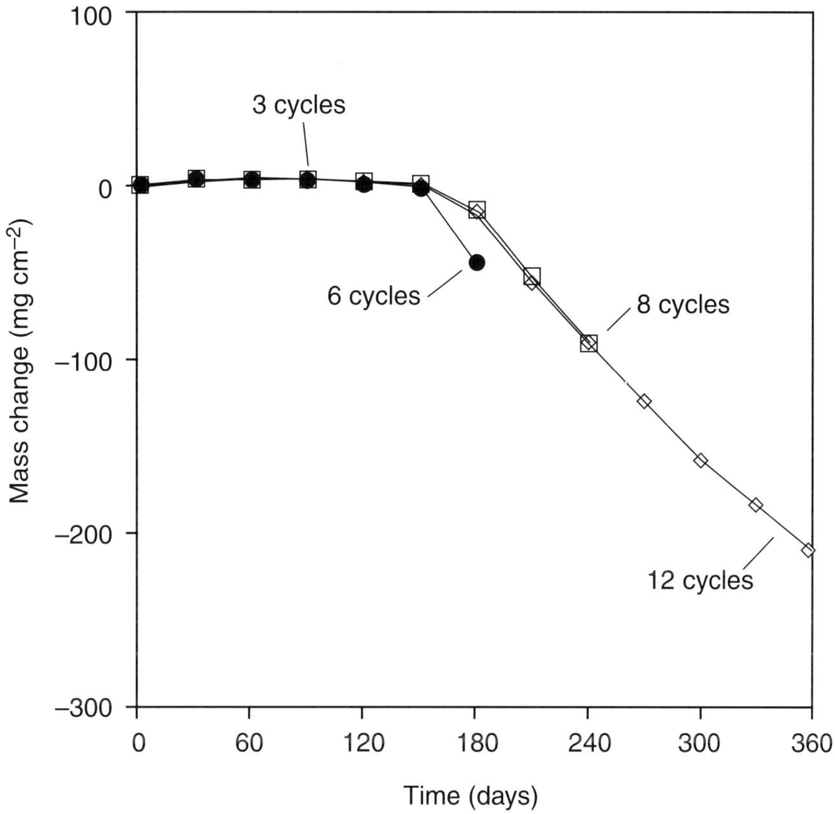

***Fig. 2** Mass change vs time plots for the cyclic oxidation in still air at 982°C of four different samples of 800HT alloy.*

titanium while, deeper into the alloy, they were rich in aluminium. Previous studies [11,12] on the oxidation behaviour of 800-type alloys reported similar results to those found here. The 230 and HR-120 alloys also showed internal attack, with the oxide precipitates typically found to be rich in aluminium, but the depths of attack were significantly less than that found in 800HT. The total depth of attack was similar in 230 and HR-120; however, the amount of attack per unit area would have been greater in 230. In addition to internal precipitates, the subsurface regions of the alloys also contained a number of small voids.

It is seen in Fig. 3 that the morphology of the alloy/scale interface was non-planar in each of the alloys, with 800HT being the most non-planar and HR-120 the least. The inner part of the scale on 800HT contained isolated metal particles, similar to what was observed by others [11,12] and consistent with a portion of the Cr_2O_3 scale growing into the alloy (i.e. oxygen diffusion through the Cr_2O_3 scale was significant). The alloy/scale interface in the 230 alloy was also very non-planar, but apparently the amount of scale growth was not yet sufficient to cause the incorporation of metal particles.

Figure 4 compares the surface and cross-sectional microstructures of the three alloys after six thermal cycles (i.e.180 days). In this case, the cross-sectional

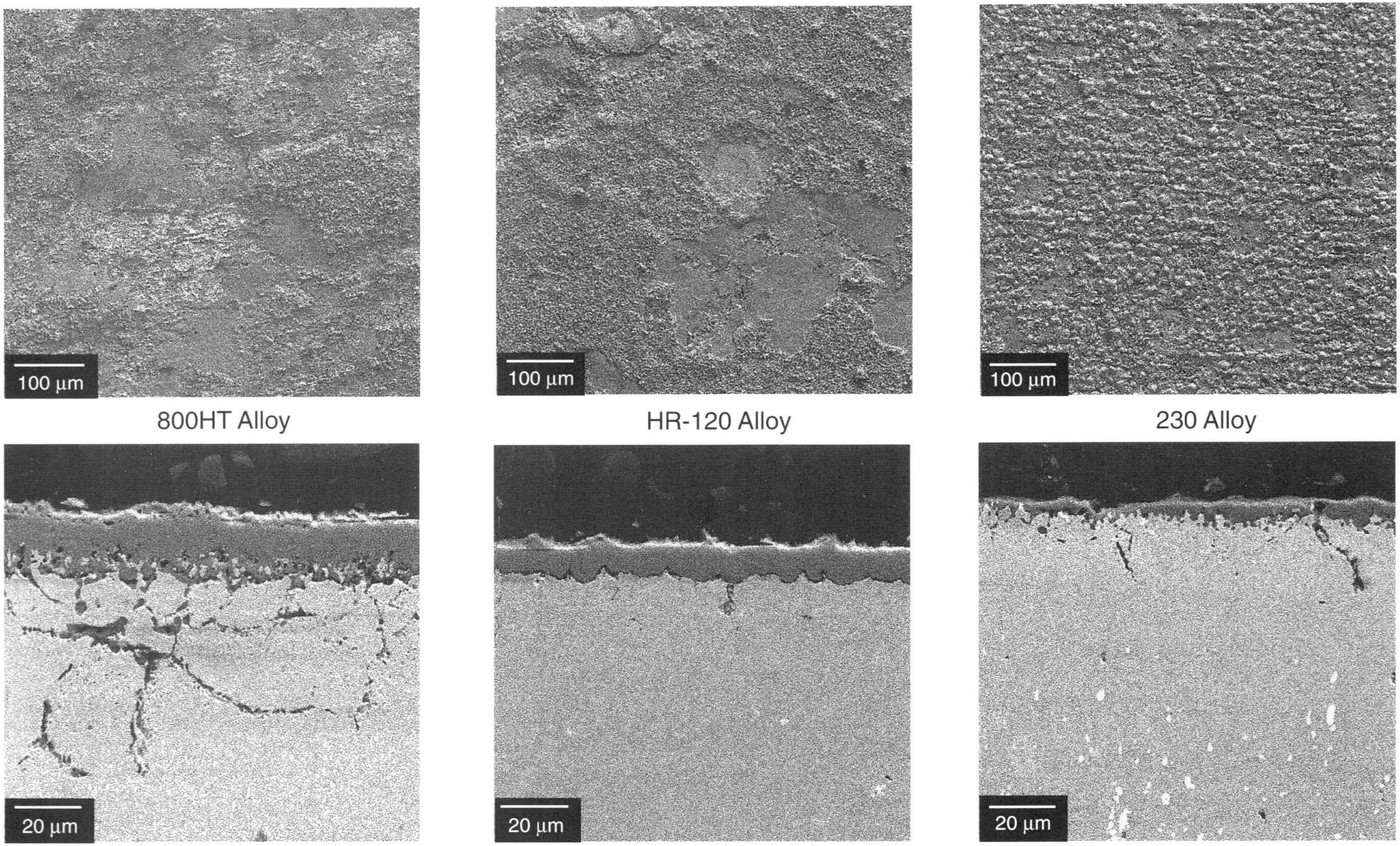

Fig. 3 *Surface and cross-sectional SEM images of the alloys 230, HR-120 and 800HT after three thermal cycles (i.e. 90 days) at 982 °C.*

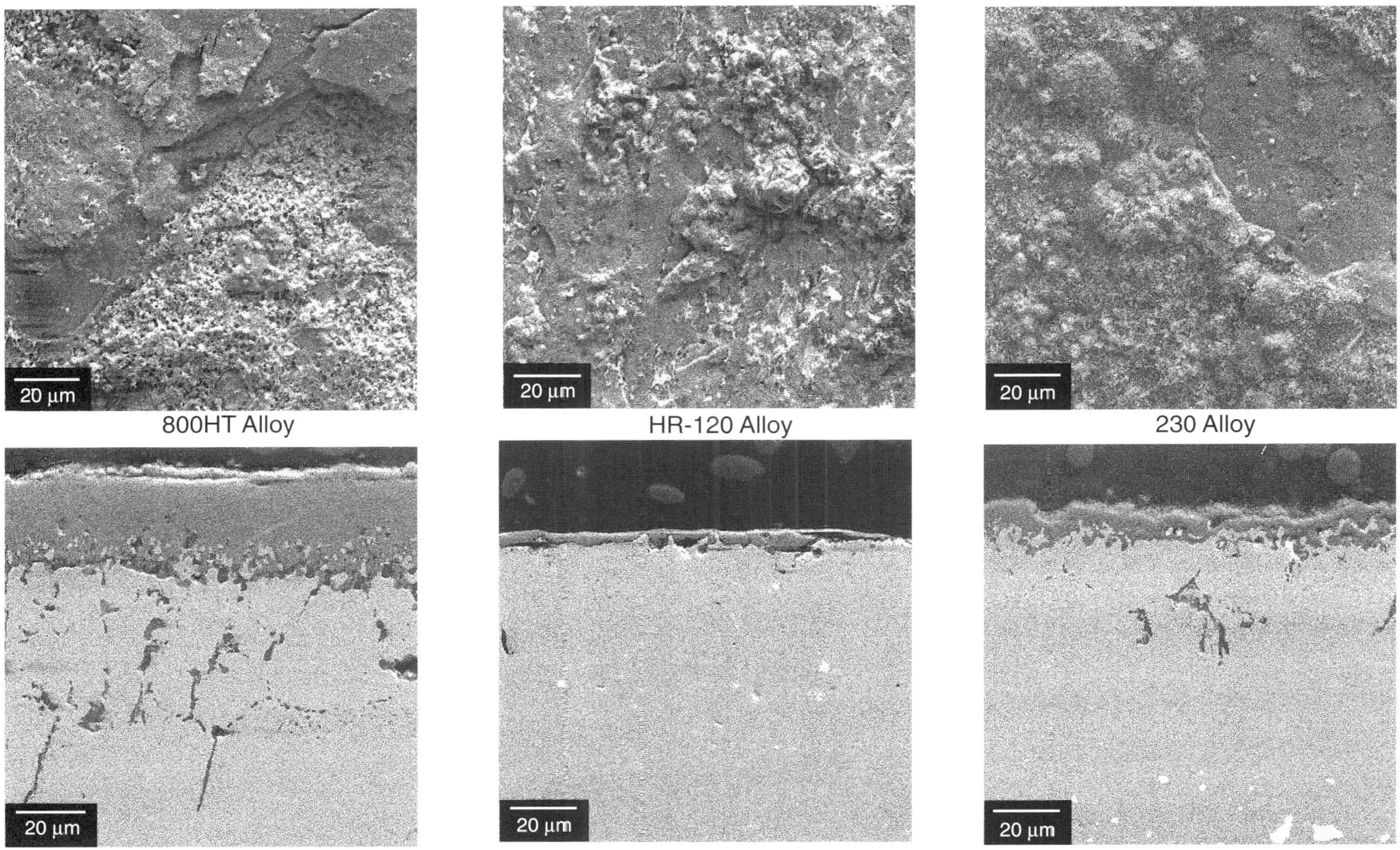

***Fig. 4** Surface and cross-sectional SEM images of the alloys 230, HR-120 and 800HT after six thermal cycles (i.e. 180 days) at 982 °C.*

microstructures are representative of unspalled or least-spalled regions. The surface images indicate that scale spallation occurred from each of the alloys, with the amount of disruption being the most on 800HT and the least on 230. X-ray diffraction and EDS analyses showed that the scales on each alloy consisted primarily of Cr_2O_3 with a thinner, outer region of $MnCr_2O_4$. The 800HT alloy showed the additional presence of $(Fe, Ni)_3O_4$. The cross-sectional images are similar to those obtained after 3 thermal cycles (Fig. 3). It is worth noting that bare alloy was seldom observed when analysing 230 and HR-120, which is indicative of the fact that scale failure occurred within the scale rather than at the alloy/scale interface. Failure within the scale is preferred from the standpoint of prolonging the service life of the alloy since, in comparison to a bare alloy, less chromium depletion occurs during subsequent oxidation.

Figure 5 compares the surface and cross-sectional microstructures of the three alloys after 12 thermal cycles (i.e. 360 days). The cross-sectional microstructures are representative of least-spalled regions. The principal constituent in the scale that remained on 800HT was found by XRD to be $(Fe,Ni)_3O_4$, with Cr_2O_3 being secondary. For HR-120 and 230, Cr_2O_3 was still the principal scale.

The variation in chromium concentration in the alloy at the alloy/scale interface is shown in Fig. 6 as a function of exposure time up to 360 days. The concentrations are presented in atomic percent (at.%) in this figure in order to discount any enrichments or depletions of the other constituent elements in the alloys. The nominal bulk atomic concentrations of chromium in 800HT, HR-120 and 230 are about 23, 26 and 27%, respectively. Alloy 800HT showed the greatest amount of surface chromium depletion, decreasing below about 11 at.% after 180 days. The surface chromium concentration in 230 remained relatively constant at about 21 at.% after the first two thermal cycles. In the case of HR-120, the surface chromium concentration remained relatively constant at 22 at.% between 3 and 8 thermal cycles, but then decreased to about 16 at.% after 12 cycles. It is noted that the errors associated with the surface concentration measurements may be large due to the limited number of measurements made. Work is currently in progress to quantify more accurately the variation in chromium surface concentrations with exposure time.

Figure 7 compares the average metal affected in each alloy after 12 thermal cycles (i.e. 360 days). The average metal affected is the summation of metal recession and internal attack, with internal attack corresponding to internal precipitation and void formation. No distinction is made here between metal recession and internal attack since, for many applications, both can be equally detrimental. It is seen in Fig. 7 that internal attack was the predominant mode of degradation in each of the alloys studied. For 800HT, internal attack accounted for about 60% of the average metal affected after 360 days of oxidation, while for 230 and HR-120 internal attack accounted for about 95%.

4. Discussion

The three alloys, 800HT, HR-120, and 230, eventually exhibited mass-loss kinetics during long-term cyclic oxidation testing in still air at 982°C. The mass-loss kinetics were different for each alloy, with 230 exhibiting the least amount of mass loss and 800HT the most, undergoing breakaway mass-loss kinetics after 150–180 days. The

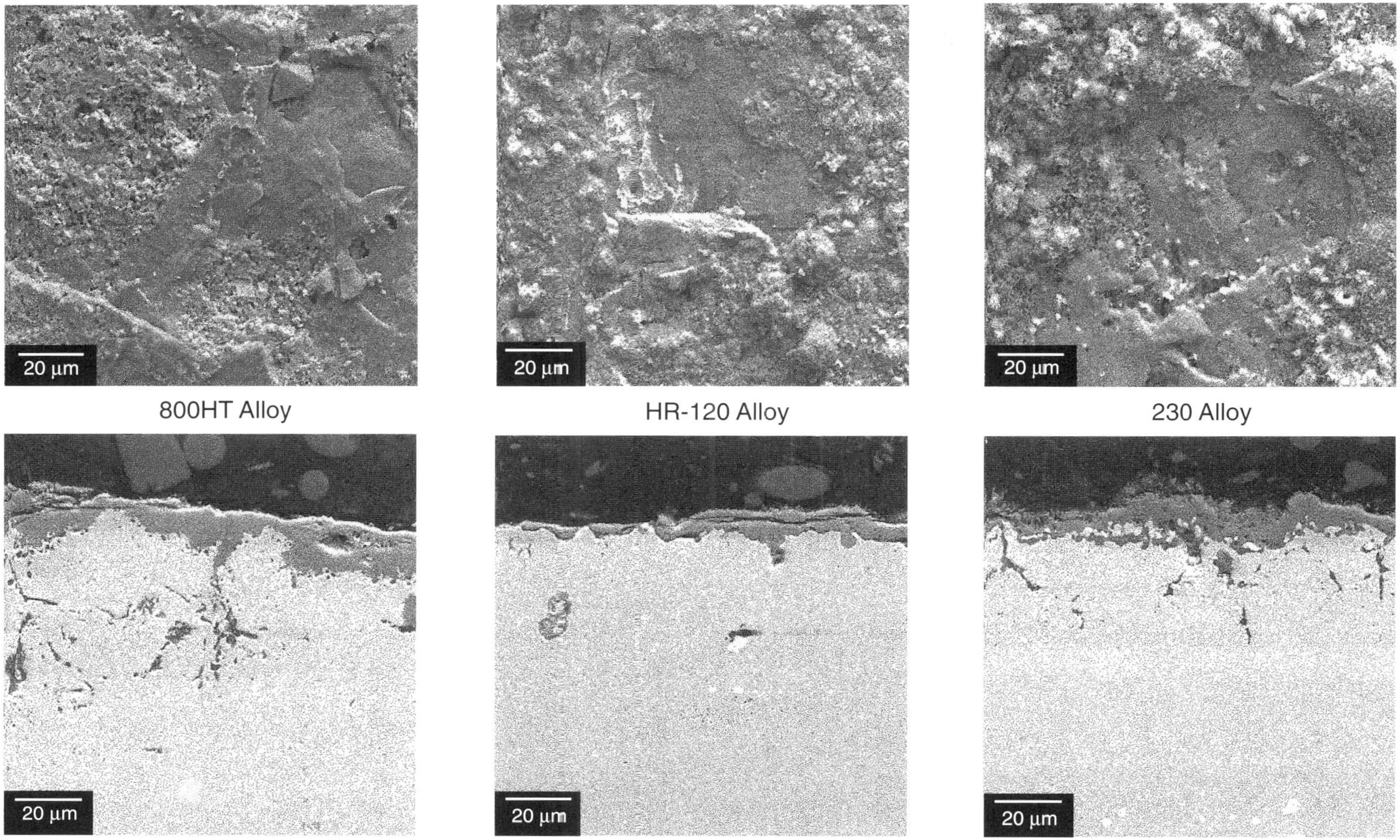

Fig. 5 *Surface and cross-sectional SEM images of the alloys 230, HR-120 and 800HT after twelve thermal cycles (i.e. 360 days) at 982°C.*

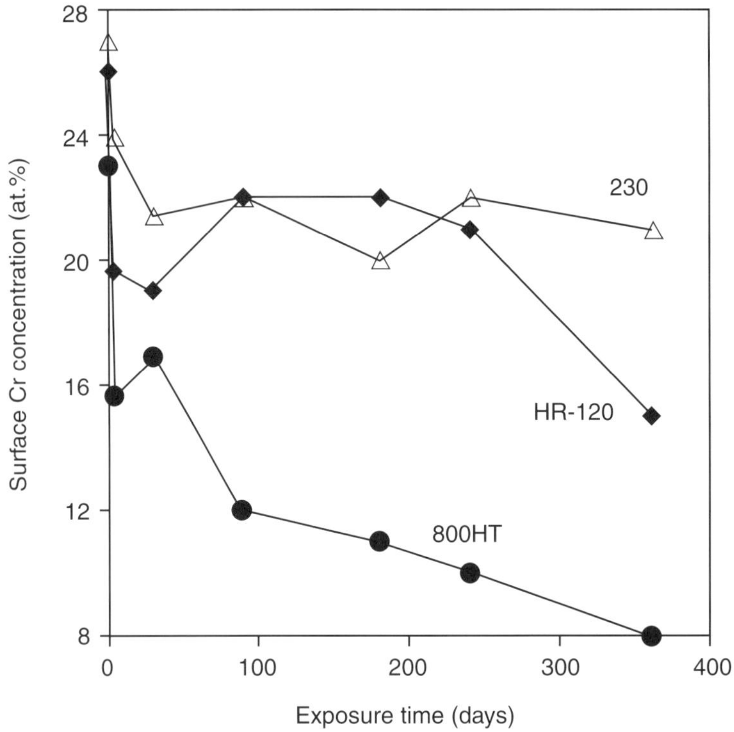

Fig. 6 *Variation in the surface chromium concentration in each of the alloys 230, HR-120 and 800 HT as a function of exposure time (30 day cycles) at 982℃.*

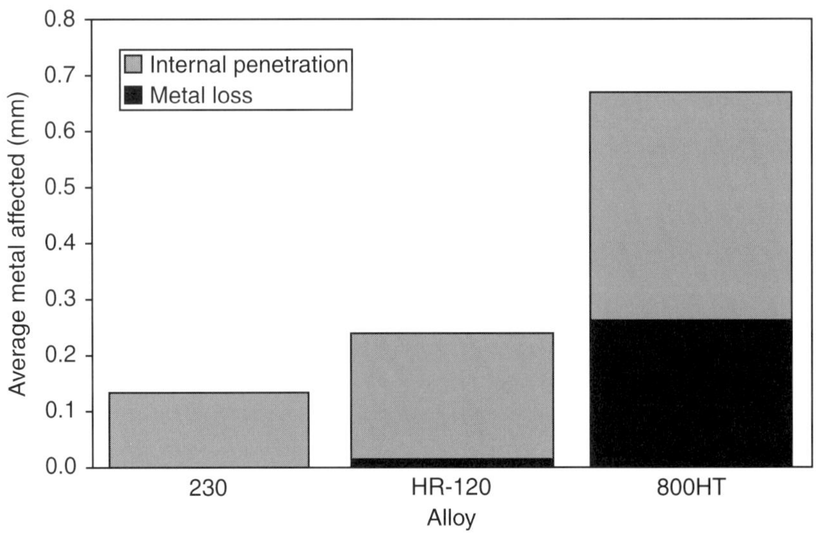

Fig. 7 *Comparison of the average metal affected (metal loss + average internal attack) for the alloys 230, HR-120 and 800HT tested for 360 days at 982℃.*

fact that the mass-loss kinetics varied between the alloys can be partly ascribed to differences in scale spallation behaviour which, in turn, can be related to differences in both scale properties and alloy compositions. For instance, it is well established that alloy composition affects the manner in which the scale spalls (i.e. partial or complete spallation) and, hence, the extents to which element enrichment and depletion occur within the underlying alloy [13].

The spallation of oxide scales generally involves the separation of cracked scale either by decohesion at the alloy/scale interface or by fracture on planes within the scale itself [14,15]. The failure mode which predominates depends on the stress state in the scale and on the degree of adhesion at the alloy/scale interface. Moon and Lee [16] quantified the fraction, F, of spalled scale resulting from a temperature drop, ΔT, by the expression

$$F = \left[\frac{\eta X (\Delta T)^2}{\beta} \right]^m \quad (1)$$

where x is the scale thickness before spalling, η is a factor that converts scale thickness to mass gain, m is a positive number which depends on the mechanical properties of the scale, and β is a constant which is dependent upon the strength of adhesion at the alloy/scale interface. The value of β increases with increasing strength of scale adhesion, while m typically varies from unity for a ductile scale to about 4 for a brittle scale [16]. Although not stated by Moon and Lee, it is clear from the present results that X, m, and β are actually cycle-dependent variables. Moreover, the values of m and β must be related in some manner to the difference in thermal-expansion coefficients between the alloy and scale (i.e. $\Delta\alpha = \alpha_{me} - \alpha_{ox}$, with α_{me} greater than α_{ox}). Since the oxidation times and temperature were fixed variables in the present study, the differences in the spallation behaviour of the alloys can be directly attributed to the cycle- and composition-dependent parameters X, m and β. The cyclic-oxidation behaviour of the three alloys studied in the context of these parameters, is discussed below.

A significant compositional difference between 230 and the other two alloys is that 230 contains nominally 0.02 mass% lanthanum, a reactive element known to decrease the extent of scale spallation at an alloying content below about 1.0 mass% [17,18]. Tawancy [19] reported that the trace lanthanum in 230 segregates to grain boundaries of the Cr_2O_3 scale, which has the consequent effects of causing the maintenance of a fine-grain scale after extended exposure and reducing the Cr_2O_3 growth rate by suppressing the flux of chromium through the scale. Thus, the superior cyclic-oxidation resistance of 230 can be largely ascribed to the presence of lanthanum decreasing X and m, while increasing β. A further factor is that 230 has the lowest CTE value of the three alloys. According to the specifications sheets of these alloys, the CTE values for 230, HR-120 and 800HT are 16×10^{-6}, 17.8×10^{-6}, and 18.7×10^{-6} K^{-1}, respectively. For comparison, the CTE for Cr_2O_3 is $7.3 - 8.5 \times 10^{-6}$ K^{-1} [2]. The lower CTE value for 230, which is due its the high refractory-metal content (14 mass% W + 2 mass% Mo), would contribute to lowering m and increasing β.

The 800HT alloy formed the thickest scale (i.e. highest X), which would enhance the occurrence of scale spallation. Moreover, according to Barrett and Lowell [20], Cr_2O_3-forming alloys which develop a distinctive outer spinel layer generally undergo a significant amount of scale spallation. In support of this conclusion, the scale on 800HT during the early stages of oxidation consisted of a relatively thick inner layer of Cr_2O_3 and a thin outer layer of the spinel $MnCr_2O_4$. The duplex scale structure possibly has the effect of decreasing the effective β and increasing *m*. The apparently faster growth rate of Cr_2O_3 on 800HT compared to the other two alloys is presumably due to one or a combination of factors, such as differences in the boundary condition at the alloy/scale interface for Cr_2O_3 and doping of the Cr_2O_3 scale. Indeed it is well established that the growth rate of Cr_2O_3 can vary over several orders of magnitude at a given temperature when the substrate composition varies [21]. Regardless of the reason for this, the faster Cr_2O_3 growth rate on 800HT, combined with more extensive scale spallation, resulted in a more accelerated subsurface depletion of chromium (Fig. 6). After 180 days the surface chromium concentration decreased to below about 11 at.%, which was insufficient for the re-formation of a continuous Cr_2O_3 scale. As a consequence, the iron and nickel oxidised to form a much less protective $(Fe,Ni)_3O_4$ scale, and this marked the commencement of the breakaway mass-loss kinetics.

The HR-120 alloy exhibited higher mass-loss kinetics than 230. This is quite possibly due to HR-120 having higher values of *X* and *m* at any given time, and a lower value of β. As discussed above, the differences in these values are believed to be largely attributable to the 0.02 mass% La in 230. The higher mass-loss kinetics exhibited by HR-120 were not sufficient to cause breakaway behaviour up to 720 days of testing. This can be ascribed to the relatively high chromium content in HR-120 (26 at.% compared to 23 at.% in 800HT) and the occurrence of scale failure within the scale rather than at the alloy/scale interface. This latter factor would particularly contribute to a lower rate of chromium depletion from the subsurface, as indicated in Fig. 6, and the ability of the alloy to sustain Cr_2O_3 growth. It is worth noting that HR-120 undergoes breakaway mass-loss kinetics after only 60 days (i.e. two thermal cycles) when the oxidation temperature is increased to 1093°C [8,9].

All the alloys underwent subsurface degradation in the form of internal oxidation and void formation. John [22] studied the oxidation behaviour of a variety of commercial, heat-resistant alloys over the temperature range 760–1093°C and found that more than 80% of the sound metal loss incurred by most alloys is a result of internal oxidation and void formation. John concluded that mass-change data are likely to underestimate significantly the extent of alloy degradation. The present results strongly support John's conclusion.

Void formation within the alloys was most probably a consequence of vacancy injection [23] and/or the internal oxidation of carbon [24,25]. Vacancy injection results from the fact that the outward diffusive flux of alloying elements undergoing selective oxidation (e.g. chromium) must be counterbalanced by an inward flux of vacancies. This is generally referred to as the Kirkendall effect [26], and it is often observed that the injected vacancies eventually coalesce to form voids. Alternatively, the carbon in the alloy can react with the inward diffusing oxygen to form CO and CO_2 gases which, if of sufficiently high pressures, can induce the nucleation and growth of cavities. Both these mechanisms for subsurface void formation are viable, and it is beyond the scope of this analysis to determine which one predominates. What is perhaps more important is the fact that the subsurface voids can act as crack-initiation sites, thus reducing the load-bearing capacity of the alloy [27].

Under equilibrium conditions, internal oxidation will occur only if

$$N_b^O N_o^s > K_{sp} \quad (2)$$

where N_b^O and N_o^s are the bulk content of constituent *B* and surface concentration of oxygen in the alloy, respectively, and K_{sp} is the solubility product of oxide BO in the alloy matrix. In general, the greater the stability of an oxide, the lower will be its K_{sp}. It is for this reason that internal precipitates of the stable oxides Al_2O_3 and TiO_2 are commonly observed in oxidised superalloys [22,28,29]. The results from the EDS analysis indicate that internal precipitates of these oxides formed in the present alloys. The greater amount of internal precipitation in 800HT can be ascribed to the higher amounts of Al and Ti in this alloy compared to 230 and HR-120.

The kinetics of internal oxidation under the conditions of cyclic oxidation have not yet been treated theoretically; however, it is easy to infer that the process is greatly complicated by scale spallation, the subsurface enrichment and depletion of alloying elements, and changes in the phase constitution of the scale as a function of oxidation cycles. The combined effect of these factors is alloy specific, and may act to either increase or decrease the extent of internal oxidation.

The final point to consider is the change in the surface chromium concentration with exposure time. As shown by Nesbitt [5], the surface concentration is predicted to decrease steadily with exposure time if it is assumed that the same proportion of scale spallation occurs after each thermal cycle and that the same scale reforms. Clearly, these assumptions are not valid for the alloys and test conditions used here. A modified model would have to account for the time dependence of the spallation behaviour and the corresponding interdependence between the subsurface changes which occur in the alloy and the type of scale which forms.

5. Conclusions

The relative performance of the alloys studied was 230>HR-120>>800HT for a testing duration of 720 days and an oxidising temperature of 982°C. The superior performance of 230 can be ascribed to its relatively low CTE and nominal content of 0.02 mass% La effecting improved scale adhesion and slower rate of Cr_2O_3 growth. The 800HT alloy formed a duplex scale structure consisting of a relatively fast-growing inner layer and a thin $MnCr_2O_4$ outer layer. The scaling behaviour of 800HT, combined with the occurrence of extensive spallation during thermal cycling, caused the rapid subsurface depletion of chromium concentration below that necessary for Cr_2O_3 reformation after about 150–180 days. This had the consequent effect of causing 800HT to undergo breakaway mass-loss kinetics. The higher chromium content and lower CTE value for HR-120 compared to 800HT helped to prelude HR-120 from undergoing breakaway kinetics within 720 days of testing.

Each of the alloys studied underwent internal attack in the form of internal oxidation and void formation. In all cases, internal attack was the principal mode of degradation. Such results demonstrate the importance of reporting the extent of internal attack in addition to metal loss or mass change data.

The present results further show that modelling of the long-term cyclic oxidation behaviour of complex alloys requires input variables which are time dependent (i.e. cycle dependent). This is because the scale and subsurface compositions change with the duration and number of cycles. Future work is necessary to correlate the interdependencies of these changes before the cyclic oxidation behaviour of complex alloys can be accurately modelled.

5. Acknowledgements

The authors gratefully acknowledge the assistance of J. E. Barnes, J. P. Cotner, M. A. Richeson and J. A. Ryan in conducting the experiments and analysing the samples.

References

1. G. C. Wood and D. P. Whittle, *Corros. Sci.*, 1967, **7**, 763.
2. M. Schütze, *Oxid. Met.*, 1995, **44**, 29.
3. D. P. Whittle, *Oxid. Met.*, 1972, **4**, 171.
4. G. Wahl, *Thin Solid Films*, 1983, **107**, 417.
5. J. A. Nesbitt, *J. Electrochem. Soc.*, 1989, **136**, 1518.
6. C. E. Lowell, J. L. Smialek and C. A. Barrett, in *High Temperature Corrosion*. (R. A. Rapp, ed.). National Association of Corrosion Engineers, Houston, TX, USA, 1983, p.219.
7. C. E. Lowell, C. A. Barrett, R. W. Palmer, J. V. Auping and H. B. Probst, *Oxid. Met.*, 1991, **36**, 81.
8. M. A. Harper, J. E. Barnes and G. Y. Lai, *Corrosion '97*, Paper No. 132, NACE International, Houston, TX, USA, 1997.
9. B. Gleeson and M. A. Harper, *Oxid. Met.*, 1998, **49**, 373.
10. J. R. Nicholls, in EFC working party report on *Guidelines for Methods of Testing and Research in High Temperature Corrosion*. (H. J. Grabke and D. B. Meadowcroft, eds). Publication No.14 in the European Federation of Corrosion Series published by The Institute of Materials, London, 1994, pp.11–36.
11. F. I. Wei and F. H. Stott, *High Temp. Technol.*, 1989, **7**, 59.
12. F. Delaunay, C. Berthier, D. Galy, J. M. Lameille, and M. Lenglet, *Microscopy of Oxidation – 3*, (S. B. Newcomb and J. A. Little, eds). The Institute of Materials, London, 1997, pp.587–596.
13. C. A. Barrett, *A Statistical Analysis of Elevated Temperature Gravimetric Cyclic Oxidation Data of 36 Ni- and Co-base Superalloys Based on an Oxidation Attack Parameter*, NASA TM-105934, NASA Lewis Research Center, Dec. 1992.
14. K. S. Chan, *Metall. Mater. Trans. A*, 1997, **28A**, 411.
15. H. E. Evans, *Int. Mater. Rev.*, 1995, **40**, 1.
16. C-O. Moon and S-B. Lee, *Oxid. Met.*, 1993, **39**, 1.
17. D. P. Whittle and J. Stringer, *Phil. Trans. Roy. Soc.*, 1980, **295A**, 309.
18. S. Mrowec, J. Jedlinski and A. Gil, *Mater. Sci. Eng.*, 1989, **A120**, 169.
19. H. M. Tawancy, *Oxid. Met.*, 1996, **45**, 323.
20. C. A. Barrett and C. E. Lowell, *Oxid. Met.*, 1975, **9**, 307.
21. H. Hindam and D. P. Whittle, *Oxid. Met.*, 1982, **18**, 245.
22. R. C. John, *Corrosion '96*, Paper No. 171, NACE International, Houston, TX, USA, 1996.
23. A. H. Rosenstein, J. K. Tien and W. D. Nix, *Metall. Trans. A*, 1986, **17A**, 151.
24. R. H. Bricknell and D. A. Woodford, *Acta Metall.*, 1982, **30**, 257.
25. D. Gan, *Metall. Trans. A*, 1983, **14A**, 1518.
26. A. Smigelskas and E. Kirkendall, *Trans. AIME*, 1947, **171**, 130.
27. R. H. Bricknell, R. A. Mulford and D. A. Woodford, *Met. Trans. A*, 1982, **13A**, 1223.
28. C. L. Angerman, *Oxid. Met.*, 1972, **5**, 149.
29. C. H. Lund and H. J. Wagner, *Oxidation of Nickel- and Cobalt-Base Superalloys*, Defence Metals Information Center (DMIC) Report 214, Battelle Memorial Institute, 1965.

18
Oxidation Limited Life of FeCrAl Based Alloys During Thermal Cycling

D. NAUMENKO, L. SINGHEISER and W. J. QUADAKKERS
Forschungszentrum Jülich, IWV-2, 52425 Jülich, Germany

ABSTRACT
The oxidation limited life of FeCrAl alloys during high temperature service is determined by depletion of the Al reservoir within the alloy matrix. Based on a recently developed model which allows quantitative prediction of the oxidation limited lives, the results of high temperature exposures for commercial FeCrAl materials are evaluated with special emphasis on the effect of thermocycling. Growth rates and adhesion of the alumina scales on thin foils of FeCrAl alloys were found to be unaffected by temperature cycling. Consequently, the life time is mainly determined by oxide growth rates, bulk aluminium content (C_0) and critical content for selective oxidation of Al (C_B). C_B values were found to be higher for the ODS material than for wrought alloys of similar composition. On the contrary, for components of 1–2 mm thickness, scale spallation, induced by thermal cycling has a crucial effect on life time. The tendency for oxide scales to spall is affected by minor alloying additions but also by alloy creep strength and component geometry.

1. Introduction

FeCrAl alloys are being used in high temperature applications because of their ability to form protective alumina scales in air and exhaust gas environments. For improving oxide scale properties with respect to growth rate and adherence the alloys, which are usually of the composition Fe + 16–20Cr +5Al (mass%), are doped with small amounts of reactive elements (e.g. Y, Ce) either in metallic form or as oxide dispersions. Two main product forms of such alloys presently exist. Foils with a thickness of 30–100 μm are, for example, being applied for car catalyst carriers whereas materials of a few millimetre thickness are being used in a large variety of applications such as heating elements, furnace parts and heat exchangers. For the last application, oxide dispersion strengthened (ODS) alloys seem to be especially suitable because of the combination of excellent oxidation behaviour with superior creep strength at temperatures above 1000°C [1,2].

Formation of the protective alumina scales relies on the aluminium reservoir in the alloy matrix. Depletion of Al due to the oxidation at elevated temperatures results in formation of rapidly growing Fe-based oxides. This event, termed 'breakaway', leads to eventual rapid failure of the component. Breakaway, a failure mode for thin foils as well as for parts of several millimetre thickness, has been subject to extensive research during the last decade [3,4]. After long term exposures the alumina scale formed on thick components usually spalls but re-heals thus depleting the aluminium from the bulk alloy. In this respect scale adhesion and resistance of the surface oxide to spallation, which can be affected by cyclic parameters, are of great concern [4]. In contrast, for the thin foils seemingly no spallation has been observed [4,5]. It was proposed that this is due to the fact that the scale does not reach the critical thickness for spallation and that the alumina growth rate is the only life limiting factor [4]. It is, however, not known whether thermal cycling can lead to scale microcracking and consequently enhanced oxide growth rates and whether the scale, formed on thin foils becomes prone to spallation under severe cyclic conditions.

In this paper results of the isothermal and cyclic oxidation tests of thin foils and thick coupons of several FeCrAl-based alloys at 1200°C are presented. Measured times to breakaway (t_B) are compared with the values calculated using the recently developed model [4].

2. Experimental

Commercial wrought alloys Aluchrom YHf and Kanthal AF, the powder metallurgical alloy Kanthal APM, the commercial ODS alloy PM2000 and ODS model alloys of different thickness have been studied. The materials investigated are listed in Table 1. It is important to note that the actual thickness of the Aluchrom YHf foils, designated by the manufacturer as 50 μm, appeared to be 57–58 μm as measured with a digital micrometer, i.e. about 15% higher than that of Kanthal AF. Coupons 20 × 10 mm were machined from the sheet material and 8 mm dia. disk coupons were cut from the hot extruded rods of the model ODS alloys. The foils of the wrought alloys were oxidised in the as-received state. The foils and 0.5–2 mm thick coupons of the ODS alloys were ground to 1200 grit surface finish. All specimens were degreased in a detergent before oxidation. Cyclic oxidation tests (20 h heating/2 h cooling) were performed in a resistance heated furnace (Fig. 1) at 1200°C in laboratory air. For determining the oxide scale growth rates during isothermal exposures

Table 1. *Data sheet for materials investigated*

Material	Batch	Thickness	Composition (mass%)	Dimensions
[a]Kanthal AF	FSF	50, 70 μm	Fe + 21.7Cr + 5.4Al + RE	20 × 10 mm
[b]Aluchrom YHf	FTF	30, 58, 80 μm	Fe + 20.3Cr + 5. 6Al + 0.05Y + 0.05Hf	20 × 10 mm
Kanthal APM	CST	1–2 mm	Fe + 20.0Cr + 5.5Al + 0.03Ti + 0.1Zr	20 × 10 mm
[c]PM 2000	FSK	100 μm	Fe + 18/8Cr + 5.8 Al + 0.50Y_2O_3 + 0.5Ti	20 × 10 mm
[d]ODM 751	CCC	1–2 mm	Fe + 16.3Cr + 4.7Al + 0.42Y_2O_3 + 0.6Ti	20 × 10 mm
[e]MA 956	CBL	1–2 mm	Fe + 19.4Cr + 4.6Al + 0.56Y_2O_3 + 0.4Ti	12 mm dia.
Model Alloy	DAG	1–2 mm	Fe + 16.0Cr + 6.5Al + 0.50Y_2O_3	8 mm dia.
Model Alloy	DAH	1–2 mm	Fe + 16.0Cr + 6.5Al + 0.50Y_2O_3 + 0.4Ti	8 mm dia.
Model Alloy	DAJ	1–2 mm	Fe + 16.0Cr + 6.5Al + 0.20Y_2O_3 + 0.4Ti	8 mm dia.

List of manufacturers: [a]Kanthal; [b]Krupp/VDM; [c]Plansee; [d]Dour Metal; [e]Inco Int.

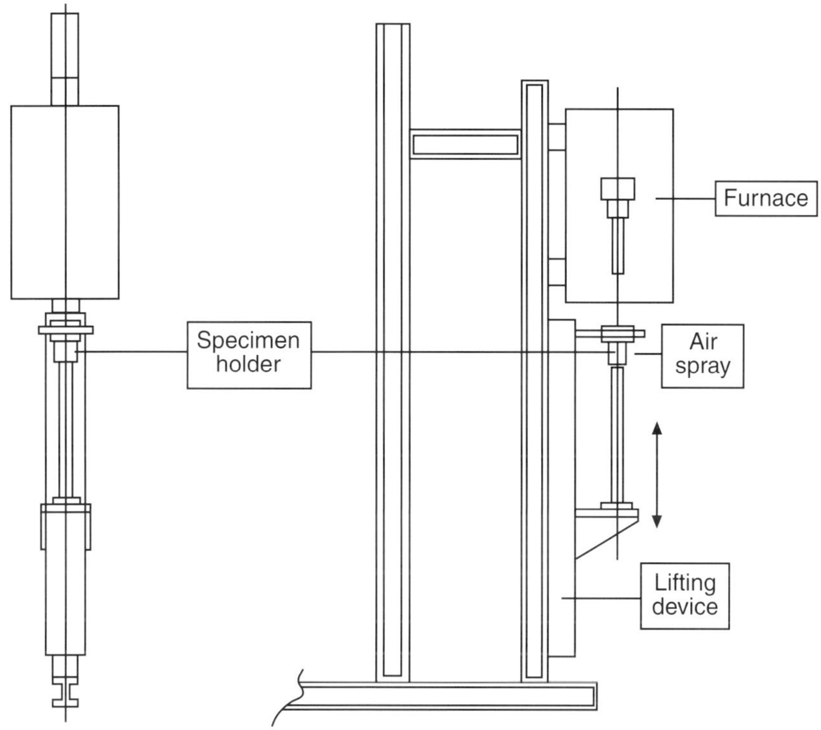

Fig. 1 Schematic illustration of laboratory equipment for cyclic oxidation testing.

thermogravimetric (TG) experiments were carried out at 1200°C in synthetic air. The oxidised samples were investigated using optical metallography and scanning electron microscopy (SEM) with energy dispersive X-ray analysis (EDX).

3. Oxidation Limited Life

A detailed description of the life-prediction model for FeCrAl alloys for different specimen geometries has been given in reference [4]. It was shown that for a flat uniform specimen of infinite length and width, assuming no spallation of the oxide scale to occur, the time to breakaway (t_B) is given by:

$$t_B = [4.4 \times 10^{-3} \times \Delta C_B \times \rho \times d / k]^{1/n} \quad (1)$$

where $\Delta C_B = C_0 - C_B$ is the aluminium reservoir in mass%, C_0 and C_B are the initial and breakaway Al contents respectively, d is the specimen thickness in cm and ρ the alloy density in mg cm^{-3}. The parameters k and n are power law oxidation rate constants [4]. Considering a specimen of given dimensions the thickness d has to be replaced by a modification factor f [6], defined as:

$$f = 2 \times Volume / Surface\ area \quad (2)$$

The factor f can be easily calculated for different component shapes:

- for a rectangular specimen $f = d/(1+d/l+d/b)$ (3)

where l is the specimen length and b the width in cm;

- for a disk shape specimen of thickness d and diameter r
$$f = d/(1+2d/r) \quad (4)$$

- for a cylinder of diameter δ and length l
$$f = \delta/(2+\delta/l) \quad (5)$$

- for a wire of diameter δ
$$f = \delta/2 \quad (6)$$

Taking spallation into account and assuming that the oxide scale spalls after reaching a critical thickness (which corresponds to a critical weight gain Δm^*), eqn. (1) changes into:

$$t_B = 4.4 \times 10^{-3} \times \Delta C_B \times \rho \times d \times (k)^{-1/n} \times (\Delta m^*)^{1/n-1} \quad (7)$$

For real components with finite dimensions the factor f rather than thickness d has to be used in this equation. Usually during long term exposures quite linear gross mass gains (i.e. total oxygen uptake including mass of the spalled oxide) for thick coupons of ODS alloys have been observed [4] which allows the life-time prediction using eqn. (7) to be made with reasonable accuracy. For calculation of t_B in eqns (1) and (7) the value of the breakaway aluminium concentration (C_B) is required. It was shown for thick components of commercial ODS alloys that the value of C_B is usually 1.2–1.3 mass% [4,6].

4.1. Life Time of Thin Foils

Figure 2 and 3 show mass change data for foils of Kanthal AF and Aluchrom YHf at 1200°C during isothermal and cyclic (20 h heating/2 h cooling) oxidation. No significant difference in mass gain between cyclic and isothermal oxidation is observed until breakaway. After about 100 h exposure the mass change curves of the 58 μm thick samples of Aluchrom YHf start to exhibit a decreased slope. Also, for the 100 μm thick foils of the ODS alloy PM2000 the isothermal and cyclic mass change data (Fig. 4) look very similar. A comparison of the isothermal oxidation data for the three alloys investigated is shown in Fig. 5. The growth rate of the 30 μm thick Aluchrom foil is slightly lower than that of the other materials.

The results of the isothermal oxidation tests were used to determine the oxide growth rate constants. An example of a double logarithmic plot of mass change data used for this purpose is presented for the 58 μm thick Aluchrom YHf in Fig. 6. The determined values of k and n for the various materials are shown in Table 2. The data exhibit the lowest growth rates for the 30 μm specimens of Aluchrom YHf and the highest for the 100 μm thick specimens of PM2000.

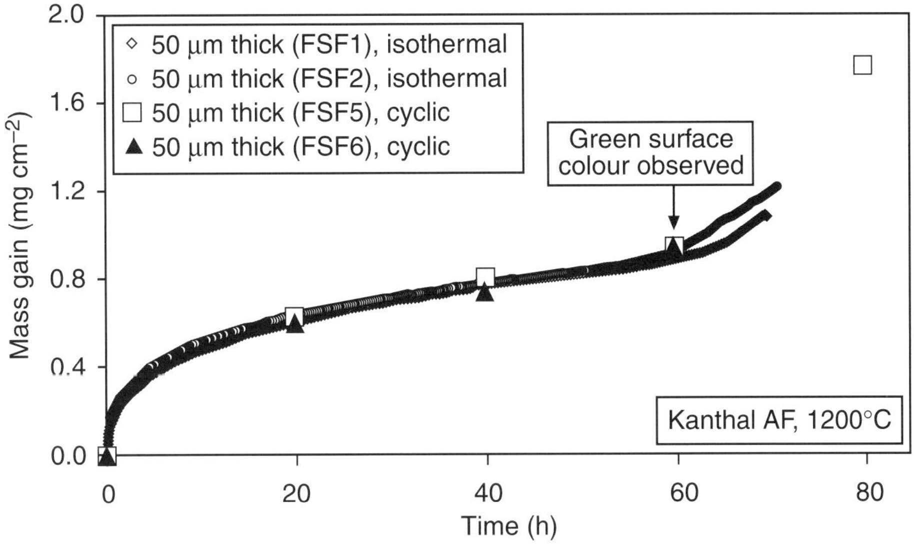

Fig. 2 *Mass change data for Kanthal AF foils (50 μm thick) during isothermal and cyclic oxidation (20-h cycles) at 1200°C in air. Arrow indicates first observation of surface colour change during cyclic oxidation.*

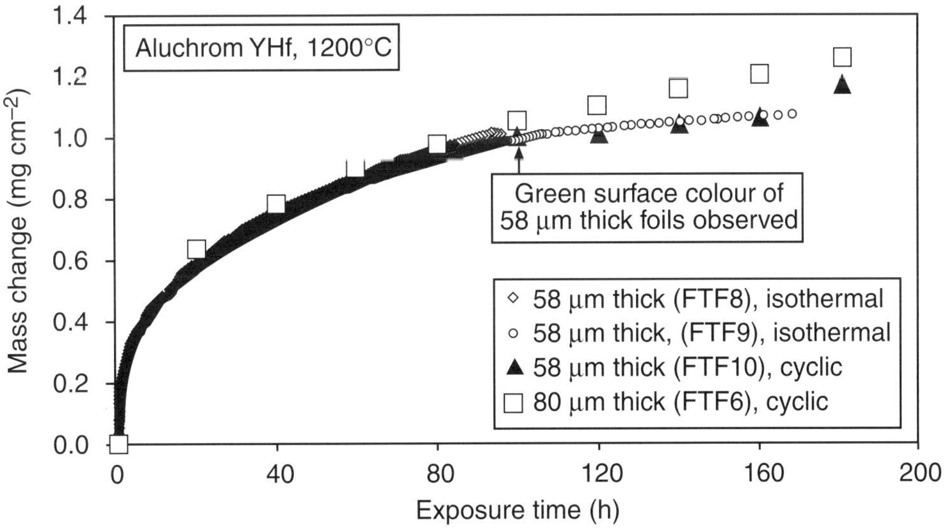

Fig. 3 *Mass change data for Aluchrom YHf foils (58 μm thick) during isothermal and cyclic oxidation (20-h cycles) at 1200°C in air. Arrow indicates first observation of surface colour change during cyclic oxidation.*

The ODS foils showed a type of breakaway oxidation which is different from that of the wrought alloy foils (Fig. 7). For the ODS material the breakaway front starts at the specimen corner propagating to the centre (Fig. 7a). Metallographic cross-sections

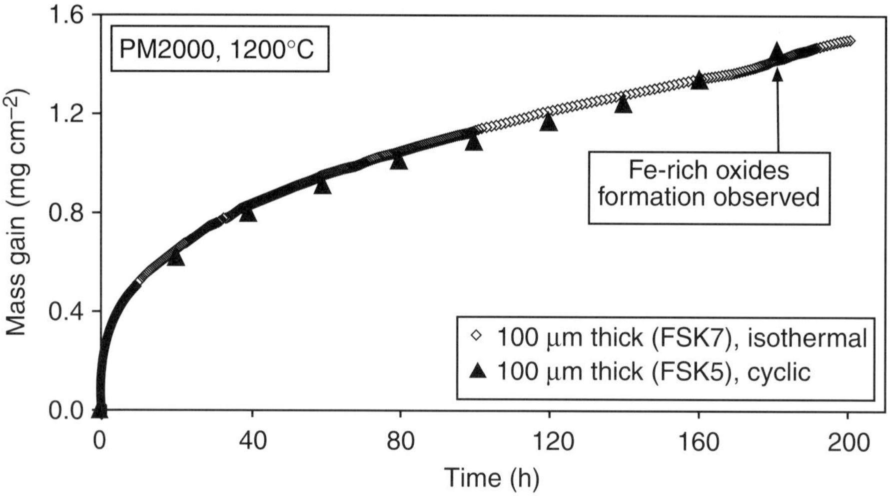

Fig. 4 *Mass change data for PM2000 foils (100 μm thick) during isothermal and cyclic oxidation (20-h cycles) at 1200°C in air. Arrow indicates first observation of surface colour change during cyclic oxidation.*

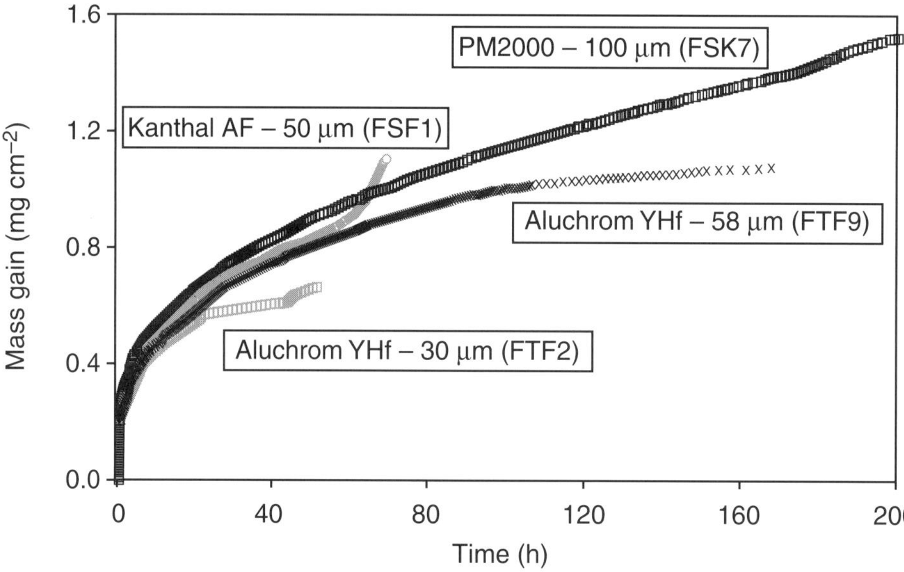

Fig. 5 *Comparison of mass change data for foils of commercial alloys during isothermal oxidation at 1200°C in synthetic air.*

show the typical appearance of commencement and progress of breakaway as previously observed for thick coupons [3,4]. The breakaway front is preceded by a region of internal oxidation (Fig. 8a) and further away from the front, TiN precipitates are found within the alloy matrix. In the case of the wrought alloy foils breakaway is accompanied by changes in surface colour and deformation of the specimen (Fig. 7b).

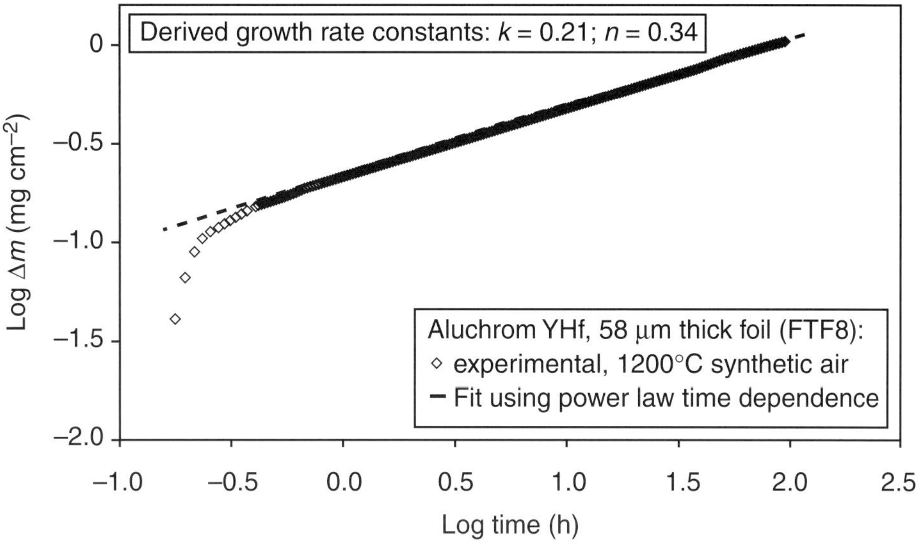

***Fig. 6** Example of double logarithmic plot of mass change data for Aluchrom YHf (58 μm thick foil) during isothermal oxidation at 1200°C to illustrate determination of oxide growth rate parameters* k *and* n.

The metallographic cross-sections reveal that this is caused by formation of chromium-rich oxide beneath the alumina scale (Fig. 8b). This double-layered scale, which is macroscopically apparent by a greenish colouring, covers practically the whole surface of the foil before rapidly growing Fe-rich oxides start to form. After longer exposure times severe scale and substrate deformation occur and often the formation of large rectangular pores is observed (Fig. 8c).

Away from the breakaway front in the matrix of the PM2000 ODS foil an aluminium content of 1.2–1.5mass% was determined by EDX analysis. This is in agreement with literature data for thick ODS coupons [4,6]. The EDX measurements in the region of internal oxidation next to the breakaway front showed, however, much lower values i.e. less than 0.5mass%. Only 0-0.3mass% of aluminium was determined by EDX in the Kanthal AF and Aluchrom YHf foils. This leads to the conclusion that in the latter case almost all aluminium had been consumed before breakaway, i.e. in this case formation of a chromia sub-scale, started. It is important to stress that EDX or EPMA point analyses may not always give relevant values of C_B because Al can be

***Table 2.** Values of power law constants* k *and* n *describing oxidation kinetics of studied alloys during isothermal oxidation at 1200°C (*k *and* n *defined as* Am = kt^n *where* Δm *is the mass change in mg* cm^{-2} *and* t *is time in hours)*

Material	Thickness	*k*	*n*
Aluchrom YHf	30 μm	0.20	0.29
Aluchrom YHf	58 μm	0.21	0.34
Kanthal AF	50 μm	0.20	0.38
PM2000	100 μm	0.24	0.34

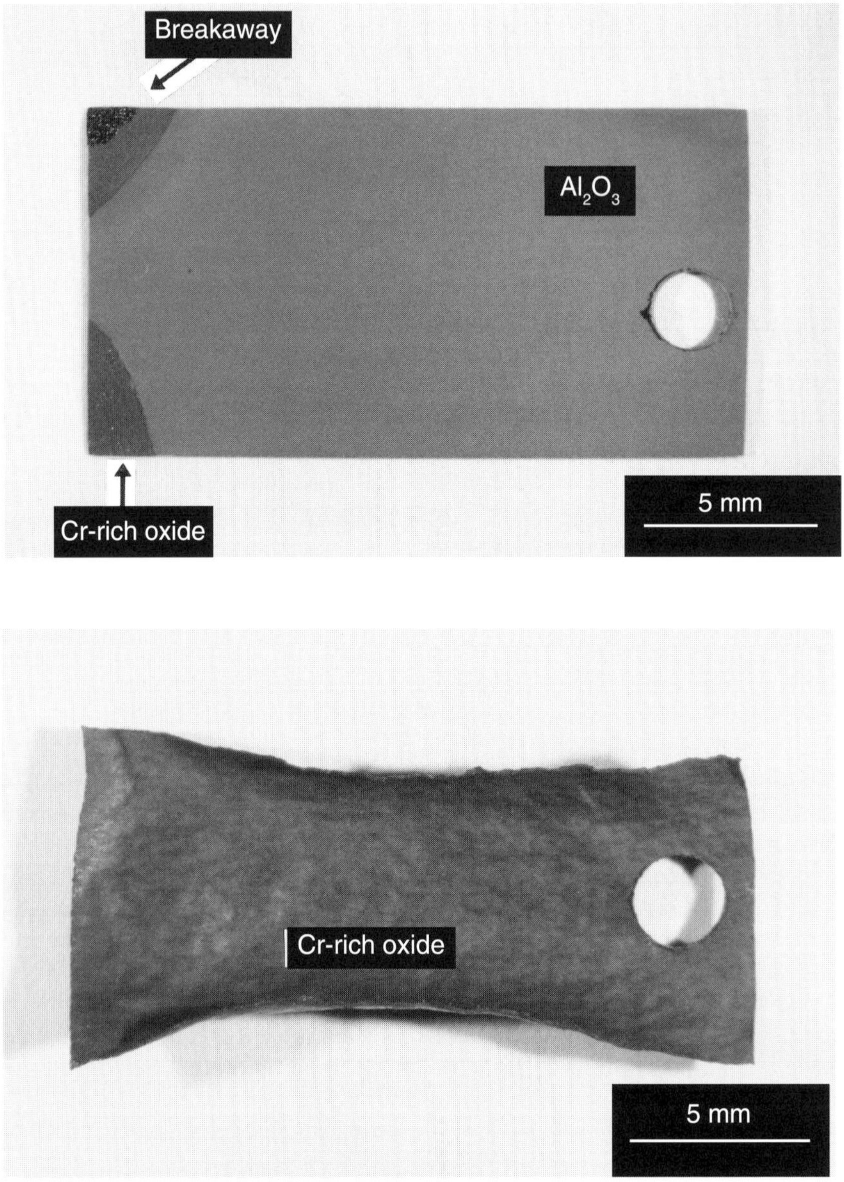

Fig. 7 *Macrographs of breakaway features for foils of commercial FeCrAl alloys after isothermal oxidation at 1200°C in synthetic air: (a) 100 μm thick foil of PM2000 after 200 h of exposure; (b) 58 μm thick foil of Aluchrom YHf after 170 h of exposure.*

tied up in internal nitride/oxide precipitates, resulting in an extremely low matrix Al concentration [3,6].

Times to breakaway were assessed from the mass change curves using eqn (1). C_B values of 1.3mass% for PM2000 foils and 0.3mass% for the wrought alloy foils were used. The calculated values of t_B are plotted in Fig. 9 in comparison with the experimental data.

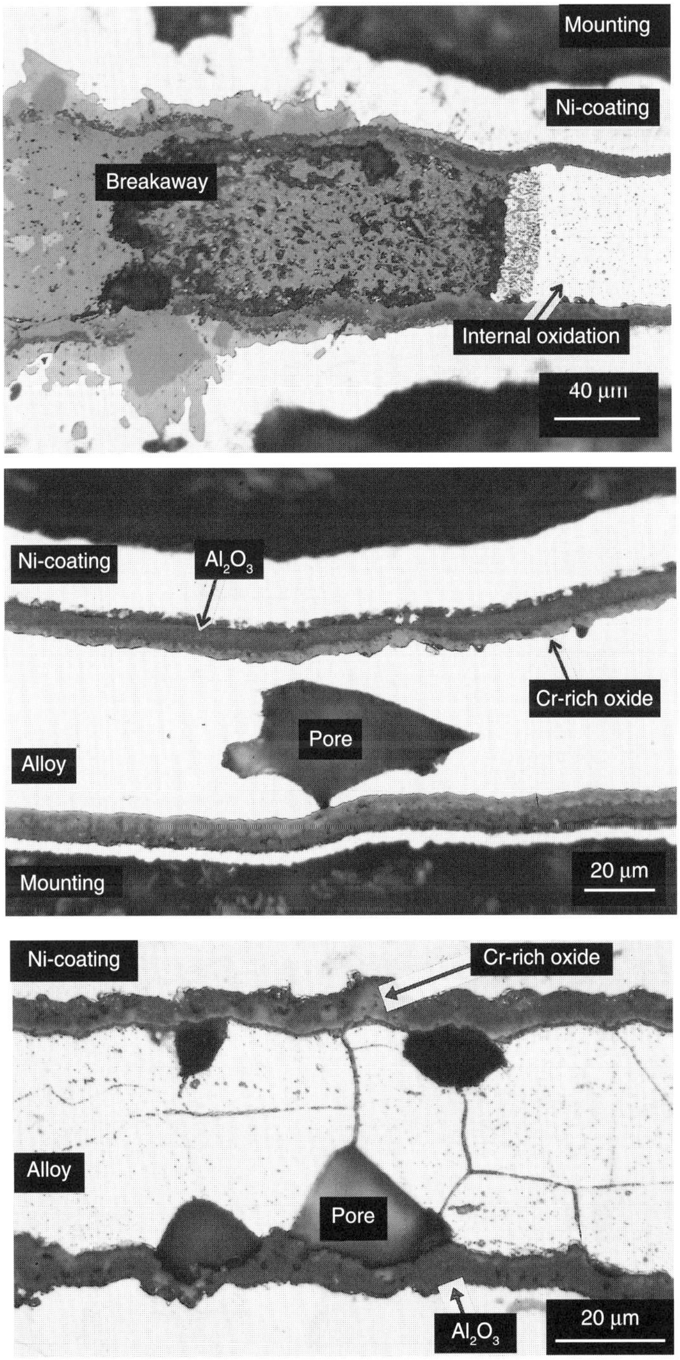

Fig. 8 *Metallographic cross-sections of FeCrAl foils after cyclic oxidation (20-h cycles) until breakaway at 1200°C in air: (a) 100 μm thick foil of PM2000 after 180 h of exposure; (b) 58 μm thick foil of Aluchrom YHf after 240 h of exposure; (c) 50 μm thick foil of Kanthal AF after 80 h of exposure (etched).*

Actually, it appeared to be a problem to derive from the original mass change data whether thermal cycling has an effect on the occurrence of breakaway in thin foils of the FeCrAl alloys. The data for isothermal and cyclic exposures show excellent agreement and no clear effect of temperature cycling on the oxidation rates was found (Figs 2–4). During the cyclic exposures changes in the colour of the specimen surface after cooling and deformation rather than significant mass gain unequivocally indicated the onset of breakaway which in case of the wrought alloys is characterised by sub-scale chromium oxide formation. It was, however, difficult to determine accurately the value of t_B during isothermal exposures. For practical reasons it was impossible to interrupt the TG tests exactly at the moment when breakaway started, since in the beginning no rapid mass gain could be detected. A green colouring , i.e. a proof for the occurrence of breakaway, was found after finishing the TG experiment with e.g. specimen FTF9 (Fig. 7b), while no clear evidence for onset of Cr_2O_3 formation could be derived from the mass change curve (Fig. 3). Another example is the 50 μm thick foil of Kanthal AF for which the value of t_B calculated using eqn (1) is only about 44 h i.e. a much shorter time than one would expect from the mass change data in Fig. 2.

For a more accurate determination of the life time it was decided to use a plot of parabolic rate constant (K_p) as a function of time to check whether slight changes in oxide scale growth rate occurred as soon as sub-scale chromia formation commenced. Fig. 10(a) shows the typical appearance of a $K_p(t)$-plot for alumina forming alloys [8]

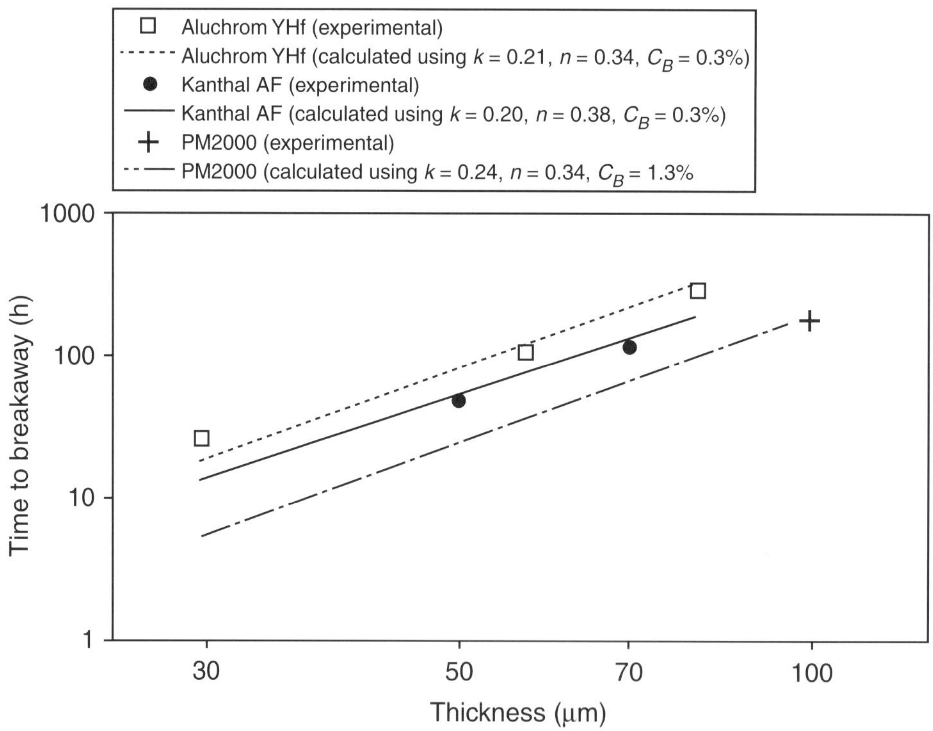

Fig. 9 Time to breakaway vs thickness plot for commercial FeCrAl foils oxidised at 1200°C (comparison of experimental and predicted results); calculation parameters are given in the Figure.

i.e. K_p slowly decreases with time during the first 40 h of oxidation of the Kanthal AF foil specimen. However, after approximately 43–44 h K_p exhibits a slow increase. At this stage apparently the chromium-rich oxide starts to form. The dominating formation of Fe-rich oxides occurs after around 57 h leading to a stronger increase of K_p. In the case of the isothermally oxidised 58 µm thick Aluchrom foil an apparent decrease in K_p was found after about 95 h (Fig. 10b). This was related to the chromia formation and confirmed by the specimens oxidised during 20-h cycles which showed greenish surface colouring after the fifth cycle, i.e.100 h exposure.

The presented results show that generally for thin foils of the wrought FeCrAl alloys, as was also reported elsewhere [5,7], three stages of oxidation can be distinguished:

1. Alumina formation;

2. Sub-scale chromia formation, and

3. Formation of iron-based oxides.

The duration of each stage depends seemingly on the type and/or amount of reactive elements and perhaps the foil thickness. In some cases the chromia stage appeared to be extremely long (e.g. more than 70 h at 1200°C for 58 µm thick Aluchrom YHf foils, Fig. 3) and no rapid mass gain during this stage was detected.

The results also confirm that for thin foils the time at which sub-scale chromia formation occurs can actually be considered as t_B because it is accompanied by significant deformation of the foil (see Fig. 7b and ref. [5]). A possible explanation for the latter effect is stresses induced by the chromia formation and growth. It is well established that both aluminium and chromium oxides on Fe-based RE-doped alloys grow mainly by inward oxygen diffusion [8,9]. In this case compressive growth stresses within the scale can be related to the larger volume of the formed oxide compared to that of the metal consumed [8] (Pilling-Bedworth ratios for chromia and alumina on FeCrAl alloys were calculated to be 2.1 and 1.8 respectively). In the growing Cr_2O_3 sub-scale constrained by the initially formed Al_2O_3 (Fig. 8b) much higher growth stresses must arise than those in the uniform and non-constrained alumina. This will result in tensile stresses and plastic deformation in the thin, mechanically weak substrate which in turn can lead to the formation of large pores with the characteristic triangular or rectangular shape as e.g. seen in the etched cross-section of the 50 µm thick Kanthal AF foil in Fig. 8(c). Of course, for the growth of pores, coalescence of vacancies left by Al and Cr atoms within the alloy matrix may contribute.

It seems that the 58 µm thick Aluchrom YHf foil behaves in a slightly different manner to that of Kanthal AF following the onset of chromia formation (compare Figs 10a and 10b). In the case of Aluchrom the second stage of oxidation is very extended and the K_p values are even lower than during the first stage. Similar observations by Ishii *et al.* [5] were explained by the fact that the growth of chromia is limited by the oxygen transport through the initially formed Al_2O_3. However, another reason for this drop in oxidation rate may be partial loss of Cr_2O_3 due to evaporation of volatile CrO_3 species [8].

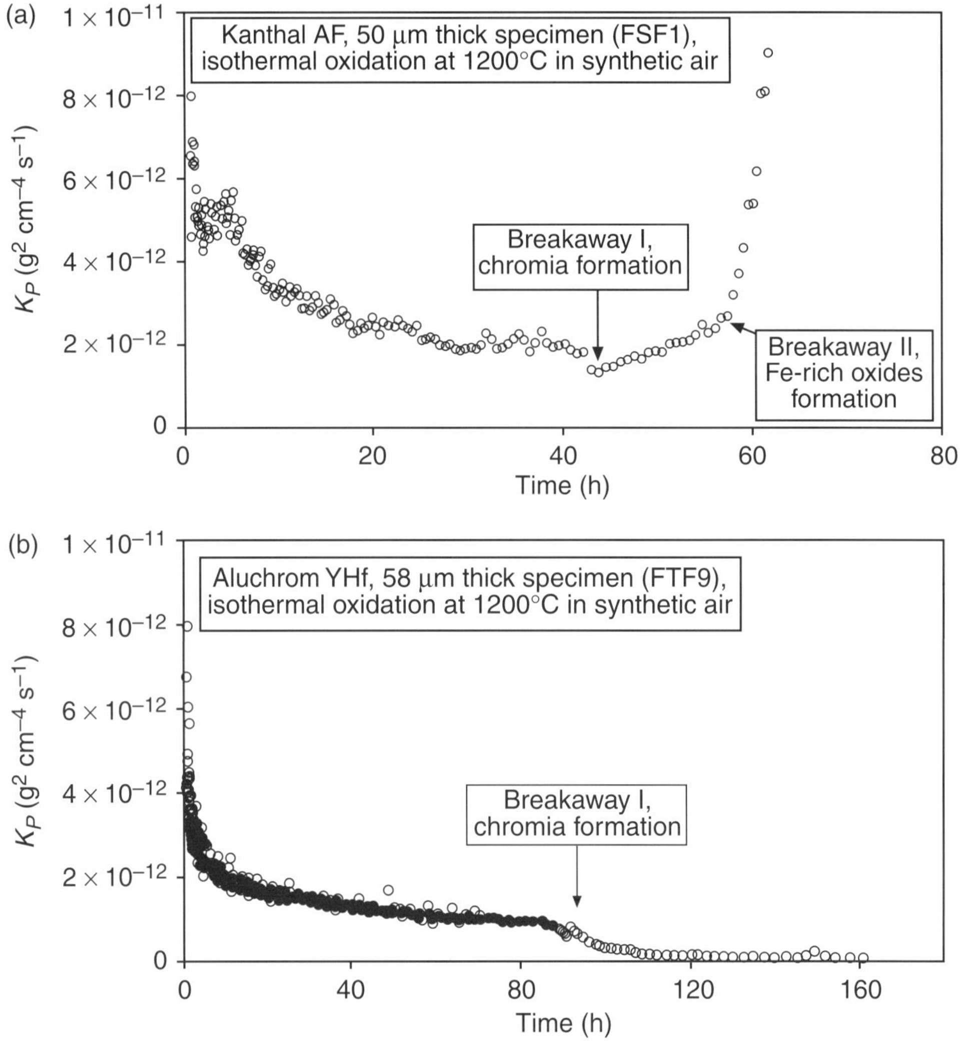

Fig. 10 *Plots of parabolic growth rate constants vs time for isothermal oxidation at 1200°C illustrating changes in growth rate due to chromia formation: (a) 50 μm thick foil of Kanthal AF; (b) 58 μm thick foil of Aluchrom YHf foil.*

In contrast to the wrought alloys the mechanism of breakaway oxidation for PM2000 foils (see Figs 7a, 8a) is very similar to that of 1–2 mm thick ODS specimens previously described [3,4]. Formation of Fe-rich oxides starts at the corner of the specimen and progresses towards the centre. This indicates that the second oxidation stage, accompanied by sub-scale chromia growth is extremely short. The time to breakaway assuming C_B to be 1.3mass% calculated using eqn (1) appeared to be around 170 h which is in excellent agreement with the experimental data (Figs 4, 9). A large region in the alloy matrix is affected by breakaway: an internal oxidation zone appears and further from the breakaway front an area with TiN precipitates is visible. In the middle of the specimen no microstructural changes due to occurrence of breakaway were observed. In the region of internal oxidation chromia was found beneath and apparently embedded in the alumina scale. The TiN precipitates were probably formed as a result of nitrogen diffusing from the breakaway area into the alloy [6]. These nitrides do not exist in the region of internal oxidation probably because of the local high oxygen activity. The area of TiN precipitates and the unaffected area can lead to the conclusion that the Al concentration and activity are still high enough in these places to maintain the protective alumina scale.

Although the higher C_B value of 1.3mass% for the ODS foils may also be attributed to a specimen thickness effect (100 µm thick PM2000 versus 58 µm thick Aluchrom YHf and 50 µm thick Kanthal AF) it seems more reasonable to assume that it is related to the other form in which the reactive element prevails in the ODS alloy (Y_2O_3 rather than Y). The easiest explanation might be that during alloy production and high-temperature exposure yttria ties up some aluminium in the form of a $Y_3Al_5O_{12}$ compound [10]. However, a simple calculation shows that less than 0.2mass% Al can completely react with 0.5mass% yttria in this way. It was found that in case of ODS Fe–Al alloys the breakaway aluminium concentration does not depend on the type of oxide dispersions (yttria or alumina) [11]. It was proposed that this is because the diffusion of aluminium towards the scale/metal interface takes place mainly through the alloy grain boundaries and thus mainly depends on the alloy microstructure. That is obviously not the case for the FeCrAl alloys where, in spite of the large grain size, the diffusion of aluminium was determined in the studied cases to be very fast [4,6] and thus unlikely to be significantly affected by differences in alloy grain boundary density. A more likely explanation for the high C_B value is that the reactive effect element in the form of a dispersion has a different effect on selective oxidation of Al in the ODS foils from that of the metallic reactive element in the wrought foils in which a breakaway aluminium content close to zero was determined. The exact mechanism by which reactive elements can affect the selective oxidation is not completely understood [8]. It might be, for instance, that after significant depletion, the Al flux towards the scale / metal interface, especially on the corners of an ODS alloy specimen, is inhibited by a rather large volume fraction of yttria dispersions.

4. 2. Life Time of Thick Components

As mentioned above, on components of 1 to 2 mm the oxide scales usually grow to a thickness at which the growth and thermally induced stresses are relaxed by scale spalling. The critical mass gain (Δm^* in eqn (7) in section 3) at which spallation starts depends on a number of factors including the intrinsic adherence of the scale to the substrate, the substrate material creep strength [14], oxidation test parameters (cooling rates, cycle frequency) etc. As illustrated by Figs 11 and 12, spallation behaviour can differ from material to material and cannot always be related directly to a value of Δm^* derived from short term mass change data [6].

For the ODS model alloy DAH the value of Δm^* is much lower than that of the MA956 specimen of the same thickness i.e. the scale spallation starts earlier (Fig. 12). Despite this fact the life of the DAH sample is almost twice that of MA956 apparently due to its larger aluminium reservoir. It is important to mention however, that an attempt to significantly increase the life time of an FeCrAl component simply by

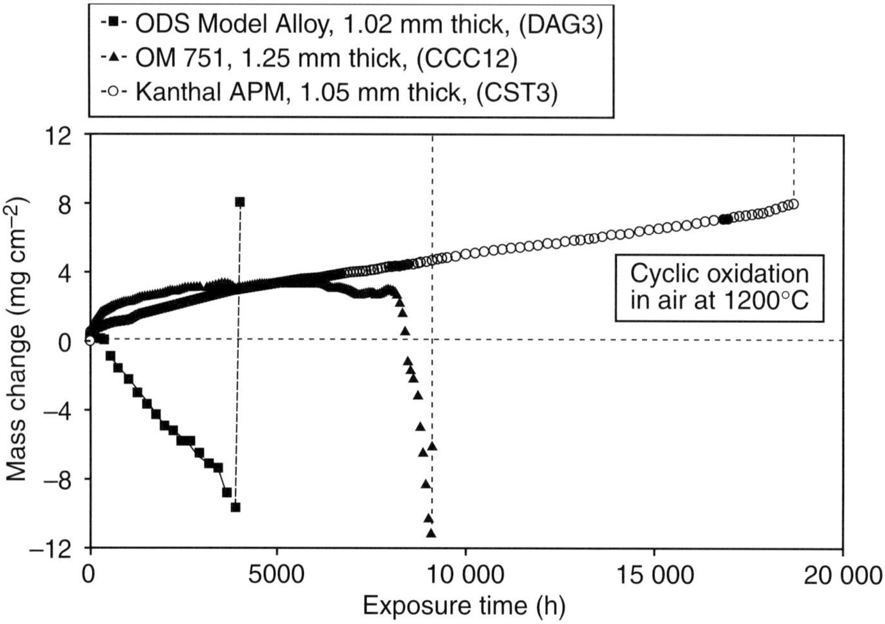

Fig. 11 Mass change data for thick coupons of FeCrAl alloys during cyclic oxidation at 1200°C in air, showing different types of spalling behaviour.

increasing C_0 will in most cases not be practical. First, the restrictions imparted by the low ductility of alloys with high Al contents are encountered [6]. Secondly, it was not possible to extend the life time of ODS model alloys by increasing C_0 from 6.5 to 12mass% with a parallel decrease in the Cr content [13]. Thus it is still to be established

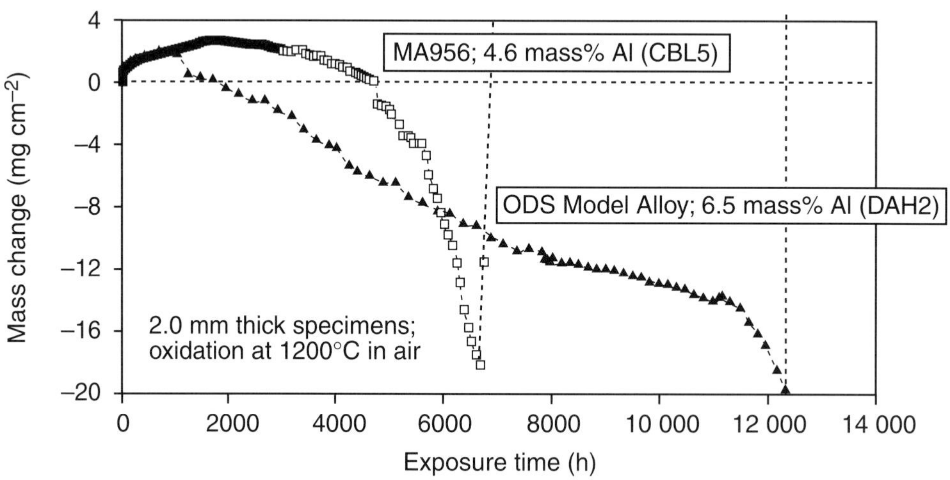

Fig. 12 Mass change data for thick coupons of MA956 (batch CBL) and ODS model alloy (batch DAH) during cyclic oxidation at 1200°C in air illustrating differences in Δm and effect of initial Al content on life time.*

which concentrations of major alloying elements (i.e. Cr, Al) give the optimum results with respect to a combination of oxidation limited life and suitable mechanical properties.

As in case of thin foils, the alumina growth rate also for thicker components, i.e. the temperature and composition dependent constants *k* and *n* in eqn (7), have a crucial effect on life time. Results published in reference [4] show that the growth rate of the alumina scale on ODS alloys increases and consequently the life time decreases if the yttria content in an MA956 type alloy is increased from 0.17 to 0.7mass%. This effect of exact yttria content on life time also occurs in case of model ODS alloys with 6.5mass% Al (Fig. 13). Comparison of the very small differences in the growth rates of the two alloys indicates that an additional factor might play a role in the large differences in life time. A possible explanation would be that a higher volume fraction of yttria shortens the life time by increasing the strength of the material, because the resistance to spallation was previously shown to be affected by the intrinsic alloy creep strength [14]. It was found [14] that in mechanically weaker alloys the growth and thermally induced stresses in alumina can be relaxed by creep of the substrate. In Fig. 13 spalling of the alumina scale is shown to start after about 1000 h in the case of alloy DAH (0.5mass%Y_2O_3) and after around 2000 h for alloy DAJ (0.2mass%Y_2O_3) This difference in spalling resistance may be correlated with the slight difference in oxide growth rate [4], but can also be affected by the difference in creep strength of the materials with the weaker material possessing the better resistance to spalling. The latter effect might also explain the excellent oxide adherence and consequently the extremely long life of the 1.05 mm thick specimen of Kanthal APM in comparison with ODS alloys of similar composition (see Fig. 11).

Resistance to spallation not only depends on the intrinsic creep strength of the material but apparently also on the specimen geometry (e.g. thickness) [6]. The life time of specimens from the same ODS model alloy (Fig. 14) is not increased by a factor of 1.85 with doubling the thickness as one would expect from eqn (7). It was proposed [6] that in a thinner (and thus mechanically weaker) coupon the stresses

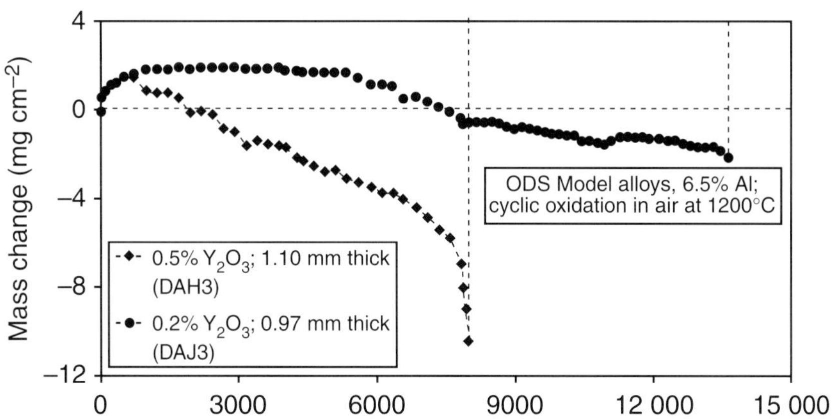

Fig. 13 *Mass change data for ODS model alloys (batch DAH and DAJ) during cyclic oxidation at 1200°C illustrating effect of yttria content on life time.*

generated in the oxide scale can be relaxed by the plastic deformation of the substrate more easily than in the case of thicker coupons.

Minor elements (e.g. titanium, which is a common addition in most of the commercial ODS alloys) can have a significant effect on the time to breakaway (Figs 15, 16). Severe spallation of the scale formed on the Ti-free ODS model alloy (batch DAG) is observed after only 100 h of oxidation at 1200°C (Fig. 15a). This results in a much shorter life time of alloy DAG than that of a Ti-doped alloy (DAH) of similar composition (Fig. 15b). Oxide scale morphologies on the two alloys after 1000 h exposure at 1200°C appeared to be clearly different. Large areas of bare metal were exposed in the case of the Ti-free alloy. In the regions where alumina was still adherent the oxide scale was found to be rather convoluted and porous (Fig. 16a), whereas addition of 0.4mass% Ti apparently promoted formation of a flat, adherent alumina scale and no significant spallation was detected (Fig. 16b). It is important to note that for thin wrought foils of FeCrAl alloys Ti additions were found to increase the scale growth rates due to 'overdoping' [12], mainly because of enhancement of internal oxidation. Further studies are necessary to elucidate under which conditions small additions of Ti are beneficial and when they are detrimental in respect to oxidation limited life.

5. Conclusions

The life time of thin ODS and wrought foils at 1200°C during cyclic oxidation is not determined by spallation of the alumina scale. In all the studied cases, the materials reached breakaway before clear oxide scale damage occurred. The oxide growth rate appeared to be virtually unaffected by thermal cycling. The onset of breakaway oxidation of wrought FeCrAl foils was not accompanied by a rapid mass gain

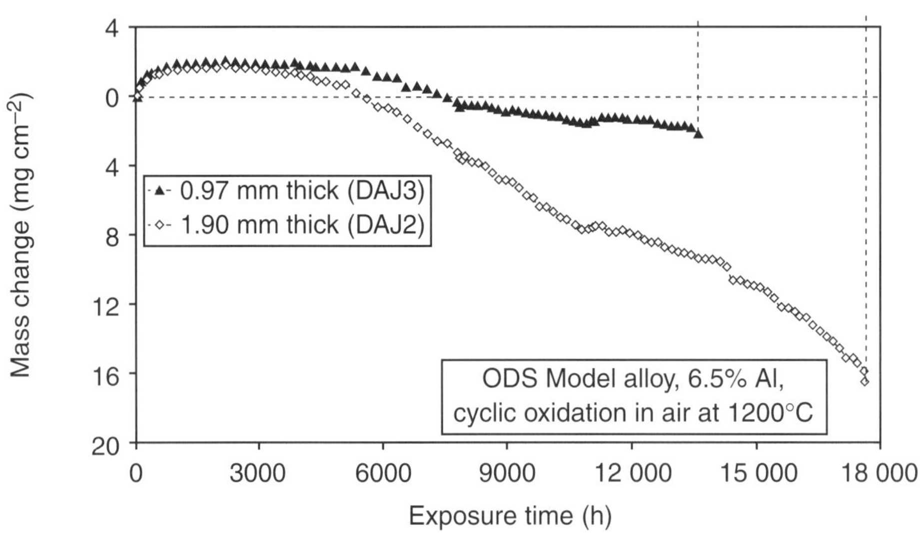

Fig. 14 *Mass change data for ODS model alloy (batch DAJ) during cyclic oxidation at 1200°C illustrating effect of specimen thickness on resistance to oxide scale spallation and life time.*

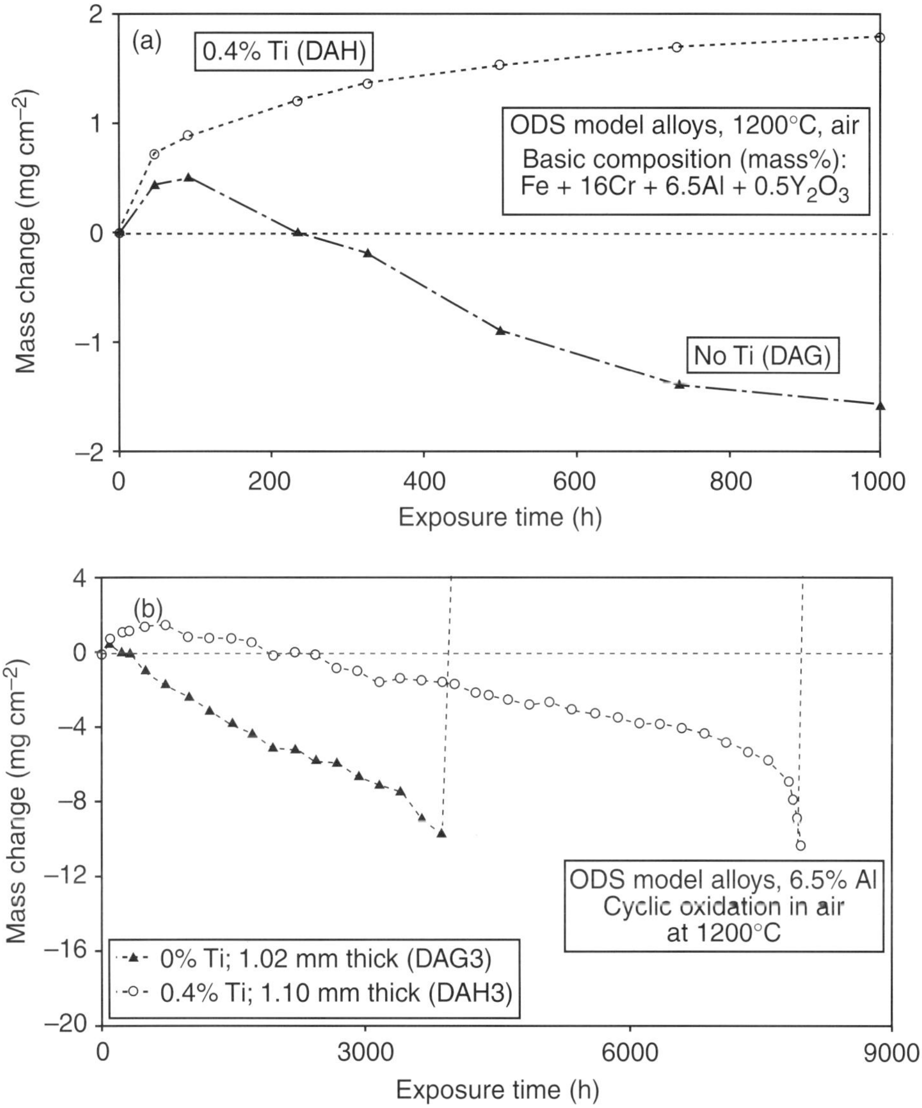

Fig. 15 Mass change data for ODS model alloys (batches DAG and DAH) during cyclic oxidation at 1200°C illustrating effect of Ti additions on (a) resistance to oxide spallation and (b) life time.

resulting from formation of Fe-rich oxides but by a rather long time period of chromia formation beneath the alumina scale. This sub-scale Cr_2O_3 formation results in only minor changes in oxide growth rate and is therefore difficult to determine during isothermal oxidation. On the contrary, the breakaway features of 100 μm thick ODS foils are similar to those of thick (1–2 mm) ODS coupons.

Wrought alloy foils showed almost total consumption of Al (C_B less than 0.3mass%) before breakaway occurred. The higher C_B values of 1.3mass% Al for ODS foils may be attributed to the other form in which the reactive element prevailed (oxide

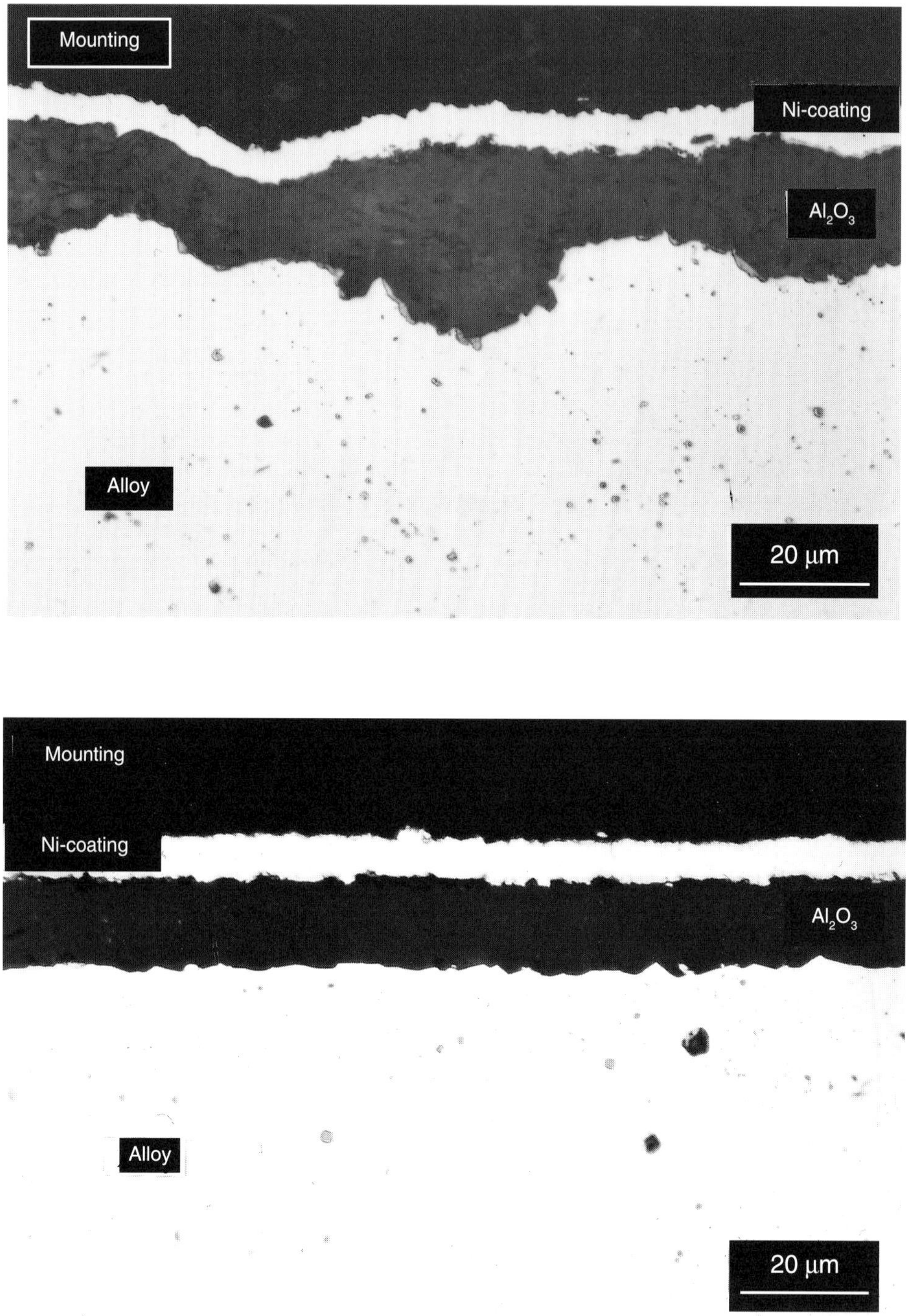

Fig. 16 *Metallographic cross-sections of ODS model alloys after 1000 h cyclic oxidation at 1200°C in air: (a) Ti-free batch DAG; (b) batch DAH containing 0.4mass% Ti.*

dispersion instead of metallic Y, Hf, etc.). Apparently the dispersion has a less positive effect on the selective oxidation of aluminium and chromium than the metallic reactive element additions.

In the case of thermal cycling of 1–2 mm thick FeCrAl components the resistance to aluminium oxide spallation plays a major role in the oxidation limited life time. Onset and rate of spallation can be affected not only by the intrinsic adhesion of the oxide to the metallic substrate but also by other factors, such as:

(a) alumina scale growth rate,

(b) creep strength of the material and consequently the amount and distribution of yttria dispersions in the case of the ODS alloys,

(c) component or specimen geometry, scales on thinner specimens of the same material being more resistant to spallation than those on thicker ones.

The life time prediction model for FeCrAl alloys proposed in refs [4,6] gives values of times to breakaway which are in reasonable agreement with experimental results.

6. Acknowledgements

The authors are grateful to their colleagues in Forschungszentrum Jülich for their assistance in carrying out the experimental work: Mrs Felden and Mr Gutzeit for optical metallography and Mr Olefs for the oxidation studies. Part of this research was supported by the European Community in the framework of the LEAFA project (Project No. BE-97-4491).

References

1. H. Bode, Developmental status of materials for metal supported automotive catalysts, in *Proc. Int. Conf. On Metal Supported Automotive Catalytic Converters*, Wuppertal, Germany, 1997, p.17–31. (H. Bode, ed.). Werkstoff-Informations gesellschaft mbH, Frankfurt, 1997.
2. F. Starr, *Mater. High Temp.*, 1995, **13**, (2), 75.
3. W. J. Quadakkers and M. J. Bennett, *Mater. Sci. Technol.*, 1994, **10**, (2), 126–131.
4. W. J. Quadakkers and K. Bongartz, *Werkst. Korros.*, 1994, **45**, 232–241.
5. K. Ishii, M. Kohno, S. Ishikawa and S. Satoh, *Mater. Trans., JIM*, 1997, **38**, (9), 787–992.
6. I. Gurappa, S. Weinbruch, D. Naumenko and W. J. Quadakkers, submitted for publication in *Materials and Corrosion*.
7. G. Strehl *et al.*, Cyclic oxidation of FeCrAl alloys: SEM and EPMA investigations, this volume, pp. 95–108.
8. P. Kofstad, *High Temperature Corrosion*, 1988, Elsevier Applied Science, London, UK, 1988.
9. W. J. Quadakkers, *Werkst. Korros.*, 1990, **41**, 659–668.
10. A. Czyrska-Filemonowicz *et al.*, AFM and SNMS studies of alumina scales formed during early stages of oxidation on INCOLOY MA956, *Proc. 3rd Int. Conf. on the Microscopy of Oxidation*, Cambridge, 1996. (S. B. Newcomb and J. A. Little, eds). The Institute of Materials, London, 1997, pp.185–194.

11. B. A. Pint, J. Leibowitz and J. H. DeVan, *Oxid. Met.*, 1999, **51**, (1/2), 181–197.
12. J. Klöwer and A. Kolb-Telieps, Effect of aluminium and reactive elements on the oxidation behaviour of thin FeCrAl foils, in *Proc. Int. Conf. on Metal Supported Automotive Catalytic Converters*, Wuppertal, Germany, 1997, (H. Bode, ed.) Werkstoff-Informations gesellscaft mbH, Frankfurt, 1997, pp.33–47.
13. W. J. Quadakkers, Factors affecting the oxidation limited life of components fabricated from FeCrAl-Based alloys containing RE- or RE-oxide additions, in *Proc. Int. Conf. On Metal Supported Automotive Catalytic Converters*, Wuppertal, Germany, 1997, (H. Bode, ed.). Werkstoff-Informations gesellschaft mbH, Frankfurt, 1997, pp. 149–159.
14. H. E. Evans, *Mater. High Temp.*, 1994, **12**, (2–3), 219–227.

19
Characterisation of Oxidation Resistance of Heat-resistant Cast Alloys in the 5 × 24-h Test

W. STEINKUSCH
POSE-MARRE, Edelstahlwerk GmbH, Erkrath, Germany

ABSTRACT
For determination of the oxidation resistance of heat-resistant cast steels small cylindrical samples were annealed in the 5 × 24-h test. To find the maximum allowable temperature the definition in SEW 471-60 (Stahl-Eisen-Werkstoffblatt) was used. The test results prove the usefulness of this test procedure.

1. Introduction

The maximum operation temperature permitted for heat-resistant cast-steel materials, which were in use about 40 years ago, was stated in the standards that were then valid, for example SEW 471-60, March 1960 [1].

Laboratory tests and practical experience were considered in these standards which were revised in 1976 [2].

Approximately 60 years ago Bandel and Volk [3] described a method to find the oxidation resistance. The method became known as the 5 × 24-h test. The method was developed due to the absence of a standard test. The mass loss ($gm^{-2}h^{-1}$) after annealing was used to measure the amount of scaling. It was determined that the maximum operation temperature should be defined to be that giving a medium oxidation rate of 1 or 1–2 $gm^{-2}h^{-1}$ in 120 h (= 5 × 24-h). The temperature up to which cast steel materials are resistant to oxidation has been defined in SEW 471-60 as follows: Steel is oxidation resistant up to temperature X at which the mass loss is 1 $gm^{-2}h^{-1}$ and at 50°C higher temperature, i.e. X + 50°C, the mass loss does not exceed 2$gm^{-2}h^{-1}$.This definition has again been incorporated in the draft of EN 10295 [4].

Development of cast steel material requires determinations of oxidation resistance for the new alloy types. Therefore numerous oxidation tests were conducted between 1968–1973 [5] using the 5 × 24-h test.

2. Experimental

2.1. Annealing

The machined and degreased samples are mounted on a metallic specimen holder (see Fig. 1), the design of which ensures minimum contact between the sample and the holder. The holder is then inserted into a chamber furnace pre-heated to a given

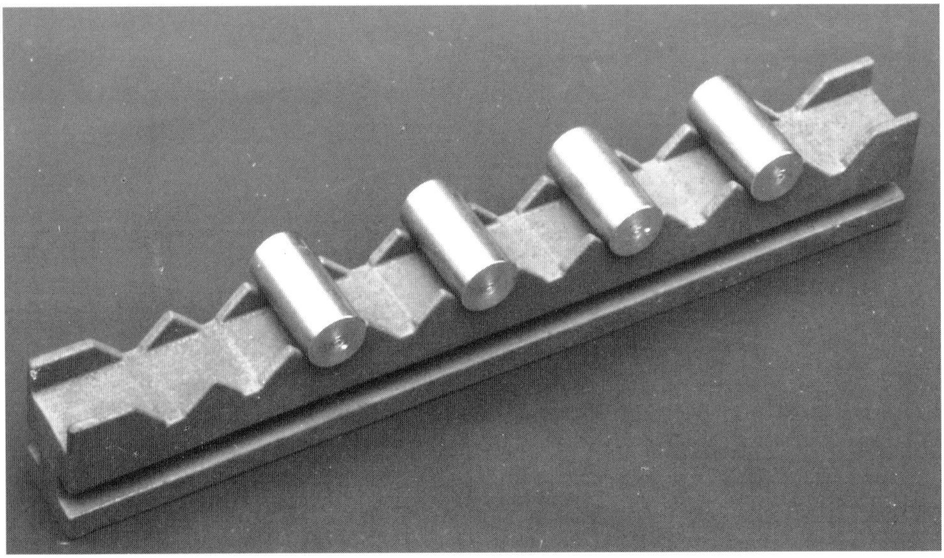

Fig. 1 Specimen holders for oxidation tests.

test temperature. The atmosphere in the furnace is air. When the samples have reached the test temperature it is maintained for 24 h. Afterwards the holder is removed from the chamber furnace and the samples are cooled down in air to room temperature. This oxidation and cooling process is repeated 5 times.

2.2. Determination of Scaling Loss

The scaling loss as a function of temperature was determined. The scaling loss was measured in $gm^{-2}h^{-1}$ which necessitates the utmost accuracy regarding the sample mass before and after the oxidation test.

Since scaling loss depends on the size of the sample surface, the sample dimensions had to be measured before the test to be able to calculate the surface area. When the test was completed the scale had to be removed.

2.3. Shape of the Samples

The oxide scale caused by annealing can spall from the sample during cooling from the test temperature. Cubic-type samples are obviously more sensitive to spallation than spherical samples. Cylindrical samples were chosen for the tests with 15 mm dia. × 30 mm height. Tests with both cubic and cylindrical samples in different sizes showed a scatter band of + 12% from the average with respect to the mass loss ($gm^{-2}h^{-1}$) for both types of samples.

2.4. Descaling

Scale caused by oxidation can be removed by either a mechanical procedure (i.e. wire brushes) or by a chemical procedure. With the chemical procedure (i.e. pickling) the cast-steel material should not dissolve. Since scaling layers differ in different areas as regards thickness, the layers were first loosened in a salt bath (NaOH at 350°C), brushed off and then the sample was pickled in a 20% H_2SO_4+2% HCl solution at room temperature. However, the amount of cast steel loss during the pickling process is not known. In NaOH at 350°C it is approximately 0.01 $gm^{-2}h^{-1}$. The chemical procedure is unreliable and complex. The mechanical procedure is more favorable and therefore wire brushing was preferred.

3. Results of the 5 × 24-h Test

3.1. Determination of the Permissible Temperature

The calculated values of mass loss for some heat resistant steels are plotted vs the test temperatures in Figs 2–13 (pp. 310–321). According to the definition in SEW 471-60 steel is oxidation resistant up to temperature *X* at which the loss of mass is 1 $gm^{-2}h^{-1}$, and at a 50°C higher temperature, the value of mass loss does not exceed 2 $gm^{-2}h^{-1}$. From Fig. 2 (a) it is seen that steel PG 25/12 (1.4837, GX40CrNiSi25–12) is oxidation resistant according to this requirement. This is also true for PG 25/35H (1.4857, GX40NiCrSi35–25) shown in Fig. 3 with a moderate increase of mass loss. However, for cast material PG24/24Nb (1.4855, GX40CrNiSiNb24–24) which shows sudden rise in mass loss (Fig. 4), the mass definition has to be changed; in this case 50°C must be deducted from the temperature at which a mass loss of 2 gm^{-2} h^{-1} is achieved. This temperature is then considered the new oxidation resistance limit. In cases like this the field information and the observations made in our stress rupture test facilities are helpful to find the right temperature.

3.2. Evaluation of Reliability

Not all test results could be demonstrated in parabolic functions as has been the case so far. The function of PG 25/20H (1.4848, GX40CrNiSi25–20) in Fig. 5, for example, is discontinuous which makes it difficult to determine the temperature exactly.

Parallel tests which run separately and independently with one material only are more reliable in such cases. Our tests did, not however, involve such parallel tests. The diagram which demonstrates the mass loss as a function of the annealing temperature is best provided with the cast material PG25/20H (1.4848, GX40CrNiSi25–20) since the amount of tests conducted is higher, each with small analytical variations. The results show that the scatter band of the average value is +30% as can be seen in Fig. 6.

3.3. Comments on Test Results

Figures 7–13 show some results from tests with 73 materials consisting of 60 heat resistant cast materials, 7 heat resistant wrought alloys and 6 filler metals. Typical cast materials with different functions are shown in Figs 7–10. The lowering of the oxidation resistance by niobium is shown in Fig. 11, the temperature decrease is between 50 and 100°C. There is no influence of grain size on the cast and forged Co-base alloy UMCo 50 (Fig. 12).

Testing time 5 × 24 h with intermediate cooling								
Specimen no.:				Material grade: PG 25/12				
Analysis:	C	Si	Mn	Cr	Ni			
	0.37	1.33	0.74	25.70	13.60			
Oxidation resistance: 1050°C				x°C = 1 $gm^{-2}h^{-1}$ x°C + 50°C = <2 $gm^{-2}h^{-1}$				

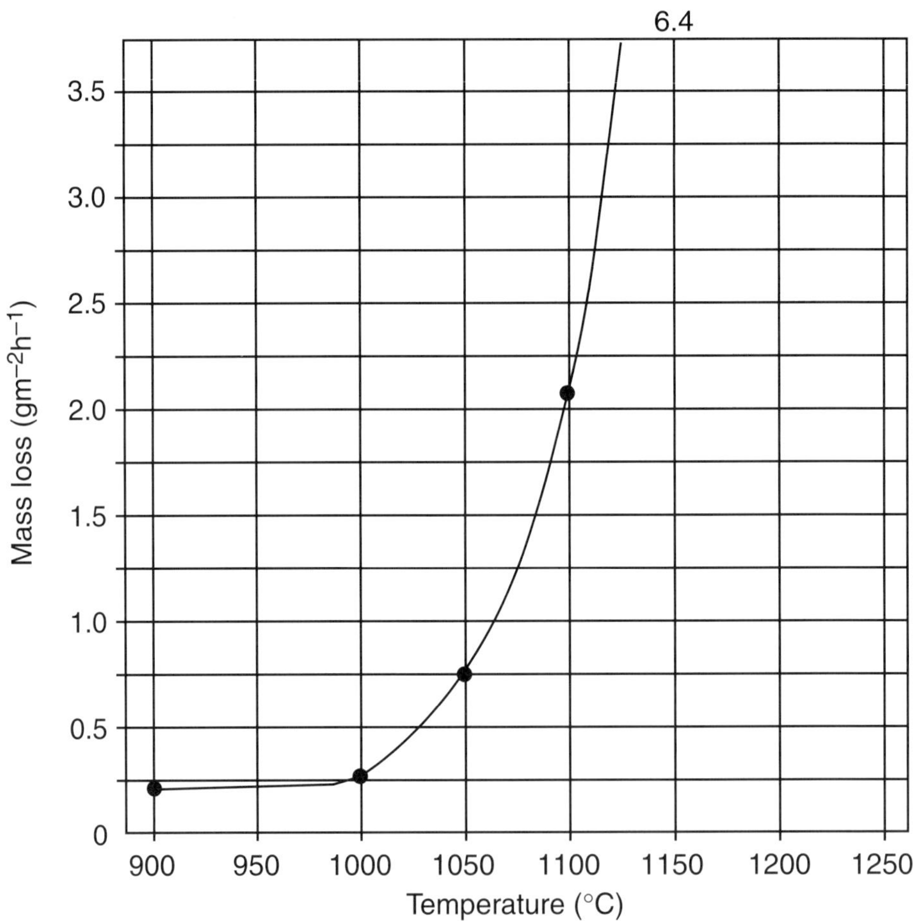

Fig. 2 *Oxidation test in air.*

Testing time 5 × 24 h with intermediate cooling								
Specimen no.: 10				Material grade: PG 25/35 H				
Analysis:	C	Si	Mn	Cr	Ni			
	0.40	1.64	0.78	25.00	37.30			
Oxidation resistance: 1125°C				x°C = 1 $gm^{-2}h^{-1}$ x°C + 50°C =< 2 $gm^{-2}h^{-1}$				

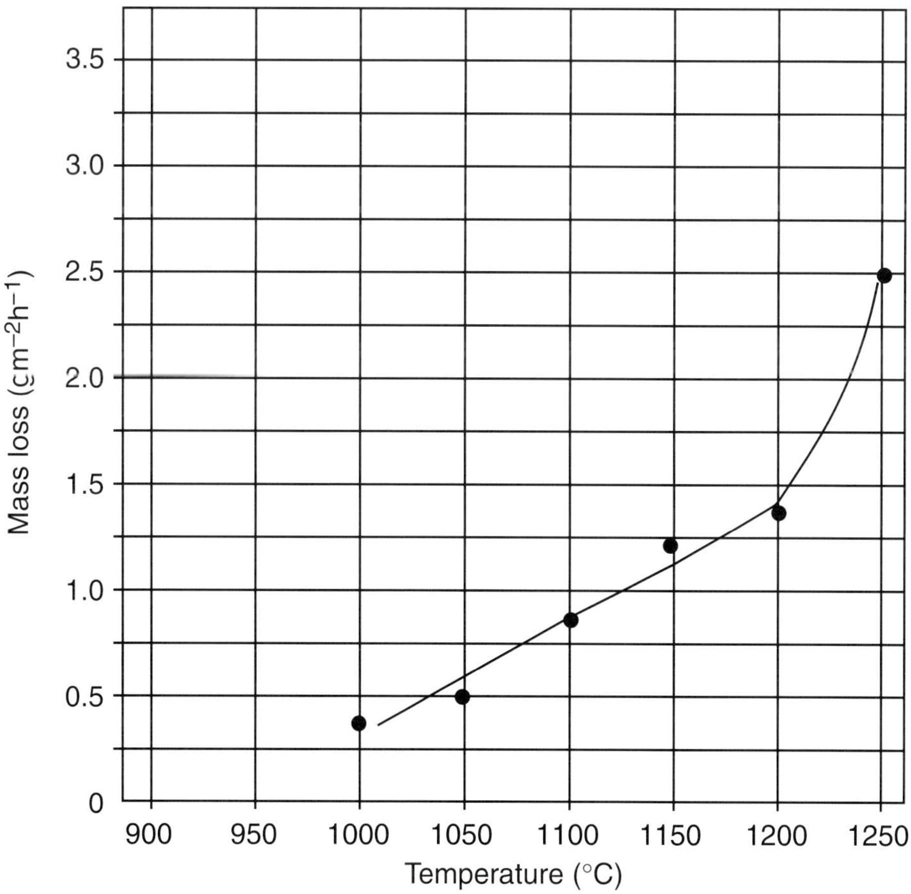

Fig. 3 *Oxidation test in air.*

Testing time 5 × 24 h with intermediate cooling								
Specimen no.: 72				Material grade: PG 24/24 Nb C				
Analysis:	C	Si	Mn	Cr	Ni	Nb		
	0.38	1.21	0.84	24.80	24.50	1.55		
Oxidation resistance: 1040°C				x°C = 1 $gm^{-2}h^{-1}$ x°C + 50°C = <2 $gm^{-2}h^{-1}$				

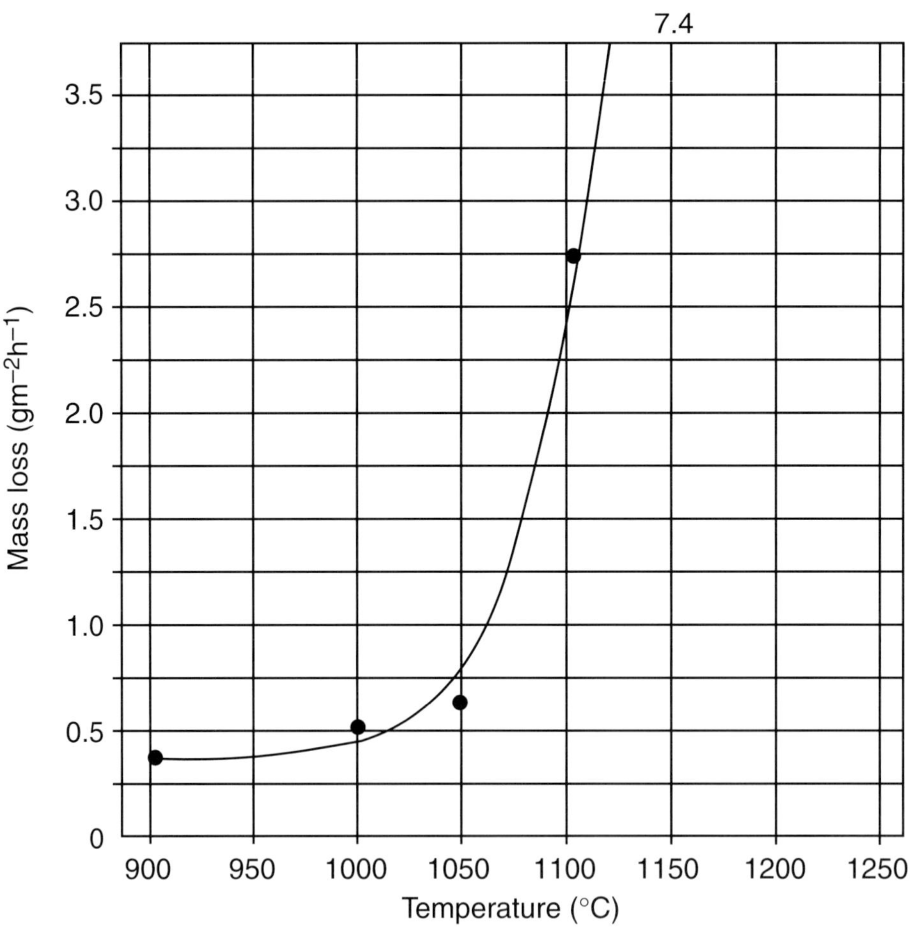

Fig. 4 Oxidation test in air.

Testing time 5 × 24 h with intermediate cooling								
Specimen no.:				Material grade: PG 25/20 H				
Analysis:	C	Si	Mn	Cr	Ni			
	0.42	1.38	1.10	25.00	20.15			
Oxidation resistance: 1045°C				x°C		= 1 $gm^{-2}h^{-1}$		
				x°C + 50°C		= <2 $gm^{-2}h^{-1}$		

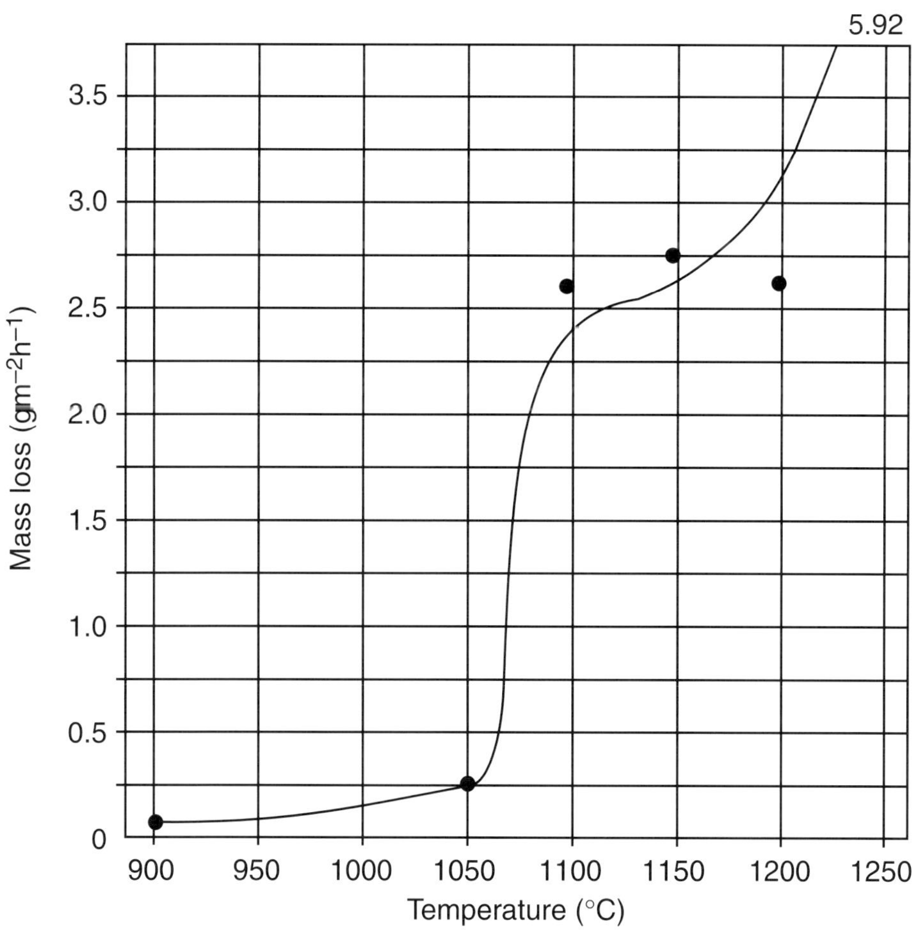

Fig. 5 *Oxidation test in air.*

Testing time 5 × 24 h with intermediate cooling								
Specimen no.:			Material grade: 25%Cr/20% Ni					
Analysis:	C	Si	Mn	Cr	Ni			
● 14	0.24	1.34	1.31	24.45	19.45			
× 15	0.54	1.51	0.63	24.95	19.40			
△ 16	0.38	0.72	0.70	23.55	21.25			
□ 20	0.34	1.09	0.98	24.40	20.25			
Oxidation resistance:			x°C = 1 $gm^{-2}h^{-1}$ x°C + 50°C = <2 $gm^{-2}h^{-1}$					

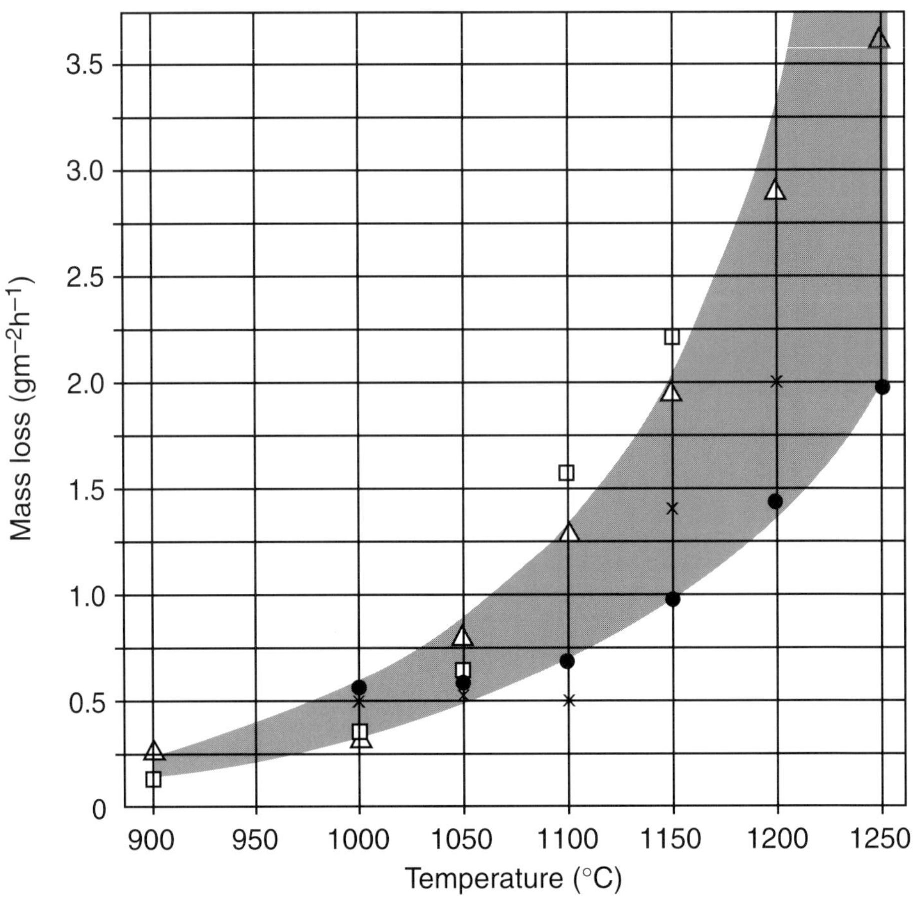

Fig. 6 *Oxidation test in air.*

Testing time 5 × 24 h with intermediate cooling								
Specimen no.: 3			Material grade: PG 18					
Analysis:	C	Si	Mn	Cr	Ni			
	0.30	1.68	0.64	15.30	0.10			
Oxidation resistance: 910°C			x°C = 1 gm^{-2}h^{-1}					
			x°C + 50°C = <2 gm^{-2}h^{-1}					

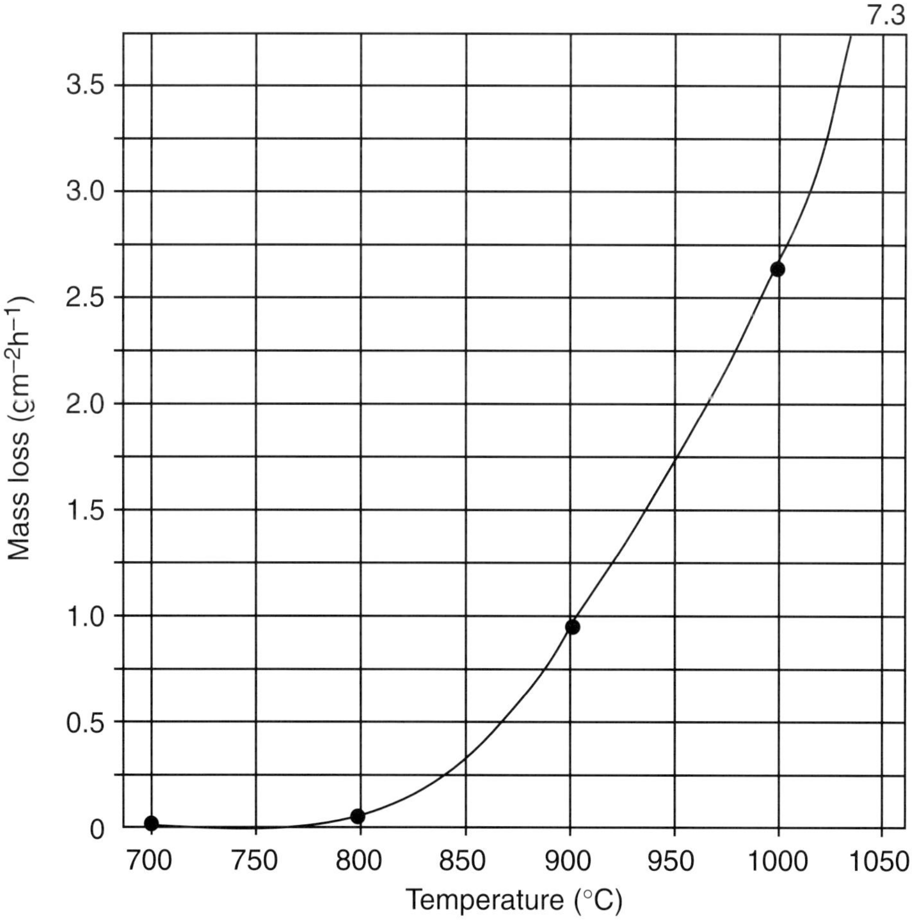

Fig. 7 *Oxidation test in air.*

Testing time 5 × 24 h with intermediate cooling								
Specimen no.: 11				Material grade: PG 29/7				
Analysis:	C	Si	Mn	Cr	Ni			
	0.61	1.65	0.59	28.50	7.85			
Oxidation resistance: 1200°C				x°C	= 1 $gm^{-2}h^{-1}$			
				x°C + 50°C	=<2 $gm^{-2}h^{-1}$			

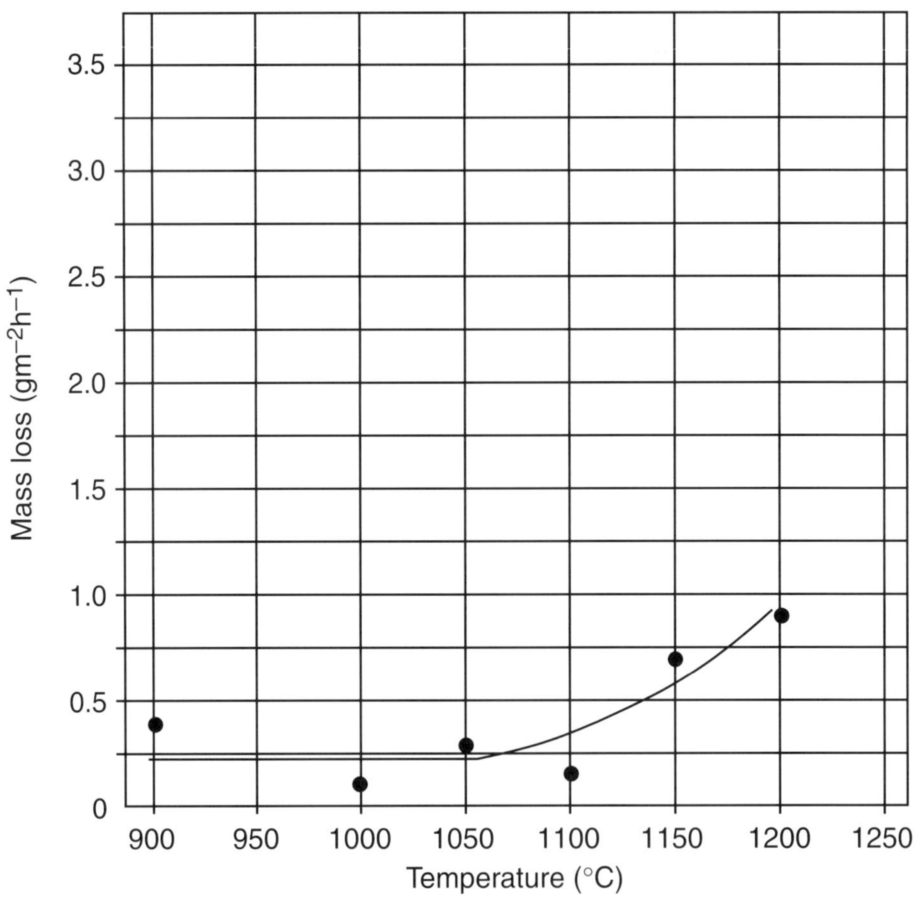

Fig. 8 *Oxidation test in air.*

Testing time 5 × 24 h with intermediate cooling								
Specimen no.: 4			Material grade: PG 18/8					
Analysis:	C	Si	Mn	Cr	Ni			
	0.16	1.69	0.69	18.90	8.30			
Oxidation resistance: 990°C			x°C = 1 gm^{-2}h^{-1} x°C + 50°C = <2 gm^{-2}h^{-1}					

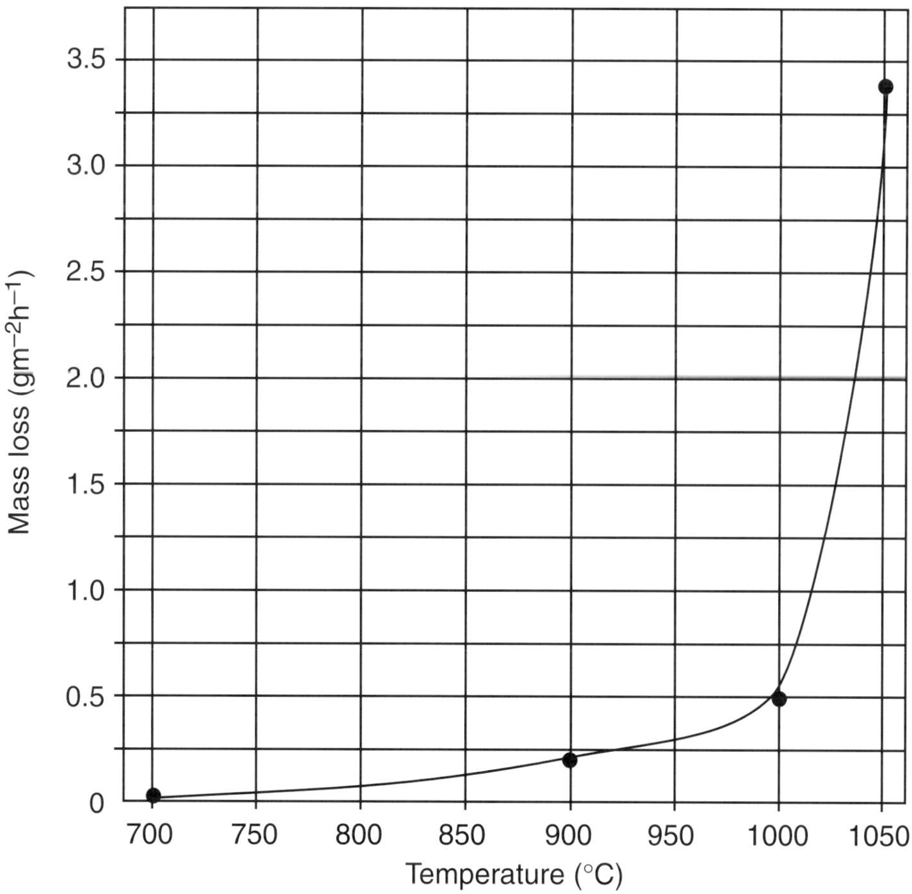

Fig. 9 *Oxidation test in air.*

Testing time 5 × 24 h with intermediate cooling								
Specimen no.: 22			Material grade: PG 25/20 H					
Analysis:	C	Si	Mn	Cr	Ni			
	0.54	1.51	0.63	24.95	19.40			
Oxidation resistance: 1130°C			x°C = 1 $gm^{-2}h^{-1}$ x°C + 50°C = <2 $gm^{-2}h^{-1}$					

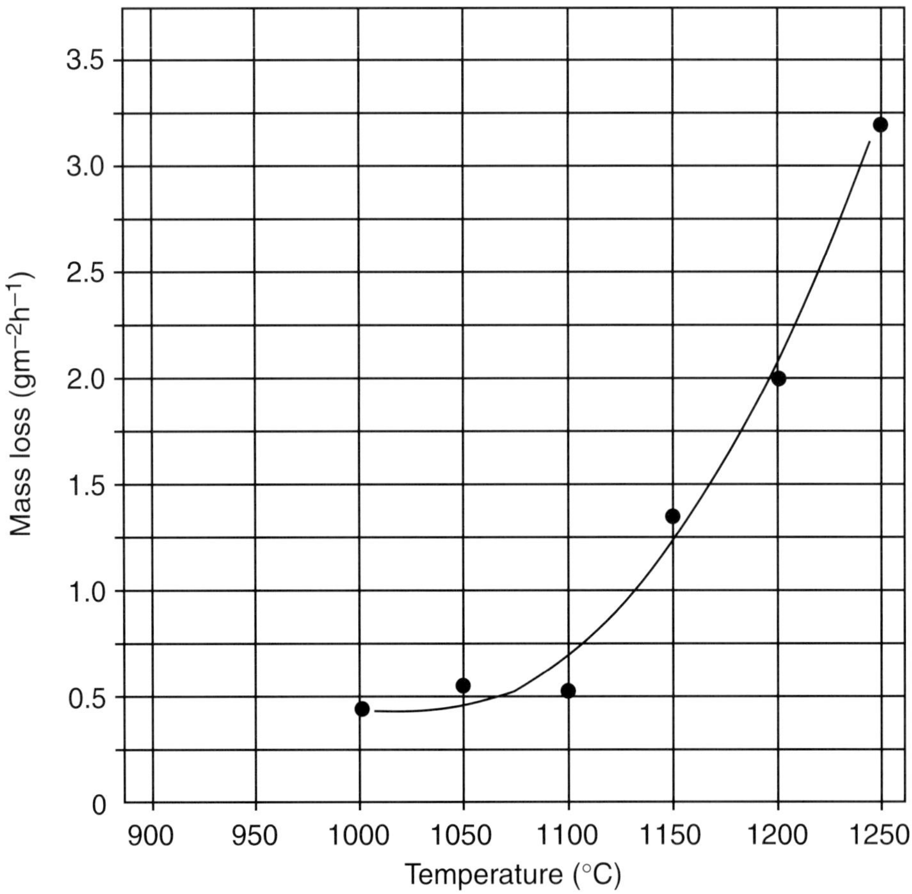

Fig. 10 *Oxidation test in air.*

Testing time 5 × 24 h with intermediate cooling								
Specimen no.: 15/17/19	Material grade: PG 25/45 H + 1.8 and 3.0 Nb							
Analysis:	C	Si	Mn	Cr	Ni	Nb		
○ ohne Nb	0.42	1.88	1.29	27.50	46.20	–		
◐ 1.8% Nb	0.39	1.94	1.05	26.40	45.50	1.73		
● 3.0% Nb	0.44	1.83	1.14	25.40	43.55	3.24		
Oxidation resistance:	x°C = 1 gm^{-2}h^{-1} x°C + 50°C = <2 gm^{-2}h^{-1}							

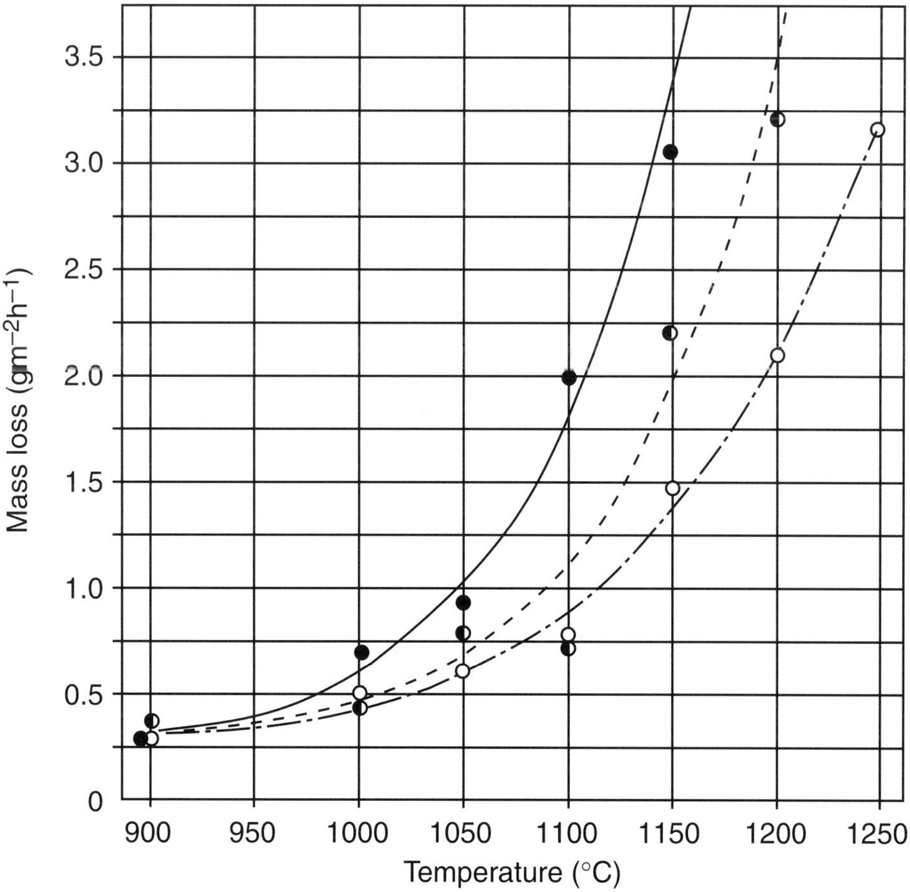

Fig. 11 *Oxidation test in air.*

Testing time 5 × 24 h with intermediate cooling								
Specimen no.:				Material grade: Forged and as–cast				
Analysis:	C	Si	Mn	Cr	Ni	Co		
● 37 = P UMCo 50	0.07	0.72	0.73	28.90		49.20		
× 39 = PG UMCo 50	0.21	0.56	0.76	29.20		49.70		
Oxidation resistance:				x°C = 1 gm^{-2}h^{-1}				
				x°C + 50°C = <2 gm^{-2}h^{-1}				

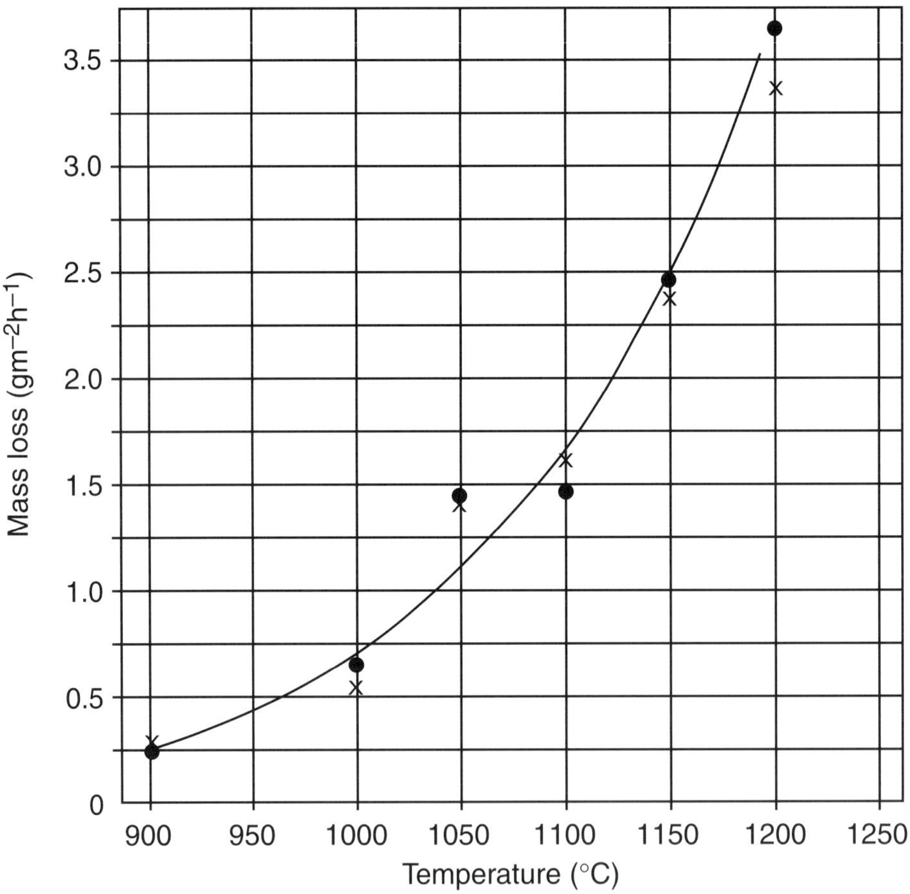

Fig. 12 *Oxidation test in air.*

Testing time 5 × 24 h with intermediate cooling								
Specimen no.:			Material grade:					
Analysis:	C	Si	Mn	Cr	Ni			
● 15 = PG25/20H	0.54	1.51	0.63	24.95	19.40			
× PG30/30	0.54	2.48	1.26	30.10	30.40			
△ 44 = PG35/65	0.02	0.19	0.30	35.05	64.25			
□ 45 = PG50/50	0.03	0.54	0.50	48.70	49.60			
+ 48 = PG60/40	0.02	0.30	0.48	57.90	40.40			
Oxidation resistance:				x°C = 1 $gm^{-2}h^{-1}$ x°C + 50°C = <2 $gm^{-2}h^{-1}$				

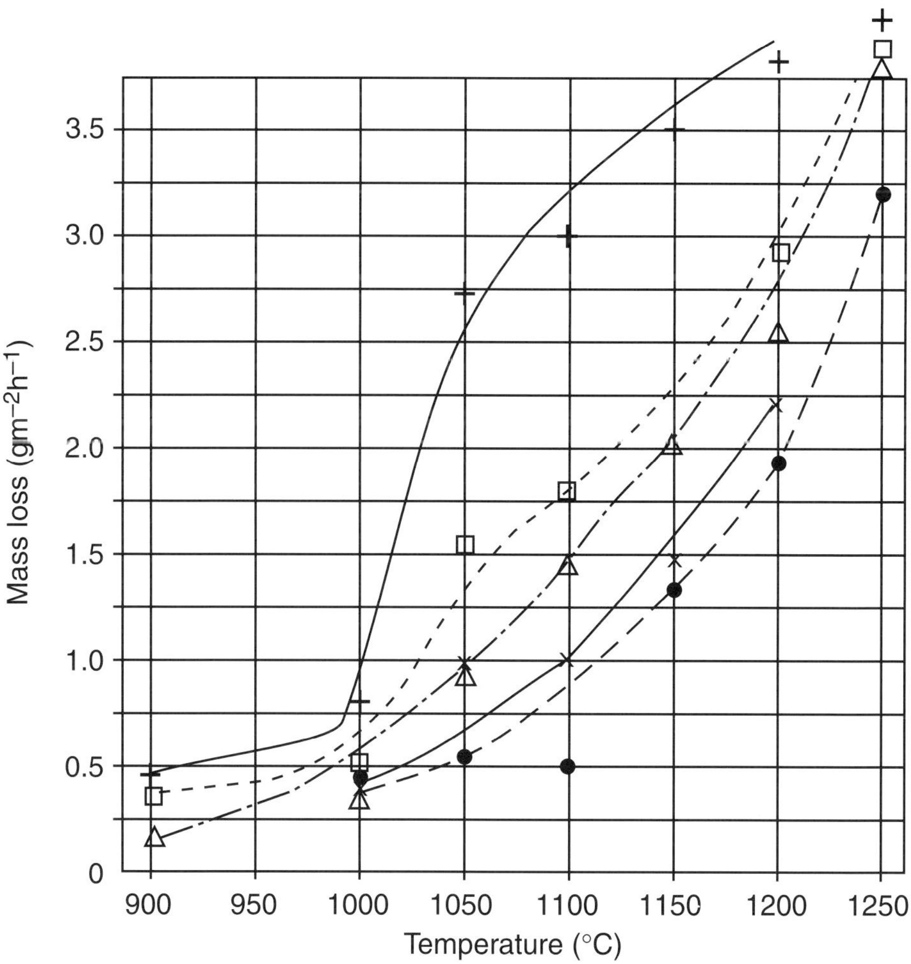

***Fig. 13** Oxidation test in air.*

There is no proof that a cast-steel material with a high Cr-content provides high oxidation resistance. If the Cr-content is increased to above 30% the oxidation resistance (Fig. 13) is obviously decreased. It is possible that there is an additional effect from the low Si-content of these materials.

4. Evaluation of the 5 × 24-h Test

The first results from this so-called 5 × 24-h test method were described almost 60 years ago by Bandel and Volk and showed that this test method is effective in finding the maximum permissible operation temperature for the used materials. With frequent temperature changes the oxidation resistance is lower than with constant temperature. This is shown in Figs 14 and 15. This means that the thermal stress in the 5 × 24-h test is very high and the determination of the permissible temperature is reliable.

5. Conclusion

The results are presented of tests made 25-30 years ago and which have been successfully used ever since. This simple method of 5 × 24-h test proves that a reliable maximum operation temperature for cast wrought and weld material can be guaranteed to the customer.

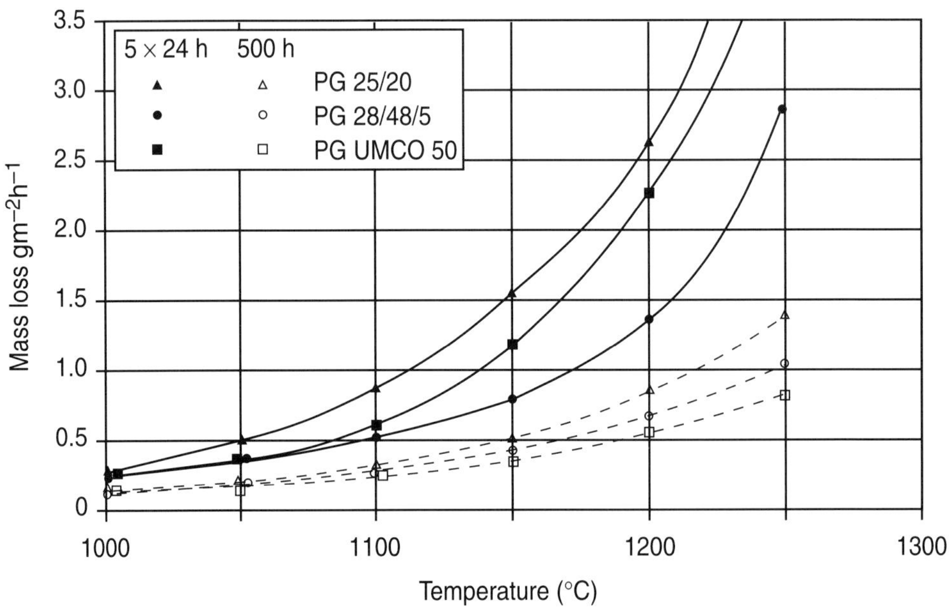

Fig. 14 *Mass loss in the 5 × 24-h test with intermediate cooling and 500 h continuous test.*

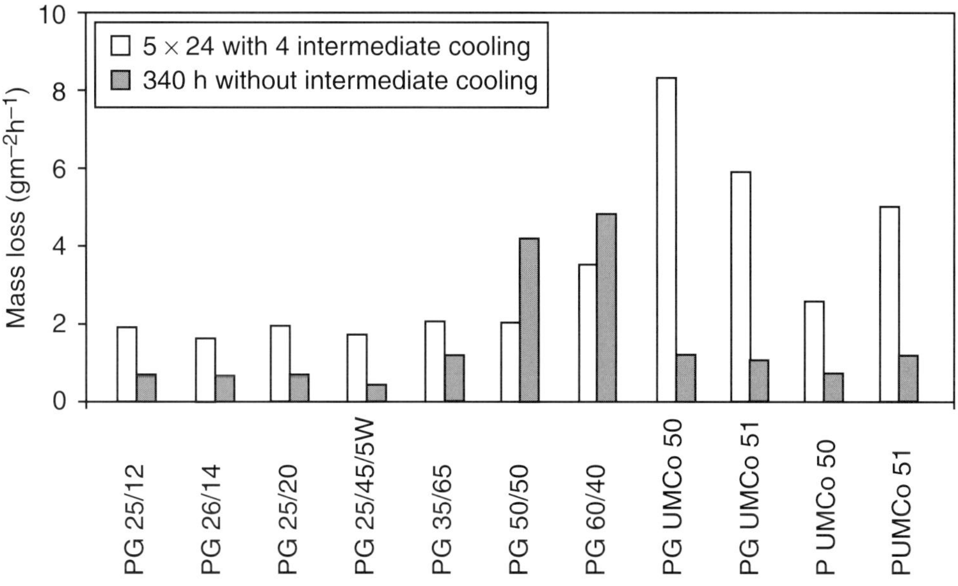

Fig. 15 *Comparison of mass losses at 1100°C.*

References

1. Stahl-Eisen-Werkstoffblatt SEW 471-60, 4th issue, March 1960. Verein Deutscher Eisenhuttenleute, Dusseldorf, Germany.

2. Stahl-Eisen-Werkstoffblatt SEW 471, 5th issue, August 1976. Verein Deutscher Eisenhuttenleute, Dusseldorf, Germany.

3. G. Bandel and K. E. Volk Die Prufung der Zunderbestandigkeit von legierten Stahlen (The testing of oxidation resistance of alloyed steels), *Arch Eisenhuttenwes.*, 1942, **8** (2), 369.

4. "Heat resistant steel castings" pr EN 10295, draft for European Standard, issued May 1998, chapter 7.2.3.1.

5. POSE-MARRE internal reports (unpublished) N. 28/68 and 40/73.

20
Cyclic Oxidation Testing by Resistance Heating

B. JÖNSSON, A. WESTERLUND and G. LANDOR
Kanthal AB, P.O. Box 502, S-734, 27 Hallstahammar, Sweden

ABSTRACT
Cyclic oxidation tests on thin strip, 70 and 90 μm, and wire 0.7 mm dia. are presented for seven experimental Fe–20Cr–5Al alloys and two commercial alloys, Kanthal®, AF and Kanthal® AE. The test method involves heating the samples by passage of electrical current with measurement and interpretation of the changes of electrical resistance during the test. It is demonstrated that the test method gives reproducible and relevant qualitative information on the oxidation process, and effects from sample geometry and chemical composition can be distinguished. It is shown that vanadium has a negative effect and zirconium a positive effect on the oxidation of the experimental alloys. Phosphorus is represented at three levels in the alloys, from 10 to 200 ppm, and is shown to have a positive effect under the test conditions.

1. Introduction

Life testing of wire samples by electric resistance heating according to the method developed by Bash and Harsch [1,2] has been used in the characterisation of electric resistance alloys (Fe–Cr–Al and NiCr(Fe)) since the 1930s. The sample, located in a chamber kept at room temperature, is heated by the Joule effect and dissipates its energy to a large extent by radiation. Used mainly as a comparative test for control of product uniformity, the test method has several desirable features. Thus, it somewhat resembles the application of the alloys in electric appliances, the test times are comparatively short, it delivers at least one quantitative result (lifetime in hours) and it allows for a relatively simple test set-up mainly due to the fact that the only hot part is the sample itself. The test is standardised by ASTM [3,4] and it has been used extensively by manufacturers of resistance materials as one of the tools for quality control.

In the last few years, a modern version of this life test has been used for material control and product development at Kanthal R&D. Computer controlled test parameters and data acquisition and storage makes the test a very efficient and reproducible tool. The set-up is easily modified to suit various sample geometries and power levels, as will be exemplified in this paper.

This paper will present and discuss some results achieved on experimental Fe–Cr–Al alloys investigated within the BriteEuRam programme LEAFA, Life Extension of Alumina Forming Alloys, and the results will be discussed with respect to material composition and sample geometry.

2. Experimental

2.1. Temperature Control

Temperature, which was the main test parameter in the old test, has disadvantages as the controlling parameter from physical, cost and practical points of view. Temperature measurement was a manual and time consuming task and it gave room for human error and suffered from poor reproducibility for a number of reasons. It has proved difficult and costly to automate temperature measurements on the samples in question. The alternative to temperature control is to use electrical parameters for control. The changes as compared to the old test are therefore mainly in the way of regulating and measuring the electrical parameters of the samples. The main parameters of the modern life test are given in Section 2.5.

Electric current, voltage or power may all be easily controlled by means of computer and thyristor control units. As the physical characteristics of the sample change during the test, it is necessary to keep one of the parameters constant. Although the choice is in principle arbitrary, power has been chosen as the constant parameter for a standard procedure, for reasons that are explained below.

Constant current has an obvious disadvantage in that any increase in resistance in the material means increase in power, which is unrealistic in most applications. Voltage is clearly an alternative, since this is the actual case in many electric appliances where resistance alloys are used and it also used to be common in industrial applications. However, voltage control tends to give 'unfair' advantage to materials that increase their electrical resistance with time since increased resistance means in this case a power loss, which can only be tolerated to a limited extent in real applications.

Power as the controlling parameter in a standard test is therefore preferred as it gives the best resemblance to heating applications in which the elements, whether gas heated or electric, are designed to deliver a specified power — not a specified temperature. Material changes that occur during the test, in which the power is kept constant, may then of course mean that temperature on a sample varies during the test time. However, this drawback is not as large as one would tend to believe as in most cases the changes are more rapid in the beginning of the test, and a representative steady state temperature on a sample can usually be given. An overall view of the experimental set-up for the life test is given in Fig. 1.

The options to use current or voltage or even (manually) temperature as the controlling parameter may, however, give important information in special investigations.

2.2. Temperature Cycling

Two minutes on/two minutes off has been chosen as the standard cycling condition although the experimental set-up allows for any cycling frequency between 30 s on/30 s off and continuous on. The choice is based on experiments with varied cycle frequencies in which it was found that this gave the shortest life and obviously the largest test throughput for the sample geometry and experimental set-up in question [1,2]. Schematically, the variation of lifetime vs cycling frequency is given by Fig. 2.

Fig. 1 *Experimental set-up for the Life test.*

2.3. Resistance Changes

Resistance in the hot condition, *Rv*, and in the cold condition, *Rc*, are monitored during the test. Both are affected by geometrical changes of the sample i.e. by reduction of cross-section and elongation that often, but not always, occur in this test.

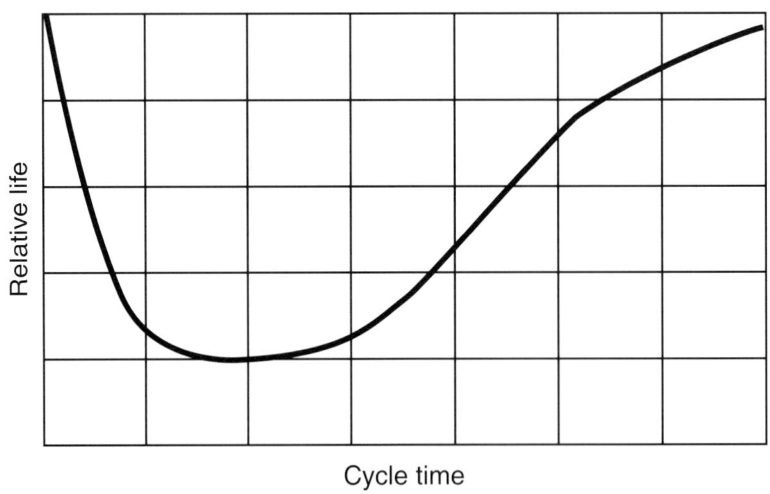

Fig. 2 *Variation of lifetime with cycling frequency. From [2].*

2.4. Interpretation of Change in Hot/Cold Resistance Ratio

Both hot and cold resistance are affected by the chemical composition of the sample and for otherwise similar Fe–Cr–Al alloys the Al content is the main influence. The impact from chemical composition is, however, smaller on Rv than on Rc, which makes it possible qualitatively to interpret changes in the ratio $Ct = Rv/Rc$ as evidence of Al loss in the alloy.

By studying Ct, the impact of geometry changes are effectively eliminated, and thus Ct provides a simple way of qualitatively monitoring Al consumption and thus the oxidation process as Al is removed from the alloy to form alumina on the surface. Quantitative conversion of the described electrical parameters are quite involved and require several second order effects to be taken into account, such as localised inhomogeneity, temperature variations along the sample, etc.

2.5. Samples and Test Conditions

The LEAFA programme with its aim of investigating the possibilities of improving the alumina based protection of Fe–Cr–Al alloys involves production and testing of seven experimental alloys, produced under extremely carefully controlled conditions [7]. Designation and chemical composition of the alloys are given in Table 1. For

Table 1. *Designation and chemical composition of experimental alloys*

Melt no.	**1 766**	**2 767**	**3 772**	**4 769**	**5 771**	**6 770**	**7 768**	**8 Ref. Kanthal AF AE**
Designation	MRef	M + P	M + V	M + Ca	M + Ti	M + Zr	M – P	
Cr (mass%)	19.9	19.9	19.7	19.7	19.7	19.7	19.8	(21.7, 21.1)
Al (mass%)	5.0	5.0	(5.1)	4.8	(4.8)	(5.0)	4.9	(5.4, 5.4)
Y (mass%)	0.05	0.05	0.05	0.05	0.05	0.05	0.05	
C (mass%)	0.009	0.010	0.010	0.010	0.009	0.009	0.007	(0.026, 0.030)
S (ppm)	<10	<10 (12)	<10	<10	<10 (34)	<10		
O (ppm)	<10	<10		<10			<10	
N (ppm)	<10 (6)	<10 (5)	(32)	<10 (5)	(23)	(20)	<10 (17)	
P (ppm)	86	236	83	64	78	78	(13)	
Zr (mass%)	—	—	—	—	—	(0.052)	—	
V (ppm)	<10	<10	83 (60)	<10	<10	<10	<10	
Ti (mass%)	—	—	—	—	0.022	—	—	
Ca (ppm)	(11)	(<10)	(<10)	<5 (26)	(43)	(10)	(<10)	

Note: Numbers in brackets are analysed by Kanthal AB by ICP OES. All other analysis made at [1].

reference, the commercial alloys Kanthal AF and Kanthal AE* were included in the tests.

The alloys were hot rolled, cold drawn/cold rolled to thin wire/strip: sample dimensions are given below.

Strip 70 μm: Thickness, 68–73 μm (thickness variation within one sample < 1μm), width ≈ 2 mm, length ≈ 135 mm, U-shape. Power, 18 Wcm^{-2}.

Strip 90 μm: Thickness, 91–94 μm, (thickness variation within one sample < 1 μm), width ≈ 2 mm, length ≈ 135 mm, U-shape. Power, 18 Wcm^{-2}.

Wire 0.7 mm dia.: Diameter 0.70 mm, length ≈ 224 mm, U-shape, Power 155 W.

The general test parameters are given below.

Life test parameters

Sample geometry: Wire 0.7 mm dia., free radiating length 224 mm, total length 250 mm, U-shape.

Temperature cycling: 2 min on/2 min off.

Regulation: Power kept constant during the test, 155 W for Fe–Cr–Al alloys.

Set-up geometry: Cabin dimensions and terminal design according to ASTM, B78-81.

Measurements: Sample hot and cold electrical resistance, *Rv*, *Rc* and ratio *Rv*/*Rc*. Lifetime in hours, given as accumulated hours at temperature.

Temperature: Varies during the test time, optional recording by two-colour pyrometer on the hottest part of the sample. Values recorded after 15 min and 48 h, or with regular intervals. Typical steady state value of 1265°C for the above conditions.

3. Results

Changes in hot resistance, *Rv*, and the *Ct* ratio as well as temperature changes as measured with a two-colour pyrometer are given in Fig. 3 (a–c) and Fig. 4 (a–c) for the three sample geometries.

---

*Kanthal and Nikrothal are registered trademarks in Sweden and other countries.

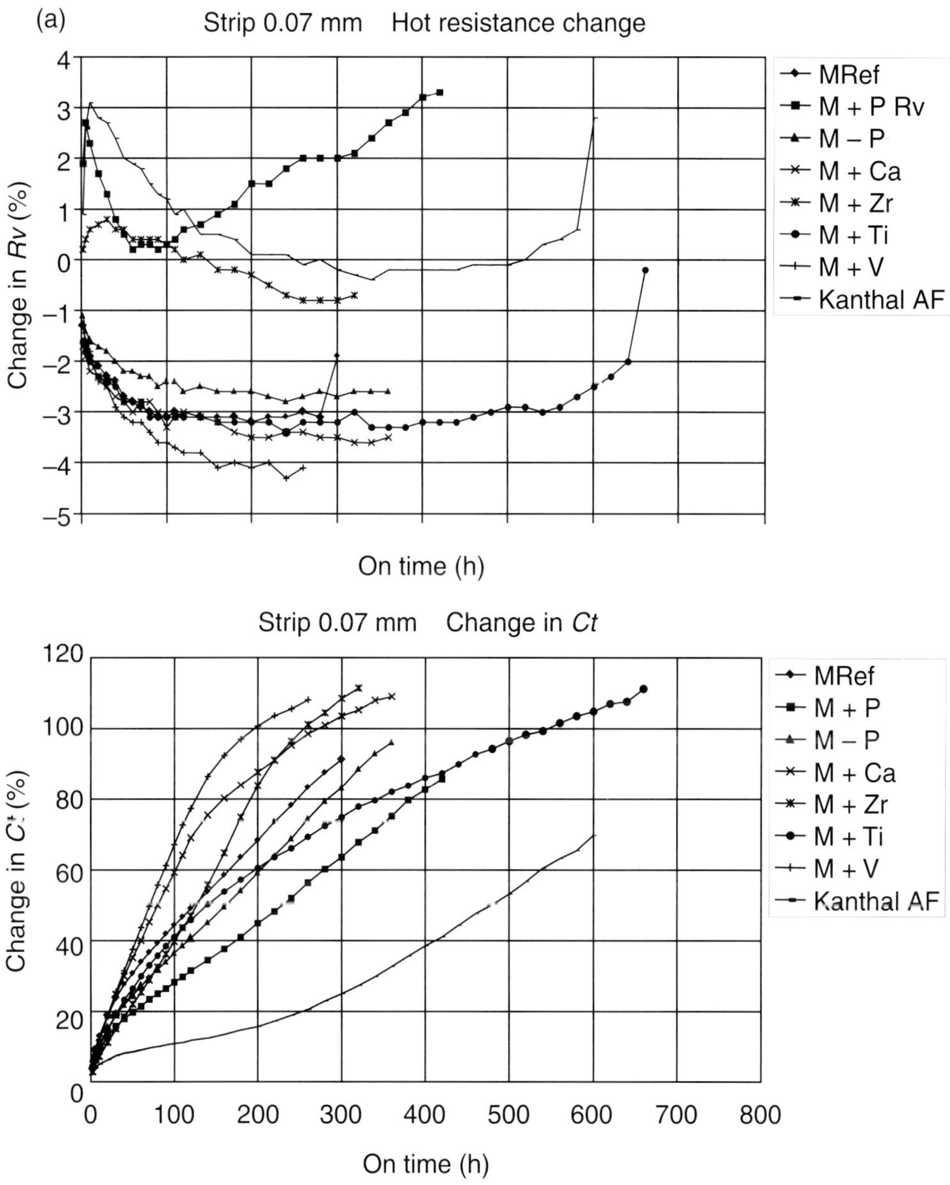

Fig. 3 (a) *Change in* Rv *and* Ct *for a strip of 70 μm.*

4. Discussion

Measured changes in electrical resistance are caused by local and general effects along the sample. These, from the electrical point of view, are connected in series, and it is therefore obvious that local disturbances are difficult to interpret from the data. A detailed model to describe this is relatively complex and is beyond the scope

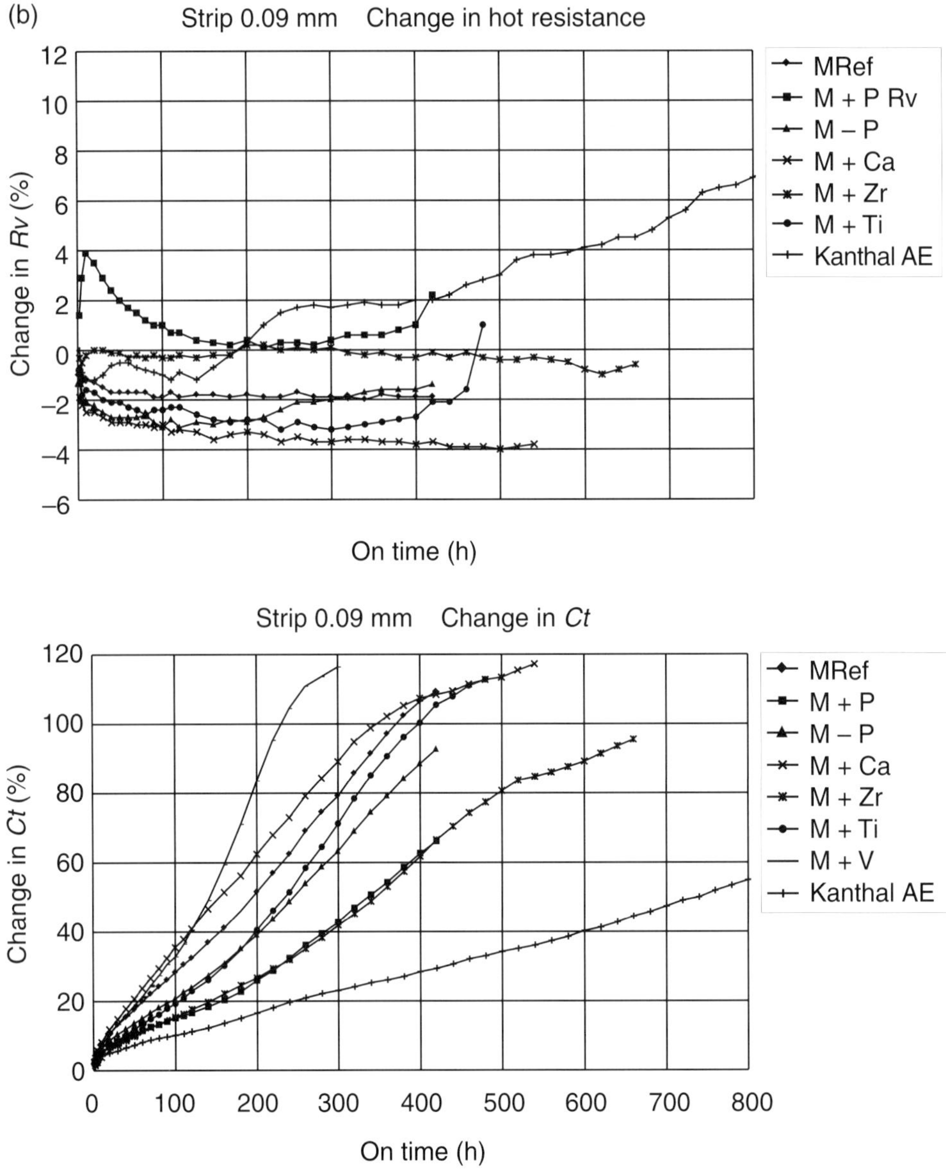

Fig. 3 (b) Change in Rv *and* Ct *for strip of 90 μm.*

of this paper. From the practical point of view, the sample can in most cases be regarded as homogeneous along the length and especially so in the earlier stages of oxidation. In the interpretation, it should be remembered , however, that failure most often is localised at a point, and close to failure effects occur that have a substantial impact on the overall measurements.

Changes in hot resistance, *Rv*, may be attributed to both geometrical and chemical effects. Elongation of the sample and reduction of metallic cross-section obviously

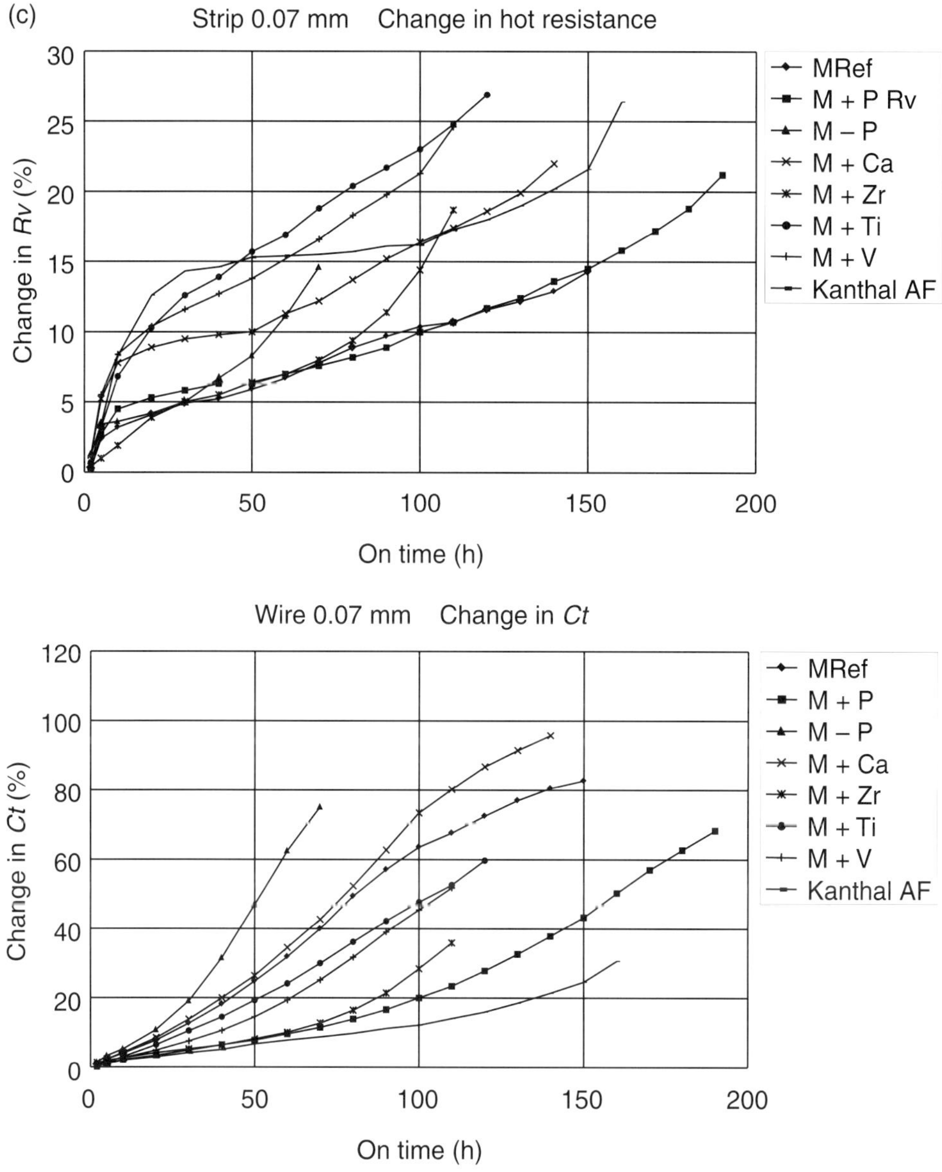

Fig. 3 (c) *Change in* Rv *and* Ct *for wire 0.7 mm dia.*

act to increase resistance, whereas shrinkage and decreased Al-content act in the opposite direction. The graphs in Fig. 3. indicate that elongation or shrinkage may occur in the case of thin strip geometry, whereas substantial elongation is indicated by the relatively large resistance increase on 0.7 dia. mm wire.

The *Ct* factor is more insensitive to local and general geometrical effects since these will affect hot and cold resistance in approximately the same way. There are several observations to be made on studying the *Ct* curves. The first, and most important, is that relevant information regarding the oxidation process from

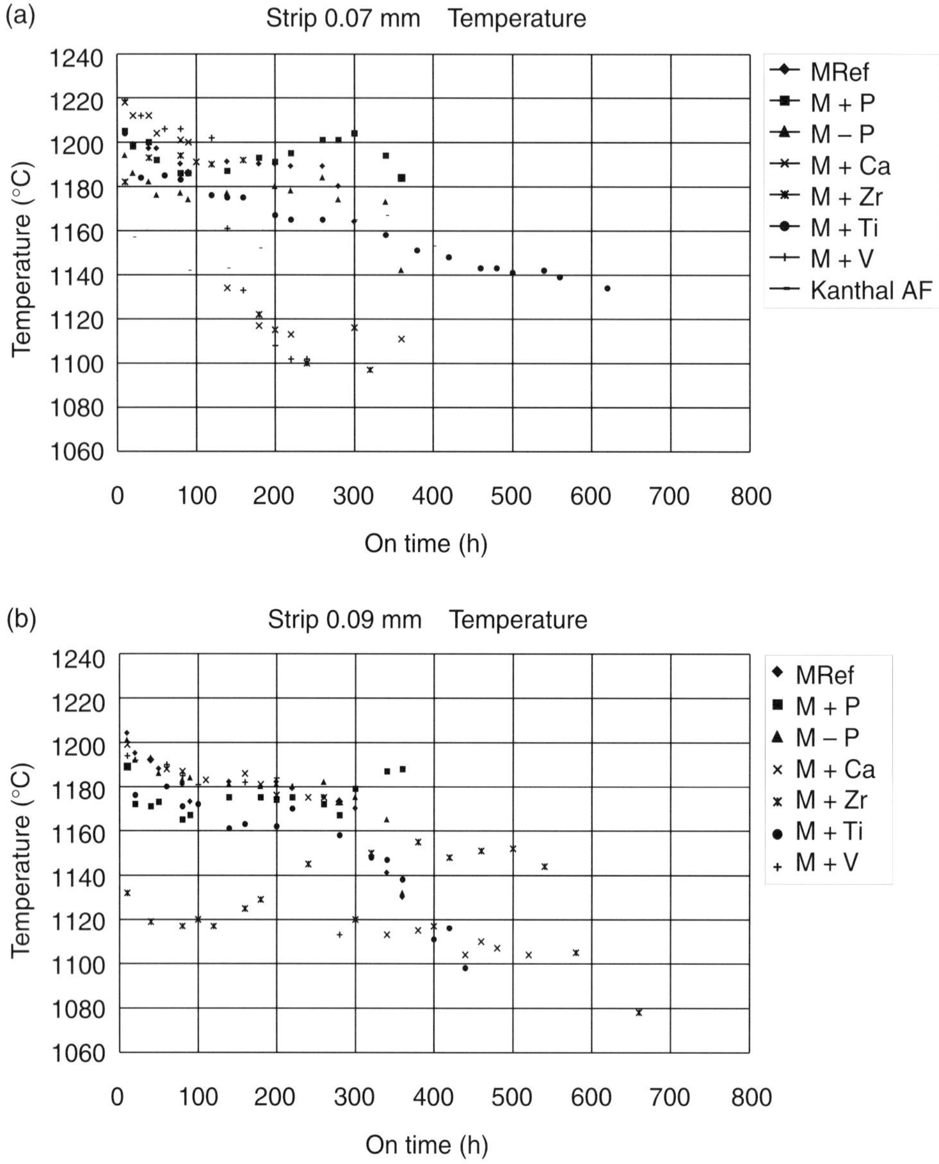

Fig. 4 Measured temperature for (a) strip 70 μm, (b) strip 90 μm and (c) (opposite) wire 0.7 mm dia.

differences in chemical composition can be made. This is evident from the fact that ranking of the alloys is similar in the tests of different sample geometries and in repeated experiments.

There is, however, a noticeable spread that is derived both from sample imperfections and from experimental parameters. Nevertheless, a summary table regarding changes in Ct, i.e. qualitatively the oxidation rate, can be prepared from the diagrams in Fig. 3 (see Table 3).

6. Acknowledgements

The tests on experimental Fe–Cr–Al alloys are part of the Brite/EuRam project BE-97-4491, Life Extension of Alumina Forming Alloys in High Temperature Corrosion Environments, LEAFA.5.

References

1. F. E. Bash and J. W. Harsch, Life test on metallic resistor materials for electrical heating, *Proc. Amer. Soc. Test. Mater.*, 1929, **29**, 506.
2. I. Zawadzka, Untersuchungen über eine optimale Verkürzung der Lebensdauer – Prüfung von Heizleitern, Elektrowärme international, Bd 26, Nr.4, April 1968.
3. Accelerated life test of Iron–Chromium–Aluminium alloys for electrical heating, ASTM standard, B 78-81.
4. Accelerated life test of Nickel–Chromium and Nickel–Chromium–Iron alloys for electrical heating, ASTM standard, B 76-81.
5. A. Seitser, Ein rechnergeführtes Verfaren zur Lebensdauerbestimmung und Qualitätssicherung von Heizdraht, Dissertation RWTH Aachen 1992.
6. B. Zimmermann, personal communication, RWTH Aachen, 1996.
7. J. Le Coze, R. Tardy, A. Kobylanski and M. Biscondi, Preparation of high purity metals and metallurgical studies performed at the Ecole des Mines from 1970. Review, *Proc. 1st Int. Conf. on Ultra High Purity Base Metals* (UHPM-94), Japan (1994). The Japan Institute of Metals, Sendai, Japan, 1995.

Part 4

The Behaviour of Coatings under Cycling Conditions

21
Thermal Cycling Testing of Thermal Barrier Coatings

R. ANTON*, S. K. JHA†, D. CLEMENS, W. MALLENER, L. SINGHEISER and W. J. QUADAKKERS
Forschungszentrum Jülich GmbH, IWV, D-52425 Jülich, Germany
*Siemens AG, D-45473 Mülheim/Ruhr, Germany
†Indian Institute of Technology, Powai, 4000 076 Bombay, India

ABSTRACT

The behaviour of TBC (Thermal Barrier Coatings) systems during thermal cycling at 1000°C and 1100°C is discussed. The typical failure modes of EB-PVD (electron beam-physical vapour deposition) and APS (air plasma sprayed) TBCs are demonstrated. The main emphasis is placed on the significance of BC (bond coat) oxidation for TBC life. Generally, failure of EB-PVD coatings occurs by spallation of the TGO (thermally grown oxide). Indications were found that this failure mode also dominates in the case of APS coatings although metallographic investigations indicate crack initiation to occur within the TBC, as a result of the rough BC surface, in the vicinity of the TGO. The significance of BC (bond coat) roughness on TGO formation and growth was shown by comparing the behaviour of TBCs with that of free standing MCrAlY BCs. In many cases TBC failure occurred earlier than would be expected from the TGO spallation measured for the free standing BCs. This is probably related to the effect of BC surface roughness on transient oxide formation and stress generation in the TGO. Specimen geometry can have a significant effect on TBC life during cyclic oxidation testing. Therefore, the choice of a suitable specimen shape is necessary to obtain reliable and reproducible data in respect to TBC life.

1. Introduction

Increased gas inlet temperatures in high efficiency, industrial gas turbines lead to an increase of component surface temperature, resulting in enhanced oxidation attack of the blading material [1]. The high operating temperatures require the use of high strength directionally solidified or single crystal Ni-base superalloys in combination with advanced cooling techniques of the blades and vanes as well as an additional protection of the components by a ceramic thermal barrier coating (TBC) [2–4]. This ceramic coating significantly reduces the temperature of the MCrAlY-type bond coat (M = Ni, Co, NiCo) due to its low thermal conductivity. The ceramic top coat usually consists of yttria-stabilised zirconia and is commonly applied by air plasma spraying (APS) or by electron beam physical vapour deposition (EB-PVD) [2]. The MCrAlY-bond coat (BC) between the Ni-base superalloy and the zirconia based TBC is usually applied by low pressure plasma spraying or vacuum plasma spraying (LPPS or VPS).

Due to the oxidation attack during service a thermally grown oxide (TGO), usually Al_2O_3, grows at the bond coat/TBC interface. The factors which limit the lifetime of this four-component system (Ni-base superalloy + BC + TGO + TBC) depend on the actual components, coating system and service conditions. A factor which frequently contributes to TBC failure is thermomechanical fatigue (TMF) because the combination of externally applied compressive stresses with stresses resulting from the thermal expansion mismatch between the various components in the TBC system can degrade the integrity of the TBC [2]. Further, sintering of the ceramic top coat, resulting in a change of the physical and/or mechanical properties of the coating, is frequently claimed to limit TBC life [2]. Which factor eventually is the dominating cause for failure depends on the actual system under consideration and is still discussed controversially in the literature. However, nearly all authors agree that one of the major damage mechanisms in practice is oxidation of the MCrAlY-bond coat [5–8]. Although different failure modes are being discussed for different systems, cracking or spalling of the coating in nearly all practical cases occurs in or near the TGO [5,6,9–11].

In the present paper the behaviour of TBC systems under thermal cycling conditions is discussed. The main emphasis will be put on the interaction between BC oxidation and TBC failure. The significance of the choice of specimen geometry during coating testing will be addressed.

2. Experimental

For the cyclic oxidation experiments at 1000 and 1100°C in air a vertical furnace with an automated specimen cycling unit was used. Each cycle consisted of 120 min heating and 15 min cooling. Cooling of the specimen was enhanced by flushing the specimens with pressurised air after their removal from the furnace. Figure 1 shows the heating and cooling curves for two typical cycles. The actual heating and cooling rates will obviously depend on the prevailing specimen shapes. In order to determine the heating and cooling rates during the cyclic testing, typical cylindrical TBC-coated and uncoated specimens with a diameter of 10 mm were used. Thermocouples were placed in the centre of each specimen and the temperatures were compared with those measured by a thermocouple attached to the specimen surface. Figure 2 (a,b) shows the temperature characteristics during heating and cooling. The heating/cooling rate of the specimen surface appears to be extremely fast, i.e. during heating the actual exposure temperature is reached at the specimen surface within *ca.* 4 min and during cooling a temperature of 28°C is reached within 1 min. The temperature changes in the specimen centre are obviously slower with the centre of the specimen without TBC coating showing a higher heating/cooling rate than the specimen with TBC.

Different specimen geometries were used for thermal cycling experiments (Fig. 3). Apart from flat, disc shaped specimens with tapered edges, three types of cylindrical geometries were used. The first cylindrical specimen (Fig. 3a) was completely coated with MCrAlY but only the central region of the cylinder was coated

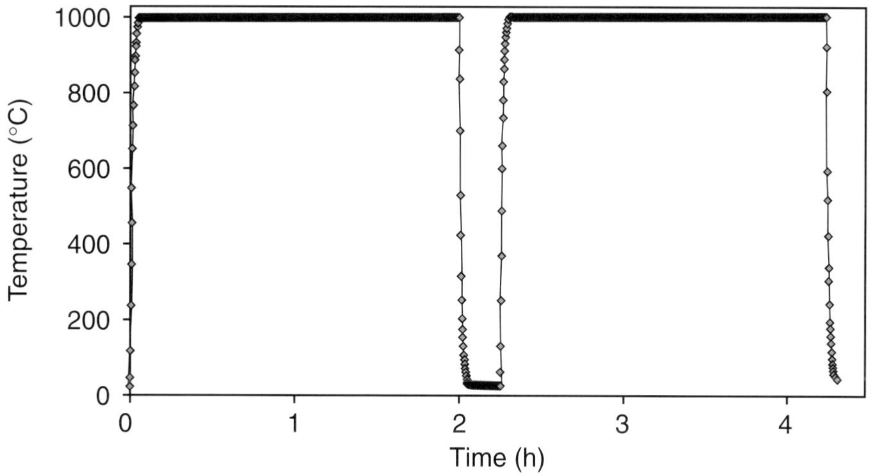

Fig. 1 *Temperature vs time for a cylindrical specimen (dia. 10 mm) during two typical cycles (120 min heating and 15 min cooling) of standard thermal cycling testing.*

with TBC. The second type of cylindrical specimen type was similar, however, only one part of the cylinder surface was free of TBC coating (Fig. 3b). The third type was completely covered with TBC, including the end-faces (Fig. 3c); this third type also had cone-like cylinder ends.

3. Results and Discussion

3.1. Comparison between PVD and APS Coating

Figure 4 shows typical examples of the macroscopical appearances of failure modes of TBCs applied by APS and EB-PVD. Typically APS-TBCs show a small crack, initiated in the TBC which propagates with time, eventually leading to spallation of large parts of the ceramic top coat after longer times (Fig. 4a,b). In contrast, EB-PVD-TBCs show a more sudden spallation behaviour, i.e. spallation of large, elongated ceramic patches occurs (Fig. 4c) without showing initiation of small cracks which would announce the actual occurrence of TBC spalling well in advance.

Failure of EB-PVD coatings in service is generally accepted to occur by spallation of the TGO [5]. This is confirmed by the results illustrated in Fig. 5, showing the cross section of a TBC applied by EB-PVD (cylindrical specimen) after 1629 h of cyclic oxidation at 1000°C. Separation of the top coat has clearly occurred at the BC/TGO interface with the alumina scale adhering to the ceramic top coat. In contrast, Fig. 6 shows the cross-section of a TBC applied by APS (cylindrical specimen) after 144 h of cyclic oxidation at 1100°C. Here, spallation seems to occur within the TBC because over nearly the whole specimen surface, patches of TBC are still adhering to the bond coat. However, a more careful investigation of the crack pattern of a number of specimens indicated, that also in the case of the APS coatings the crack initiation probably occurs at the TGO/BC interface. Whereas the TGO is still adhering in valleys of the rough bond coat surface, spallation of the TGO has in many cases occurred at

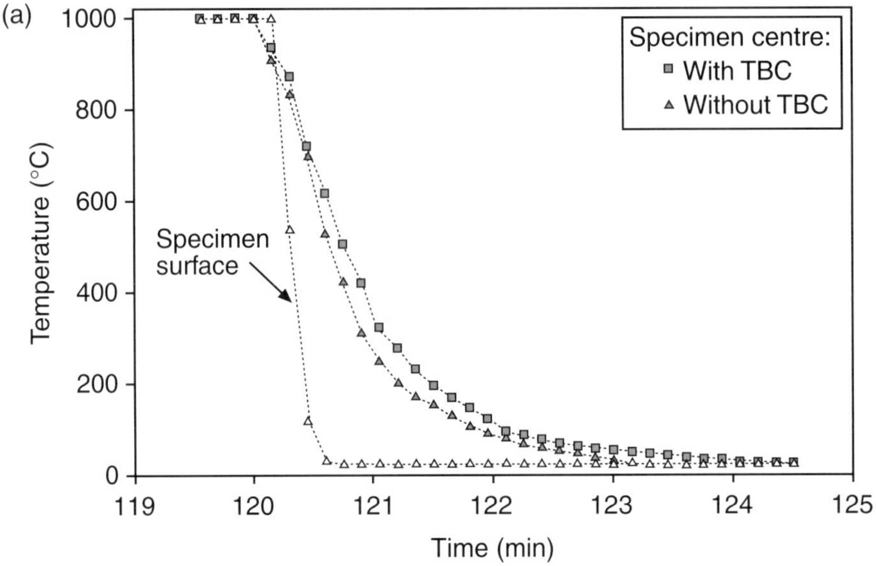

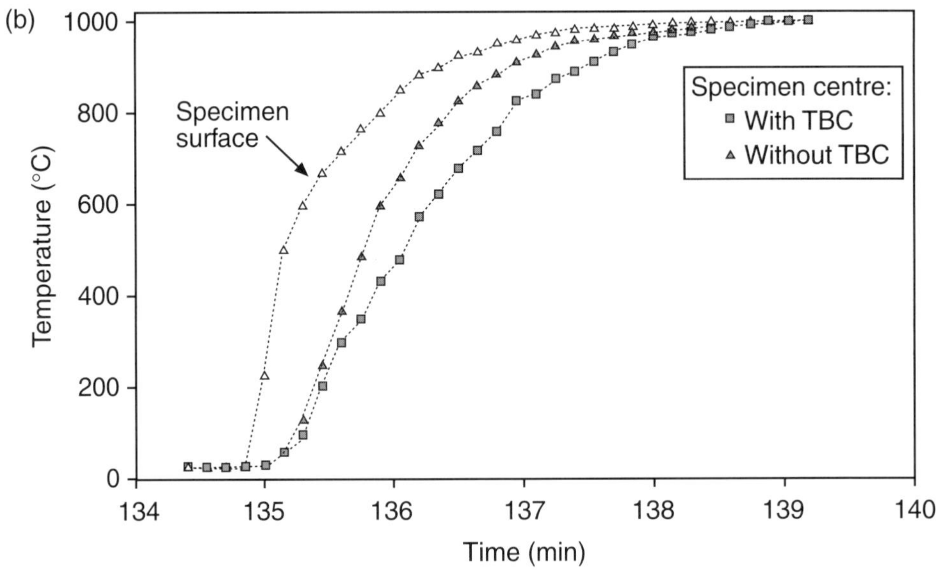

Fig. 2 *Comparison between temperature measurement at specimen surface and specimen interior during thermal cycling at 1000°C in air showing effect of TBC on heating/cooling rate.*

the surface tips. In these places the TGO is adhering to the TBC (Fig. 6). A possible failure mechanism, which leads to the observed crack patterns for APS-TBCs is illustrated schematically in Fig. 7 [5]. In the valleys of the rough BC-surface, high compressive stresses are initiated in the TGO during specimen cooling. The TGO/BC strength is apparently sufficiently high to withstand these stresses and consequently the TGO in the valley areas possesses a high resistance against spallation. On surface tips the compressive stresses in the TGO can be transformed

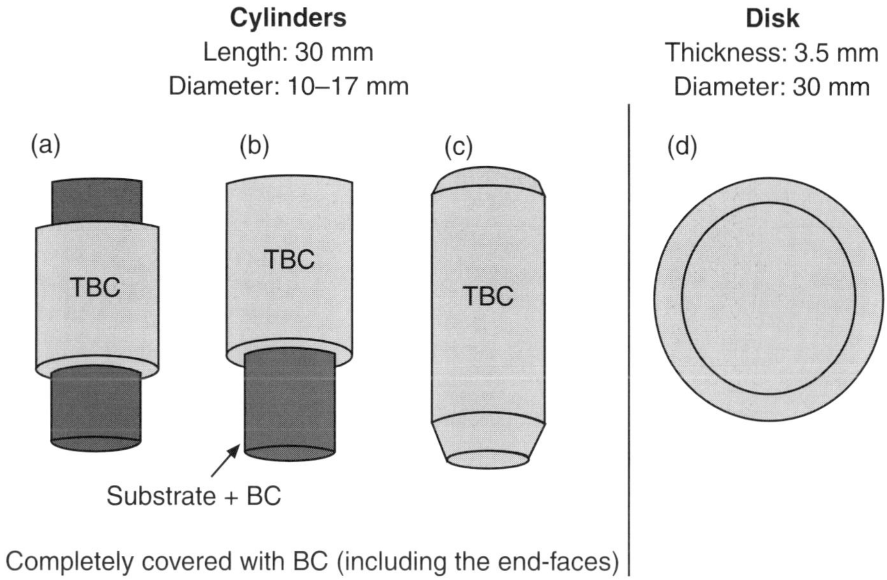

Fig. 3 Various shapes of specimens used during thermal cycling testing of TBCs.

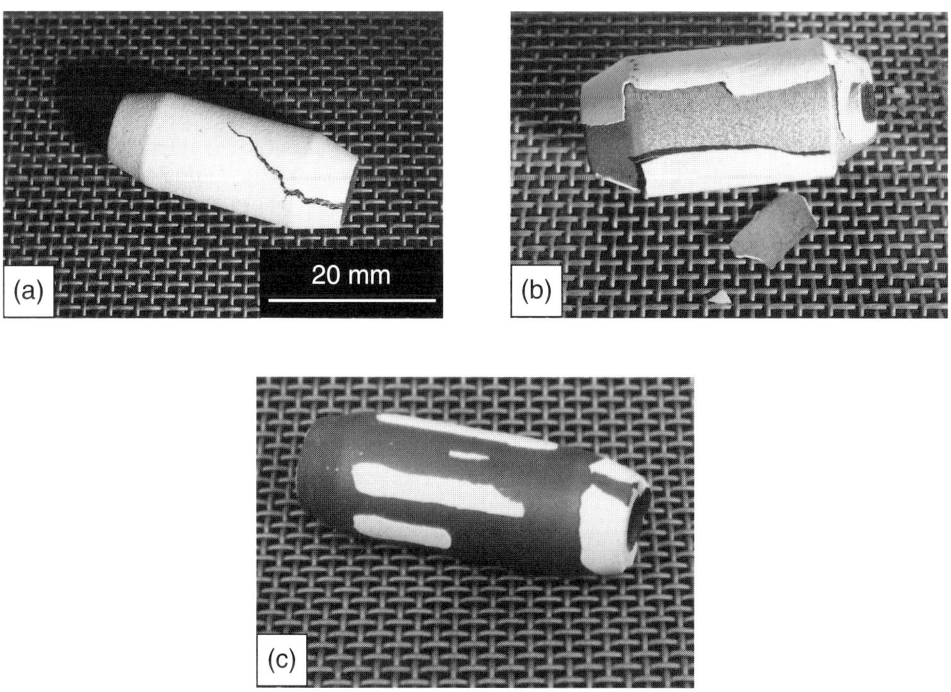

Fig. 4 Macroscopical failure appearance of specimens coated by (a, b) APS and (c) EB-PVD after long term thermal cycling.

Fig. 5 *Typical failure of an EB-PVD coated specimen at the BC/TBC interface after 1629 h thermal cycling at 1000°C in air.*

into tensile stresses leading to easy crack formation in the TGO and/or at the TGO/BC interface. This crack subsequently propagates approximately parallel to the BC/TBC interface, leaving the TGO and the adhering TBC in the valleys unaffected. This failure mechanism macroscopically gives the impression that the crack actually occurs within the inner part of the TBC. However, the present results strongly indicate that during thermal cycling not only EB-PVD but also APS-TBCs primarily fail due to crack initiation in the TGO or at the TGO/BC interface.

The question of whether TBC failure occurs at the BC/TBC interface, the TBC/TGO interface or within the TBC is in many cases difficult to answer. If macroscopically visible crack formation occurs as shown, for example, in Fig. 4(a), the place at which a metallurgical cross-section is made does not necessarily coincide with the place of crack initiation. Consequently the metallographic investigation can lead to erroneous conclusions in respect to the location of failure initiation. An additional aspect which should be borne in mind, is that the frequently described ideal case for occurrence of failure at the BC/TGO interface only prevails if a homogeneous, practically single phase TGO is present. Frequently, however, the thickness of the TGO is not exactly the same over the whole BC surface. Protrusions, related to reactive element-containing internal oxides, are growing from the TGO/BC interface into the BC leading to local differences in scale thickness. This results in a complex stress situation both within the TGO and at the interface of the TGO with TBC and BC. Consequently, mixed failure modes occur whereby in one specimen cracking can occur at both these interfaces as well as within the TGO (Fig. 8).

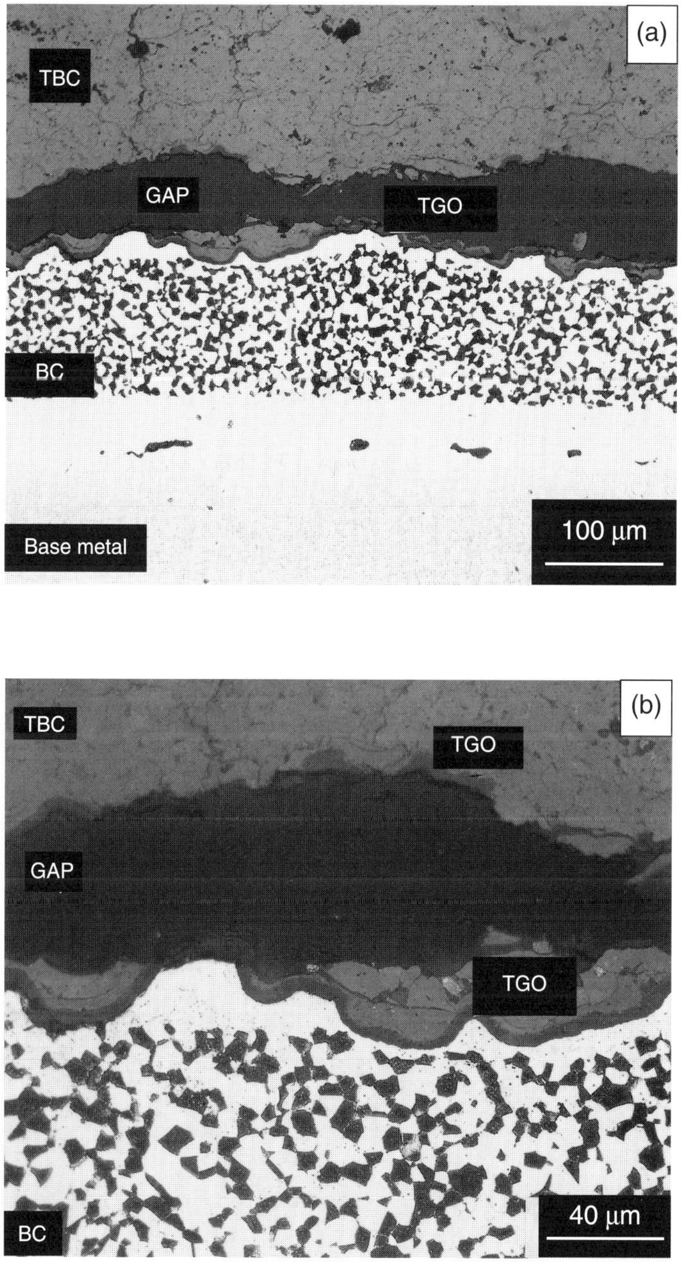

***Fig. 6** Spalling morphology of an APS-TBC after 144 h of thermal cycling at 1100°C in air.*

3.2. Transient Oxides in the TGO

MCrAlY-bond coat compositions are designed to form protective, single phase alumina surface scales during high temperature exposure. This is generally the case

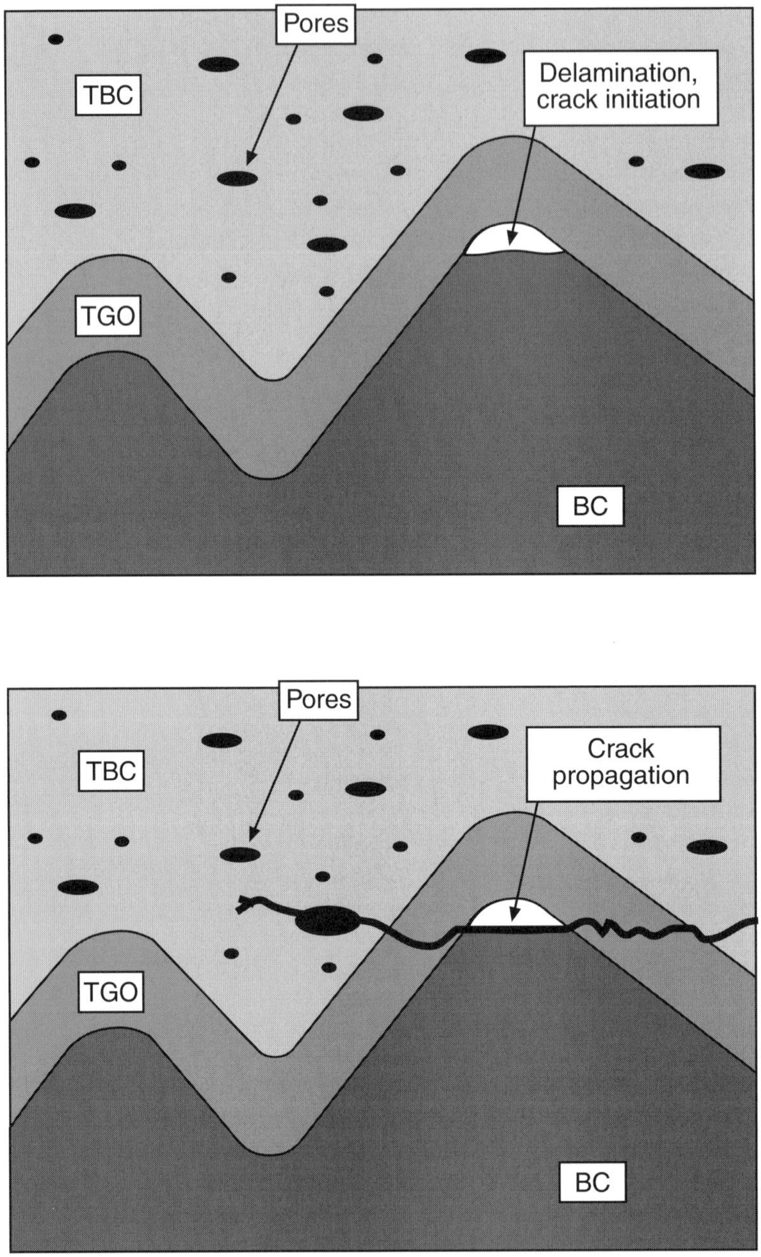

Fig. 7 *Schematic illustration of APS-TBC failure during long term thermal cycling showing local delamination at the BC/TBC interface on tips of the rough BC surface resulting in subsequent crack propagation in the TBC.*

if the bond coats contain a sufficiently high amount of aluminium. It has, however, to be considered, that MCrAlY-bond coats are multiphase alloys. Depending on their exact composition, the coatings can consist of γ-Ni, β-NiAl, γ'-Ni_3Al, α-Cr, σ-(Co,

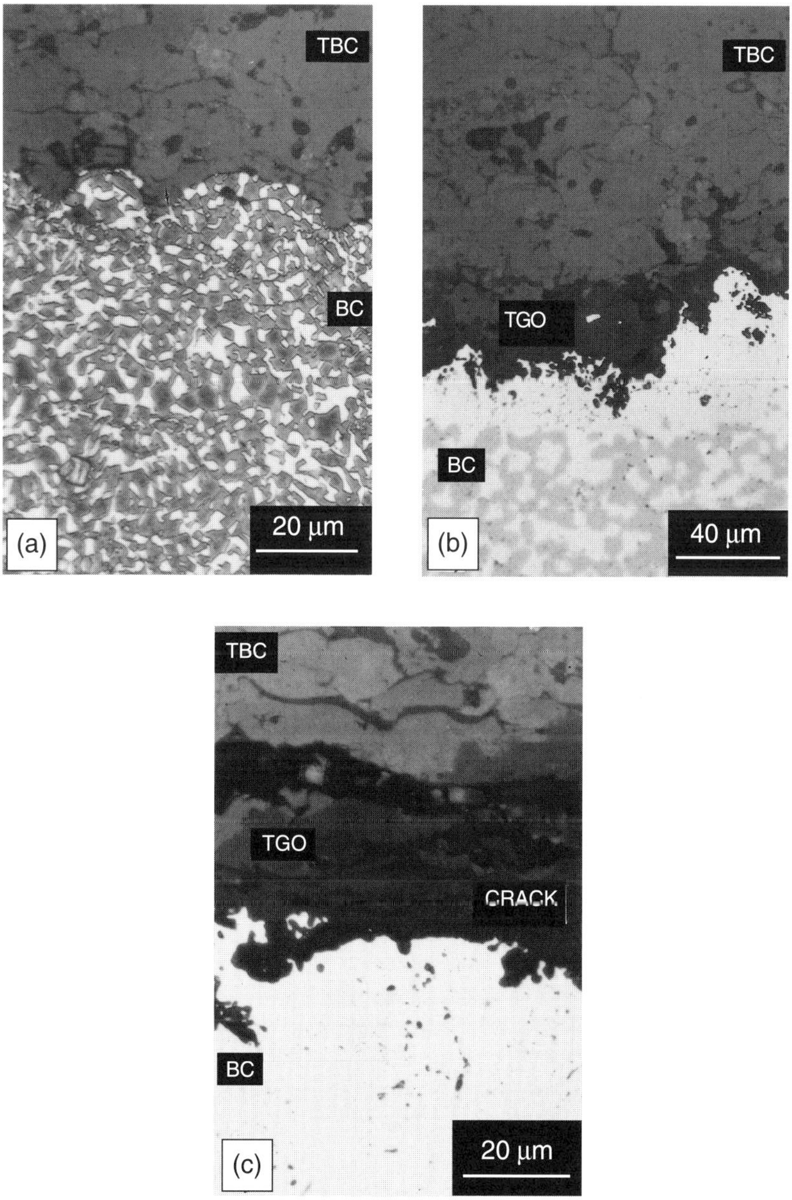

Fig. 8 *Metallographic cross-section of APS-TBC; (a) as-sprayed, (b) after 1108 h thermal cycling at 1000 °C, (c) after failure (3900 h).*

Cr), etc. [1,12]. Due to the strong differences in Al and Cr contents, for example, in the transient oxidation stage all these phases will tend to form different types of oxides, ranging from pure alumina to chromia or e.g. (Ni, Co)(Cr, Al)-spinels. In addition it has been shown, that although the γ- and β-phases both tend to form alumina based scales, the transient oxides formed on the two phases can prevail in different modifications [13–15]. After longer exposure times, the various oxide types formed in the early stages of oxidation will eventually be undergrown by the stable α-alumina; however, the protective properties of this final scale will certainly be affected by stresses generated due to transformations of the transient oxides into the phase(s) which are stable after longer times. The shape, morphology, and stability of the transient oxides will depend on the composition, size and distribution of the various compounds in the bond coat matrix at the moment at which it is exposed to high temperature. If, for example, coarse Cr-rich phases are present, relatively large patches of Cr-rich oxide will be formed during the transient oxidation stage. In some cases the chromia patches still persist up to very long exposure times. This is illustrated in Fig. 9 showing the cross section of an EB-PVD coating after 1692 h of cyclic oxidation at 1000°C. In this failure case patches of Cr-rich oxides were formed in the TGO due to the oxidation of Cr-rich σ-phase precipitates [1,12] at the BC surface during the transient oxidation stages. After longer exposure times alumina is the dominant oxide, however, the large chromia precipitates still remain. It is believed, that stresses occurring due to the heterogeneous character of the TGO lead in this case to early damage in the TGO and consequently to failure of the TBC.

The fact that heterogeneous scales tend to be prone to early crack formation can also be observed in cases where, after long term exposure, spinel type oxide phases are being formed on top of the alumina due to aluminium depletion in the sub-surface layer [16]. This is illustrated in Fig. 10, showing a cross-section of an EB-PVD-coating after 1000 h oxidation at 1000°C. Although failure at the alumina/BC interface is observed in some places, in-scale cracking is additionally detected due to local formation of (Ni, Co)(Cr, Al)-spinel.

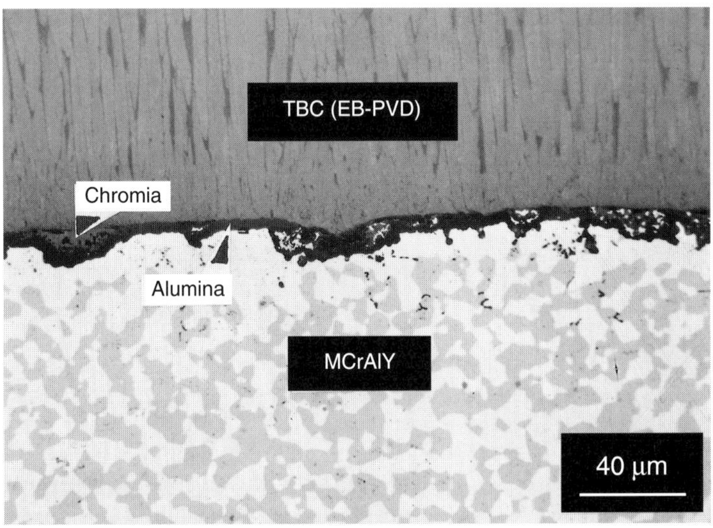

Fig. 9 *Failure of TBC (applied by EB-PVD) after 1692 h of thermal cycling at 1000°C in air due to formation of Cr-rich oxides during transient oxidation.*

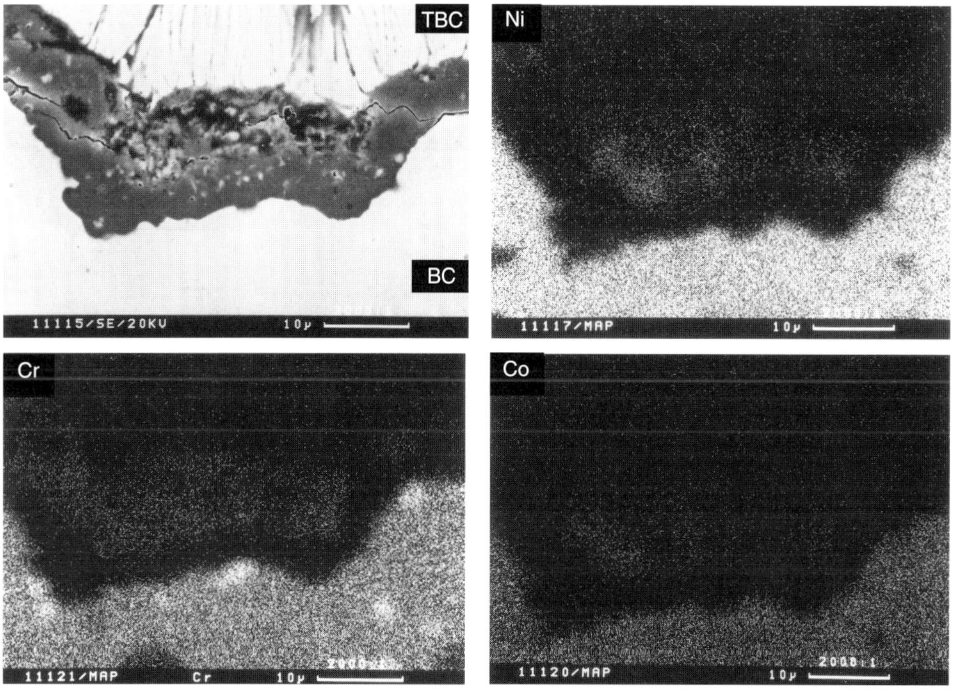

Fig. 10 *Effect of spinel phases in TGO on crack start and/or propagation of an EB-PVD coated specimen after 1000 h oxidation at 1000°C in air.*

3.3. Effect of TBC on TGO Growth and Adherence

Figures 11 and 12 show a comparison of the oxidation rates of free standing MCrAlY bond coats and a TBC coated cylindrical specimen during 1000 h of cyclic oxidation at 1000°C. The cylinder was completely covered (including end faces) with bond coat and partly covered with TBC (see Fig. 3a). For both specimens the same BC powder batch was used. In the interpretation of the mass change results, it should be taken into account that the free standing BCs have a ground surface (1200 grit surface finish), whereas the BC on the TBC coated specimen prevails in the as-sprayed condition and consequently has a surface roughness which is common to VPS coatings (see e.g. Fig. 6). On the free standing bond coats the scales grown at 1000°C generally show no clear indication of spalling up to the maximum exposure time of 2000 h. At 1100°C onset of scale spallation starts between a few hundred and up to around one thousand hours, depending on coating composition. Considering the larger surface area of the as-sprayed BCs on the TBC-coated specimens, their oxidation behaviour with respect to oxide growth rate is very similar to that of the free standing BCs (Fig. 12). However, in many cases TBC spallation occurs earlier than would be expected from the initiation of the TGO spallation measured for the free standing bond coats (Fig. 12). This could be the result of a number of factors:

1. The TGO-growth rate beneath a TBC differs from that on an uncoated specimen.

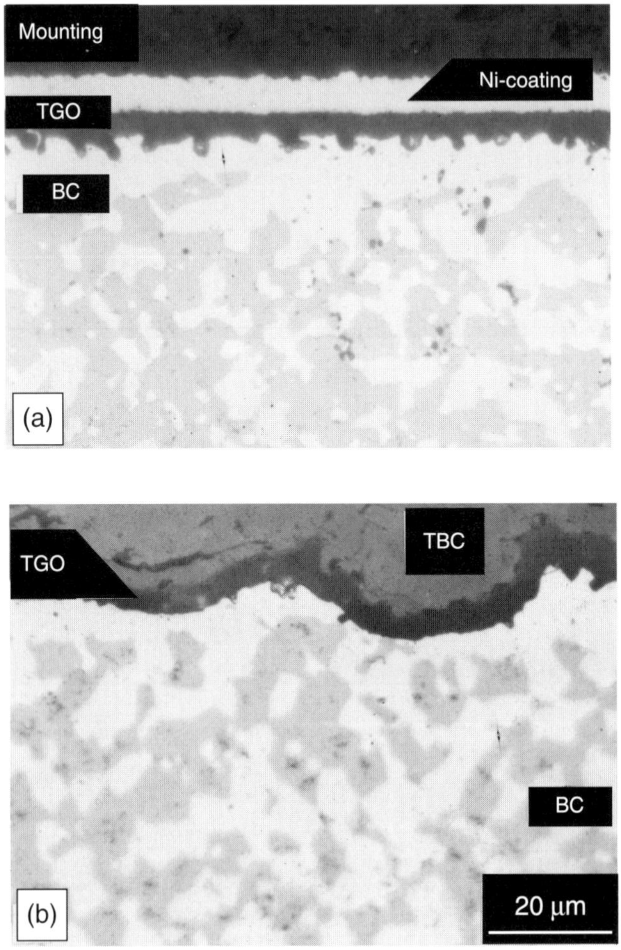

Fig. 11 *Comparison between oxide scale on (a) free standing BC and (b) at TBC/BC interface after 1000 h of thermal cycling at 1000°C in air.*

2. Impurities (such as S) diffuse from the base metal through the bond coat to the BC/TGO interface, thereby lowering the bonding of the TGO with the BC.

3. The transient oxidation on a rough surface differs from that of a ground surface, e.g. due to geometrical effects.

4. The surface roughness of the as-sprayed bond coat affects the adherence of the TGO. The compressive stresses initiated in the scale during specimen cooling are transformed into tensile at surface tips (compare Fig. 6).

5. The TBC on the cylindrical specimen is applied after the BC was subjected to a diffusion heat treatment. Although this pre-treatment is carried out in vacuum or inert gas, some oxidation of the oxygen active elements such as Al

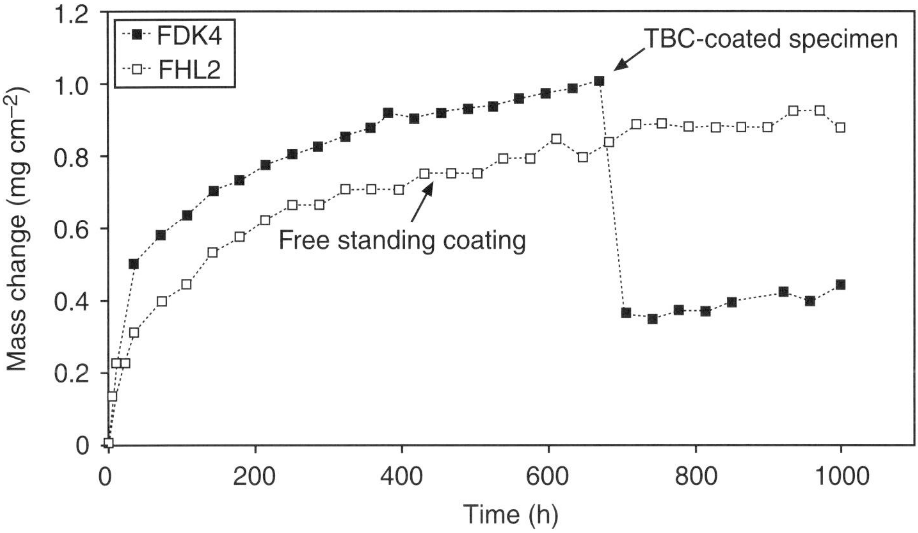

Fig. 12 *Mass change of free standing BC and TBC-coated cylindrical specimen (Fig. 3a) during cyclic oxidation at 1100℃.*

and Y in the BC cannot be prevented. Therefore the transient oxidation stage will differ from that of the free standing bond coat which is tested in the as-ground condition.

6. The oxide at a BC/TBC interface grows under stress conditions which are different from those on a free standing bond coat surface. The oxygen uptake due to scale growth is accompanied by a volume increase, leading to radial tensile stresses in the TBC. If the bonding between TBC and TGO is larger than that between TGO and bond coat, then this stress increase would enhance spallation of the TGO.

Although the exact mechanisms have not yet been completely established the prevailing results indicate that the TGO adherence and perhaps also growth on the ground surface on a free standing BC clearly differs from that on an as-sprayed coating surface. This is illustrated in Fig. 13, which compares cyclic oxidation data of free standing BCs (ground surface) with those of cylindrical specimens which were completely covered with a 200 µm thick BC of the same powder batch. In both cases spallation of the oxide scale occurs much earlier on the coated specimen than on the free standing BC. This result clearly shows that the observed early spallation of some of the TBCs is not necessarily related to an effect of the TBC on the growth and/or adherence of the TGO (item 1, mentioned above). Although the different stress conditions under which the TGO grows beneath the TBC play an important role for oxide adherence, the results in Fig. 13 indicate that earlier spallation is not necessarily enhanced by the presence of a TBC (item 6). Therefore, it is more likely that the early spallation of some of the TBCs is related to the effect of specimen geometry, e.g. surface roughness, on the transient oxidation and/or the stresses initiated in the TGO (items 3 and 4).

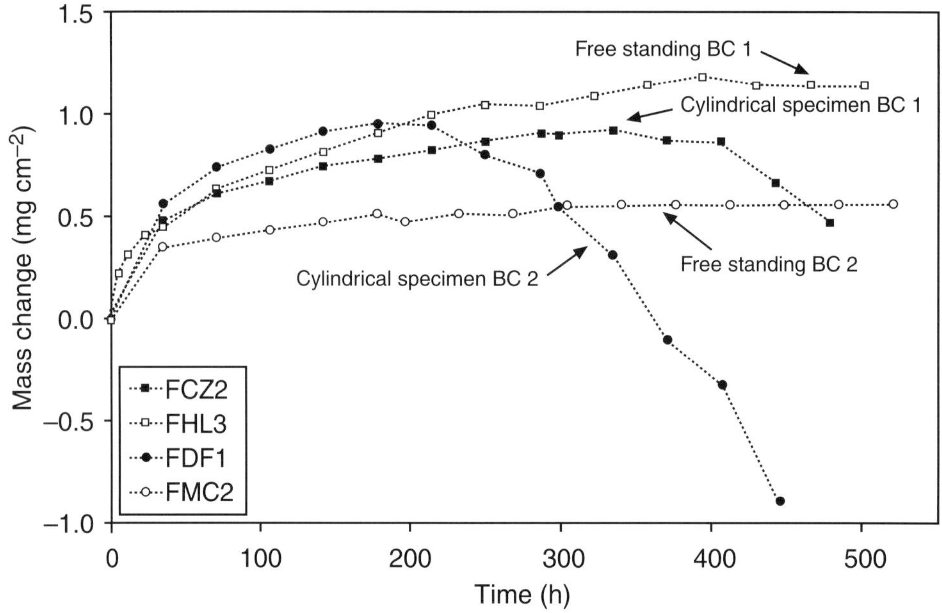

Fig. 13 *Mass change of free standing BCs (ground surface) and cylindrical specimens (Fig. 3a) during cyclic oxidation at 1100°C air. Cylindrical specimens (without TBC) coated with a 200 mm thick BC of the same powder batch as the free standing BC.*

3.4. Failure at the Interface between Bond Coat and Base Metal

In all cases previously discussed, the failure after long term thermal cycling appeared near the BC/TBC interface. Although this was the main failure location in nearly all cases studied, in some specimens failure occurred in other locations of the TBC coated specimens. Figure 14(a) shows an optical macrograph of a TBC-coated cylindrical specimen after 4754 h of cyclic oxidation at 1000°C. No macroscopical failure such as cracking or spalling of the TBC is apparent. However, after cross-sectioning, a gap between base-material and bond coat is clearly visible (Fig. 14b). As expected from the macroscopical observation, excellent bonding between TBC, alumina scale and MCrAlY bond coat still exists. The failure of this specimen can be related to a chemical incompatibility between the bond coat and the base metal. Interdiffusion processes resulted in the formation of a brittle phase at the BC/base metal interface, which eventually cracked during long term thermal cycling. It was, however, remarkable that this failure did not lead to a macroscopically visible cracking of the TBC. This result illustrates the necessity of developing methods for reliable failure detection during testing of TBC systems.

A further example of a TBC coated system which failed at the BC/base metal interface is shown in Fig. 15. It illustrates the cross section of a disc-shaped specimen after 396 h of cyclic oxidation at 1100°C in air. The image in Fig. 15(a) was taken in the middle of the specimen, revealing a crack in the ceramic top coat parallel to the BC/TBC interface. The image in Fig. 15(b), which was taken near the edge of the specimen, reveals the actual reason for the TBC failure. As the edges of the disc-

Fig. 14 *TBC-coated specimen (applied by APS) after 4754 h thermal cycling at 1000°C in air; (a) macroscopical appearance of the specimen, (b) metallographic cross-section showing crack formation at interface between base metal and bond coat.*

shaped specimen were, for experimental reasons, not completely covered with MCrAlY, oxidation of the unprotected base-material occurred. Upon further cyclic exposure the oxidation front propagated along the interface between base metal and bond coat, leading to a volume increase and subsequently to crack initiation in the ceramic top coat.

4. Conclusions

During thermal cycling the macroscopical failure appearance of TBCs applied by APS differs from that of PVD-TBCs. Usually APS-TBCs show a small crack, initiated in the TBC, which propagates with time, eventually leading to spallation of large irregular parts of the ceramic top coat. In contrast, EB-PVD-TBCs show a more sudden spallation behaviour, typically of large, elongated ceramic patches, without any apparent warning of the onset of failure by the formation of small cracks.

Failure of EB-PVD coatings generally occurs by spallation of the TGO although in case of multiphase TGOs in-scale failure is sometimes observed. Failure interpretation of APS-TBCs, based only on metallographic cross-sections, can be misleading due to the rough BC-surface. Although in most cases the TGO is still adhering in valleys of the rough surface, indications were also found that failure initiation can occur at

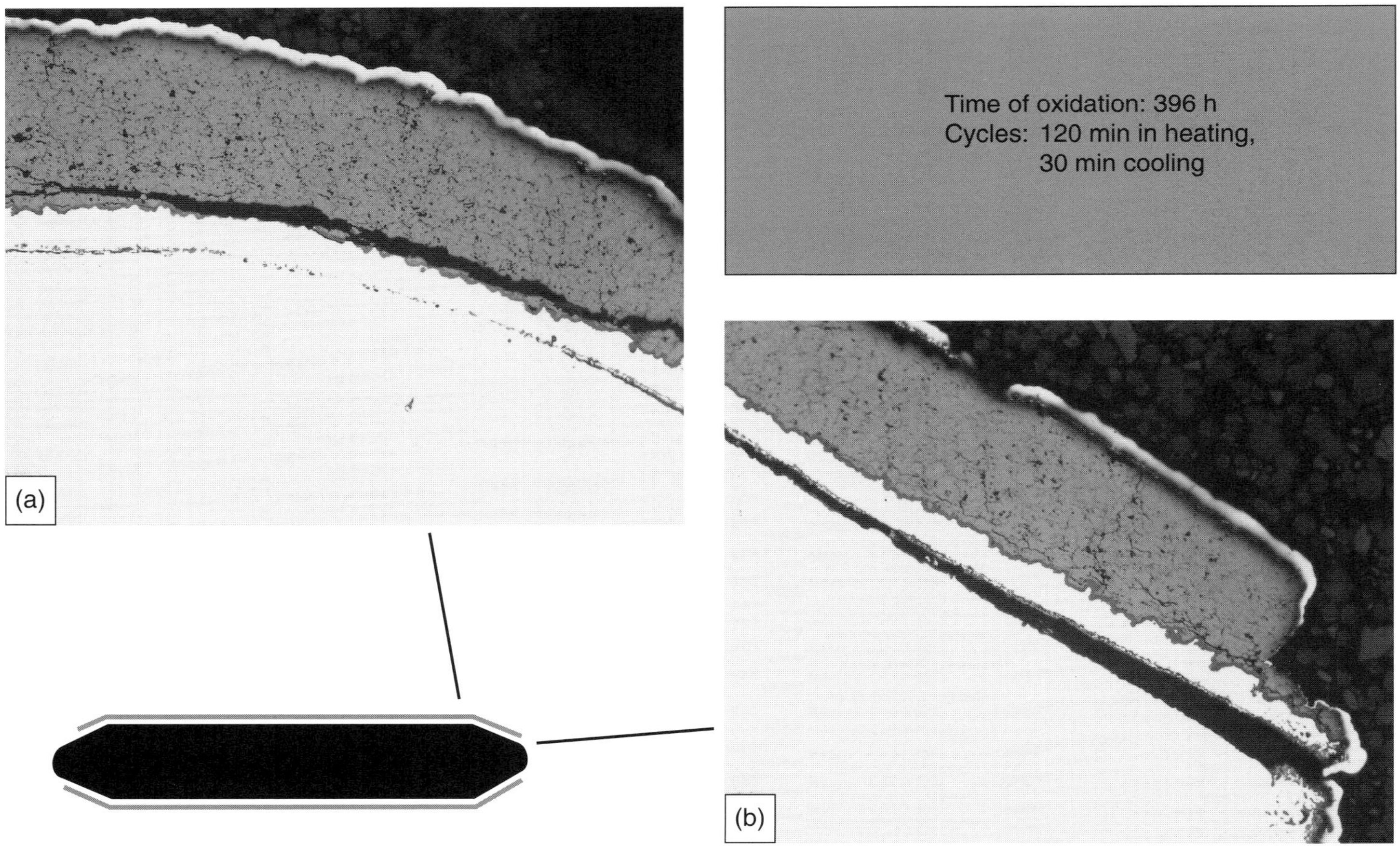

Fig. 15 *Cross-section of disk-shaped TBC-coated specimen (applied by APS) after 396 h of thermal cycling at 1100°C in air; picture taken (a) in the middle and (b) at the edge of the specimen showing TBC failure due to oxidation at interface between base metal and BC.*

surface tips of the TGO/BC interface. The initiated crack propagates horizontally in the TBC leaving the TGO in the valleys unaffected. This leads to a macroscopically visible damage type, frequently called 'white failure'.

Due to their multiphase character, MCrAlY-BCs tend to form heterogeneous oxide scales during transient oxidation. Stresses will be generated in the scale due to transformation of the transient oxides into those phases that are thermodynamically stable after longer times. Such phase transformations usually imply volume changes in the TGO resulting in an enhanced tendency to scale cracking and/or spalling. In many cases TBC failure appeared to occur earlier than one would expect from the mass change data obtained during cyclic oxidation of free-standing MCrAlY-BCs. Indications were found that this is related to the effect of BC roughness on transient scale formation and/or stress generation in the scale.

In rare cases, TBC failure occurs due to reactions at the interface between BC and base metal. This can, for example, be caused by oxidation of an incompletely coated base metal or by brittle phase formation due to interdiffusion processes between BC and base metal. These results highlight the significance of a suitable choice of specimen geometry during cyclic oxidation testing if a misleading interpretation of results concerning the actual TBC adherence is to be avoided.

References

1. W. Beele, N. Czech, W. J. Quadakkers and W. Stamm, *Surf. Coat Technol.*, 1997, **94/95**, 41–45.

2. L. Singheiser, R. Steinbrech, W. J. Quadakkers, D. Clemens and B. Siebert, 'Thermal barrier coatings for gas turbine applications – failure mechanisms and life prediction', *6th Liége Conf. on Materials for Advanced Power Engineering*, 1998, Liége, B, Proc. European Commission, (J. Lecomte-Beckers, F. Schubert and P. J. Ennis, eds), 1998, Vol. II, pp.977–996.

3. D. Wortmann, B. Nagaraj and E. Duderstadt, *Mater. Sci. and Eng. A* , 1989, **121**, 433–440.

4. S. Bose and J. de Masi-Maran, *J. Therm. Spray Technol.*, 1997, **6** (1), 99.

5. W. J. Quadakkers, A. K. Tyagi, D. Clemens, R. Anton and L. Singheiser, The significance of bond coat oxidation for the life of TBC coatings, TMS Symposium: San Diego, 28 Feb.–4 March, 1999. *Elevated Temperature Coatings: Science and Technology — III* (J. M. Hampikian and N. B. Dahotre, eds). The Minerals, Metals & Materials Society, 1999, 119–130.

6. P. Hancock and J. R. Nicholls, *Mater. High Temp.*, 1994, **12**, (2/3), 209–218.

7. D. Gupta and D. S. Duvall, in *Superalloys 1984*. Metallurgical Soc. of AIME, Warrendale, PA, 1994, p. 711.

8. L. Singheiser, Proc. *1st Plasmatechnik Symp.*, Lucerne, Switzerland, 1988, Vol. 2, p. 77.

9. A. Evans, J. Wand and D. Mum, 'Mechanism-based life prediction issues for thermal barrier coatings', NASA TBC Workshop, Cleveland, Ohio, *Proc.*, 1997, p. 45.

10. P. K. Wright, NASA TBC Workshop, Cleveland, Ohio, *Proc.*, 1997, p. 73.

11. J. C. Schäffer, NASA TBC Workshop, Cleveland, Ohio, *Proc.*, 1997, p. 99.

12. W. J. Quadakkers, W. Stamm, N. Czech, E. Wallura and H. Hoven, Rhenium containing plasma sprayed NiCoCrAlY coatings with improved oxidation and interdiffusion properties, *EUROMAT '97* (L. Sarton and H. Zeedijk, eds). Maastricht, The Netherlands, 1997 Vol. 3, pp. 209–215.

13. D. Clemens, V. Vosberg, L. W. Hobbs, U. Breuer, W. J. Quadakkers, and H. Nickel, *XXIX Colloquium Spectroscopicum Internationale CSI*, 1995, *Fresenius' J. Analyt. Chem.*, 1996, **355**, 703–706.

14. D. Clemens, V. Vosberg, J. Penkalla, E. Schumann, A. Czyrska-Filemonowicz and W. J. Quadakkers, TEM and SNMS studies on the oxidation of Ti- and Si-containing NiCrAlY alloys; *IX Conf. Electr. Micr. of Solids*, 6–9 May, 1996, Zakopane, Poland (A. Czyrska-Filemonowicz *et al.*, eds), 1996, pp.435–440.
15. D. Clemens, V. R. Vosberg, H. J. Penkalla, U. Breuer, W. J. Quadakkers and H. Nickel, AOFA, 9. Arbeitstagung für Angewandte Oberflächenanalytik, Aachen, 1996, *Fresenius' J. Analyt. Chem.*, 1997, **358**, 122–126.
16. D. Clarke, Precursor to TBC failure caused by constrained phase transformation in the thermally grown oxide, in *TMS Symp. 1999, Elevated Temperature Coatings: Science and Technology III* (J. M. Hampikian and N. B. Dahotre, eds). The Minerals, Metals & Materials Society, 1999, pp.67–78.

22
Investigations of the Microstructure and Stress State in Oxide Scales and Buried Layers with Diffraction Methods

T. SCHMACKERS and W. REIMERS
Hahn-Meitner-Institut, Berlin

ABSTRACT

For a detailed understanding of the oxidation and fatigue behaviour of high temperature materials, the knowledge of the interaction between the microstructure and residual stresses, which arise during material processing, is of practical importance. Because non-destructive diffraction techniques are sensitive to the different crystalline components of materials, they offer significant advantages in terms of the phase specific analysis of the microstructure as well as the residual stress state within the material. The paper gives some practical examples for the analysis of the temperature induced changes in both stresses and microstructure.

Conventional X-ray techniques were applied to analyse the oxidation of a TiAl substrate and the development of the stress state in a NiO scale on fine grained, coarse grained and single crystalline Ni substrates during isothermal growth and cooling. These investigations reveal that the oxidation of the substrate is always accompanied by compressive stresses in the oxide scale. The absolute amount of the stress state depends on the thickness of the oxide scale, the oxidation temperature and the microstructure of the substrate. From the relation between the stress state and the microstructure conclusions may be drawn concerning the oxidation behaviour.

In addition, investigations of the oxidation behaviour and the temperature dependence of the stress state in a NiCoCrAlY layer of a NiCoCrAlY/INCONEL alloy 718 layer compound have also been conducted. The results show an increase of compressive stresses even after one thermal cycle.

In order to investigate the stress state of this layer in a thermal barrier coating system where it is buried under an additional ZrO_2–7mass%Y_2O_3 ceramic layer, high energy synchrotron radiation providing greater penetration depths has been used. These investigations show that it is possible to analyse *in situ* the stress distribution in the ceramic layer as well as in the bond coat, the latter having a thickness of only 150 μm and deposited between the ceramic layer and the substrate.

1. Introduction

One of the main topics for the lifetime evaluation of components subjected to high temperatures is a sufficient characterisation of their behaviour under real service conditions. In most of the cases the building and performance of protective oxide scale plays an important role in the corrosion of high temperature materials.

The protective character of the oxide films is maintained under certain conditions, in which the growth induced residual stresses are of particular interest because they can induce scale fracture and a loss of the adherence to the substrate material [1–4]. Another aspect of the oxide film stress behaviour is the thermal part of the stresses which are built up at temperatures different from the temperature on which the oxide film was formed. During cooling strong compressive stresses are usually generated which, above critical scale thicknesses, lead to scale fracture and a detrimental degradation of the material.

In thermal barrier coating systems the stress state of the bond coat which is also used for oxidation protection of the substrate becomes even more complex due to the overlying ceramic layer. The performance of the bond layer and the thermally grown oxide that is built between the ceramic and the bond layer influences the adherence of the thermal barrier coating and therefore to a large extent its lifetime.

In this paper the results of investigations are presented, which were performed *in situ* by X-ray diffraction methods on growing oxide scales on nickel and titanium aluminium substrates.

The substrate materials represent in the case of nickel a main component of many high temperature materials and in the case of titanium aluminium an independent material.

In addition, the oxidation behaviour and stress state of a NiCoCrAlY bond layer on a In718* substrate was investigated *in situ*. With high energy synchrotron radiation a new tool has been used which allows the stress state of the bond coat in a thermal barrier coating system to be analysed even if it is buried between the ceramic layer and the substrate [5].

2. Experimental

The specimens were cut from bulk material into pieces of approximately 10×10 mm^2. They were polished by a 6 μm diamond suspension. The chemical composition of the substrate materials is summarised in Tables 1 (a) and 1(b).

The purity of the Ni-single crystal was 99.999%.

Table 1(a). *Chemical composition of the polycrystalline Ni-specimen [6]*

Chemical element	Ni	C	Mn	Si	Co	Mg	Cu	Fe	S
(Mass%)	Bal.	0.0027	0.23	0.02	0.01	0.030	0.01	0.04	0.002

Table 1(b). *Chemical composition of the titanium aluminum specimen [7]*

Chemical element	Ti	Al	H	N	O
(Mass %)	Bal.	35.90	0.001	0.011	0.12

*INCONEL alloy 718 (UNS N 07718) referred to in the paper as In 718.

The investigated duplex thermal barrier coating system consists of a plasma sprayed ceramic layer of 7mass% Y_2O_3 stabilised zirconia with a thickness of 0.5 mm and a low pressure plasma sprayed (LPPS) NiCoCrAlY bond layer with a thickness of 0.15 mm, both deposited on the Ni-based superalloy In718 with a thickness of 2 mm. The dimensions of the specimen were $50 \times 10 \times 2.65$ mm^3. For the investigations of the oxidation behaviour of the bond coat samples without the ceramic layer were used.

In order to analyse the diffraction elastic constants (DEC) of the plasma sprayed zirconia, a bulk sample with the dimensions $50 \times 10 \times 3$ mm^3 was also investigated.

The high temperature measurements were conducted using a mirror oven device as shown in Fig. 1.

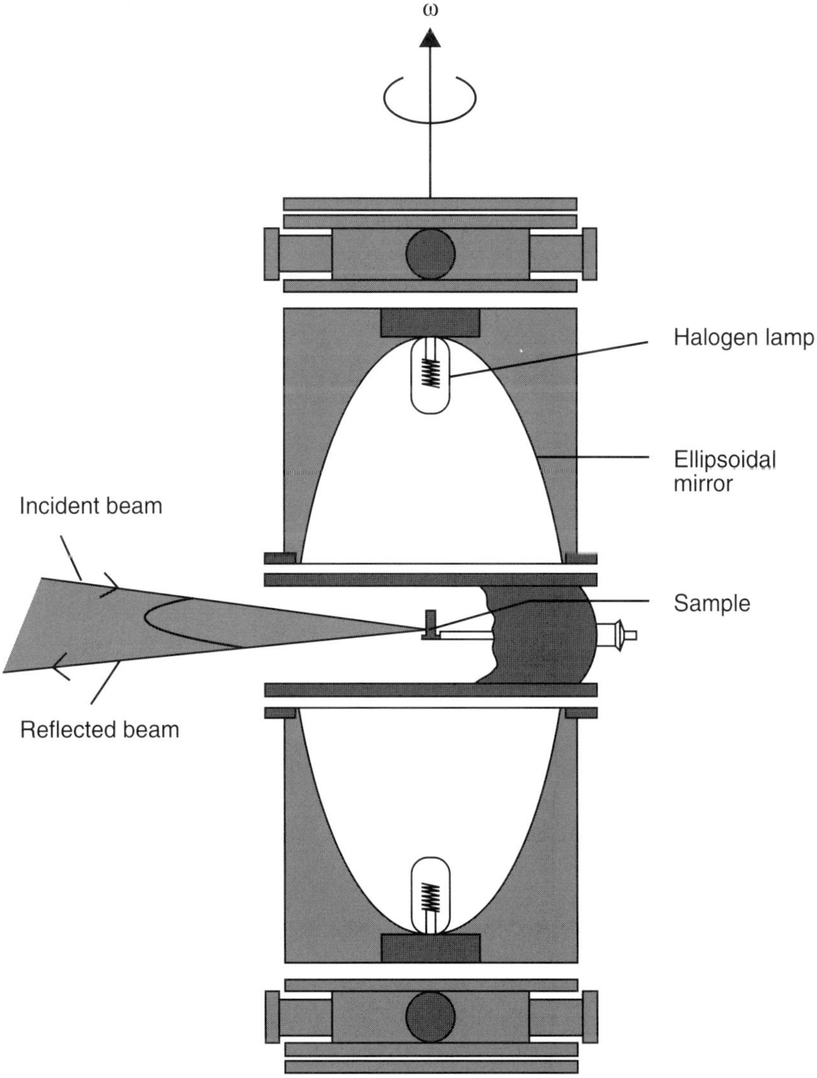

Fig. 1 *Mirror oven device used for high temperature measurements.*

To achieve a more realistic temperature distribution across the sample thickness, a new oven device has been developed. The schematic view of it is shown in Fig. 2.

The sample is mounted on a water cooled sample holder. Its cooling rate can be regulated by the flow rate of the water within narrow limits. While the back side of the sample is cooled, the surface of the ceramic layer is heated by four special halogen lamps up to temperatures of 1200°C. Sheets of alumina around the sample ensure that only the surface is heated.

The temperatures are measured with thermocouples, fixed at the surface and the back side of the sample.

This device is mounted on an eulerian cradle, which allows an *in situ* analysis of the thermal stresses by means of diffraction methods using X-ray, neutron or high energy synchrotron radiation.

3. Principles of Measuring Techniques

The principles of X-ray stress analysis using conventional X-rays are based on eqn (1), which represents the relation between the measured strains $\varepsilon_{\phi\psi}$ and the stress components σ_{ij}. The factors s_2(hkl) and s_1(hkl) are the DEC.

$$\varepsilon_{\phi\psi} = \tfrac{1}{2}s_2(\text{hkl})\begin{bmatrix}(\sigma_{11}\cos^2\phi + \sigma_{22}\sin^2\phi + \sigma_{12}\sin 2\phi)\sin^2\psi \\ +\sigma_{33}\cos^2\psi + (\sigma_{13}\cos\phi + \sigma_{23}\sin\phi)\sin 2\psi\end{bmatrix} + s_1(\text{hkl})(\sigma_{11} + \sigma_{22} + \sigma_{33}) \tag{1}$$

The strain and stress components are determined in the axes system shown in Fig. 3.

The formula (1) can be simplified for a biaxial stress state with rotational symmetry to the well known $\sin^2\psi$-law:

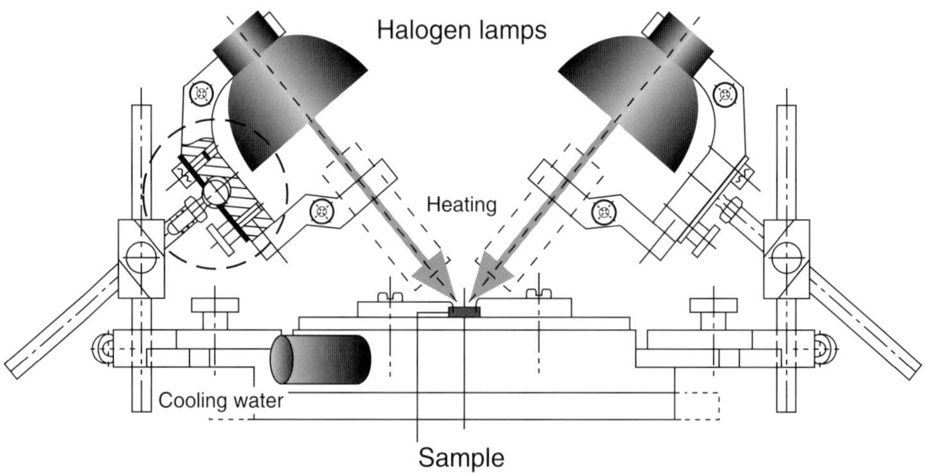

Fig. 2 *Oven device to apply defined temperature gradients to thermal barrier coating systems.*

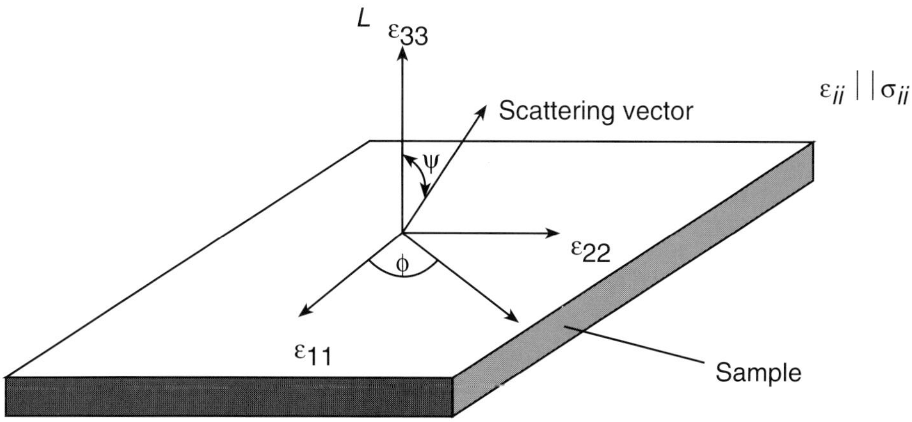

Fig. 3 Coordinate system for the strain measurement.

$$\frac{\partial \varepsilon_{\phi\psi}}{\partial \sin^2 \psi} = \tfrac{1}{2} s_2(\mathrm{hkl})\sigma_\phi \tag{2}$$

$$\sigma_\phi = \sigma_{11}\cos^2\phi + \sigma_{22}\sin^2\phi \tag{3}$$

The different ψ-angles were obtained by tilting the coupled sample-furnace system in the so called *ω-mode,* in which the ψ-angle is given by the difference between the Bragg angle θ and the angle ω. The stress determination by eqn (2) is achieved by a linear regression of the analysed strains.

For the investigations that have been performed with high energy synchrotron radiation and an energy dispersive detector the lattice strains (eqn. (1)) can be obtained from the shifts in the detected energy E_x with eqn (4) and eqn (5) [8]:

$$E_x = \frac{hc}{\lambda} \tag{4}$$

$$d^{hkl} = \frac{hc}{2\sin\theta} \cdot \frac{1}{E_x^{hkl}} = const. \cdot \frac{1}{E_x^{hkl}} \tag{5}$$

Here h denotes the Planck constant, c the velocity of light and λ the wavelength. The energy dispersive experiments were carried out at the High Energy Diffraction Beamline (ID 15) of the ESRF in Grenoble, France [9]. This beamline allows investigations up to an energy of 300 keV. The measurements were performed at a diffraction angle of 2θ = 10 deg. The gauge volume was defined by a slit in the primary beam and two slits in the diffracted beam with an opening width of 80 μm in the diffraction plane. It has the form of a parallelepiped with the dimensions 1.3 mm perpendicular to the scattering vector and 0.13 mm parallel to it (Fig. 4).

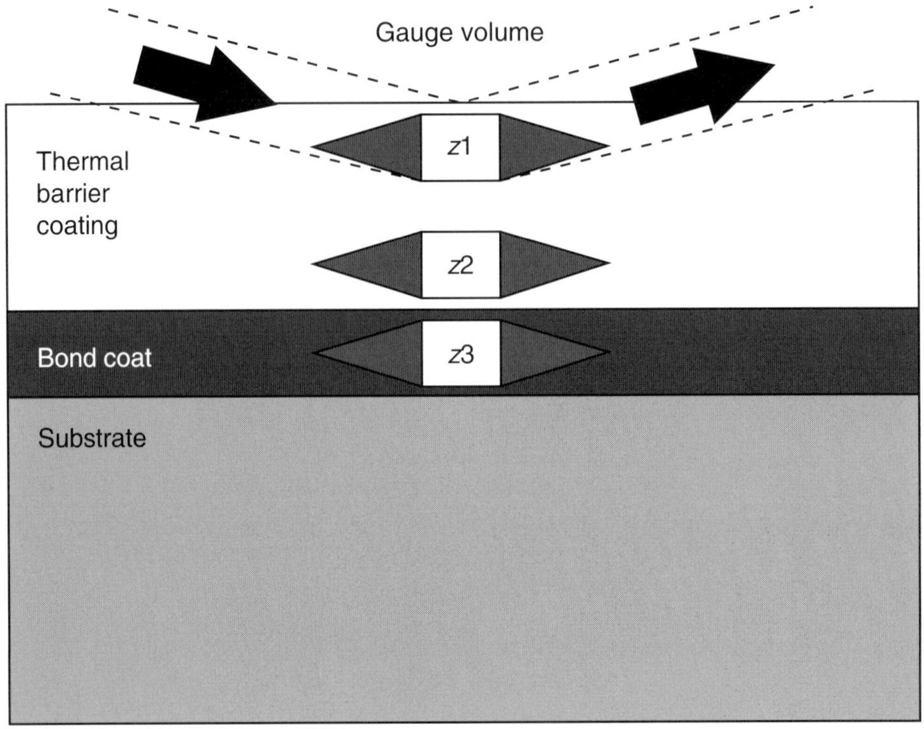

***Fig. 4** Geometric form of the gauge volume and its positions z1, z2 and z3 for the investigations of the stress state.*

4. Results

4.1. NiO on Nickel

4.1.1. Microstructural characterisation

The microstructure of the NiO layers is shown for three cases in Fig. 5(a–c).

The micrographs in Fig. 5(a,b) demonstrate a duplex oxide film growth on either polycrystalline substrate but the fraction of the inner layer is smaller for the coarse grained material. The oxide film formed on the single crystal exhibits only a single layer structure. These findings are due to the mechanisms of the inner and outer oxidation forming an inward and an outward growing layer, respectively. The horizontal line separating both layers is nearly identical with the substrate surface at the beginning of the oxidation. The structure of both layers is different in terms of size and orientation of the oxide grains. The inner layer consists of fine grained, isotropic oriented NiO-crystallites and is generated by the oxidation of the nickel grains. Hence on the single crystalline nickel substrate no inner oxide layer is found. The outer oxide layer shows a more coarse structure with a columnar like growth. Its growth mechanism is the transport of Ni-cations to the oxide surface where new oxide material is formed [1,2].

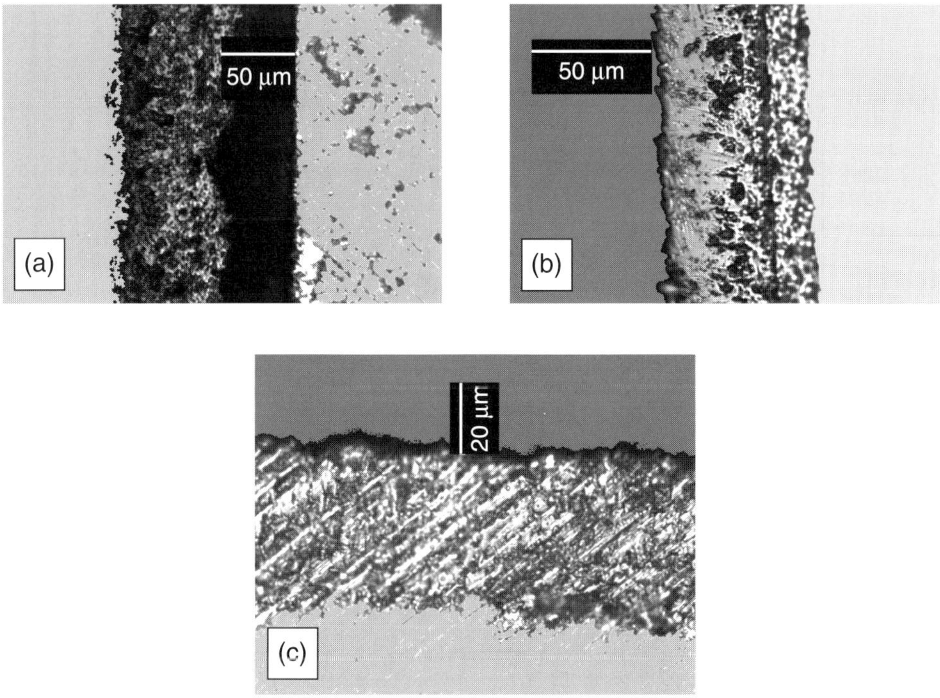

Fig. 5 *Microstructure of the NiO oxide layers on (a) a fine grained and (b) a coarse grained nickel substrate after 70h at 900°C. The grain sizes of the polycrystalline substrates were approximately 10 μm for the fine grained and 100 μm for the coarse grained nickel; (c) shows the microstructure of a NiO layer on a single crystal nickel substrate after 120h at 900°C.*

At this oxidation temperature of 900°C an examination was made of how the substrate microstructure affects the stress state in the oxide film. The results are shown in Fig. 6.

The oxide films on the fine grained as well as on the coarse grained nickel substrate exhibit the same distribution of growth stresses. Hence, the differences in the substrate grain structure for polycrystalline materials are not sufficient to affect a difference between both stress distributions. Both distributions become nearly time independent after a short incubation period. With respect to other investigations [10] it can be concluded that the stress contributions of inner and near surface zones having different mechanisms of oxide formation is nearly balanced out in the sense of a dynamic equilibrium. This is schematically shown in Fig. 7.

This discussion can be enhanced by consideration of the single crystalline substrate. If there is no inner oxidation then the stress generating processes of the outward growing scale will dominate completely. It would lead to a steady increase of compressive stresses as was found here.

4.2. Oxidation of TiAl

4.2.1. Microstructure and composition of the oxide scale

Figure 8 shows a micrograph of an oxide scale formed on TiAl (1). At least four sublayers can be distinguished having a volume fraction high enough for the detection

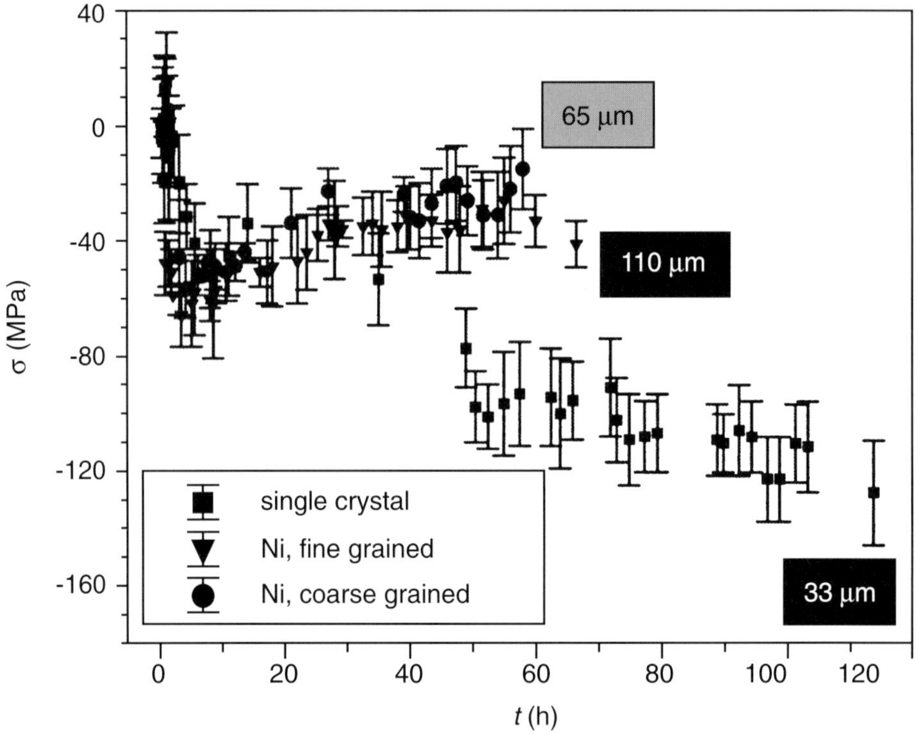

Fig. 6 *Residual stresses vs oxidation time for different substrate microstructures. The numerical values represent the thickness of the oxide layers at the end of the experiment.*

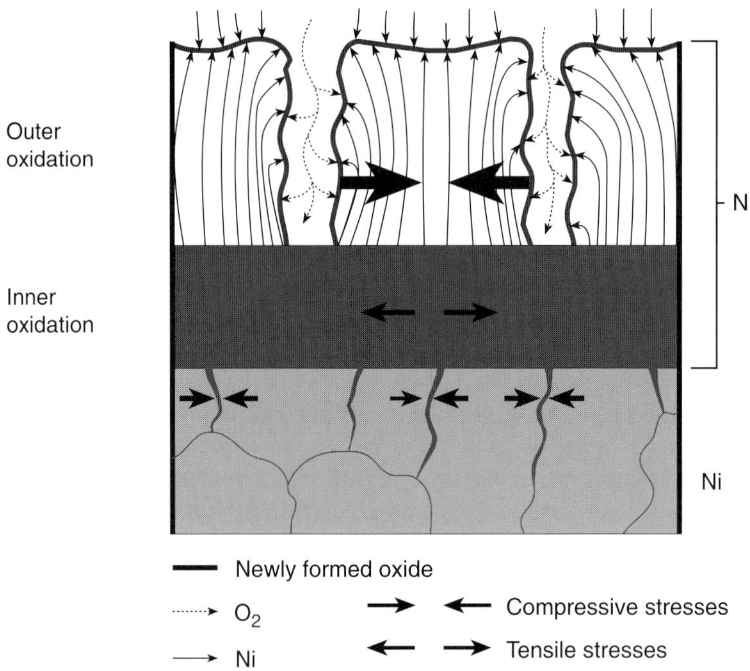

Fig. 7 *Oxidation processes of fine and coarse grained Ni and the stress state within the different layers (schematic).*

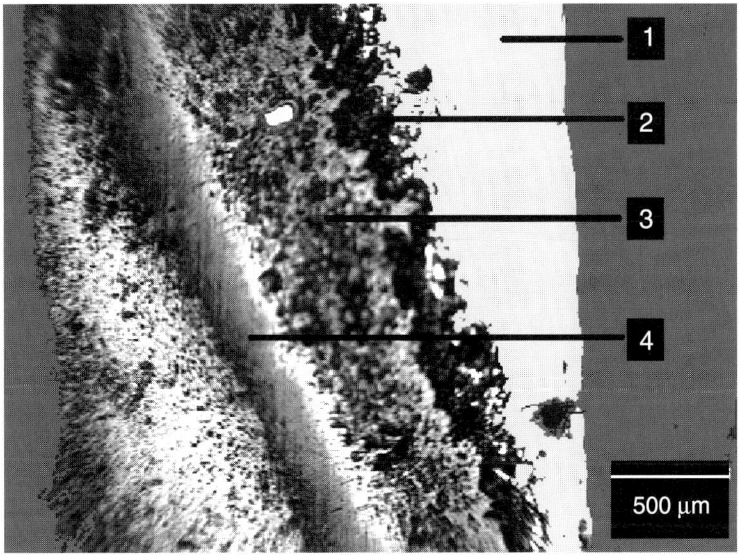

Fig. 8 Cross-section of the oxide scale on TiAl prepared by grinding under an angle of approximately 10 deg. The numbers given in the image are to mark the different subscales.

in a X-ray pattern. However, because of the large thickness of the outermost layer 4 the inner layers cannot be detected due to the attenuation of the X-rays by the layer 4.

The outermost layer relevant for X-ray diffraction consists of TiO_2 which is demonstrated by the diffraction pattern in Fig. 9. The inner layers consist mostly of a mixture of TiO_2 and Al_2O_3 with different fractions of each [11].

4.2.2 Stress analysis

The strain measurements were carried out using Cu-Kα radiation as providing the best peak-background ratio for the strong fluorescence of Ti. Figure 10 shows the results of two independent measurements on two specimen at 900°C.

Obviously there is an oscillating fluctuation of the stress evolution rather than a monotonous development as was found for nickel oxide. If a critical amount of compressive stresses of nearly –100 MPa is reached, a relaxation of the stress state takes place. The period of the oscillation is approximately 40 h. Due to the rapidly increasing layer thickness the deeper layers with a different phase composition could not be investigated. Therefore the discussion of the stress inducing processes will be limited to the outermost titanium oxide layer. Due to the complex structure of the entire oxide scale, there is also a different distribution of oxidation induced stresses in these sublayers. Hence a microstructural based model is required that takes the interaction of the sublayers into consideration. As a result of SEM studies and further investigations [10] a model of the oxidation behaviour can be developed which is schematically represented in Fig. 11.

Taking into account that the interfaces between the different sublayers are very irregular and each of them has large bulges similar to cavities filled with newly formed oxide material this interpretation is based on a localised loss of adherence

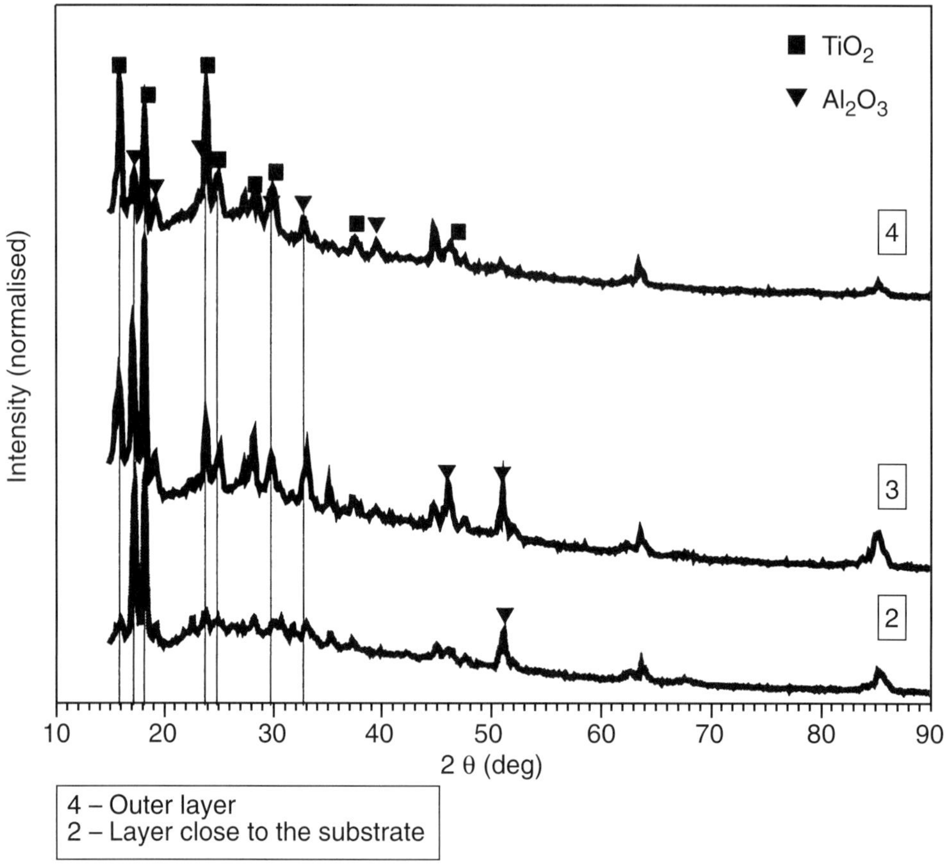

Fig. 9 *X-ray diffractograms of the different sublayers of the oxide scale on TiAl.*

between the TiO_2 layer and the inner oxide layers. The loss of adherence is randomly distributed over the entire interface. It is correlated with the analysed critical amount of compressive stresses (Fig. 10) that decrease by buckling of the TiO_2 layer. A healing of the cavity defects by newly formed oxide material restores the initial state so that compressive stresses in the TiO_2 can again accumulate.

4.3. Thermal Barrier Coating System

4.3.1. NiCoCrAlY/In718

The results of the phase analysis of the NiCoCrAlY bond coat in the as-sprayed condition and after an annealing of 60 h at 900°C carried out with Cu-Kα radiation are shown in Fig. 12.

It can be seen that the bond layer consists of two phases with different symmetries denoted by F and P in connection with the Miller indices of the lattice planes. After annealing the 116-Al_2O_3 reflection can be detected. The increase of the scattered intensity of this reflection during annealing has also been investigated *in situ*. The

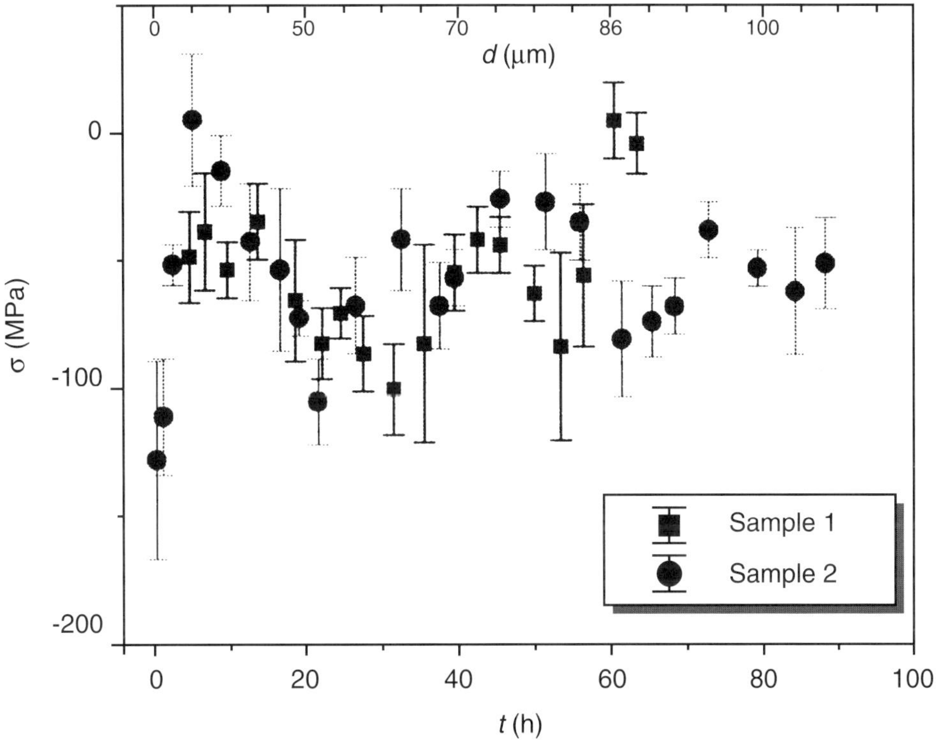

Fig. 10 Residual stresses in the outer TiO_2 layer on TiAl at 900°C.

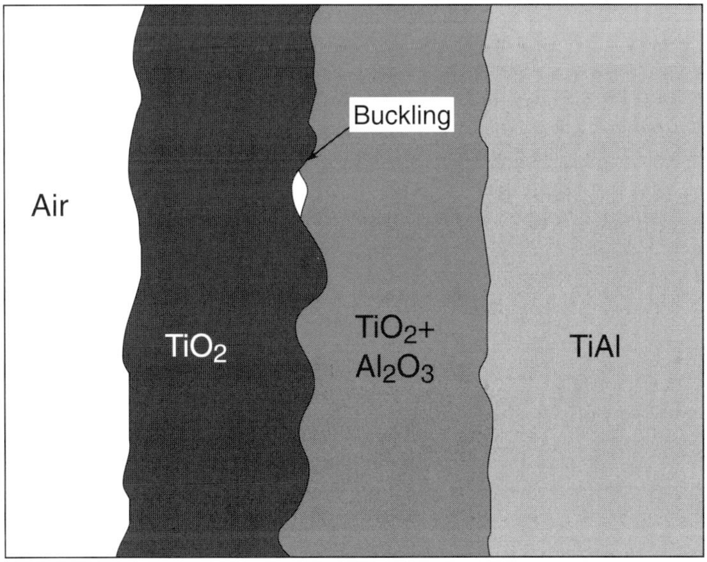

Fig. 11 Oxidation behaviour of TiAl (schematic).

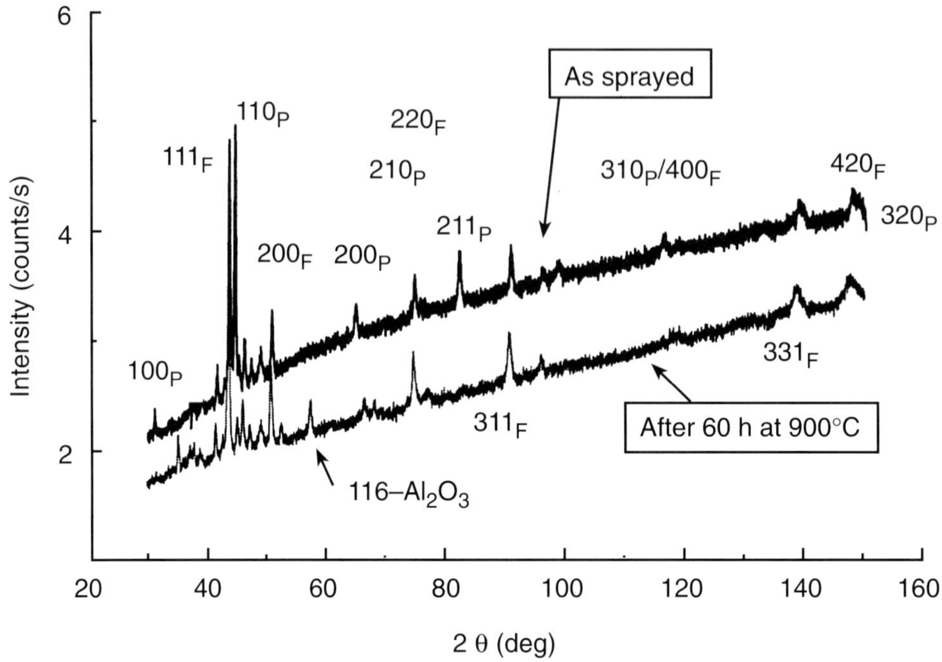

Fig. 12 *X-ray diffractograms of the NiCoCrAlY layer in the as-sprayed condition and after annealing for 60 h at 900°C.*

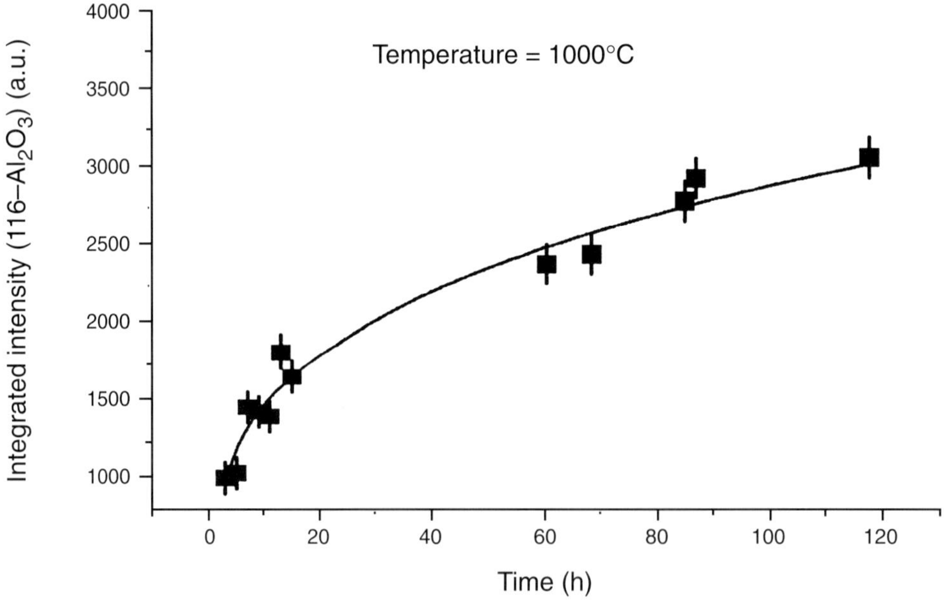

Fig. 13 *Increase of the integrated intensity of the 116-Al_2O_3 reflection.*

results are shown in Fig. 13.

After *ca*. 5 h at 1000°C the 116-Al_2O_3 reflection becomes detectable. After 120 h there is still an increase in the scattered intensity. This shows that the oxide scale has not reached a thickness comparable to the penetration depth of the X-rays. A SEM investigation confirms this. The Al_2O_3-layer has a thickness of up to 1–2 μm.

The *in situ* stress analysis of the NiCoCrAlY bond coat in the bond coat/substrate composite (Fig. 14) reveals no significant variation of the stress state during the first heating. During cooling of the sample an increase of compressive stresses can be revealed. The second heating of the layer compound leads to a relaxation of the stresses at 1000°C. After the second cooling compressive stresses up to –65 MPa are built up again.

Due to the temperature dependence of the ductility and Young's modulus of the NiCoCrAlY a relaxation of the stress state at temperatures above 700°C occurs. The *in situ* analysis of the development of the stress state at 1000°C (Fig. 15) confirms this. Though there is oxidation no change of the stress state can be analysed.

An investigation of the oxide scale after cooling to room temperature reveals compressive stresses of $\sigma_{Al_2O_3} = -555 \pm 27$ MPa.

The results of the stress analysis of the thermal barrier coating system which were performed with high energy synchrotron radiation are shown in Fig. 16 and Fig. 17.

From Fig. 16 it can be seen that a temperature gradient across the sample thickness as achieved with the gradient oven device causes compressive stresses in the ceramic layer. These stresses increase with increasing temperature differences between the sample surface and its backside.

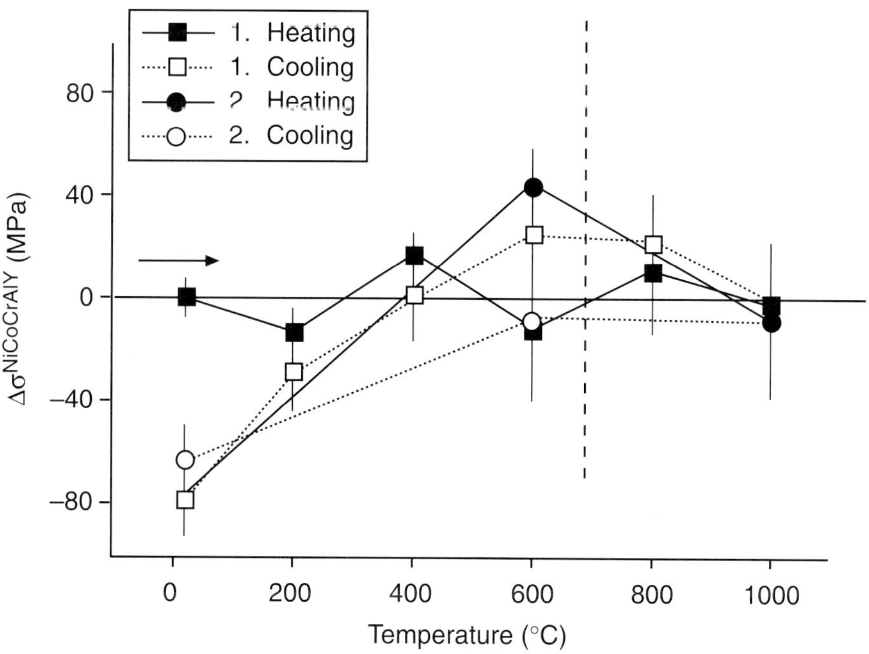

Fig. 14 In situ *stress analysis of the NiCoCrAlY bond coat in the bond coat–substrate composite.*

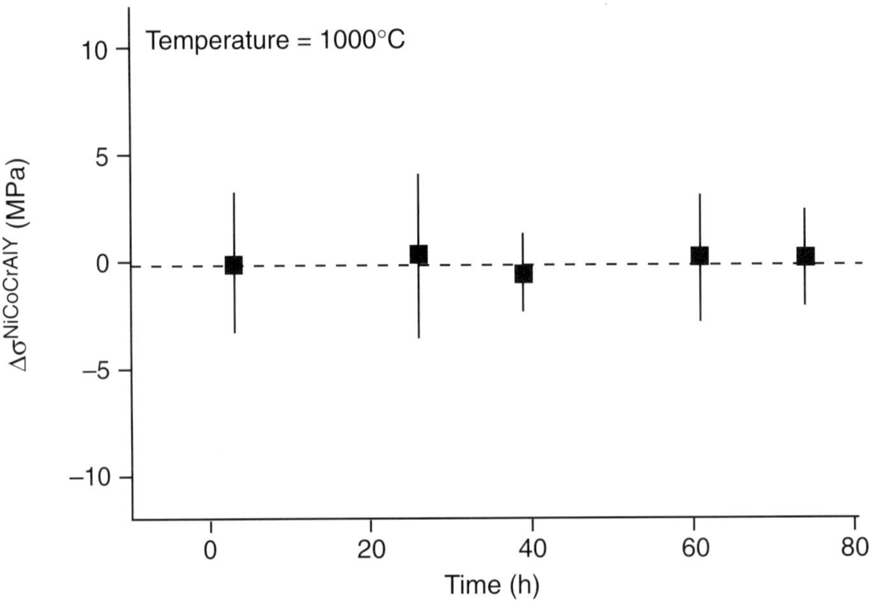

Fig. 15 In situ *stress analysis of the NiCoCrAlY bond coat in the bond coat–substrate composite*

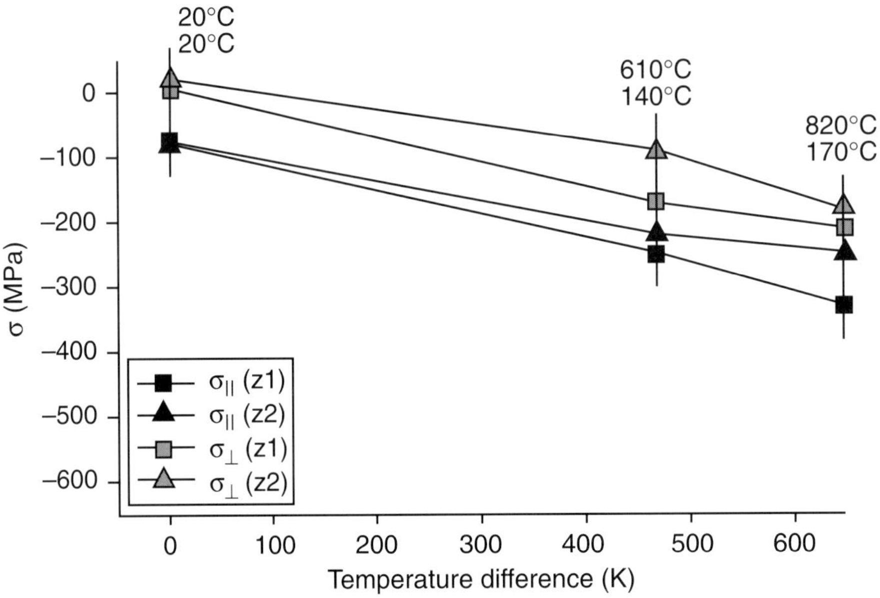

Fig. 16 In situ *stress analysis in the ZrO_2–$7Y_2O_3$ layer of the thermal barrier coating system for different temperature gradients. The numbers indicate the temperatures of the surface of the ceramic layer and the backside of the substrate, respectively.*

The stress state in the bond layer does not change significantly by applying these temperature distributions (Fig. 17). The main reason for this behaviour is the low thermal conductivity of the ceramic layer. This leads to a steep decrease of the temperature across the ceramic layer in such a way that the temperature of the bond coat does not change that much though the temperature difference across the sample varies significantly.

5. Conclusions

The analysis of the oxidation of different Ni substrates as well as of TiAl reveals that the oxidation induces compressive stresses in the oxide layer. From the investigations of the microstructure in connection with *in situ* stress analysis conclusions can be drawn concerning the oxidation process. The experiments performed with the NiCoCrAlY/In718 layer compound show an oxidation of the bond coat material together with compressive stresses after two thermal cycles. *In situ* stress analyses performed with high energy synchrotron radiation reveal an increase in compressive stresses in the ceramic layer of the thermal barrier coating system while the compressive stress in the bond coat does not change for the applied temperature distributions.

The examples given in this paper confirm that non-destructive diffraction methods which allow phase specific investigations of the microstructure and stress state are an appropriate tool for the analysis of oxidation processes. They complement commonly used methods and lead to a detailed understanding of the oxidation processes.

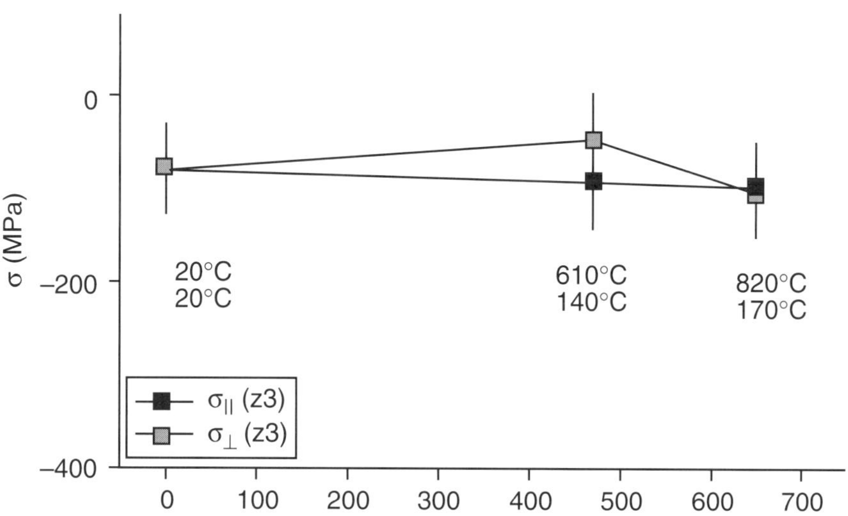

Fig. 17 In situ *stress analysis of the NiCoCrAlY bond layer of the thermal barrier coating system for different temperature gradients. The numbers indicate the temperatures of the surface of the ceramic layer and the back side of the substrate, respectively.*

References

1. A. Atkinson, *Rev. Mod. Phys.*, 1985, **57**, (2) 437–469.
2. F. N. Rhines and I. S. Wolf, *Met. Trans.*, 1970, **1**, 1701.
3. C. Courty, J. L. Lebrun, F. Armanet, G. Beranger and C. Fayoux, Mechanical and microstructural investigation by X-ray diffraction of nickel oxide formed at 900°C, 360–366, in *Proc. 2nd Int. Conf. on Residual Stresses* (G. Beck and S. Denis, eds). Elsevier Applied Science, London, 1988.
4. J. C. Pivin, J. Morvan, D. Mairey and J. Mignot, *Scr. Metall.*, 1983, **17**, 179–182.
5. T. Schmackers, W. Reimers, H. U. Baron and T. Cosack, Residual stresses in zirconia thermal barrier coatings due to strongly inhomogeneous temperature distributions, in *Proc. Materials for Advanced Power Engineering, 1998,* Liège. (J. Lecomte-Beckers, F. Schubert and P. J. Ennis eds.), III, 1998, pp.1647–1656.
6. Data from the manufacturer Krupp VDM GmbH, Werdohl.
7. Data from the manufacturer Gesellschaft für Elektrometallurgie mbH Nürnberg.
8. W. Reimers *et al.*, *J. Nondestr. Eval.*, 1998, **17**, (3), 129–140.
9. P. Suortti and T. Tschenscher, *Rev. Sc. Instr.*, 1995, **66**, (2), 1798.
10. T. Gnäupel-Herold and W. Reimers, Residual stresses in oxide scales on nickel- and TiAl-substrates, in Proceedings of ICRS-5, 1997, pp.952–957 (T. Ericsson, M. Odén and A. Andersson eds.), Linköping, Sweden Vol. I (1997).
11. S. Becker, A. Kahmel. M. Schorr and M. Schütze. Metall., 1992, **38**, 425.

23
Numerically Calculated Oxidation Induced Stresses in Thermal Barrier Coatings on Cylindrical Substrates

G. KERKHOFF, R. VAßEN and D. STÖVER
Institut für Werkstoffe und Verfahren der Energietechnik I, Forschungszentrum Jülich GmbH, 52425 Jülich, Germany

ABSTRACT
Thermal barrier coating (TBC) systems are often produced by vacuum plasma spraying of a highly dense bondcoat layer and atmospheric plasma spraying (APS) of a porous (10–15%) Y_2O_3-stabilised zirconia top coat. The main failure mechanisms in such systems are crack propagation due to oxidation of the bondcoat and to thermal stresses. A finite element method (FEM) is used to calculate stress development during thermal cycles. The investigated system consists of a cylindrical substrate (Ni-base-superalloy), a bondcoat (CoCrAlY) and the APS-TBC (ZrO_2+7%Y_2O_3). The viscoelastic properties of the materials are taken into account. The influence of a thermally grown oxide layer between the bondcoat and the TBC is shown. The oxidation induced stresses have been calculated and compared to thermal stresses.

1. Introduction

The efficiency of gas turbines can be improved by increasing the turbine inlet temperatures [1]. However, the temperature of the hot gas path component, e. g. the nickel alloy turbine blades, should remain constant. This can be achieved by the application of protective thermal barrier coatings (TBCs). Gas turbines with TBC coated blades can operate with hot gas temperatures up to 1400°C [2]. Increasingly demanding operating conditions require more precise predictions of the guaranteed life time of the TBCs. Several life time models have been proposed [3–8], but a reliable model that correlates TBC life with thermomechanical properties and microstructural features of the coatings is still not available.

Failure of TBC is essentially initiated by stresses which are induced during thermal cycling loading as a result of thermal expansion mismatch and oxidation of the rough bondcoat [9,10]. Once the stresses and their development during loading-time are known, it should be possible to calculate the time until delamination by taking into account fracture mechanics considerations on crack initiation and crack propagation [3].

In gas turbines spallation often occurs at the ends of blades, where radii of curvature are very small [1]. In the present paper this influence of the curvature on thermal stresses is taken into consideration by using cylindrical geometries. The investigated TBC-system is a superalloy (IN738)* coated with a vacuum-plasma-sprayed corrosion-resistant bondcoat (BC) layer (MCrAlY) and an atmospheric-plasma-sprayed ZrO_2–7mass%Y_2O_3 ceramic top layer. Time-dependent creep has been taken into account for the bondcoat and the TBC.

The main focus has been set on the calculation of the stresses in the TBC due to oxidation of the bondcoat. This is done for a triple-layered cylindrical system by two different approaches: (i) an analytical calculation of the stresses due to the thermal expansion mismatch, and (ii) by a numerical calculation of the growth stresses at high temperatures and subsequently taking into account thermal stress due to thermal cycling. The numerical method also includes BC roughness.

2. Finite Element Model

2.1. Mesh Properties

All calculations in this paper have been made with the same geometrical parameters of a hollow cylinder. The inner radius r_0 has been set to 10 mm, the thickness of the substrate d_1 to 1 mm, the thickness of the bondcoat d_2 to 0.15 mm and the thickness of the TBC d_3 to 0.3 mm. The influence of different interface curvatures and different TBC-thicknesses has been investigated elsewhere [11–13]. The initial thickness of the thermally grown oxide scale (TGO) is set to 0.1 μm.

The finite element analysis has been made by using the FEM-code ABAQUS [14]. The three-dimensional problem has been reduced to a two-dimensional mesh by restricting the calculations to the case of a long cylinder. Such a system can be simulated by a two-dimensional mesh using general plain strain elements. The mesh is generated in a plane perpendicular to the cylindrical axis. The profile of the bondcoat roughness is taken to be sinusoidal with an amplitude of 20 μm and a wavelength of 50 μm. Only one wavelength of the roughness is simulated. Such a simple model can be used when fixing the translational degrees of freedom of all nodes in a tangential direction. With this constraint the results of calculated displacements of the nodes are the same as the results of a completely discretised cross section.

The translational degrees of freedom of the interface nodes on each side are set equal to each other to simulate adhesion of the different coatings.

2.2. Material Properties

The material properties assigned to the substrate-, bondcoat-, TGO- and TBC-elements are listed in Table 1. A variation of these values has been investigated by means of finite element methods in [18]. The creep strain rates of these elements have been calculated using modified Norton's creep law

$$\dot{\varepsilon}_{cr} = A(T)\sigma^n t^m \tag{1}$$

with the parameters given in Table 1.

* INCO trade mark.

Table 1. *Material data taken from [7,15–17]*

	E (GPa)	ν	α (°C⁻¹)	*A* (hMPa^n)$^{-1}$	*n*	*m*
IN 738						
T = 200°C	191	0.3	12.8 10^{-6}			
T = 500°C	180	0.3	13.7 10^{-6}	6.8×10^{-29}	8.37	0
T = 900·C	140	0.3	15.8 10^{-6}	1.2×10^{-15}	4.7	0
MCrAIY						
T = 200°C	220	0.3	13.4 10^{-6}	6.4×10^{-10}	1.8	–0.19
T = 500°C	200	0.3	14.4 10^{-6}	2.2×10^{-9}	1.9	–0.21
T = 900·C	160	0.3	16.6 10^{-6}	9.7×10^{-7}	3.4	–0.31
ZrO_2 + 7mass%Y_2O_3						
T = 200°C	40	0.3	10 10^{-6}	2.3×10^{-16}	1.49	
T = 500°C	40	0.3	10 10^{-6}	2×10^{-10}	1.49	
T = 900·C	40	0.3	10 10^{-6}	1.3×10^{-6}	1.49	
Al_2O_3						
T = 200°C	380	0.3	5.9 10^{-6}	6.8×10^{-29}		
T = 500°C	380	0.3	7.4 10^{-6}	6.8×10^{-29}		
T = 900·C	380	0.3	8.0 10^{-6}	6.8×10^{-29}		

E = Young's modulus; ν = Poisson's ratio; α = Coefficient of thermal expansion (CTE)

2.3. Thermal Loading

Each thermal cycle calculated in this model consists of a heating period, in which the temperature of each node is raised from room temperature up to 950°C in 3 min, a high-temperature period which lasts 2 hours and a subsequent cooling period of 6 min.

2.4. Thermally Grown Oxide Scale Growth

As the TGO-elements in the FE-Model cannot be created during the calculation, the simulation starts with a very thin TGO-scale of about 0.1 μm at the beginning of the high temperature phase of the second thermal cycle. During this phase the temperature values of the nodes belonging to the TGO-elements are raised to the level needed to impose the desired strain on the TGO-elements. In such a procedure the TGO-elements expand exclusively and they reach the desired strain level at the end of the heating phase. This procedure seems to be a good simulation of the TGO-growth. At the end of this phase, the stress-state is conserved and the temperature of the TGO-nodes is set back to 950°C. Subsequently, cooling to room temperature and another heating to 950°C is calculated with the thicker TGO-scale.

3. Analytical Stress Calculations

The heating and the cooling periods of one thermal cycle are relatively short compared to the high temperature phase. During the dwell time at 950°C creep effects have to be taken into account, whereas the stress development during heating and cooling should mainly be elastic, as considerable creep starts in the bondcoat only at temperatures above 600°C. The temperature for the onset of creep in the TBC-material is even higher [11]. This means that it should be possible to calculate the difference of the stress levels at the beginning and at the end of the heating or cooling period using an elastic approach.

The formula for radial and tangential thermal stresses in a system of *i* concentric cylinders tied to each other has been derived in detail in [11]. For a homogenous temperature distribution they are given as

$$\sigma_{ir} = \frac{E_i}{1+\nu_i}\left(\frac{C_i 1}{1-2\nu_i} - \frac{C_i 2}{r^2}\right) - \frac{\alpha_i E_i}{1-\nu_i}\frac{\Delta T}{2r^2}\left(r^2 - r_{i-1}^2\right) \tag{2}$$

$$\sigma_{it} = \frac{E_i}{1+\nu_i}\left(\frac{C_i 1}{1-2\nu_i} + \frac{C_i 2}{r^2}\right) - \frac{\alpha_i E_i \Delta T}{1-\nu_i} + \frac{\alpha_i E_i}{1-\nu_i}\frac{\Delta T}{2r^2}\left(r^2 - r_{i-1}^2\right) \tag{3}$$

with ΔT being the temperature difference (925°C) and with *i* indexing the stresses in the different coatings. The origin of *r* is the centre of the cylinder. The constants C_{i1} and C_{i2} can be calculated by applying appropriate boundary conditions

$$\begin{matrix} \sigma_{1r}(r_0) = 0 & \sigma_{ir}(r_i) = \sigma_{(i+1)r}\;(r_i) \\ \sigma_{kr}(r_k) = 0 & u_{ir}(r_i) = u_{(i+1)r}\;(r_i) \end{matrix}, i = 1,2,\ldots,k-1 \tag{4}$$

Setting up the equations for a system with four concentric cylinders, one can calculate the cooling stresses in the system for different TGO-thicknesses. As temperature dependent material data are not taken into account in this model, averaged material data should be used so as to maintain comparability with numerically calculated stress levels.

4. Results and Discussion

First, analytical calculations will be presented which reveal the influence of an Al_2O_3-scale between the bondcoat and the TBC. Assuming a stress-free state at 950°C resulting from stress-relaxation caused by creep-effects, elastic stress-distributions have been calculated for cooling down to room temperature. The calculations have been made for different TGO-thicknesses. The radial stress-distribution over the whole TBC-system is plotted in Fig. 1. The location of the interfaces between the coatings is obvious. *L* is the sum of substrate, bondcoat, TGO and TBC thickness.

In a TBC-system without an oxide-scale, the TBC is under tensile stress of 40 MPa in a radial direction and under compressive stress of about 350 MPa in a tangential direction after a homogeneous temperature change of –920°C. This stress state is caused by the thermal expansion mismatch between the ceramic top layer and the metallic bondcoat and substrate.

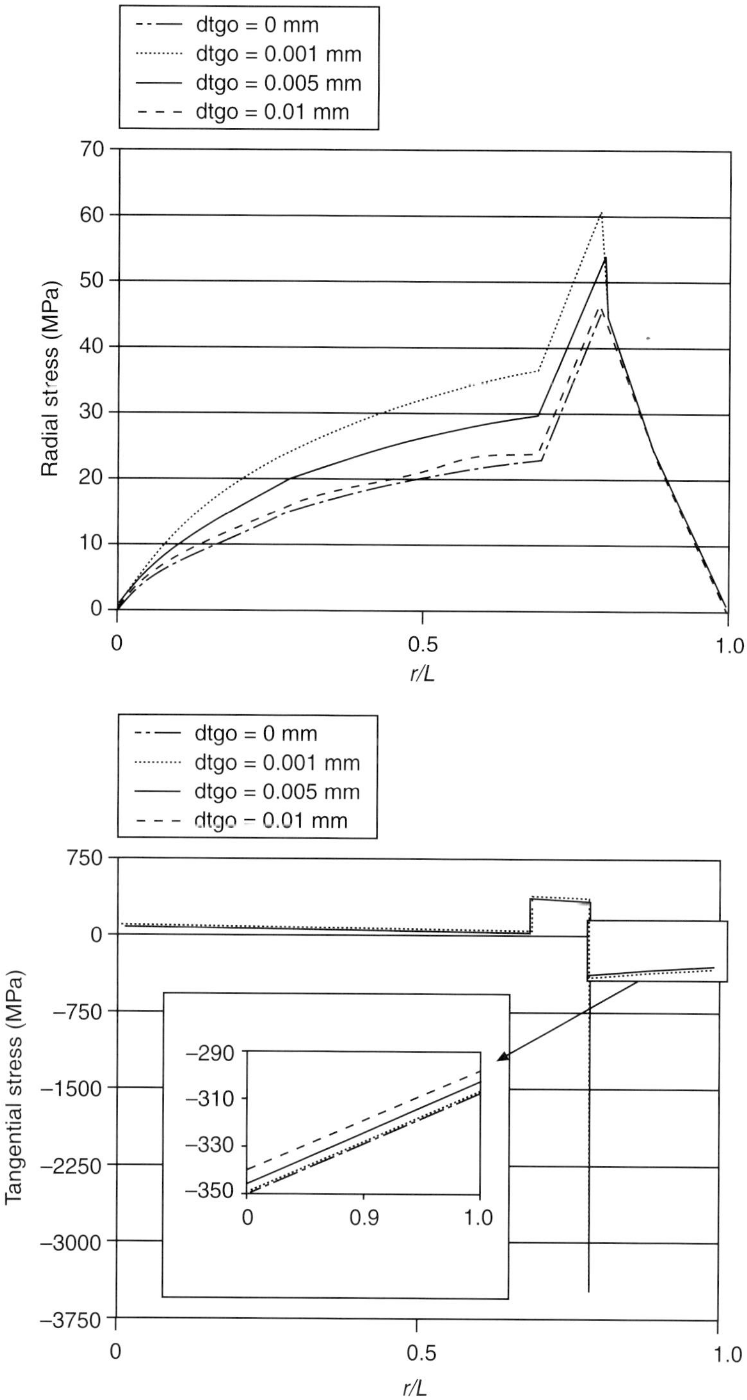

Fig. 1 *Results of the analytical calculation of the radial and tangential stress-distribution in the radial direction in a TBC-system with an oxide scale after cooling down from a stress-free state at 950°C (dtgo = TGO thickness).*

Implementing an oxide scale of 1 μm does not affect the stress state in the TBC. It only leads to high compressive stresses in the TGO itself in a tangential direction. This high stress level is also a result of the thermal expansion mismatch and the very small thickness of the oxide layer. The conservation of momentum forces lead to the high stress levels in the oxide layer. Raising the thickness to 5 or 10 μm does slightly affect the radial stress distribution. The broadening of the oxide scale leads to a reduction of about 50 MPa in the compressive stress level in the oxide scale itself and to a reduction of about 10 MPa in the compressive stress levels in the TBC.

These results indicate, that there is no significant effect of an oxide-scale on the stresses within the TBC during thermal cycling. However, high compressive stresses in the TGO are built up.

In the analytical model the stresses due to the TGO-growth have not been taken into account. In order to examine the influence of the growth of the oxide-scale during the high-temperature periods, finite element simulations have been carried out (Fig. 2).

The system is set to be stress-free at the beginning of the first cycle. The TGO-elements are not activated during the first cycle. Fast and homogeneous heating lead to compressive TBC-stress in the radial direction and to tensile TBC stress in the circumferencial direction. This is the typical elastic behaviour of a ceramic coated metallic cylindrical substrate. Due to the thermal expansion mismatch the substrate and bondcoat tend to expand more than the TBC. At temperatures above 600°C the bondcoat and at higher temperatures the TBC start to creep. During the dwell time at 950°C the TBC stresses relax and a nearly stress-free state is reached at the end of the dwell time. Subsequent cooling leads to tensile TBC-stresses in the radial direction and to compressive TBC stresses in the tangential direction. After the heating period of the second cycle the same high temperature stress levels as at the end of the dwell time of the first cycle are reached. These stress levels are very low and will approach zero for a couple of thermal cycles. For this reason it is reasonable to start the calculation from a stress-free situation. After a few thermal cycles the high-temperature stress levels approach zero and the stress development during the following thermal cycles is always the same. In this calculation however only one thermal cycle has been calculated to let the system relax during dwell time. In the high temperature period of the second cycle, the TGO-elements are activated and an artificial strain of 1500% is imposed on these elements exclusively. Thus, an oxide scale growth from 0.1 μm to 1.5 μm has been simulated. As the TGO-elements are surrounded by the bondcoat- and TBC-elements, high compressive stresses are calculated for the TGO-elements both in radial and tangential directions. The influence on the TBC-stresses are shown in the second part of Fig. 2. During the second dwell time, when the TGO-elements grow, an additional compressive stress is built up in the radial direction. As the cylindrical TBC is pushed in the radial direction, tensile stresses are built up in the tangential direction. In the high temperature period the creep strain rates of the TBC elements are significantly high. Due to relaxation effects, the stress development is not linear.

The growth rate for the oxide scale used in this calculation is not very realistic. In fact only two cycles have been calculated, to minimise calculation time. But one can easily deduce the development of the TBC-stresses over several hundreds of cycles by taking into account a realistic growth rate. During the first thermal cycles bondcoat oxidation can be taken to be negligible. At the same time TBC-creep leads to stress-

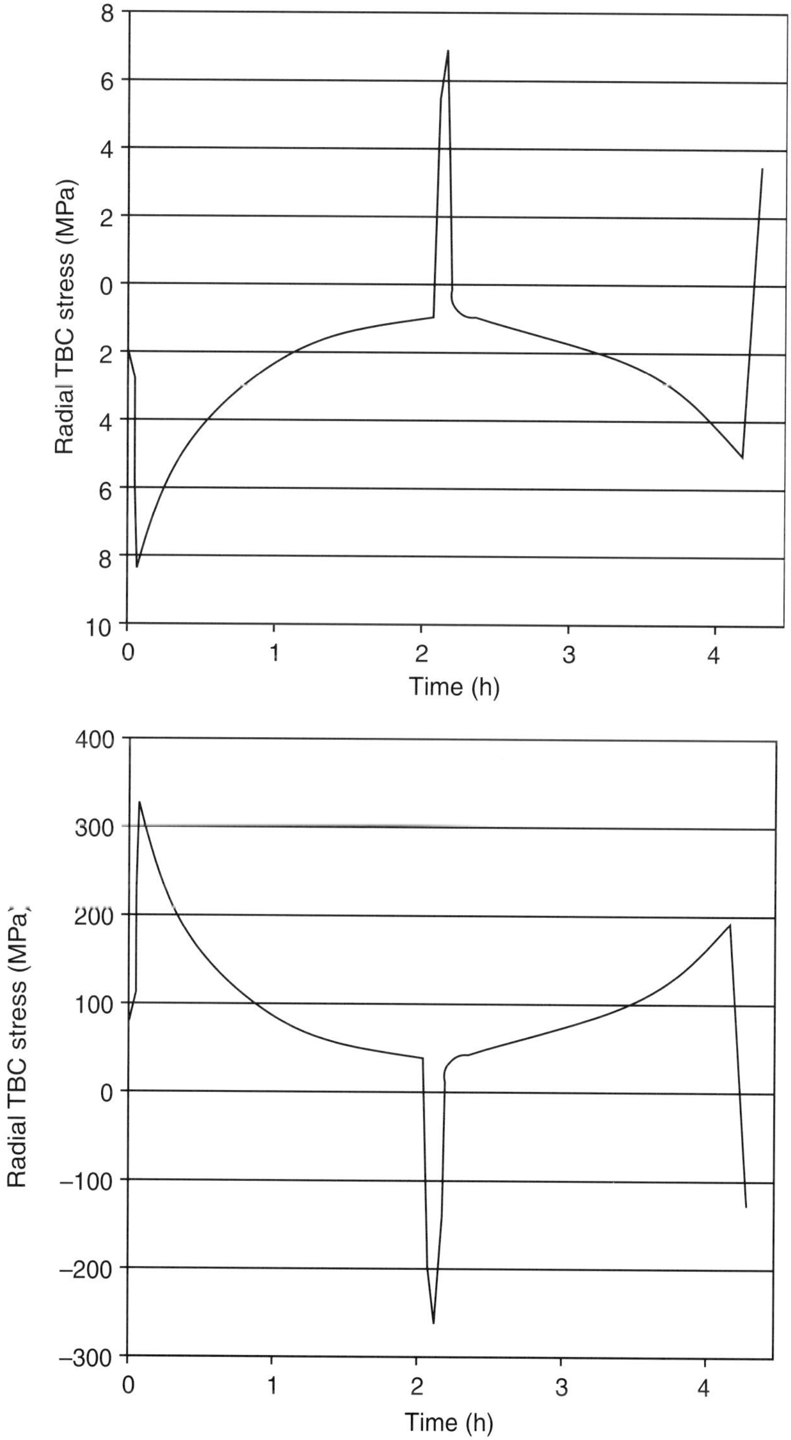

Fig. 2 *Numerical calculation of the radial and tangential TBC stress at the interface for two thermal cycles, the first without TGO-scale, the second with TGO-growth.*

relaxation during the high temperature phases of the first cycles. TBC-stresses reach very low values within a few cycles and consequently tensile stresses occur at room temperature in the radial direction and compressive stresses occur in the tangential direction. This means that during the first cycles, where TGO-formation can be neglected, the combination of the stresses in both directions at room-temperature favours crack propagation in the horizontal direction.

Further thermal cycling leads to the formation of an oxide scale between the bondcoat and the TBC during the high temperature phases. With growth during the high temperature phases of the thermal cycles, the oxide scale imposes supplementary stresses on the TBC. This leads to a reduced stress state in the radial and the tangential directions at room temperature. This is contradictory to the fact that TGO-formation is a life time limiting factor. The situation changes if a sinusoidal BC roughness is introduced into the model. Again two thermal cycles have been calculated, the first without a TGO-scale and the second with a TGO-scale growing over the whole high temperature period. Figure 3 shows the radial distribution for the radial stresses in the TBC above the roughness peak and above the roughness valley.

At the end of the first cycle the TBC is under tensile radial stress above the roughness peak at room temperature. In the valley of the roughness profile the TBC is under compression in the radial direction. That means, that during the first cycles, during which BC oxidation can be neglected, the stress fields around the roughness peak favour horizontal crack initiation.

At the end of the second cycle TGO-growth leads to a zone above the roughness peak and above the roughness valley, where the stress states switch their signs. A similar behaviour has been observed in the literature [19]. Above the peak a region of compressive radial stress occurs and above the valley, at a distance from the interface of the order of the amplitude of the roughness, a region of tensile radial stress occurs. This tensile stress region could be an explanation for the crack propagation of cracks initiated above the roughness peaks.

5. Conclusion and Perspective

Thermal stresses and oxidation induced stresses have been calculated for cylindrical TBC systems for thermal cycling loading. Creep effects and the growth of the TGO-scale during high temperature dwell time have been taken into consideration. Additionally, the roughness of the interface between the bondcoat and the thermal barrier coating has been simulated by a sinusoidal roughness profile. It can be shown, that by calculating the time dependent stress distribution in such a model, qualitative explanations can be made for the influence of BC oxidation and BC roughness on the life time of TBC systems.

On the basis of these stress calculations a fracture mechanics approach could lead to a general life time model, which is based on the material properties and geometries of the system.

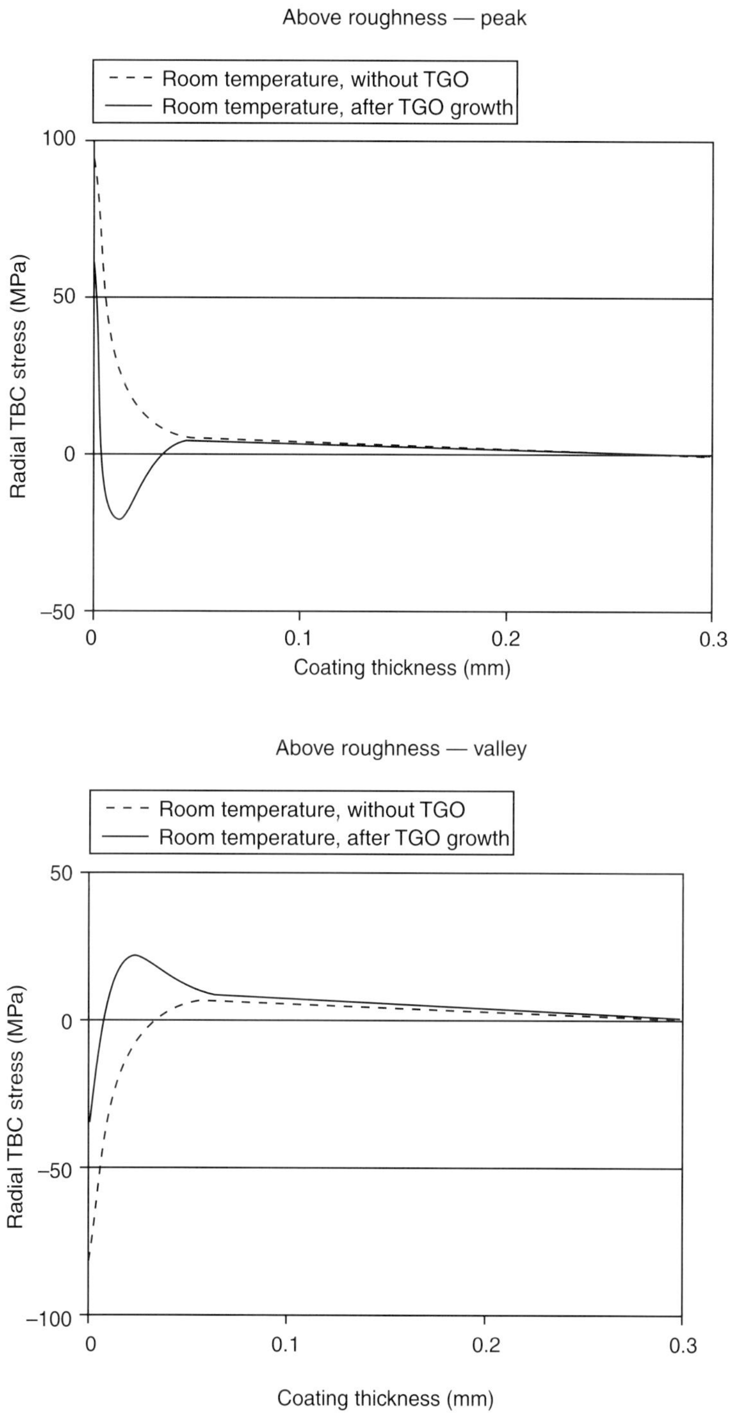

Fig. 3 *Numerically calculated radial distribution of radial stresses in a cylinder with a BC roughness and a growing TGO-scale.*

References

1. P. Hancock and M. Malik, in *Materials for Advanced Power Engineering — Part 1* (D. Coutsouradis *et al.*, eds). Kluwer Academic Publishers, Dordrech 1994, p.658.
2. W. Beele, Zuverlässigkeitsforderungen an Wärmedämmschichten in stationären Gasturbinen, Fortschrittsberichte der Deutschen Keramischen Gesellschaft 11 1996, pp.145–154.
3. C. A. Andersson, Thermal stress fracture of ceramic coatings, *Fract. Mech. Ceram.*, 1983, **6**, 497–509.
4. R. A. Miller, Oxidation based model for thermal barrier coating life, *J. Am. Ceram. Soc.*, 1984, **67**(8), 517–522.
5. R. A. Miller, Life modelling of thermal barrier coatings for aircraft gas turbine engines, *J. Eng. Gas Turbines & Power,* 1989, **111**, 301–305.
6. S. M. Meier, D. M. Nissley, K. D. Sheffler and T. A. Cruse, Thermal barrier coating life prediction model development, *J. Eng. Gas Turbines & Power,* 1992, **114**, 258–263.
7. J. T. DeMasi, K. D. Sheffler and M. Ortiz, Thermal barrier coating life prediction model development, Phase I, Final Report. Contract NAS3-23944, NASA CR 182230, 1989.
8. S. M. Meier, D. M. Nissley and K. D. Sheffler, Thermal barrier coating life prediction model development, Phase II, Final Report. Contract NAS3-23944, NASA CR 189111, 1991.
9. R. A. Miller and C. E. Lowell, Failure mechanisms of thermal barrier coatings exposed to elevated Temperatures, *Thin Solid Films,* 1982, **95**, 265–273.
10. A. H. Bartlett and R. D. Maschio, Failure mechanisms of zirconia-8wt% yttria thermal barrier coating, *J. Am. Ceram. Soc.*, 1995, **78**(4), 1018–1024.
11. G. Kerkhoff, R. Vaßen, C. Funke and D. Stöver, Finite element analysis of thermal stresses in thermal barrier coatings on cylindrical substrates, in *Proc. 6th Liège Conf. on Materials for Advanced Power Engineering,* 1998, 1669–1677.
12. D. M. Nissley, Thermal barrier coating life modelling in aircraft gas turbine engines, *J. Therm. Spray Technol.*, 1997, **6**(1), 91–98.
13. U. Szieslo, Berechnung von Eigenspannungen in metallgespritzten rotationssymmetrischen Körpern, *Z. Werkstofftech.*, 1980, **11**, 110–115.
14. ABAQUS, User's Manual, Version 5.6, Hibbit, Karlsson and Sorenson, Pawtucket, Rhode Island (1996).
15. K. H. Kloos, J. Granacher and H. Kirchner, Mechanisches Verhalten des Schutzschicht-Grundwerkstoff-Verbundes von Gasturbinen-schaufeln unter betriebsähnlicher zyklischer Beanspruchung. *Mat. wiss. Werkst. tech.*, 1994, **25**, 209–217.
16. C. Funke, J. C. Mailand, B. Siebert, R. Vaßen and D. Stöver, Characterisation of ZrO_2-7wt%Y_2O_3 thermal barrier coatings with different porosities and FEM analysis of stress redistribution during thermal cycling of TBCs, *Surf. Coat. Technol.*, 1997, **94–95**, 106–111.
17. G. Thurn, G. A. Schneider and F. Aldinger, Schädigung von Wärmebarrieren durch Kriechrelaxation, *Fortschrittsbericht der DKG* 1995, **10** (4), 41–50.
18. G. C. Chang and W. Phucharoen, Finite element thermal stress solutions for thermal barrier coatings, *Surf. Coat. Technol.*, 1987, **32**, 307–325.
19. A. M. Freborg, B. L. Ferguson, W. J. Brindley and G. J. Petrus, Modelling oxidation induced stresses in thermal barrier coatings, *Mater. Sci. Eng.*, 1998, **A245**, 182–190.

24
Electron Beam–Physical Vapour Deposited Thermal Barrier Coating Lifetime Response to Various Bond Coat Pre-treatments

K. FRITSCHER, U. SCHULZ and M. SCHMÜCKER
DLR German Aerospace Center, Institute of Materials Research, D-51170 Cologne, Germany

ABSTRACT
A test facility for thermocyclic loading of thermal barrier coatings (TBCs) and an appropriate sample geometry was designed to establish a standard test procedure for ranking purposes. IN100 samples with an electron-beam physical vapour deposited (EB-PVD) bond coat and an EB-PVD TBC were tested until spallation to assess the TBC lifetime. The bond coat was pretreated in three different aspects to investigate the influence of pretreatment procedures on TBC lives. The variants were differences in particle size diameter during shot peen densification, different shot media (glass and zircon), and heat-treatment atmospheres (Ar, vacuum, H_2 and air). Apparent deficiencies and a ranking were attributed to differences in thermally grown oxide (TGO) formation found in pretreated bond coats and in fully processed TBC systems by XRD, SEM and TEM.

1. Introduction

The application of ceramic thermal barrier coatings (TBCs) on components in the hot section of gas turbines will allow the turbine inlet gas temperatures to be raised significantly and consequently also the overall engine efficiency. The lifetime of TBCs on substrates until spallation occurs is usually the most critical property of components. Hence much effort in TBC development aims at solutions to prolong the lifetime of TBC systems. The lifetime depends on various exterior and interior influences. The main exterior effects are atmosphere, temperature and time which govern the concurrent TGO layer formation progress of the bond coat, variations in temperature and the impact on thermomechanical fatigue phenomena, gas velocity and erosion etc. Internal effects include the interdiffusion of substrate and bond coat elements, physical properties like thermal conductivity and the CTE (coefficient of thermal expansion) of the substrate and the coating system layers and their CTE mismatch, and the distribution of stresses provoked by the respective surface curvatures on the components.

The intention of this work is to present a simple test methodology installed at DLR which allows a lifetime ranking on TBC systems on arbitrary substrate compositions to be established. The ranking in this work will be carried out on IN100*

*Inco trade mark. (IN100: nominal composition 15Co, 10Cr, 5.5Al, 5Ti, 3Mo, 0.15C, 0.01B, 0.05Zr, lV max. mass%, bal. Ni.)

samples with a NiCoCrAlY bond coat and an EB-PVD PYSZ (partially yttria stabilised zirconia) TBC. The effect of different bond coat treatments on TBC life — dependency of peen shot diameter, bond coat annealing atmospheres and peen shot media — will be addressed.

2. Sample Geometry

A simple geometry of cylindrical test pins of 6 mm dia. × length 65 mm^2 was preferred since — apart from the ease of manufacture — a favourable uniformity of the development of stresses below the surface on testing is more likely to be achieved. Edge effects which lead to stress singularities are suppressed by the long cylindrical geometry. The bond coat on the cylinder walls will oxidise on furnace-rig testing and a TGO will steadily grow in thickness. This progress will alter the circumferential stress distribution in the TBC on thermal cycling until overcritical stresses accumulate and eventually lead to a buckling mode spallation of the TBC.

As the stress situation will develop rather homogeneously during the test along the cylindrical TBC layer spallation of the TBC will, as a rule, occur simultaneously at various locations. Thus, spallation on those cylindrical samples can easily be observed.

3. Furnace Test Rig

The furnaces used for thermocyclic testing of TBC samples are resistance-heated by $MoSi_2$ heating elements which allow extensive long-wave infrared back-radiation of heat from the oxide fibre brick refractory furnace walls. This aspect was judged to be of uppermost relevance for more reliable TBC testing. PYSZ TBCs which are translucent for visible light and beyond in the range of 0.5 to 6 μm wavelength are no longer transparent for infrared radiation with longer wavelength. Long wave length radiation will be absorbed at the TBC surface. So the transmitted heat from the furnace walls with a high proportion of long-wave infrared radiation will heat the surface of the TBC samples prior to the bond coat. The low heat conduction by the diffusive progress of phonons in the solid is the physical property which is highly valuable in zirconia and which makes this material so valuable as a thermal barrier material in many high temperature applications, e.g. preferentially on internally cooled turbine blades. For the hypothetical case the short-wave radiation dominates the heat transfer and the columnar zirconia crystals will act like fibre glass light conductors and will directly heat up the bond coat/TBC interface instead of the TBC surface. The resultant distribution of heat during heat-up will be opposite to the real heat distribution in a TBC system on a blade in service. This situation is considered to be unrealistic and should be avoided due to the concomitant deficiencies in stress distribution which may cause the establishment of an incorrect test procedure for TBCs.

One cycle consists of 50 min heating of the samples to 1100°C internal specimen temperature and 10 min cool-down in forced ambient air of 2 ms^{-1} below the furnace. The effect of heating a TBC sample by introducing it in the furnace is shown in Fig. 1. Two hollow cylindrical samples of 6 mm dia. are internally equipped with

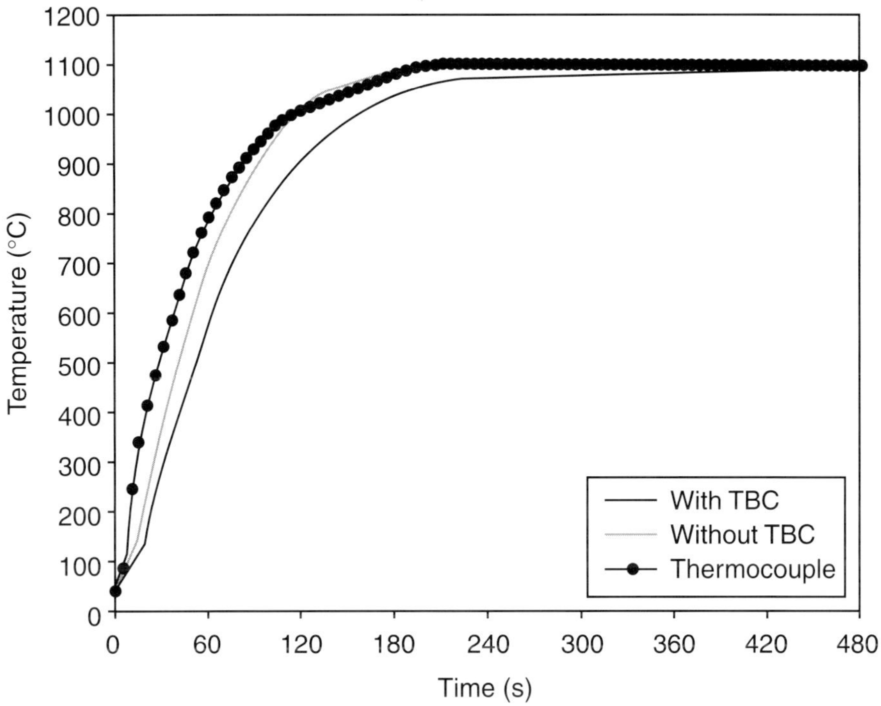

Fig. 1 *Temperature vs time of cylindrical samples with and without TBC during heat-up in the cycling furnace rig with unprotected thermocouple as reference.*

thermocouples which are in intimate contact with the substrate material. One of the two samples was applied with a 200 μm thick TBC. As soon as both samples are lifted from the ambient environment into the hot zone of the furnace they will both be heated but at significantly different rates. The difference can be attributed to the low heat conductivity via phonon transport. The terminal peak temperatures are the same for both samples. However, the effect of the TBC during rapid temperature changes on smoothing temperature alterations becomes clearly visible in the transient temperature regimes.

The cooling rate of the samples is shown in Fig. 2. It shows analogously to heating ($\cong$ 10 Ks^{-1}) the delayed cooling of the TBC-protected sample. The overall cooling time (cooling rate $\cong$ 5 Ks^{-1}) was intentionally extended to 10 min for 6 mm dia. samples in order to be at or below 40°C in the interior of the samples. The surface effectively reaches room temperature. This cooling procedure turned out to be necessary to attain reliable cycle numbers which not only represent the oxidation progress but also the effects of thermal fatigue. Additionally, the samples are cooled down in 20 h intervals for microscopic inspections of the TBCs away from the test facility. This long interval in ambient air of approximately an hour contributes some more load to the TBC system. This extra load can be attributed to the ambient humidity of the air and the obvious tendency of the TBC to absorb water vapour readily at room temperature. Partial or total loss of TBC from the samples can occasionally be observed during this inspection period.

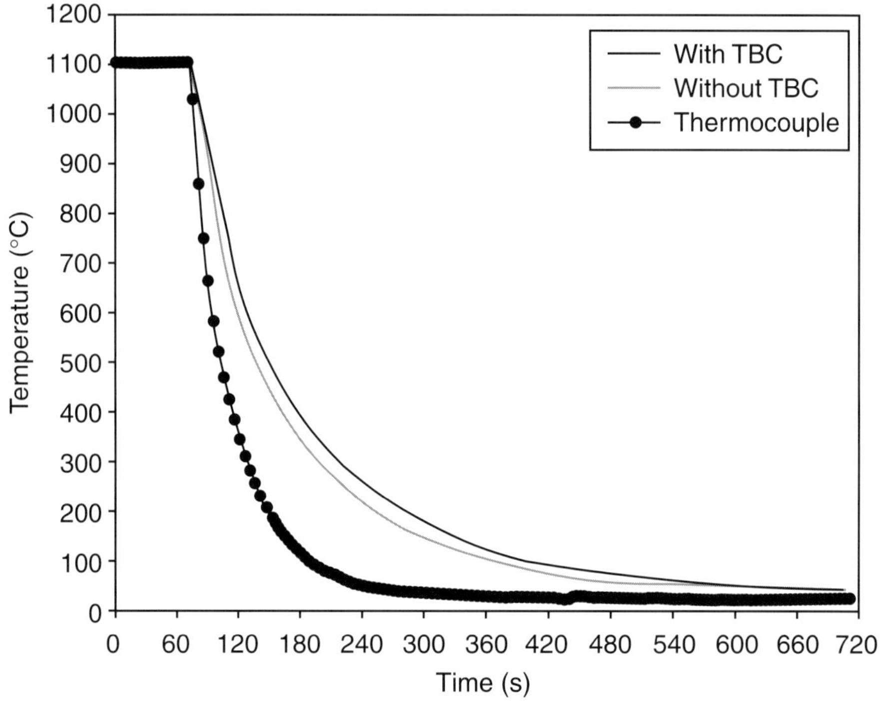

***Fig. 2** Temperature vs time of cylindrical samples with and without TBC during cooling with forced air and unprotected thermocouple as reference.*

4. Sample Preparation

Cylindrical IN100 samples of 6 mm dia. and of 0.25 μm roughness (R_a) were ultrasonically cleaned in soap water and in ethanol before a 100 to 120 μm NiCoCrAlY bond layer (Ni–22Co–20Cr–12Al–0.2Y in mass%) was deposited by EB-PVD. The layers were densified by glass bead peening at 12 N Almen intensity, cleaned as mentioned above, and heat-treated for 4 h at 1080°C in vacuum or in various atmospheres before the final EB-PVD TBC application. The impact energy of the shot stream is conveniently calibrated by means of a single-side peened spring steel specimen (Almen Strip) which curves as a result of surface plastic deformation caused by the peening impacts. The resultant height of this curvature determined in a standard Almen gauge represents the Almen intensity.

A first attempt to modify the common pretreatment procedure of the bond coat was to vary the densification procedure, i.e. by peening with distinctive glass-ball diameters. The glass bead shot was separated into four different fractions of glass-ball diameters between 160 and 450 μm, which are usually intermixed and were individually used here for peen densification at constant 12 N Almen peening intensity. A constant peening intensity was achieved by adjusting the compressed air pressure for grit ejection. Subsequent annealing was routinely carried out in a vacuum of 10^{-5}mbar (10^{-3}Pa) at 1080°C.

A second approach to varying the bond coat pretreatment was to anneal in different atmospheres: these were Ar, vacuum, H_2, and air. Annealing was at 1080°C for 4 h in all cases.

One further attempt to modify the pretreatment procedure of the bondcoat was to alter the densification procedure, that is, by peening with zircon shot instead of glass beads. The zircon shot ($ZrSiO_4$) consists of spherical beads of 200 to 250 µm dia. A slightly dimmed lustre of the peened surface compared to bright glass bead peened coating surfaces was observed. Subsequent annealing was done under vacuum of 10^{-5}mbar at 1080°C.

5. Results and Discussion

5.1. Variation of Peening Shot Diameter

The lifetime of TBC protected samples with a peening pretreatment of the bond coat by use of various glass bead diameters is shown in Fig. 3.

The lifetime is defined as the time to occurrence of a spalled surface area longer than 2 mm in one direction. In spite of the small number of samples investigated it is obvious that the lifetime results show no clear preference for the use of a distinct peening shot diameter. Instead the lifetimes can apparently be attributed to the life categories low, medium and high. Short cyclic lives were found to be synonymous of adhesive failure at the bond coat/TGO interface [1]. Intermediate cyclic lives often occur if mixed oxides have formed near the TGO/TBC interface. These categories are also to be seen in Fig. 3 indicating that the effect of the mechanically-related pretreatment of different peen grit diameters is outruled by the statistical nature of the inherent adhesion mechanisms. They still dominate the statistics which suggest that one sample has a short cyclic life, six are intermediate and one is high.

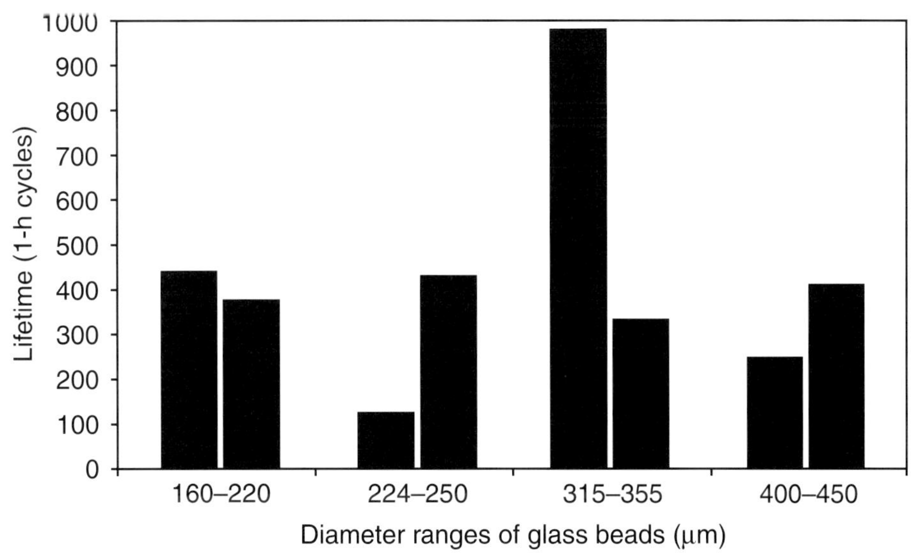

Fig. 3 *Influence of peening shot diameters on TBC lifetime (cycles) at 1100°C.*

5.2. Variation of Annealing Atmosphere

The inefficiency of the mechanical pretreatment of peening densification with peening bead diameter variation, as mentioned above, led to investigations which focus on chemistry-related pretreatments. Annealing in various atmospheres therefore was carried out. The results are given in Fig. 4. They also show a rather narrow lifetime distribution irrespective of the atmosphere. The TGO thickness is generally of the order of 5 μm after some 500 test cycles. As TGO thickness and the lifetimes seem to correlate it is assumed that a critical TGO thickness was attained to initiate spallation of the TBCs on testing. The TGO growth rate at 1100°C of NiCoCrAlY bond coats on flat samples was investigated as a function of oxygen partial pressure ranging over 20 orders of magnitude in a parallel study [2]. Exclusive growth of mainly α-Al_2O_3 at very similar rates occurred in all cases irrespective of the oxygen partial pressure. This growth mode is obviously responsible for a life expectancy which is also independent of the atmosphere of the annealing pretreatment. The thicknesses of the TGOs lead to in-plane compressive stresses and are further increased by the mismatch between the TGO and the bond coat which in turn creates a stress component perpendicular to the interface and to subsequent spallation [3].

In Fig. 4 the lifetime of the differently pretreated samples is intentionally supplemented by the time to occurrence of a first fissure. Fissures do not generally lead to instantaneous spallation but indicate an apparent status of failure. With reference to this failure mode the vacuum treated samples can be ranked worst. Comparative SEM and XRD investigations on fully treated bond coats before TBC application showed that the bond coats are strong alumina formers if subjected to any atmosphere for 4 h at 1080°C. However, under high vacuum conditions a selective coverage of the β–NiAl areas with alumina phases was observed close to metallic areas corresponding to γ phase. At other locations a spongy chromia layer was found.

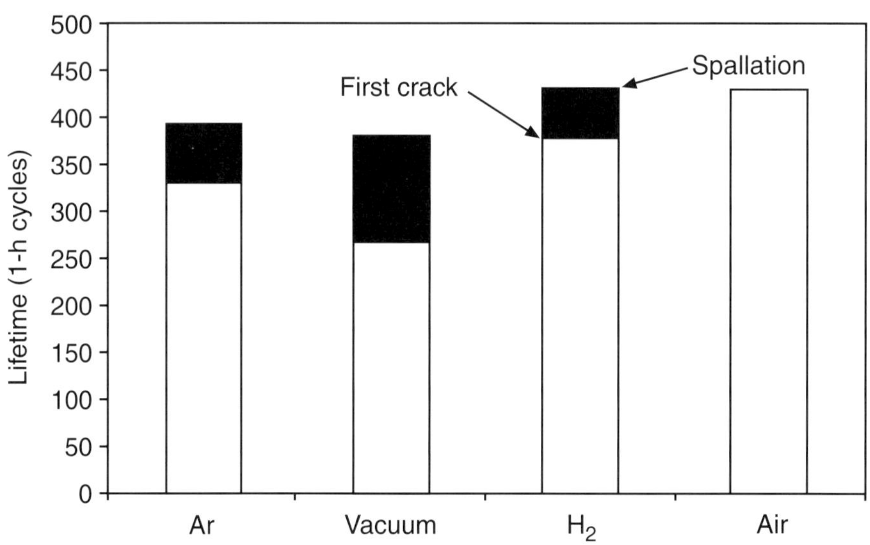

Fig. 4 *Influence of various atmospheres during bond coat annealing on the TBC cyclic lifetime at 1100 °C.*

The XRD results also suggest that the TGOs on vacuum annealed bond coats are essentially chromia and α-Al_2O_3 could not be detected. This behaviour of chromia formation on an alumina-forming composition cannot be explained in thermodynamic terms but must be related to kinetics. Extensive chromia formation on alumina-forming PM 2000 alloy (Fe–20Cr–5Al–0.5Y_2O_3) was also found if the annealing temperature was kept slightly below that required [4]. The coefficient of thermal expansion between chromia and TBC is also less favourable than between alumina and TBC and can attribute to overcritical stress build-up.

5.3. Variation of Peen Shot Material

The lifetime of samples pretreated with glass balls or with zircon shot is shown in Fig. 5 in a Weibull plot. The two intersections of the data lines and the number-of-cycles axis refer to 483 and 992 cycles. The lifetimes of zircon treated samples are represented by the steeper line shifted to longer times which indicates a smaller scatter band of the results as well as a higher potential for lifetime and lifetime predictability. The glass-ball peened samples show a more shallow lifetime distribution graph.

The wider scatter and lower lifetime of the glass-peened samples results from occasional glassy Ca-silicate inclusions introduced from the shot on glass-ball peening. These expand at high temperatures to form an interlayer along some areas at the TGO/TBC interface and represent the weakest link within a TBC system [5]. The TGO essentially consists of a corundum phase with some $YAlO_3$ dispersed in it as was found in fully-treated juvenile samples (1080°C for 4 h in vacuum, followed by thermal loading at 1000°C during evaporative application of the TBC material in a low pressure oxygen atmosphere) by TEM. In some areas, however, the corundum layer changes to isolated particles of crystalline alumina-chromia solid solution in

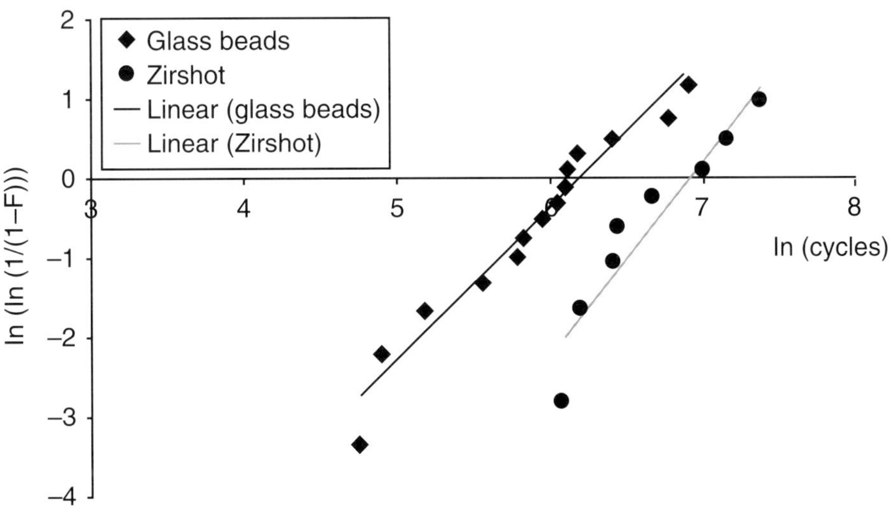

Fig. 5 Cyclic lifetime at 1100°C of samples subjected to bond coat peen densification using glass bead and alternatively zircon shot in a Weibull plot.

contact with the glassy Ca-silicate phase. Some crystalline Y-silicate, Y_2SiO_5 is occasionally present at the TGO/bond coat interface at the expense of $YAlO_3$ inclusions which suggests that the compound has formed at extremely low oxygen partial pressures below the glassy silicate phase. The low permeability of glass layers for oxygen slowed down the oxidation of Al and Cr and thus Y was selectively oxidised and synthesised with the glass phase to crystalline Y_2SiO_5. The most hazardous reaction is the liquefaction of glass phase and its extremely low surface tension. Glass can penetrate through all cracks and corners by capillary forces and will wet adjacent interfaces there. It will contract on cooling with a different thermal expansion coefficient which will result in tensile stresses in the vitreous phase. Due to the poor mechanical properties of glass it will exhibit multiple cracking.

On the other hand remarkably high lifetimes on a few particular samples were attained. They obviously contradict the expectations suggested by the low mechanical properties of the vitreous silicate phase interlayers. A mechanism which can compensate the poor properties is assumed to be operative. Here the reduced diffusivity of oxygen and probably also of some metal ions through this oxide phase is speculated to account for some benefits, e.g. lower growth rate of the TGO and avoidance of spinel oxides at the TGO/TBC interface region. The glass distribution may also have an effect. After peening it was found by EDS mapping that Si-containing inclusions showed a distribution of thin dash-like inclusions. They can be undercritical in size to account for instantaneous spallation. Instead local detachment may allow some stress relief. Taking all the aspects into account a densification treatment of the bond coat by zircon shot is favoured because of the smaller scatter and longer overall cyclic lifetimes.

6. Conclusions

Cyclic testing of cylindrical TBC coated samples is a valuable tool for the ranking of TBC systems. By microscopic observation of the failed samples failure mechanisms can be traced and valuable hints for an improvement of TBC systems can be recognised. In this study IN100 substrates with EB-PVD NiCoCrAlY bond coats subjected to various treatments and an EB-PVD PYSZ TBC were cyclically tested at 1100°C. TBC lifetime is independent of EB-PVD NiCoCrAlY bond-coat peening if different shot diameters have been used. Lifetime results indicate no clear preference for the use of a distinctive heat treatment atmosphere (vacuum, hydrogen, argon, air) before TBC deposition. Premature fissure formation, however, indicates that vacuum annealing can be critical. The use of zircon shot material instead of glass shot results in a lower scatter and in higher lifetimes.

7. Acknowledgements

The automated test facility was designed and set into operation by Thomas Krell, Uwe Leushake and Wolfdieter Zimmermann. Careful sample and coating manufacture by Jörg Brien, Claus Kröder, Helmut Mangers and Hartmut Schurmann is highly appreciated. TEM samples were skilfully prepared by Gudrun Paul.

References

1. Y. Jaslier and S. Alpérine, Electron beam physical vapour deposition thermal barrier coatings: a comparative evaluation of competing deposition technologies, in *Thermal Barrier Coatings,* AGARD-R-823, Chapter 8, pp.1–10, April 1998.
2. C. Leyens, K. Fritscher, R. Gehrling, M. Peters and W. A. Kaysser, *Surf. Coat. Technol.*, 1996, **82**, 133–144.
3. V. K. Tolpygo and D. R. Clarke, *Acta Mater.*, 1998, **46**(14), 5267–5274.
4. H.-P. Martinz, personal information with Plansee AG Reutte/Tirol.
5. K. Fritscher, M. Schmücker, C. Leyens and U. Schulz, *Mater. Sci. Forum*, 1997, **251–254**, 965–970.

25
Degradation of Aluminide, Chromium–Aluminide and Thermal Barrier Coatings in Conditions of Cyclic Thermal and Mechanical Loads

L. SWADŹBA, B. MENDALA and M. HETMAŃCZYK
Silesian Technical University, Institute of Materials Science and Engineering,
Krasi´nskiego 8, 40-019 Katowice, Poland

ABSTRACT
Results from studies of degradation processes of aircraft engine turbine blades subjected to strong cyclic thermal load during operation are presented. The blades were covered with chromoaluminide diffusion coatings (HTLA) and aluminide coatings (LTHA) produced by the pack cementation method. The HTLA coating in the initial state before operation had a thickness of 45 μm and contained the NiAl phase with a content less than 50at.% Al. The aluminide coating contained more than 50at.% Al. The coatings were formed on aircraft engine turbine blades made of Ni base alloy and after comprehensive laboratory tests they were sent for use in service. After operation the blades were dismounted and subjected to studies of the degradation mechanisms of the chromoaluminide (HTLA) and aluminide (LTHA) coatings. It was found that the degradation process takes place as a result of corrosion at the grain boundaries of the NiAl phase until the diffusion zone is reached and than there is delamination between the scale and the diffusion zone. In the aluminide coatings the corrosion takes place at precipitates which are rich in high-melting elements and at cracks formed as a result of thermal fatigue. The influence of structure and thermal treatment of the TBC (thermal barrier coatings) coatings on thermal cyclic degradation has been investigated. It has been shown that the coatings without thermal treatment undergo accelerated degradation as a consequence of thermal cycling. The TBC coatings after thermal treatment showed much higher resistance to thermal cycling when an Al_2O_3 zone was formed on the $ZrO_2xY_2O_3$–MeCrAl boundary. By selecting spraying parameters the amount of monoclinic phase may be decreased from 22% in the powder to 4–6% in the TBC coatings.

1. Introduction

Coatings on engine turbine blades are exposed to high static and variable stresses and to high temperatures. Static stresses and high temperature lead to destruction by deformation as a result of creep. In the particular loading conditions of military engine blades the influence of thermal cycles leads to crack initiation by thermal fatigue. Degradation by impact with the environment is the result of strong oxidation and high temperature corrosion processes.

Investigations after service are a necessary stage of gas turbine blade evaluation. From the investigations conducted so far it appears that in the majority of cases the degradation process of blades begins and develops in thin surface zones. Formation of defects in the coating on the blade surface can be caused by (i) high temperature corrosion and gas erosion, (ii) low resistance of blades to thermal fatigue and (iii) insufficient heat and high temperature creep resistance of the alloy used.

An analysis of defects in the coatings on engine turbine blades subjected to long term use is an important stage in evaluating the operating abilities. An analysis of damage and the reasons for damage has allowed coatings having better durability in the given operating conditions to be selected [1–6].

The covering of combustion chamber and turbine blades with TBCs makes it possible to increase the temperature of inlet gases. These coatings allow the following advantages to be obtained [7–9]: an increase in temperature of inlet gases, reduction of substrate temperature, a lowering of the amount of the cooling air and protection against thermal deformation and cracking [10].

The TBC coatings are mainly obtained by low pressure plasma spray (LPPS), electron beam physical deposition (EB-PVD) and air plasma spray (APS) processes [11–19]. These coatings consist of the ceramic $ZrO_2xY_2O_3$. outer zone and Me Cr ARY (Me=Ni, Co, Fe) bond coat. In order to increase corrosion resistance of the MeCrAlY bond coat and adhesion of ceramic coatings processes of thermal spraying are combined with diffusion processes [9,17,18].

2. Causes of Destruction of Gas Turbine Parts During Operating Conditions

Components of an aircraft engine turbine such as combustion chambers and blades can be damaged as a result of thermo-mechanical cycles of fatigue and oxidation in areas of propagating cracks. Understanding causes of destruction can bring many advantages. First, the design of engine parts can be improved in order to decrease processes in the destruction of the components second, it can be helpful in selecting materials and coatings and thirdly, it facilitates evaluations of component durability [1].

The main reasons for turbine blade damage are the following: thermal fatigue, destruction as result of creep, and hot corrosion.

2.1. Damage as a Result of Thermal Fatigue

Many blades have a system of cooling channels and cracking due to thermal fatigue initiates near these in the area of the edge of the attack. The majority of aluminide coatings are brittle at working temperatures below the ductile–brittle transition temperature (DBTT).

Depending on the crack location the oxidation process can take place dramatically along these cracks. In non-coated blades with an equiaxial grain structure thermal fatigue cracking is initiated in the subsurface layer by carbides of MC type.

2.2. Damage as a Result of Oxidation

The majority of advanced aircraft engines have turbine blades cooled by air. Cooling channels can be of different shapes from single channels to connected cooling systems. In the coated blades oxidation along non-coated outer channel walls and hole clogging lead to quick oxidation and limitation of durability in some engines.

Oxides form pits with some intercrystalline oxide wedges penetrating into the subsurface. The delaminated oxides can clog the cooling channels and limit the flow of cooling air. These oxides have lower thermal conductivity than the substrate material and serve as a barrier between cooling air and the inner surface of the blade. This leads to an increase in the blade temperature and in the oxidation rate of alloy components along the grain boundaries. In the directionally crystallised blades these oxides can penetrate deeply into the substrate material along the columnar grain boundaries and create long cracks [2].

2.3. Damage by Hot Corrosion

High temperature corrosion is of two types. The first, Type I, begins to have an effect at 800–925°C after a short period of incubation, while the second Type II, takes place at 700–775°C after a long period of incubation. The first type of hot corrosion is an accelerated form of oxidation resulting in oxide formation. The dissolution of oxides is a consequence of the molten Na_2SO_4 salt present on the components. Sulfur diffuses into the substrate where it forms sulfides. Microstructurally, chromium sulfide particles, are associated with Type I, reaction front and are visible under the microscope whereas sulfides are never visible in Type II reaction, although some differences in the sulfur content of the substrate and the reaction front may be picked up with the aid of EDX analysis [2].

Hot corrosion of the first type usually begins in the middle part of the blade from the high pressure side since sulfur contaminated enriched gas impacts this region first. The partial deposition of contamination can also take place in the area of the blade platform. Hot corrosion can destroy the blade surface to a significant extent [3].

2.4. Damage Due to Overheating

Significant changes in the structure of blades can occur depending on the degree of overheating. In extreme cases blades may be covered with blisters and coatings can undergo spalling due to diffusion zone melting [4]. In the initial stage partial melting of the grain boundaries together with formation of γ–γ' eutectics along the grain boundary areas can occur. Once γ–γ'eutectic melting has begun it cannot be removed by any kind of thermal treatment and that is why blades showing such features are rejected at inspection [5]. At the Silesian Technical University investigations of degradation of protective aluminide, chromoaluminide and TBC coatings have been conducted for many years. This article presents an analysis of the damage caused by thermal cycles, thermal fatigue and oxidation. Those processes take place simultaneously on the blade surface [6].

3. Experimental Procedures and Materials

The aim of the investigation was to determine the degradation mechanisms of a turbine blade surface covered with diffusion chromoaluminide or aluminide coatings after several hundred hours of operating in very complex conditions. The scope of the work included: selection of blades for tests, visual evaluation of the surface, making cross sections at the point of the greatest damage, metallographic and microanalytical examinations and EDX analysis. On the basis of the above tests a hypothetical description of degradation mechanisms of the Cr–Al and Al coatings during operation was made.

For the tests blades were selected with chromoaluminide and aluminide coatings after use in service on the same turbine impeller of an aircraft engine. The Cr–Al coating structure was characteristic of a low active process (HTLA) and the structure of the aluminide coating was characteristic of a high active process (LTHA). The blade material was the Ni base creep resistant alloy EI867 with the composition given in Table 1.

Quantitative and qualitative analyses were made with an X-ray microanalyser JCXA 733 of the JEOL. Quantitative analyses and microphotographs were made by wave dispersion spectrometers (WDS) whereas qualitative analyses used an energy-dispersion spectrometer (EDS) LINK 860. The electron accelerating voltage was 15kV.

Investigations of the destruction mechanisms of TBC coatings were conducted on specimens made of alloys used for combustion chambers of aircraft engines. The aim of the investigations was to determine the resistance of the coating to degradation in the cyclic oxidation conditions. NiCrAlY and $ZrO_2xY_2O_3$ powders were the materials for the coatings which were formed by the APS spraying method. Thermal cycles were based on heating to 1200°C and intensive cooling in an air stream to 250°C.

4. Investigation Results

4.1. Cr–Al Coating Degradation

The NiAl or Ni_3Al phase in which precipitates elongated in the perpendicular direction to the surface are observed forms the diffusion zone. That type of structure is a result of a growth mechanism based on Ni diffusion from the base material into the coatings. The chromoaluminide coatings were formed in the high temperature (>1000°C) low activity process (HTLA) powder mixtures. Such process conditions favour formation of NiAl phase coatings in which the aluminium concentration is from 21 to 31.6 mass%. These conditions lead to a coating formation resulting only from Ni diffusion towards the surface. The diffusion zone is relatively thick in relation

Table 1. *The chemical composition of EI 867 alloy*

Alloy	**Elements (mass%)**						
	Ni	C	Cr	Al	Mo	W	Co
EI 867	62.00	0.1	10.50	4.90	11.50	6.00	5.00

to the total thickness of the coatings and contains higher amounts of such elements as: Mo, Cr, W and Ti. The CrAl coating has higher plasticity than coatings obtained in the so-called high activity processes in which the formation mechanism is based on the dominant diffusion of aluminium to the base material. The diagram of the Cr–Al coating structure is presented in Fig. 1.

The diagram of blades with areas marked of the greatest corrosion damage and of the cutting points for metallographic examinations is presented in Fig. 2.

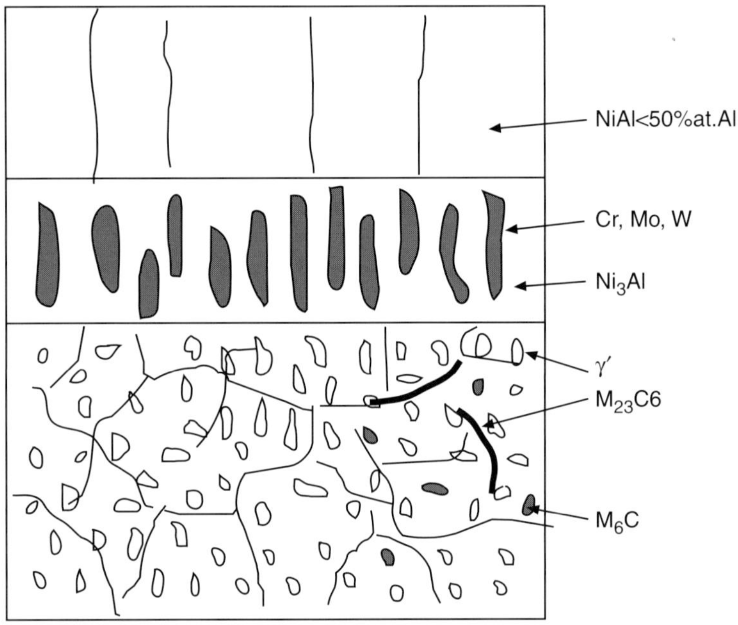

Fig. 1 Diagram showing the structure of Cr–Al coatings obtained in the low activity process.

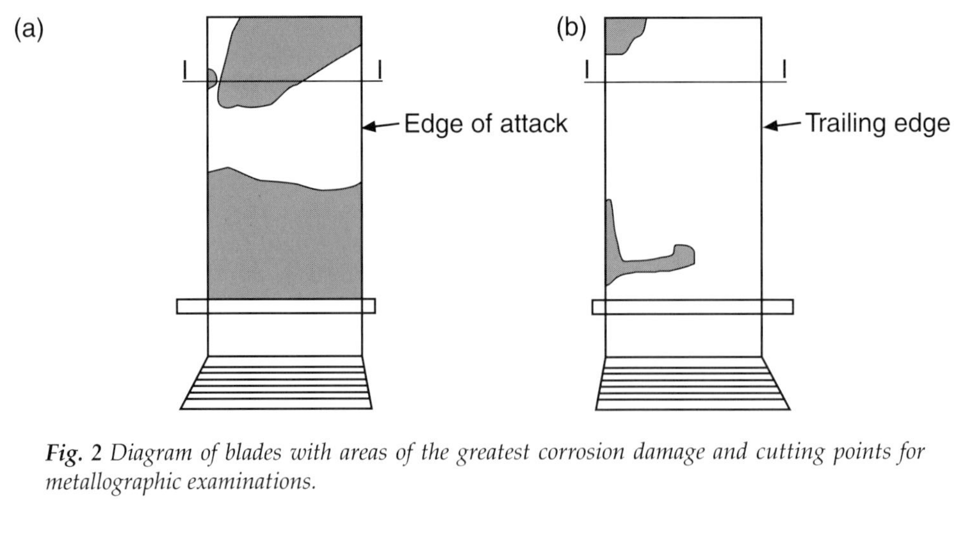

Fig. 2 Diagram of blades with areas of the greatest corrosion damage and cutting points for metallographic examinations.

The analysis of the structure presented in Fig. 3 (a–c) show that the corrosion processes are first initiated at the grain boundaries of the NiAl phase and then at the grains.

This process takes place until the corrosion process reaches the boundary with the transition zone which provides an effective barrier to the further corrosion process because of its increased chromium content. The next stage in the degradation mechanism is the propagation of the corrosion processes in the coating. The subsequent step is corrosion at the boundary with the diffusion zone and plastic deformation of the scale with loss of contact between the scale and base material.

Figure 3 (b) shows a crack between the scale and the transition zone. Figure 3 (c) shows significant plastic deformation of the scale.

This mechanism is important for determining the structure of coatings obtained in the low activity process. Destruction due to fatigue cracking was not observed in the coatings of this type. Therefore, the hypothesis was confirmed that low active coatings showed plasticity in the working conditions of this type of blade.

A diagram of the microstructure including the grain boundary with marked points of the X-ray microanalysis is presented in Fig. 4 and Table 2.

The X-ray microanalysis in the micro-areas showed that there was an increased concentration of aluminium and chromium in the corrosion products. It may be assumed that a spinel of the $NiAl_2O_4$ type constitutes the scale.

The aluminium diffusion to the scale that is being formed causes a decrease in the aluminium content in the coating matrix. As can be seen in Table 2 the Al concentration in the matrix corresponds to the Ni_3Al phase. The presence of sulfur in the corrosion products which initiates the corrosion process is characteristic of the Type II corrosion. When sulfur is detected in the corrosion product it is near the coating–corrosion product interface but always associated with oxygen [20]. This proves that in HTLA coatings the Type II mechanism takes place.

High chromium content and presence of sulfur in the scale are characteristic. The Al concentration inside the impoverished grain corresponds to the content in the Ni_3Al phase.

4.2. Degradation of the Al (LTHA) Coatings

The NiAl phase with more than 50mass% Al forms the matrix of these coatings. Precipitates of phases rich in elements such as Cr, Mo and W are present in the structure. These coatings are obtained in the high activity processes at low temperature with high Al content in the mixtures or suspension. After thermal treatment at about 950°C a diffusion zone and a structure consisting of NiAl phases with more or less 50at.% Al appear as a result of diffusion processes at the boundary with the base material.

A diagram of the structure of the aluminide coating on the blade in the LTHA process is presented in Fig. 5.

The highest degree of degradation in the form of spalling and corrosion products was found on the high pressure side in the middle of the blade height and below this, near the platform region. The surface degradation on the low pressure side of the blade takes place in the same places as on the high pressure side although the degradation intensity is lower. This picture of the surface degradation results from

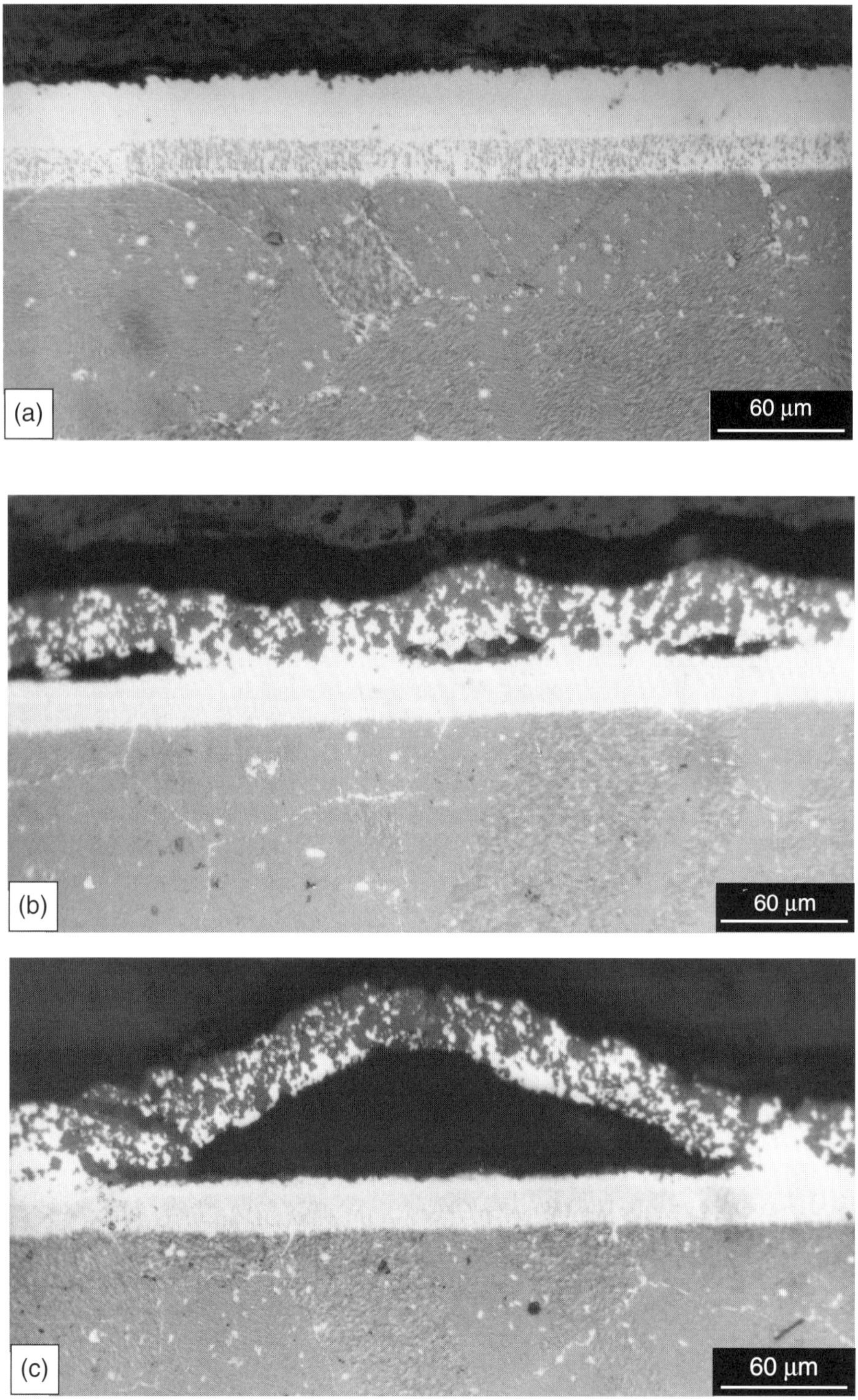

Fig. 3 *Microstructures of the Cr–Al coatings on the blade after service.*

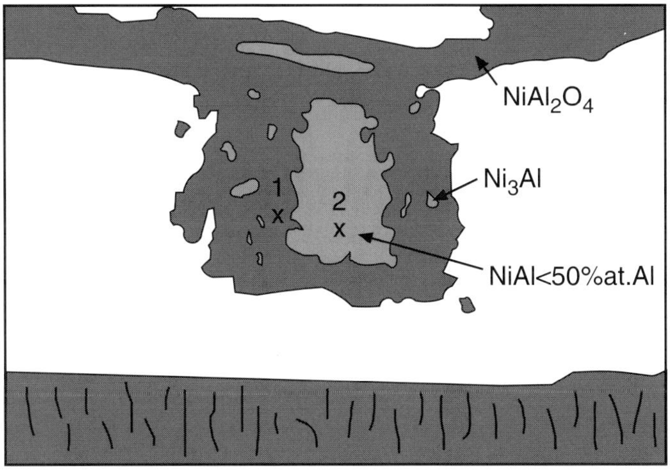

***Fig. 4** Diagram showing degradation of Cr–Al coating in NiAl grain boundary.*

differences of temperature developing on the blade height and the blade cross-section and also from varying thermal loads and thermal fatigue during operation.

The aluminide coating degradation in the high activity processes takes place in the following stages:

- initiation of the corrosion process on the surface and along the NiAl grain boundaries by precipitation of phases rich in Mo and W
- the corrosive impact of the environment and scale formation
- scale cracking due to thermal fatigue
- crack growth resulting from overlapping processes of thermal fatigue and corrosion
- crack formation between the scale and the NiAl diffusion coating.

The degradation process can also begin with coating cracking without participation of the corrosion process; cracks then go through the whole thickness of the coating.

***Table 2.** Results of the quantitative microanalysis of elements in the Cr–Al coating*

Measurement points	O		Mg		Al		S		Cr		Co		Ni	
	% mass	at.%	% mass	at.%	% mass	at.%	% mass	at.%	% mass	at.%	% mass	at.%	% mass	at.%
1	23.01	39.00	1.46	1.63	43.92	44.15	0.56	0.47	7.06	3.68	4.62	2.13	19.37	8.94
2	—	—	—	—	10.68	20.60	—	—	1.64	1.64	—	—	87.68	77.76

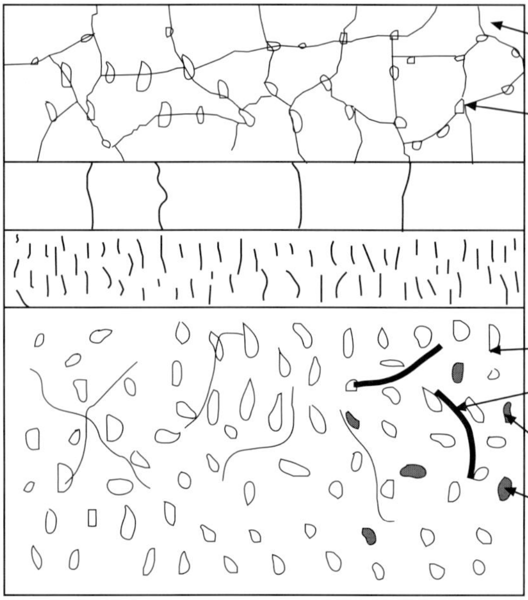

Fig. 5 *Diagram of the aluminide coating structure formed in the high active process.*

Important degradation stages of the LTHA coatings on the blades in operation are presented in Figs 6 (a–f). Cracks (light) which reach the boundary with the coating in the next stage (Fig. 6f), are visible in the scale (Fig. 6a). At this degradation stage thermal and mechanical stresses lead to decohesion between the scale and the NiAl zone.

The cracks, which grow as a result of the oxidation process can be seen; delamination then takes place at the boundary of the scale and the coating with subsequent spalling of the scale.

These results show explicitly that the corrosion processes are initiated at the grain boundaries with the higher Mo and W concentrations. These elements significantly weaken the corrosion resistance of the NiAl phase grain boundary.

The microanalysis results show that this zone consists of precipitates with high concentrations of molybdenum, tungsten and chromium, i.e. 50.35, 14 and 25at.% respectively.

4.3. Thermal Barrier Coatings

The grain morphology of the $ZrO_2xY_2O_3$ (Metco204 B-NS) powder is presented in Fig. 7. This powder is characterised by very good surface quality without so-called satellites and has grains with a spherical shape; irregular grains were effectively absent. Large and fine grains with a regular spherical shape were also present. The phase analysis showed that the $ZrO_2\,xY_2O_3$ powder is a mixture of tetragonal and monoclinic variations of the zirconium oxide ZrO_2. The powder contains 79.18vol.% of the tetragonal phase and 20.82vol.% of the monoclinic phase.

The powder phase analysis results are presented in Fig. 8. The flowability was

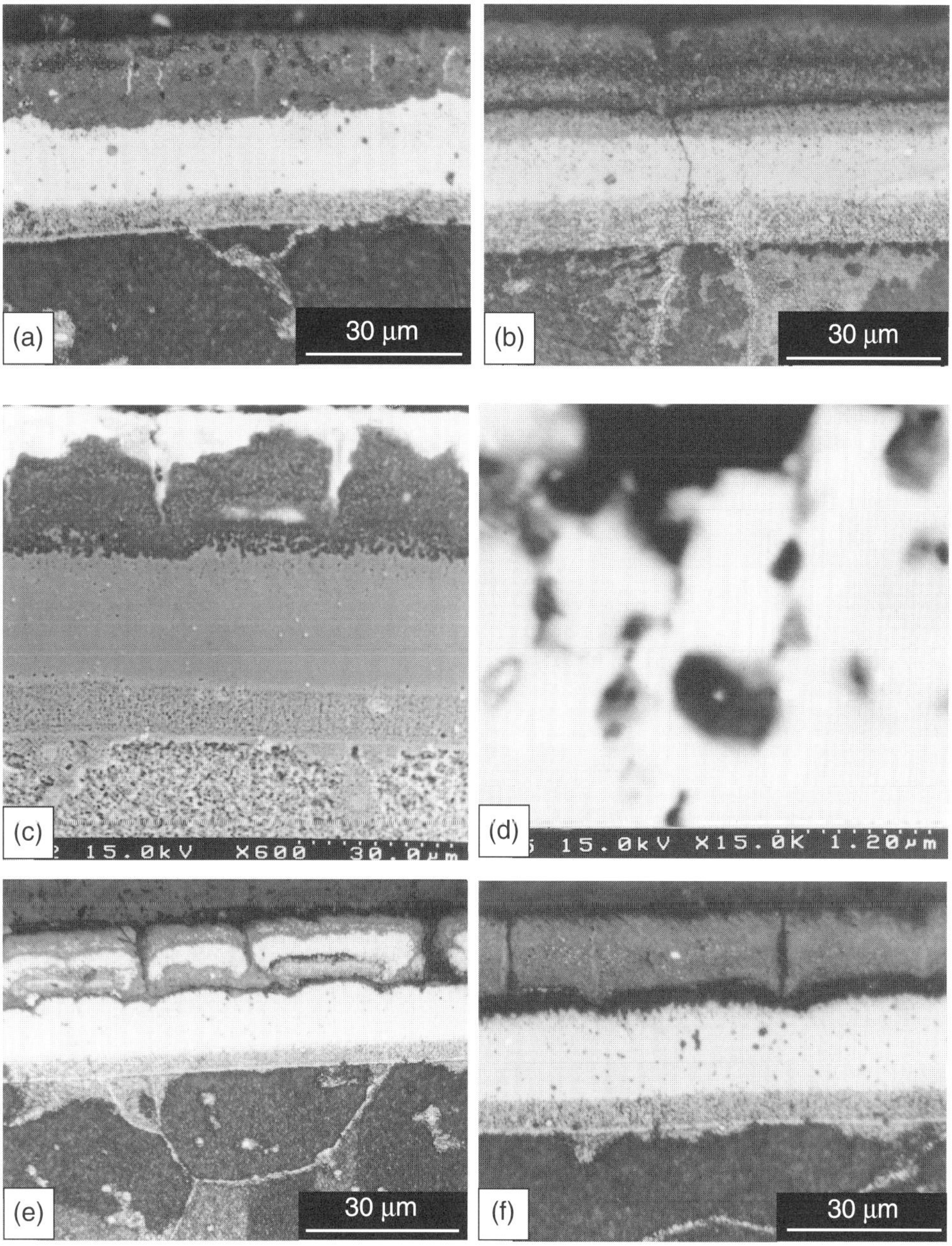

Fig. 6 Main degradation stages of the LTHA coatings on the blades in operation: (a) thermal fatigue cracks in scale, (b) cracks through coatings, (c) oxidation along thermal fatigue cracks, (d) oxidation front in Mo, W area below scale zone, (e) and (f) undercutting of scale.

measured by means of a test cup (ASTM No. 4). The flow rate of the applied powder was very good in comparison with other tested powders.

The microstructure of the TBC (MeCrAlY + $ZrO_2\,xY_2O_3$) coating obtained by the APS spraying method is presented in Fig. 9.

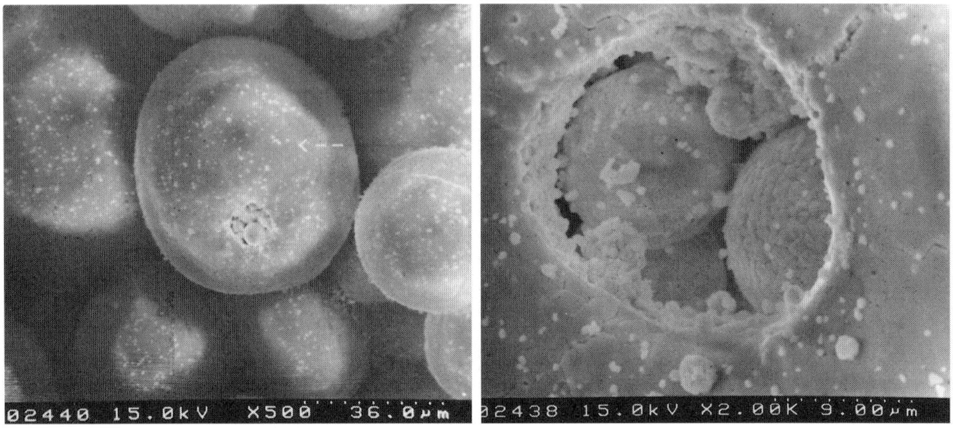

Fig. 7 *Morphology of* $ZrO_2 xY_2O_3$ *powder used for TBC spraying.*

Figure 10 presents the phase analysis results of the ceramic coating surface. As can be seen the spraying parameters significantly decreased the amount of the monoclinic phase, the presence of which has a disadvantageous influence on the properties, particularly the thermal cycling variables. The coating porosity was 10–12%. The coating thickness was about 350 µm.

TBC coatings of different chemical composition and structure after 120 thermal shocks at 1200°C–250°C cooling for 5 min are presented in Fig. 11 (a–c). As is seen from the specimen surface with ZrO_2 outer zone specimens: 0.07 mm MeCrAlY + 0.07 mm (60% $ZrO_2xY_2O_3$ + 40% MeCrAlY) + 0.07 mm $ZrO_2xY_2O_3$ (Fig. 11a) and specimens 0.05 mm MeCrAlY + $Al_{diff.}$ + 0.15 mm $ZrO_2xY_2O_3$ (Fig. 11b) underwent

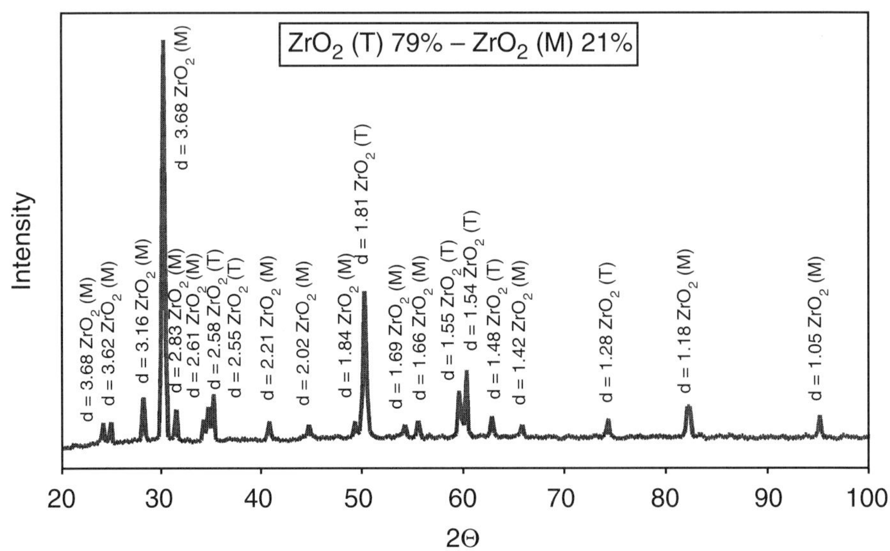

Fig. 8 *Result of phase analysis of* $ZrO_2 xY_2O_3$ *powder used for APS.*

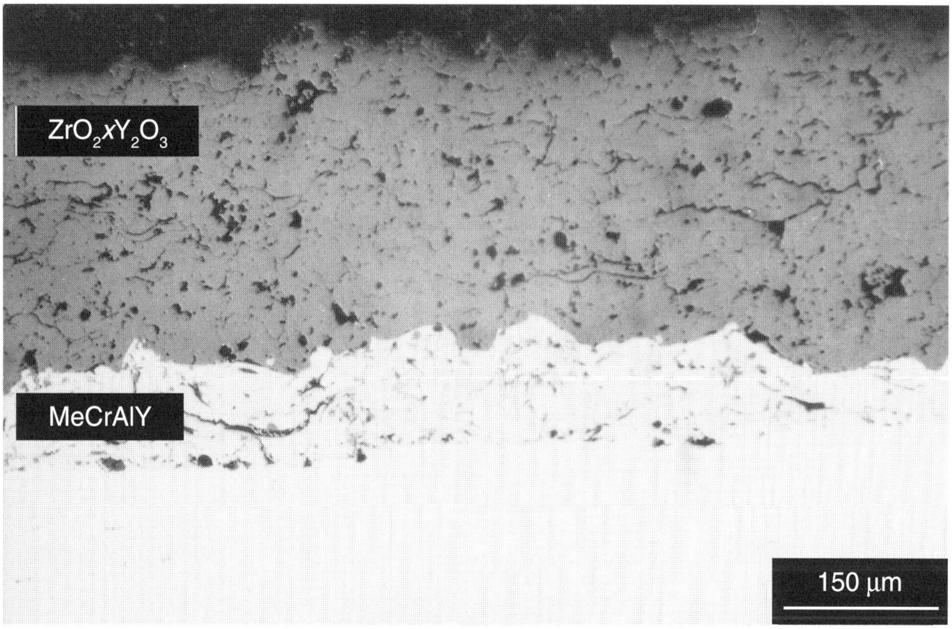

Fig. 9 *Microstructure of the TBC MeCrAlY + $ZrO_2xY_2O_3$ coating.*

the quickest spalling — after 25 cycles. The analysis of specimens after 120 shocks (Fig. 11c) showed that the best properties in test conditions were shown by the 0.1 mm MeCrAlY+ 0.2 mm $ZrO_2xY_2O_3$ specimen.

The above results show clearly that the insulation properties of coatings as determined by the maximum thickness, good adhesion to the substrate and porosity

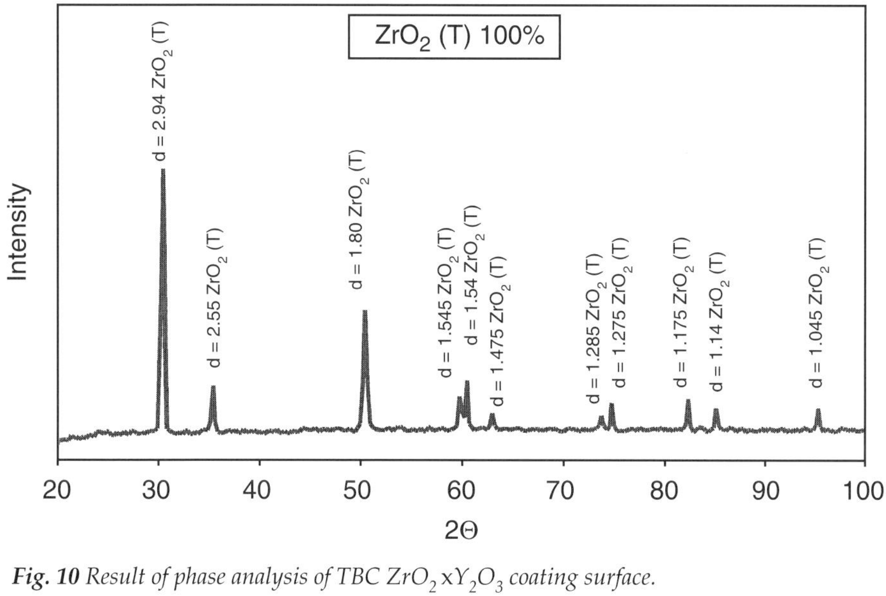

Fig. 10 *Result of phase analysis of TBC $ZrO_2xY_2O_3$ coating surface.*

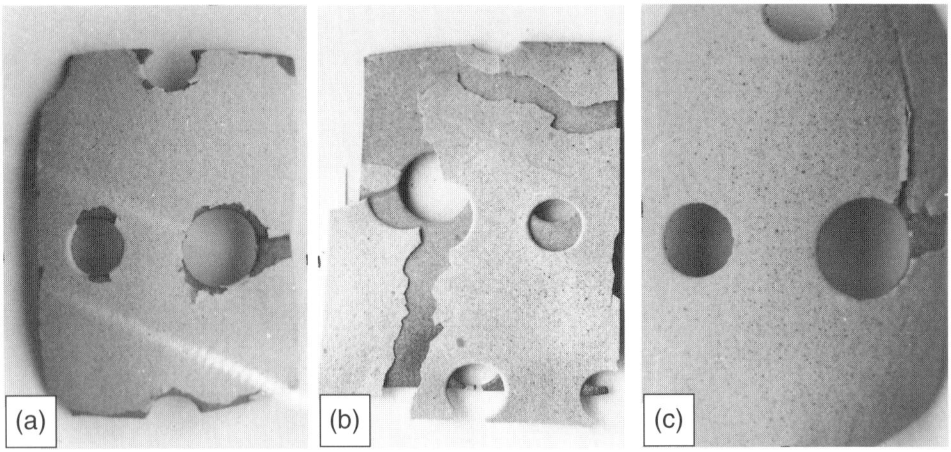

Fig. 11 Appearance of the specimens with TBC coatings after thermal shocks at 1200 °C–250 °C: (a) 0.07 mm MeCrAlY + 0.07 mm (60% $ZrO_2xY_2O_3$ + 40% MeCrAlY) + 0.07 mm $ZrO_2xY_2O_3$, 25 shocks, (b) 0.05 mm MeCrAlY + Al_{diff} + 0.15 mm $ZrO_2xY_2O_3$, 25 shocks, (c) 0.1 mm MeCrAlY + 0.2 mm $ZrO_2xY_2O_3$ 120 shocks.

are the factors ensuring their high durability. The application of multilayer MeCrAlY–$ZrO_2xY_2O_3$ coatings of which MeCrAlY layers were the intermediate layer in order to compensate stresses did not give the expected results in the conditions of the test. An overaluminising process for the MeCrAlY layer did not give positive results.

Figure 12 presents microphotographs of the zone between the bond coat and $ZrO_2xY_2O_3$ outer layer which underwent exfoliation. Although good adhesion between the bond coat and substrate was preserved very strong oxidation can be observed.

The oxidation process caused a very strong segregation of elements, particularly aluminium and yttrium, into the oxide. Good results of thermal shock resistance were obtained by the thermal pre-treatment of the TBC coatings. The thermal treatment resulted in strong bond formation between the MeCrAlY bond coat zone and the $ZrO_2xY_2O_3$ ceramic zone through the Al_2O_3 zone. The microstructure of the MeCrAlY–$ZrO_2xY_2O_3$ coating structure with the marked (see arrow) Al_2O_3 zone binding MeCrAlY with the ceramic layer and distribution of elements in the analysed zone are presented in Fig. 13.

The TBC coatings subjected to thermal treatment show thermal cyclic oxidation resistance much higher than without that treatment. On the basis of analysis of structures including different cross sections, welded joint areas, holes and gaps, TBC spraying process technology was selected for the combustion chamber of the aircraft engine. The process was conducted by a computer controlled robot system.

An evaluation of the influence of the TBC coating on thermal shock resistance showed the considerable influences of structure, chemical composition and thickness on properties. It should be noted that tests were conducted in severe conditions of thermal shock and hot corrosion. NiCrAlY–$ZrO_2xY_2O_3$ coatings with appropriate parameters of spraying and thermal treatment had the best properties in these conditions. The heat treatment of TBC coatings in vacuum has a significant influence

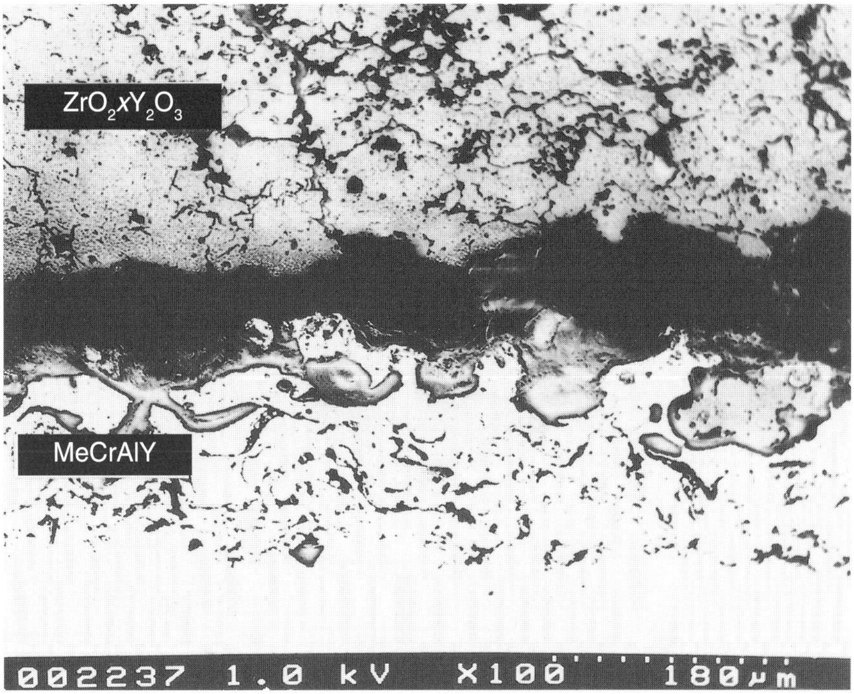

Fig. 12 *Microstructure of ceramic and bond coat delamination zone after thermal shocks.*

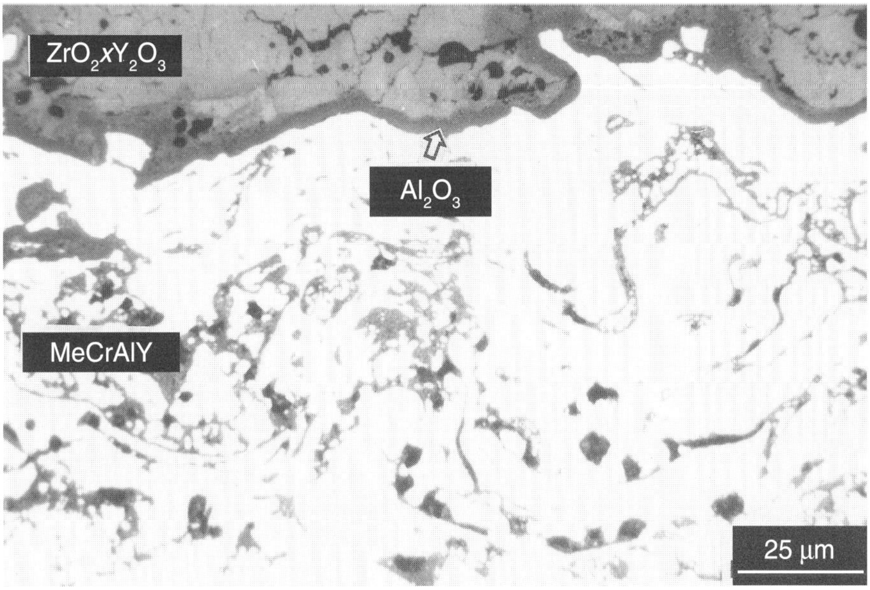

Fig. 13 *Microstructure of the MeCrAlY–$ZrO_2xY_2O_3$ coating with Al_2O_3 zone after heat treatment and thermal shock.*

on their properties. Multilayer coatings systems did not meet expectations at this stage of investigations in these severe conditions.

Conclusions

The analysis of degradation mechanisms of the LTHA, HTLA and TBC coatings leads to the following conclusions:

(1) The degradation mechanisms of LTHA and HTLA coatings in service are significantly different.

(2) The HTLA coatings have too low and Al concentration and possess a columnar structure of the NiAl phase which is unfavourable for securing high operating properties of the blades.

(3) The LTHA coatings with high Al concentration have disadvantageous properties in the service conditions of blades subjected to cyclic thermal and mechanical loads, because:

- precipitates rich in Mo and W on the NiAl phase boundaries are the source of accelerated corrosion, and

- the coatings are working in the very disadvantageous DBTT range.

(4) Overaluminising MeCrAlY coatings with more advantageous DBTT properties should be developed.

(5) Thermal cycling is a good criterion of the property evaluation of the Thermal Barrier Coatings.

References

1. G. T. Embly and E. S. Russel, *Proc. 1st Pearson Internal. Turbine Conf.*, Dublin, Ireland, 1984, pp.157–164.
2. A. K. Koul, J. P. Immarigeon, R. V. Dainty and P. C. Patnaik, Degradation of high performance aero-engine turbine blades, in *Advanced Materials and Coatings for Combustion Turbines* (V. P. Swiminathan and N. S. Cheruvu, eds). Published by ASM International, Materials Park, Ohio, 1994, pp.69–74.
3. P. Patnaik and J. Elder, in *Superalloys 88* (D. N. Duhl *et al.*, eds). TMS–AIME, Warrendale, New Jersey, 1988, p.775.
4. O. Au, A. K. Koul and J. P. Immarigeon, Effect of coatings on the mechanical properties of Ni and Co base superalloys, in *Advances in High Temperature Structural Materials and Protective Coatings* (A. K. Koul *et al.*, eds). National Research Council of Canada, 1994, pp. 237–245.

5. F. S. Petit and G. H. Meier, in *Superalloys 84* (M. Gell *et al.*, eds). TMS–AIME, Warrendale, New Jersey, 1984, p.651.
6. L. Swadzba, B. Formanek and A. Maciejny, *Mater. Sci. Eng.*, 1989, **A121**, 407–412.
7. S. Durham *et al.*, Ceramic thermal barrier coatings, in *Advances in High Temperature Structural Materials and Protective Coatings* (A. K. Koul, V. R. Parameswaran, J. P. Immarigeon and W. Wallace, eds). National Research Council of Canada, 1994. pp. 226–230.
8. M. Meier, D. K. Gupta and K. D. Sheffler, *JOM*, March (1991), 50–53.
9. J. Wortman, B. A. Nagraj *et al.*, *Mater. Sci. Eng.*, 1989, **A121**, 433–440.
10. K. D. Sheffler and D. K. Gupta, *J. Eng. Gas Turbines & Power,* 1988, **110**, 605–609.
11. T. N. Rhys-Jones, *Corros. Sci.*, 1989, **29**, (6), 623.
12. F. C. Toriz, A. B. Thakker and S. K. Gupta, *Surf. Coat. Technol.*, 1989, **39/40**, 161–172.
13. W. J. Brindley and R. A. Miler, *Adv. Mater. Proc*. 1989, **8,** 29–33.
14. G. W. Meetham, *Mater. Sci. Technol.*, 1986, **2**, March, 290–294.
15. B. J. Gill and F. L. Ridding, *Surf. Eng.*, 1986, **2**, (3), 169–175.
16. Ch. Mertens-Lecomte, D. Muck and J. Garcia, *in Materials for Advanced Power Engineering, Part II* (D. Coutsouradis, ed.). Kluwer Academic Publishers, 1994, pp.1313–1321.
17. A. Maciejny, L. Swadzba, B. Formanek and B. Mendala, *in Materials for Advanced Power Engineering, Part II* (D. Coutsouradis, ed.). Kluwer Academic Publishers, 1994, pp.1291–1302.
18. L. Swadzba, *et al.*, An influence of the thermal barrier coatings on thermal shock and hot corrosion resistance, in *Materials for Advanced Power Engineering* 1985, **5**, (III), 1709–1720, (J. Lecomte-Beckers, F. Schubert and P. J. Ennis, eds). *Proc. 6th Liège Conf. Part III*, 6th *Liège Conf. on Materials for Advanced Power Engineering*, 5–7 October 1998, Liège, Belgium.
19. J. A. M. Boogers, R. J. H. Wanhill and H. J. C. Hersbach, Thermal shock and oxidation resistance of ceramic coatings, National Aerospace Laboratory NLR The Netherlands, NRL Tech. Publ. 89085, March 1989.
20. P. Sahoo and G. W. Goward. On the suitability and application of MCrAlY coatings under various operating conditions. *Proc. 8th National Thermal Spray Conf.* Houtson, Tx, Sept. 1995, pp.539–544.

26
Thermal Cycling of NiAl Intermetallic Coatings Prepared by High Velocity Oxy-Fuel (HVOF) Thermal Spraying

J. A. HEARLEY, J. A. LITTLE and A. J. STURGEON*
Department of Materials Science and Metallurgy, University of Cambridge, Pembroke St, Cambridge CB2 3QZ, UK
*TWI, Granta Park, Abington, Cambridge CB1 6AL, UK

ABSTRACT
The thermal cycling behaviour of a HVOF NiAl IMC coating has been studied between 723–1173K for 500 h in air. Results indicated that the coating performed well between 723–1023K with no evidence of spallation or de-bonding from the substrate. However, under thermal cycling at 1173K the coating failed catastrophically with severe spallation. This may be attributed to two possible effects. Firstly, the thermal stress within the coating as a result of the cycles large temperature drop and, secondly, the phase transformation in the 2.25 Cr–1Mo steel substrate from ferritic to austenitic causing a substrate expansion during heating and cooling. Characterisation of the coating's top-surface revealed alumina (Al_2O_3) and nickel oxide (NiO). The presence of haematite (Fe_2O_3) suggested that iron cation diffusion had occurred from the substrate during thermal cycling.

1. Introduction

High Velocity Oxy-Fuel (HVOF) thermal spraying is a versatile method for applying intermetallic compounds (IMCs) from powders to form dense, adherent and homogeneous coatings with low porosity and oxide content to protect critical components used in high temperature corrosive environments. Characteristics of such compounds include high melting points and excellent resistance to environmental degradation together with mechanical and chemical stability at high temperatures [1]. Nickel aluminide IMCs, notably NiAl, show excellent resistance due to the formation of a continuous, compact and fully adherent alumina (Al_2O_3) scale [2,3].

These coatings have a potential application in waste incineration and power generation plants to protect heat exchangers. However, during operation such components may often experience severe thermal transients. Therefore, it is necessary to assess the coatings thermal shock characteristics and resistance to thermal cycling.

Commercially available HVOF thermal spray systems consist of a spray gun, a fuel and oxygen flow regulator, a heat exchanger and an automatic powder feed hopper [4]. The gaseous or liquid fuel is premixed with oxygen, at high pressures and flow rates, prior to entering an internal combustion chamber [5,6]. The oxy-fuel mixture, on injection into the combustion chamber, is continuously ignited and the combustion products accelerated along the gun barrel or nozzle to produce a supersonic flame with a velocity of approximately 2000 ms^{-1}. The powder to be sprayed is injected either axially or radially, depending on gun design, into the flame under pressure using an inert gas. Once in the flame, the powder is rapidly and uniformly heated to become molten/semi molten and is accelerated along the gun barrel or nozzle towards the substrate, reaching typical velocities of 600–1000 ms^{-1}. On impinging the substrate, the heated powder particles deform, spread laterally and rapidly solidify to form the typical microstructure of an HVOF thermally sprayed coating. This type of coating therefore consists of many overlapping and connected lenticular 'splats' or lamellae.

2. Experimental

Using a UTP TopGun™ HVOF thermal spray gun, cylindrical coupons of 2.25Cr–1Mo, dia. 10 mm, were fully coated with an inert gas atomised NiAl IMC powder, 31.5mass% Al, of spherical morphology and nominal particle size 15–45 μm. The temperature profiles of the coupons were monitored throughout spraying. Spray parameters are described in Table 1.

Thermal cycling experiments were conducted using a specially designed rig [7] between the temperatures of 723–1173K for 500 h in air. The length of each cycle was 1 h; heating for 45 min and cooling for 15 min. During each cycle the temperature profile of the coupon was recorded at the attachment of a thermocouple. The coefficient of thermal expansion (CTE) of the coating and substrate was measured individually between 323–773K using a Mettler TA 3000 thermomechanical analyser (TMA). Standard analytical techniques were used to characterise the coatings and to examine the effect of thermal cycling on the coating–substrate interface.

3. Results and Discussion

3.1. Temperature Profiles

The coupon's temperature profile throughout spraying revealed very little substrate heating. After four gun passes, the substrate temperature had reached only 460K (Fig. 1). This was attributed to the HVOF process, which uses more kinetic energy as opposed to thermal energy to deposit a coating. Overheating of the substrate during thermal spraying can be problematic especially when component geometries are

Table 1. *UTP TopGun™ HVOF spray parameters*

Fuel	Propylene
Fuel flow rate (Ls^{-1})	1.46
Oxygen flow rate (Ls^{-1})	3.93
Gun power (kW)	76.50

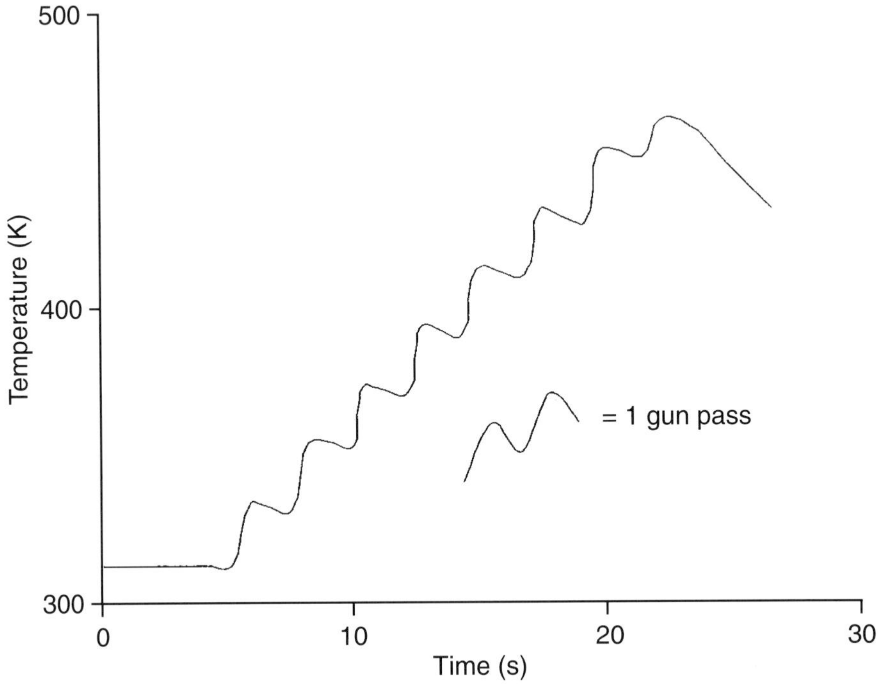

Fig. 1 Temperature profile during spraying.

complicated. One must also be aware of the effect of residual stresses within the coating as a consequence of the spraying process, and their significance during thermal cycling. The role of residual stresses in coating performance has been reviewed for both plasma and HVOF thermal spraying [8,9].

Temperature profiles during thermal cycling indicated that the coupons reached temperature quickly during the heating cycle (Fig. 2). Upon cooling an average minimum temperature of 330K was measured.

3.2. Coating Microstructure and Properties

The typical microstructure of the as-sprayed HVOF NiAl IMC coating shows the distinctive 'splats' or lamellae associated with this thermal spray process (Fig. 3). The overall coating microstructure has good homogeneity and uniformity, with some closed porosity along lamella boundaries, e.g. at A. Measured coating properties are given in Table 2 [10].

3.3. Thermophysical Properties

The CTEs of the NiAl IMC coating and substrate (Fig. 4) were measured between 323–773K. The CTE of the coating was measured as $14.7 \times 10^{-6}K^{-1}$ compared to $14.1 \times 10^{-6}K^{-1}$ for the substrate. It is immediately apparent that both the coating and substrate have a similar CTE over the temperature range measured. However, CTE measurements above 773K could not be performed due to the limitations of the

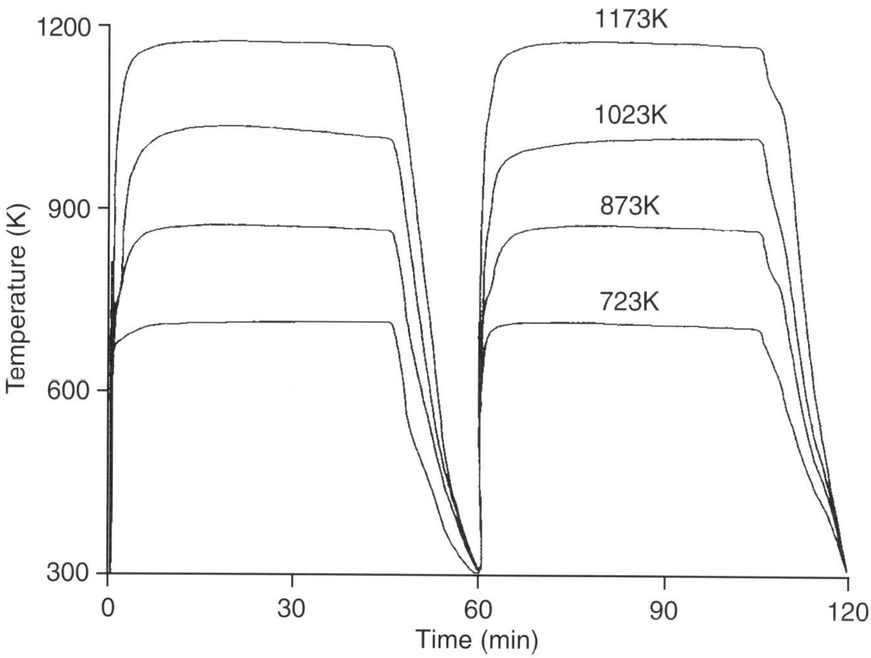

Fig. 2 *Coupon's surface temperature during thermal cycling.*

thermomechanical analyser and, from Fig. 4, the trend does suggest that the difference in CTE between the coating and substrate may increase with temperature.

Values measured for the coating compared favourably to the reported CTE value for bulk NiAl which has been measured as 15.1 and $10^{-6}K^{-1}$ between 550–1290K [11]. At lower temperatures values have been reported to range from $7.8 \times 10^{-6}K^{-1}$ to $12.6 \times 10^{-6}K^{-1}$ between 120 and 340K [12]. The texturing of the lamellae may also induce some anisotropic effects and thus deviations from linearity may affect the CTE of a thermally sprayed coating [13]. Other processes, which may influence the thermal shock characteristics, include grain growth, crack propagation, diffusion of substrate elements and oxidation.

3.4. Thermal Cycling

During thermal cycling there are two important phenomena that must be considered: growth of the protective oxide scale and its possible spallation, and the detachment

Table 2. *HVOF NiAl IMC coating properties*

Porosity (vol.%)	2.6 + 0.3
Oxygen content (mass%)	0.8 ± 0.03
Young's modulus (GPa)	249 ± 12.5
Hardness (HV)	398 ± 23.0

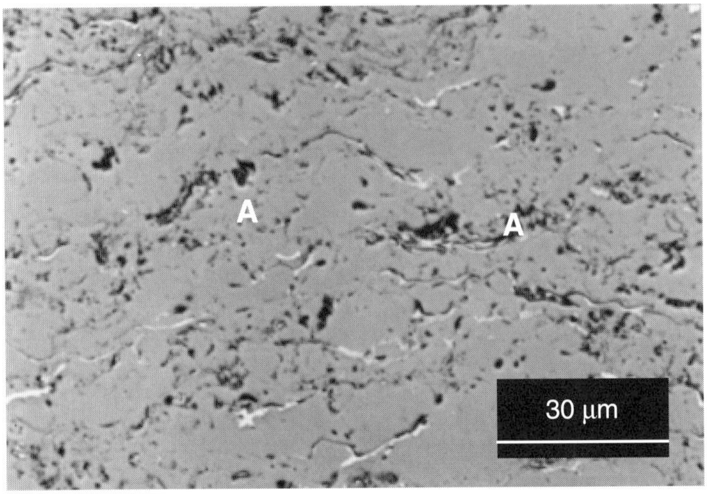

Fig. 3 *Microstructure of HVOF NiAl IMC coating.*

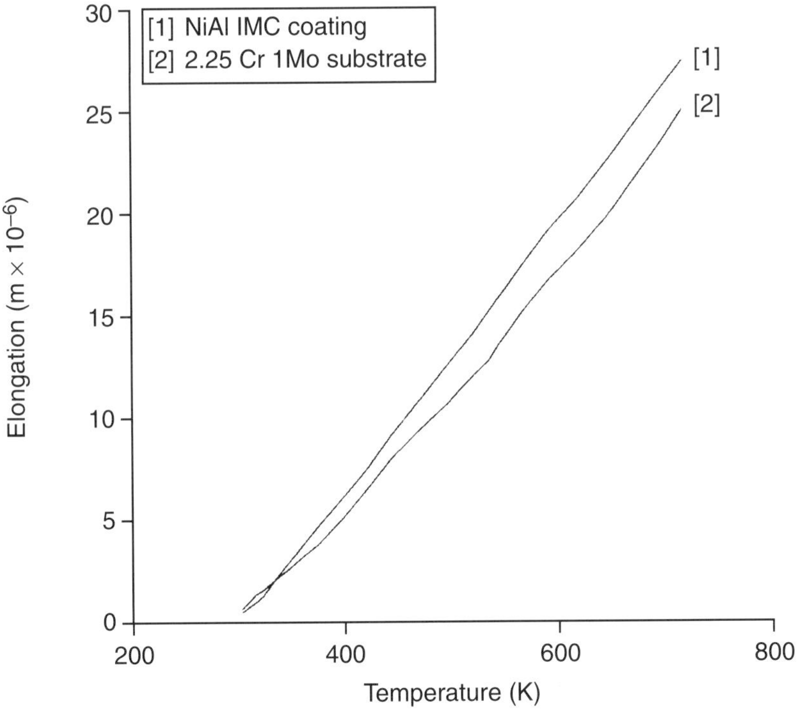

Fig. 4 *HVOF NiAl IMC coatings measured CTE.*

of the bulk coating from the substrate. Spallation arises due to stress generation either as growth stresses, which develop during the isothermal formation of the oxide scale, or thermal stresses, which arise due to differential thermal expansion or contraction between the substrate-coating and coating–oxide scale.

The mass change, revealed that those coupons cycled between 723 and 1023K performed extremely well with no cracking or evidence of coating loss by spallation (Fig. 5). However, those cycled at 1173K, failed catastrophically after only 50 cycles as can be seen in Fig. 6. Cycling at 1173K gave a greater initial mass gain compared to those coupons cycled at the lower temperatures. This was attributed to the initial oxidation of the substrate as a result of severe coating cracking after only 6 cycles. This initial large mass gain was followed by a mass loss due to coating spallation. A further mass gain was recorded due to the rapid oxidation of the newly exposed steel substrate before there was catastrophic coating spallation giving a large mass loss after which the run was stopped. This comprehensive coating failure was attributed to two possible effects. Firstly, thermal stresses induced within these coupons as a result of the cycles large temperature drop and, secondly, the phase transformation in the 2.25 Cr–1Mo steel substrate from ferritic to austenitic causing a substrate expansion during heating and cooling [14]. The thermal stress in the coating can be calculated using the following expression [15].

$$\sigma_{coating} = \frac{E_{coating}\Delta T(\alpha_{coating} - \alpha_{metal})}{1+2\left(\frac{E_{coating}}{E_{metal}}\frac{t_{coating}}{t_{metal}}\right)} \tag{1}$$

where E is the Young's modulus, α is the CTE and t is the thickness of the coating.

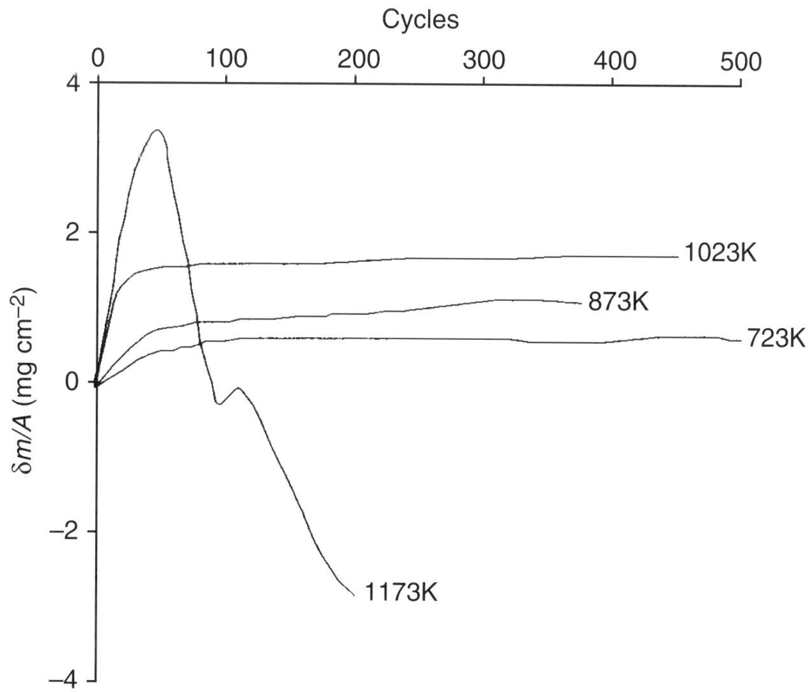

Fig. 5 *Mass change of coupons under thermal cycling.*

Fig. 6 Coupon cycled at 1173K.

The relationship between $\sigma_{coating}$ and ΔT, shows a linear dependence where a decrease in $\sigma_{coating}$ is observed for a smaller ΔT (Fig. 7a). Similarly, $\sigma_{coating}$ decreases with coating thickness (Fig. 7b). For those coupons cycled at 1173K, $\sigma_{coating}$ was calculated as 114 MPa for a temperature drop, ΔT, of 850K.

A numerical model was also used to predict the change in average residual stress within the coating for each complete cycle [16] for those coupons cycled at 1173K (Fig. 8). Initially, the coating is in slight tension due to quenching stresses originating from the spraying process. The stresses within the coating then change markedly from tensile to compressive as the heating cycle proceeds, since the CTE of the coating is slightly larger than that of the substrate. These compressive stresses reach a plateau of 160 MPa before changing once more to tensile during the cooling cycle. These tensile stresses are considerably larger than those arising from the spraying process indicating that there has been little stress relaxation on cooling. However, in practice subsequent cycles will result in these stresses being dissipated by cracking leading to coating spallation.

3.5. Characterisation

Characterisation of the coating which had remained attached to the coupons cycled at 1173K (Fig. 9) revealed a top-surface of small cubic crystallites, e.g. at B. EDS determined their composition to be of Ni, Fe, and O. Platelet crystallites were also observed e.g. at C and were found to be composed of Al and O.

X-ray diffraction (XRD) identified α-alumina, nickel oxide (NiO) and haematite (Fe_2O_3) phases to be present (Fig. 10), which showed that iron cations had either diffused through the coating from the substrate or passed through the cracks of the damaged coating. The spalled coating from these coupons revealed a top-surface of small crystallites rich in Al and O (Fig. 11). XRD identified α-alumina, NiO and Fe_2O_3 phases to be present (Fig. 12). No transient aluminas were identified.

(a)

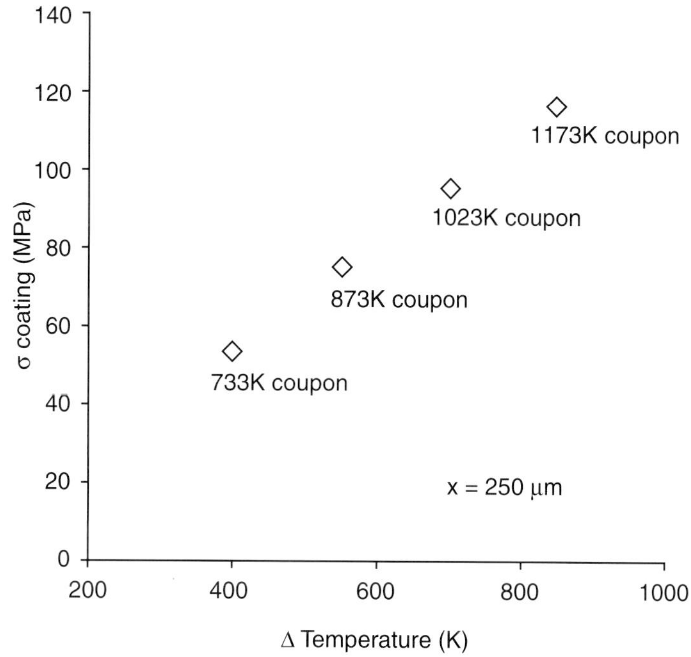

(b)

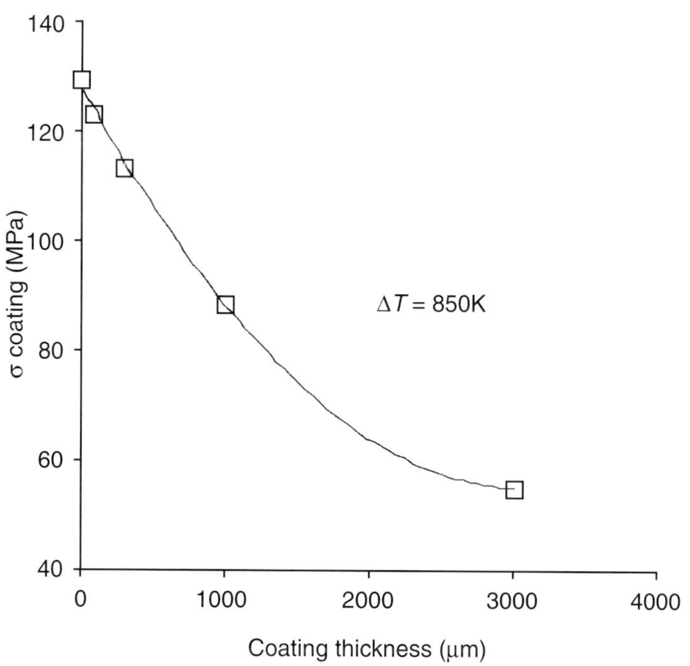

***Fig. 7** (a) Relationship between calculated $\sigma_{coating}$ and temperature change ΔT. (b) Relationship between calculated $\sigma_{coating}$ and coating thickness.*

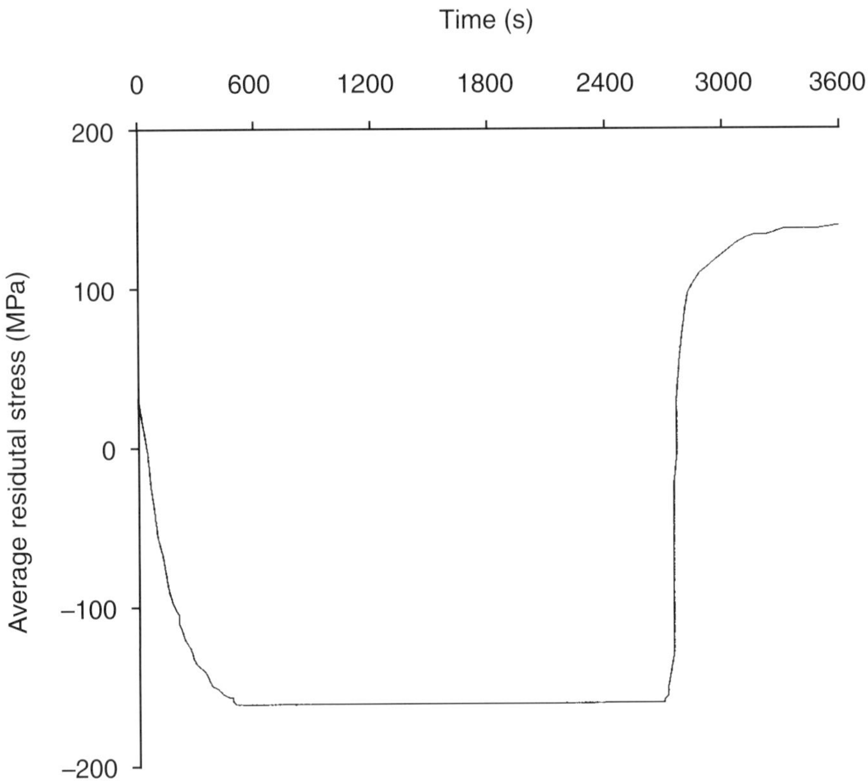

Fig. 8 Average calculated residual stress in coating cycled at 1173K, 45mm heating, 15 min cooling.

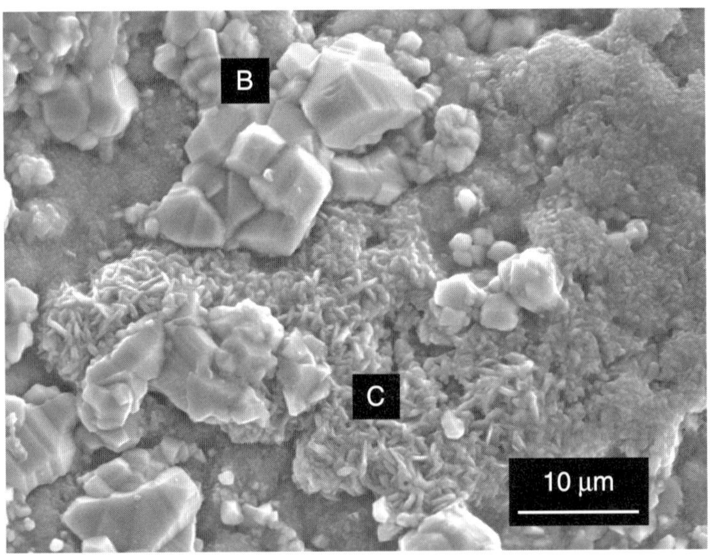

Fig. 9 Morphology of attached coating cycled at 1173K.

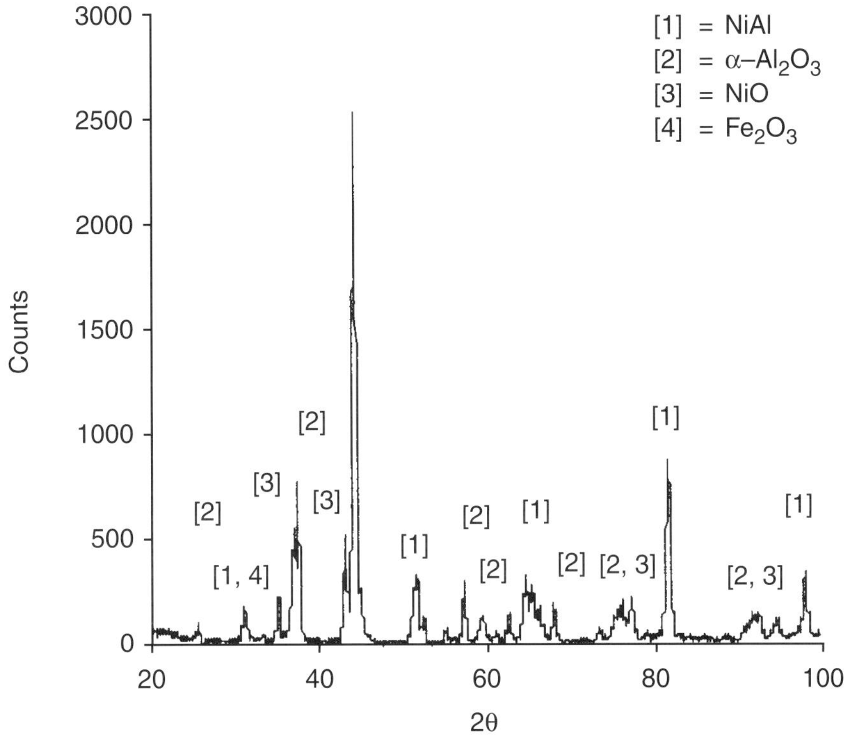

Fig. 10 *XRD pattern of attached coating cycled at 1173K.*

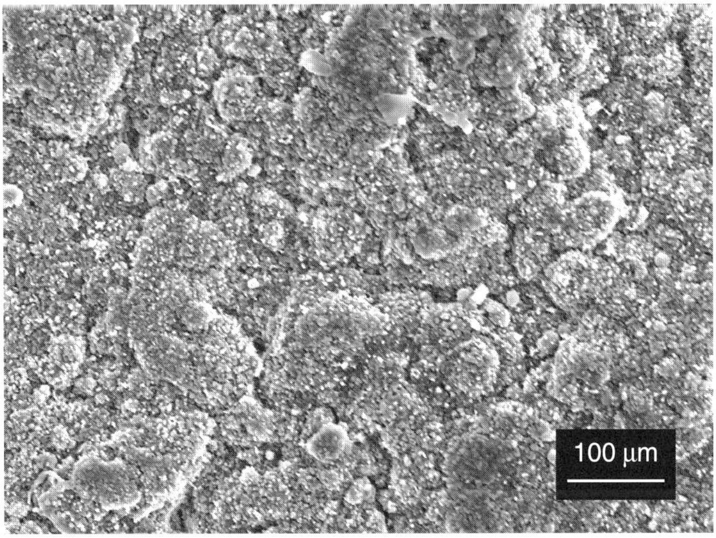

Fig. 11 *Morphology of spalled coating cycled at 1173K.*

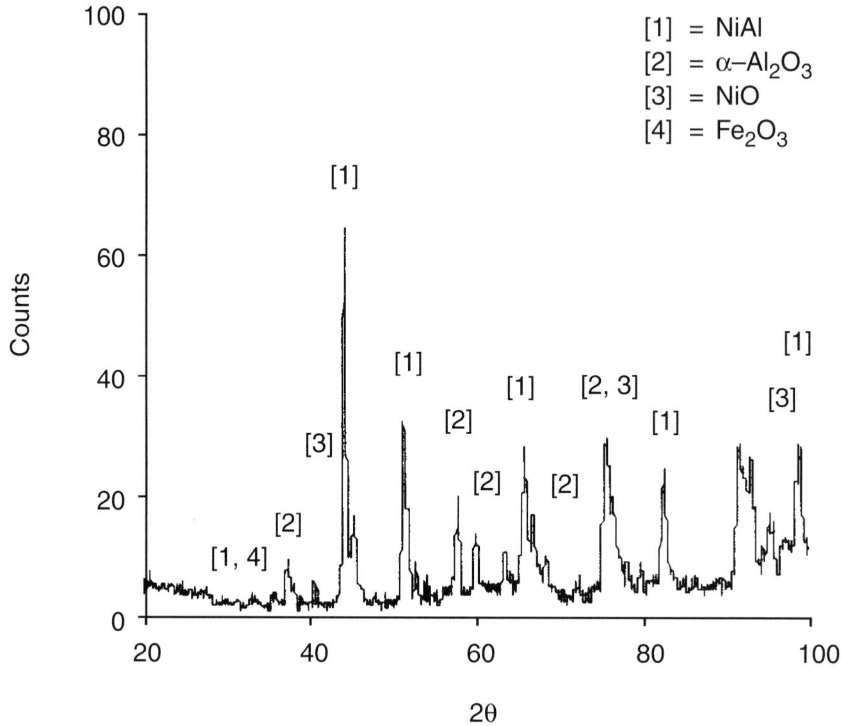

Fig. 12 XRD pattern of spalled coating cycled at 1173K.

Figure 13 is a microstructural examination of the original substrate-coating interface and its subsequent change during thermal cycling. All coatings remained attached to their substrates with the exception of those cycled at 1173K. Coupons cycled at 723–873K showed similar coating–substrate interfaces. However, small voids at this interface, e.g. at D were observed in the coupon cycled at 1023K and can be seen in greater detail in Fig. 13 (e). These voids may be formed by the coalescence of vacancies at the coating–substrate interface after prolonged iron cation diffusion through the coating. This iron cation diffusion has been confirmed using isothermal experiments between 723 and 1173K [14].

Firstly, the thermal stress generated within the coating as a result of the cycles large temperature drop and secondly, the phase transformation in the 2.25 Cr–1Mo steel substrate from ferritic to austernitic causing a substrate expansion during heating and cooling. The increased porosity visable at the substrate–coating interface may have also contributed to the ease of spallation.

4. Conclusions

1. NiAl IMC HVOF coatings exhibited excellent performance between 723 and 1023K with no evidence of spallation or de-bonding from the substrate.

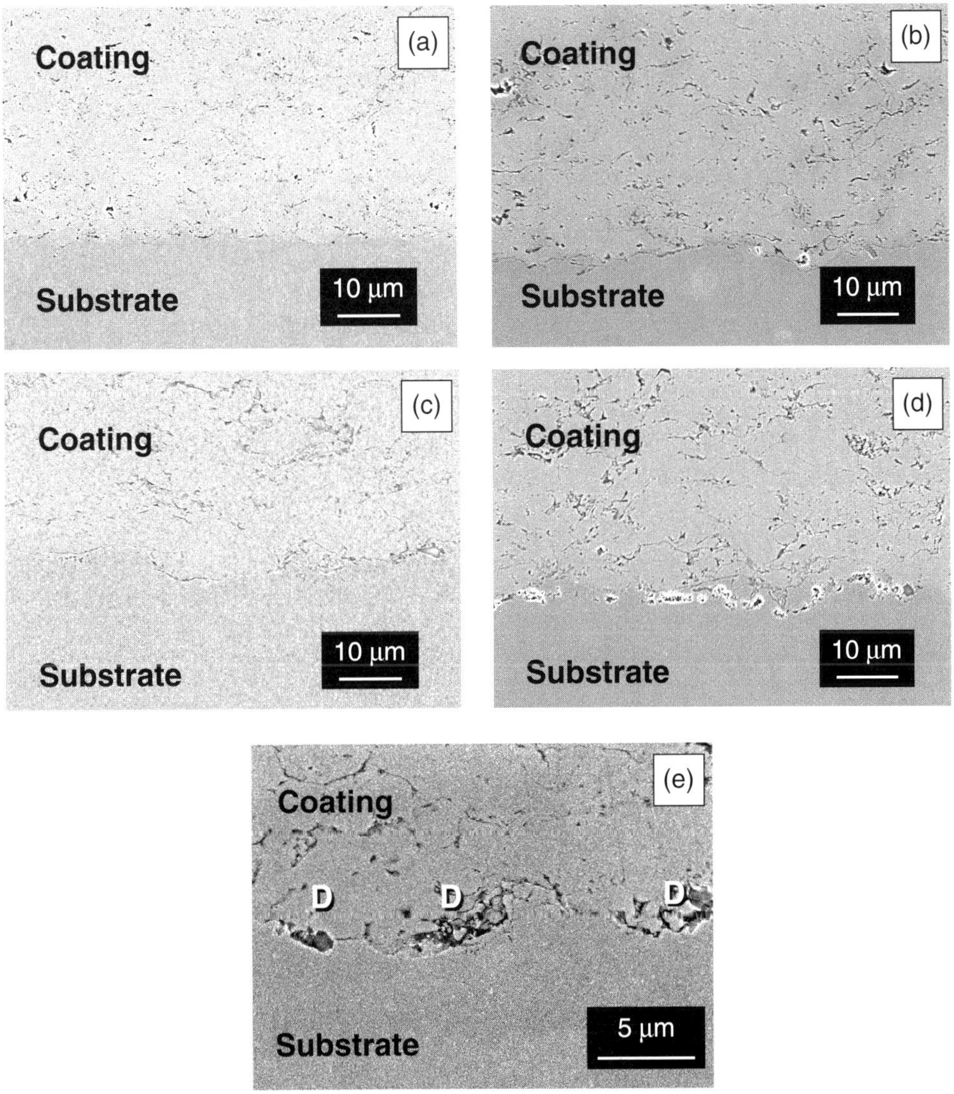

Fig. 13 *Coating–substrate interfaces of cycled coupons. (a) Cross-section of as-sprayed coupon, (b) cross-section of coupon cycled at 723K, (c) cross-section of coupon cycled at 873K, (d) cross-section of coupon cycled at 1023K, and (e) voids at coating–substrate interface, cycled at 1023K.*

2. Thermal cycling at 1173K caused severe coating spallation. This may be attributed to two possible effects. Both eqn (1) and the numerical model showed good agreement in determining the coating stress level.

3. The large initial mass gain seen in the coupons cycled at 1173K was attributed to the initial oxidation of the substrate as a result of severe coating cracking before spallation.

4. Characterisation of the coupons cycled at 1173K revealed evidence of iron cation diffusion.

References

1. C. T. Liu, J. O. Stiegler and F. H. Froes, *Ordered Intermetallics,* ASM Metals Handbook, 10th edn. 1990, pp.913–915.
2. E. A. Aitken, *Intermetallic Compounds* (J. H. Westbrook, ed). Chapter 25. John Wiley & Sons Ltd, 1967.
3. J. Doychak, *Intermetallic Compounds, Vol. 1, Principles* (J. H. Westbrook and R. L. Fleischer, eds). Chapter 43. John Wiley & Sons Ltd, 1994.
4. R. Mclntyre, *Mater. Process.,* 1996, 455–457.
5. B. Irving, R. Knight and R. W. Smith, *Weld. J.,* 1993, **72**, (7), 25–30.
6. D. W. Parker and G. L. Kutner, *Adv. Mater. Proc.,* 1994, **7**, 31–35.
7. J. H. Chen, Oxidation of Nickel-Base Superalloys for Turbine Disc and Advanced Turbine Blade Applications, 1996, Ph.D. Thesis, University of Cambridge, UK.
8. T. W. Clyne, *Key Eng. Mater.,* 1996, **116**, 307–330.
9. S. Kuroda, Y. Tashiro, H. Yumoto, S. Taira and H. Fukanuma, Peening action and residual stresses in HVOF thermal spraying of 316L stainless steel, *Proc. 15th Int. Thermal Spray Conf.,* Nice, France, 1998, 539–550.
10. J. A. Hearley, J. A. Little and A. J. Sturgeon, submitted to *Surf. Coat. Technol.*
11. D. B. Miracle, *Acta Metal.,* 1993, **41**, (3), 649–684.
12. R. D. Noebe, R. R. Bowman and M. V. Nathal, *Int. Mater. Rev.,* 1993, **38**, (4), 193–232.
13. J. Havsky and C. C. Berndt, *Surf. Coat. Technol.*, 1998, **102**, 19–24.
14. J. A. Hearley, High Temperature Corrosion of HVOF NiAL Intermetallic coatings, 1999, PhD. Thesis, University of Cambridge, UK.
15. G. D. Oxx, *Prod. Eng.,* 1958, **29**, (61).
16. Y. C. Tsui, Adhesion of Plasma Sprayed Coatings, 1996, Ph.D. Thesis, University of Cambridge, UK.

27
Development of Metal–Organic Chemical Vapour Deposition (MOCVD) Coatings for Protection of Alloys under High Temperature Cyclic Conditions

S. CHEVALIER, G. BONNET, C. VALOT, J. C. COLSON and J. P. LARPIN
Laboratoire de Recherche sur la Réactivité des Solides, UMR 5613 CNRS, 9 avenue Alain Savary, BP47870, 21078 Dijon cedex, France

ABSTRACT
The Metal–Organic Chemical Vapour (MOCVD) technique has been used to prepare Cr_2O_3 and Nd_2O_3 films on the F17Ti chromia-former alloy. Isothermal and cyclic oxidation tests at 1000 and 950°C respectively were performed on uncoated and coated steels under air at atmospheric pressure. The Cr_2O_3 coatings gave no beneficial effects during the isothermal tests, since the oxide growth rate was equal to that of the uncoated alloy. When a Nd_2O_3 film was applied, the oxidation rate and the total mass gain were decreased. Under cyclic conditions, Cr_2O_3-coated alloys spalled after 35 cycles. The scale adherence was improved compared to that on the uncoated specimen, which exhibited spallation after 3–4 cycles. A Nd_2O_3 film inhibited any spallation phenomenon, since no mass loss could be observed, even after 40 cycles.

The reactive element effect was studied by morphological analyses and residual stress determinations. The change of chromia grain morphology on coated alloys, compared to bare steels, is considered to greatly influence the mechanical properties of the oxide scale.

1. Introduction

High temperature conditions usually lead to failure of the protective chromia oxide scale and then to rapid degradation of chromia-former alloys. When spallation occurs under isothermal conditions, it reduces the thickness of the protective oxide layer and can increase the oxidation rate of the oxide-forming element. Under such circumstances, breakaway oxidation happens and promotes drastic reduction of the metallic substrate life. However, spallation is more often observed when alloys are submitted to thermal shocks or during cooling from the oxidation temperature to room temperature. Thermal cycling treatment is usually more severe than isothermal oxidation, because it causes spallation of the protective scale. Under such conditions, the formation of oxide cracks is generally followed by partial or complete decohesion of the protective oxide layer from the alloy. Thermal stresses, during the cooldown period of a thermal cycle, are assumed to cause the oxide scale fracture [1,2].

Many studies have pointed to an increase in the oxide scale adherence by introducing reactive elements (RE) such as rare-earth elements. They are able, on the one hand, to decrease the oxidation rate at high temperature and, on the other hand, to improve the oxide scale adherence to metallic substrates. The decrease of the oxidation rate is often ascribed to the inhibition of outward cation diffusion [3–7]. The blocking of cationic diffusion is able not only to annihilate vacancy condensation at the metal–oxide interface [8–10], but also to change the chromia scale mechanical properties by modifying the oxide grain morphology from large, columnar grains on RE-free alloys to small, equiaxed grains on RE containing steels [9,11–15]. Both phenomena may explain the improvement of the chromia scale adherence, since the absence of vacancies should increase the adhesion between the oxide scale and the metallic substrate, and the change in chromia grain morphology should improve the chromia scale mechanical properties.

The aim of this work was to try to increase the oxide scale adherence on a chromia-former alloy by applying Cr_2O_3 and Nd_2O_3 metal-organic chemical vapour deposition (MOCVD) coatings. The oxidation behaviour of uncoated and coated alloys was tested under isothermal and cyclic conditions. The chromia grain morphology was analysed and correlated with a previous study on chromia scale mechanical properties.

2. Experimental

A commercial chromia-former alloy, F17Ti (X3CrTi17, European designation), was oxidised under isothermal and cyclic conditions. It contains: C 0.015; Cr 17.19; Mn 0.39; Ni 0.12; Si 0.4; Ti; 0.48 mass%; balance iron. Samples were cut into $10 \times 10 \times 0.7$ mm^3 plates, polished down to 1μm with diamond paste and washed in water and alcohol. Coatins of Cr_2O_3 and Nd_2O_3 were prepared using the MOCVD technique. The deposition conditions and film characterisations have been discribed elsewhere [6,16,17]. The film thickness was 200 nm in all cases. Samples were oxidised at 1000°C in air at atmospheric pressure for 50 h. Isothermal oxidation kinetics were followed by thermogravimetry. Under cyclic conditions, specimens were tested at 950°C during 40 cycles, one cycle being equal to two hours at 950°C, followed by one hour at room temperature. They were continuously weighed in a thermobalance and spallation was noted as mass loss on the kinetic curves. Mass-gain curves only included the hot periods of the tests. Mass loss during cold periods appears as vertical drops in the kinetic curves.

Samples were analysed by SEM/EDX and X-ray diffraction after oxidation experiments.

3. Results

3.1. Isothermal Oxidation Tests

After 50 h at 1000°C, all mass gain curves exhibit a parabolic shape (Fig. 1). Parabolic rate constants were calculated by curve-fitting plots of mass gain vs. square root of time

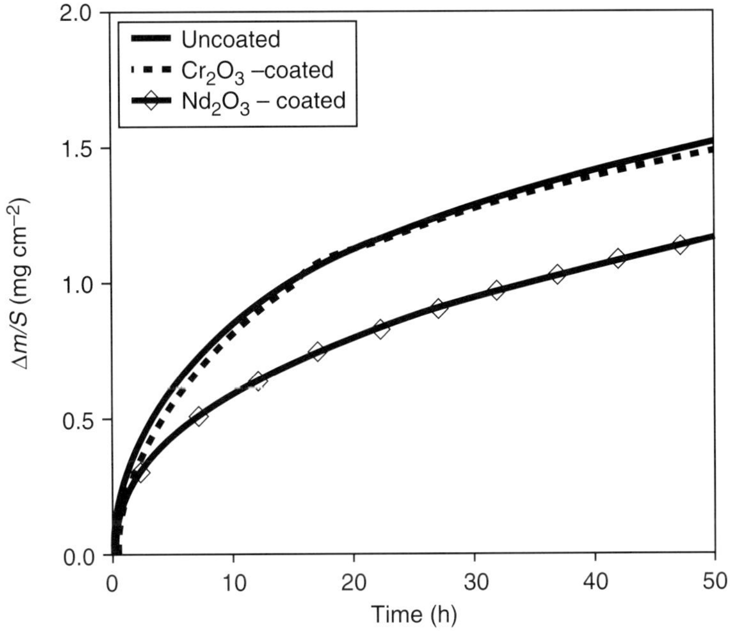

***Fig. 1** F17Ti isothermal-oxidation kinetic curves at 1000°C in air at atmospheric pressure.*

[18]. They are 1.9×10^{-11} $g^2cm^{-4}s^{-1}$ for uncoated alloys, 1.8×10^{-11} $g^2cm^{-4}s^{-1}$ for Cr_2O_3-coated alloys and 6.5×10^{-12} $g^2cm^{-4}s^{-1}$ for Nd_2O_3-coated alloys. Oxidation experiements were conducted in the temperature range 700–1050°C. Activation energies were calculated from Arrhenius plots and were equal to 223 $kJmol^{-1}$ in the case of bare steel and 188 $kJmol^{-1}$ when Nd_2O_3 films were applied to the F17Ti surface.

The Cr_2O_3 films did not appear to be effective in improving the oxide growth rate, since the parabolic rare constant and the mass gain are close to uncoated steel; Nd_2O_3 coatings decreased the oxidation rate and reduced the mass gain during the 50 h oxidation tests.

Spalled areas are detected on uncoated and Cr_2O_3-coated alloys after cooling to room temperature indicating poor adherence of the oxide scale whereas no cracks and no spallation are observed on the RE containing alloys.

Scanning electron microscope (SEM) and X-ray diffraction analyses indicate that the oxide scales are composed of duplex layers: $Mn_{1.5}Cr_{1.5}O_4$, TiO_2 mixed oxide in the outer part and Cr_2O_3 in the internal part of the scale for RE free alloys. On Nd_2O_3 coated-samples, $Mn_{1.5}Cr_{1.5}O_4$ is characterised in the outer part of the scale and Cr_2O_3 in the main part of the oxide layer. The Nd-containing phase is identified as $NdTi_{21}O_{38}$, segregated at the spinel phase grain-boundaries and distributed through the chromia scale.

3.2. Cyclic Oxidation Tests

Figure 2 shows the cyclic oxidation behaviour of uncoated and coated steels. The parabolic shape of the oxidation mass-gain curves is retained. Spallation occurs after

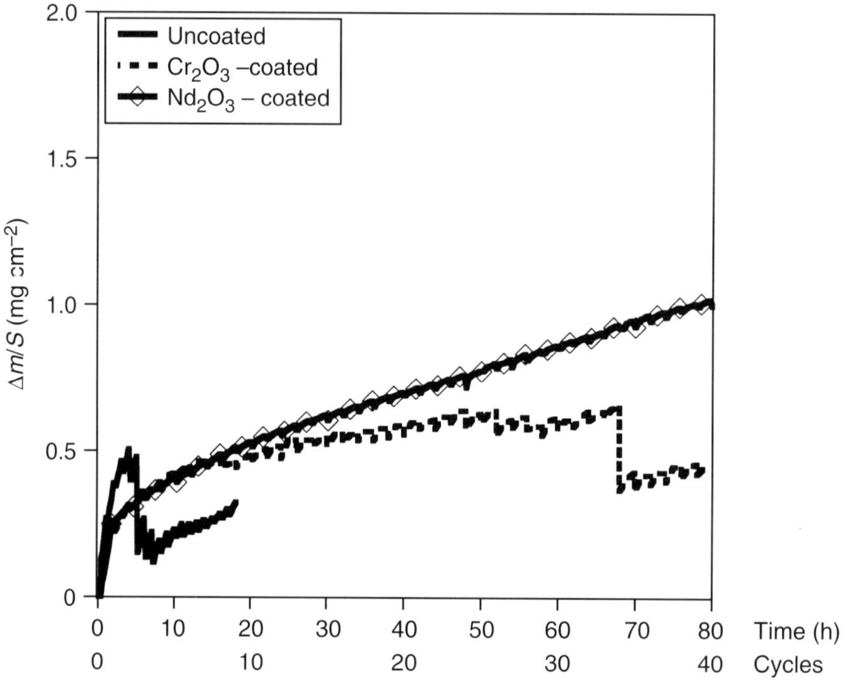

Fig. 2 F17Ti cyclic-oxidation kinetic curves at 950°C in air at atmospheric pressure.

5 cycles on uncoated samples, as indicated by mass loss on the kinetic curve. The surface morphology, shown in Fig. 3, consists of regions of unspalled oxide and exposed metal. Detachment takes place at the metal–oxide interface and not within the scale.

Cr_2O_3 coatings improve spallation resistance, since mass loss is observed after 35 cycles. Surface morphology is then similar to that of uncoated alloys. The Cr_2O_3 coatings retard oxide fracture but are not able to completely inhibit this detrimental effect.

The most beneficial effect is observed for Nd_2O_3-coated samples, since no spallation occurs during the 40 cycles. Longer experiments (100 cycles) have confirmed the beneficial effect of Nd_2O_3 since the oxide scale was still perfectly adherent to the metallic substrate in such conditions. The surface morphology (Fig. 4) shows small crystallites covering all the surface. No spalled areas are detected, indicating a great improvement in spallation resistance compared to RE-free specimens.

In each case, the oxide scales are composed of the same products as in isothermal conditions.

4. Discussion

The stresses which develop in the oxide during thermal cycling include growth stresses, due to change of volume, and thermal stresses, due to the differences of expansion coefficients between oxides and substrates [1]. Huntz *et al.* [19,20] have established that the growth stress level was lower than the thermal stress. That is

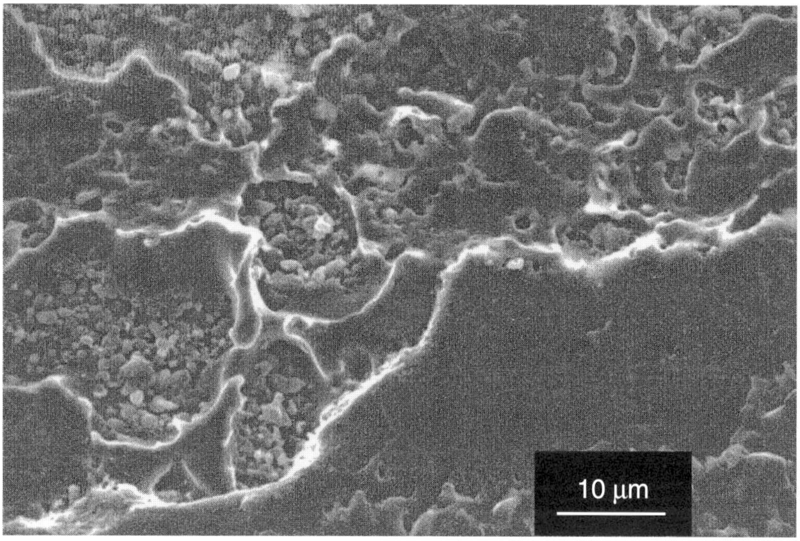

Fig. 3 Surface of F17Ti uncoated alloys after oxidation under cyclic conditions.

the reason why, residual stresses can be considered to correspond to thermal stresses. On uncoated alloys, the thermal stress is relaxed by spallation of the oxide scale. A previous study, on a Fe–30Cr alloy, showed that residual stress relaxation (Fig. 5) corresponded to the appearance of cracks through the chromia scale in the case of uncoated specimens [21]. On Nd_2O_3-coated samples stress relaxation was still observed, but without any scale detachment (Fig. 6). It was proposed that creep was the main method of stress release in that case. Fracture cross-sections indicated that the chromia scale was composed of small equiaxed grains on Nd_2O_3-coated alloys (Fig. 7), whereas they were large and columnar on RE free samples (Fig. 8). As creep

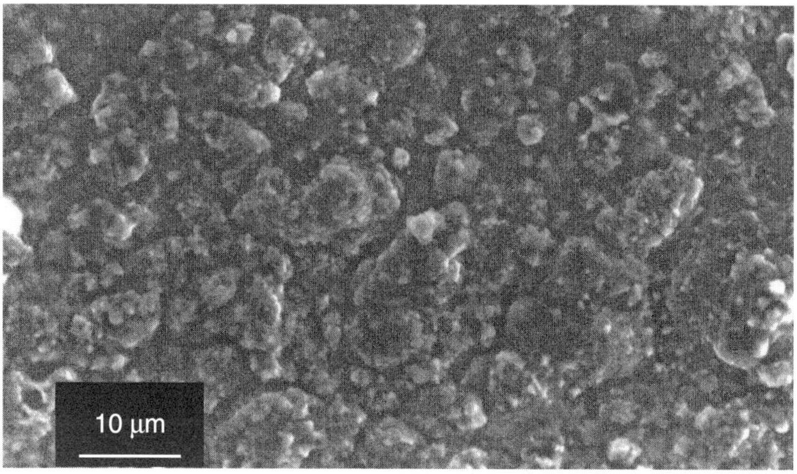

Fig. 4 Surface of F17Ti Nd_2O_3-coated alloys after oxidation under cyclic conditions.

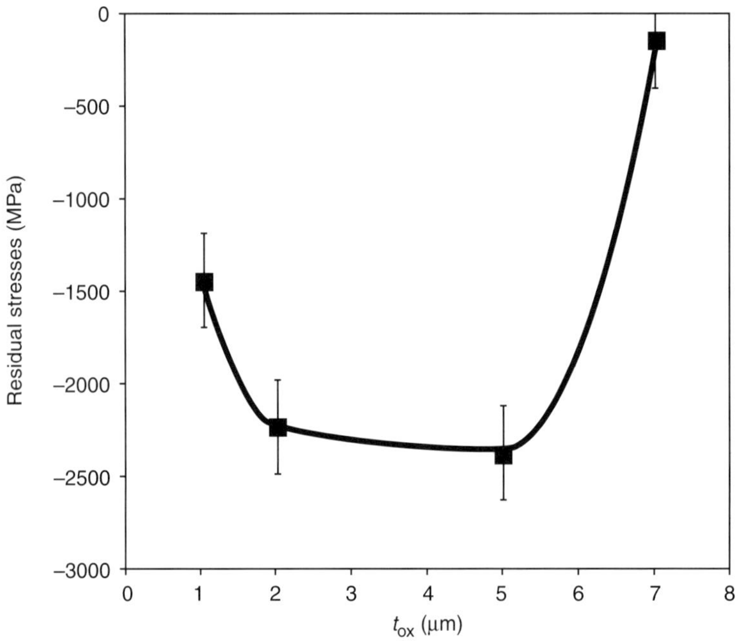

Fig. 5 *Evolution of residual stresses in chromia vs the thickness of oxide scale formed on uncoated Fe–30Cr alloys.*

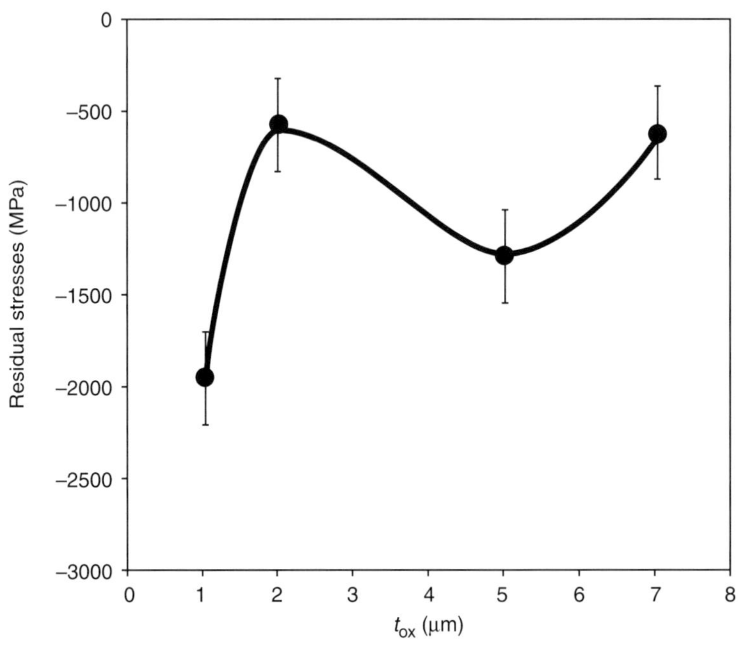

Fig. 6 *Evolution of residual stresses in chromia vs the thickness of oxide scale formed on* Nd_2O_3*-coated Fe–30Cr alloys.*

Fig. 7 *Fracture cross-section of the chromia scale formed on* Nd_2O_3*-coated Fe–30Cr alloy.*

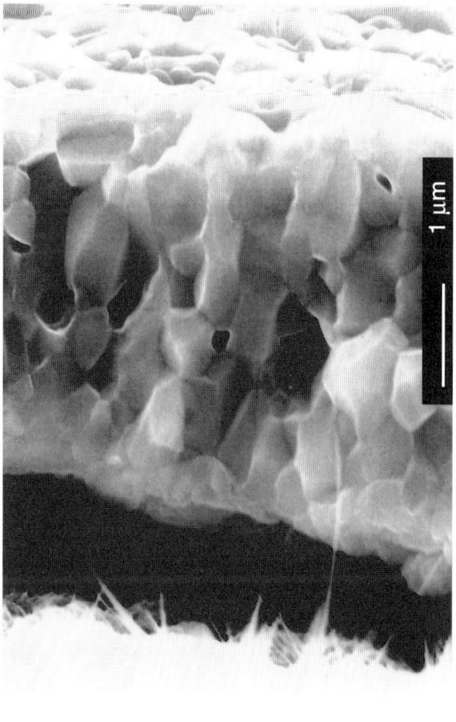

Fig. 8 *Fracture cross-section of the chromia scale formed on uncoated Fe–30Cr alloy.*

rate is promoted by small grain size, thermal stress is supposed to relax by plastic deformation, resulting in convoluted morphologies as noted by Jedlinski [22].

In the case of the F17Ti alloy, fracture cross-sections, realised on uncoated samples, show that the scale was composed of large columnar chromia grains (Fig. 9). Thermal stress is relaxed by spallation as in the Fe–30Cr case. On Nd_2O_3-coated alloys, oxide scale is composed of small, equiaxed grains, allowing stress relaxation by plastic deformation (Fig. 10). The mechanical properties of the oxide scale are then greatly improved by the presence of RE. The change of oxide grain morphology is probably due to an inversion of the chromia growth mechanism, which evolves from

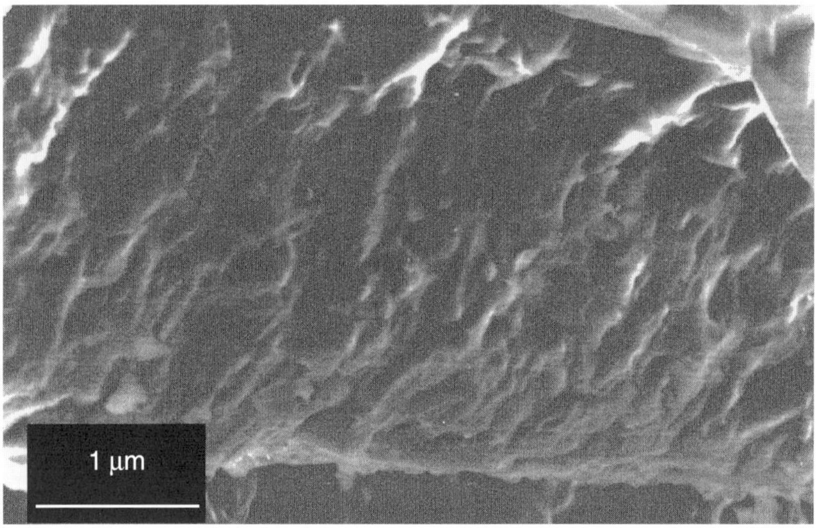

Fig. 9 *Fracture cross-section of F17Ti uncoated alloys after oxidation tests.*

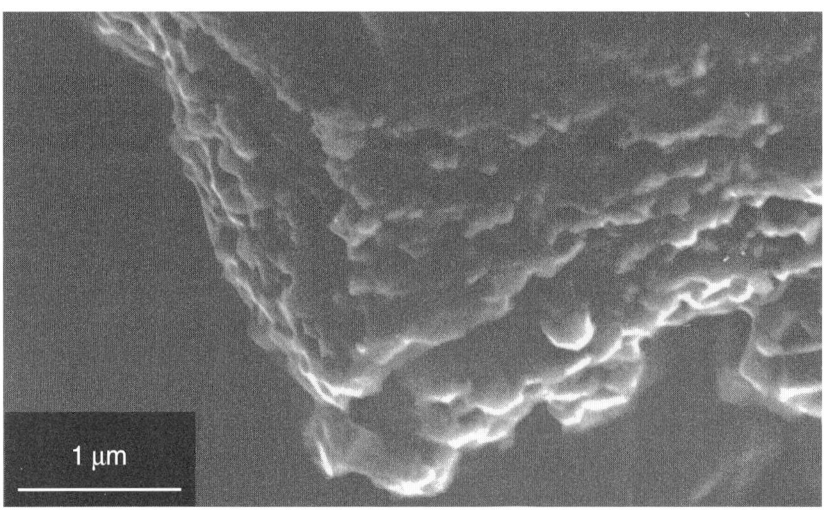

Fig. 10 *Fracture cross-section of F17Ti Nd_2O_3-coated alloys after oxidation tests.*

predominant outward cation diffusion on RE free alloys to inward anion diffusion on Nd_2O_3-coated steels [6].

Cr_2O_3-coatings appear to have little effect on the isothermal oxidation behaviour but retard spallation phenomena under cyclic conditions. One should ask for the reason for this improvement, since no modification of the chromia growth mechanism was observed in that case. Mitra *et al.* [23] proposed that Cr_2O_3 or Al_2O_3 fine particules, composing superficial coatings, could act as nucleation sites for oxide growth and then form a modified oxide layer. The chromia grains which grow during the first oxidation steps should be small and equiaxed. The presence of small equiaxed grains should allow plastic deformation in the chromia scale and, thus, thermal stress without any spallation. When the oxidation process continues, the outward cationic diffusion leads to growth of large and columnar grains. When this columnar structure reaches a critical thickness, thermal stress can no longer be relaxed by plastic deformation and cracks appear through the chromia scale leading to drastic damage which promotes the spallation phenomenon. This speculation is verified by the observation of fracture cross-sections on Cr_2O_3-coated alloys after the oxidation experiment (Fig. 11). A duplex chromia scale structure is observed: small, equiaxed grains close to the oxide–metal interface and large, columnar grains in the external part of the oxide scale. This morphology is in agreement with the proposed model for stress relaxation in chromia scales.

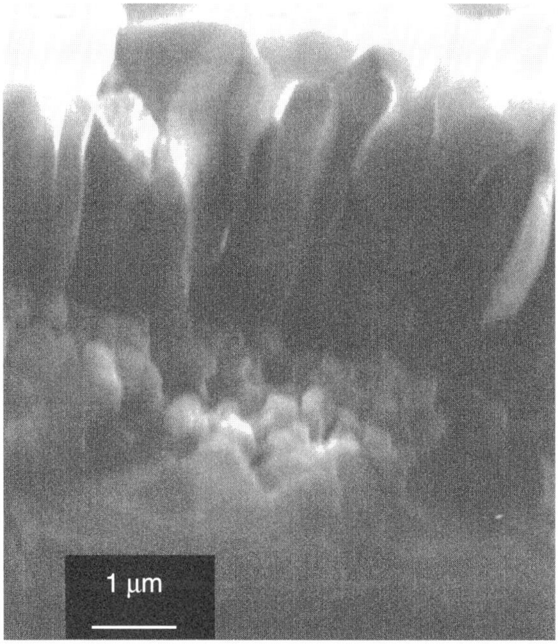

Fig. 11 *Fracture cross-section of F17Ti Cr_2O_3-coated alloys after oxidation tests.*

5. Summary

Once again, the introduction of a RE into a chromia-former alloy has decreased the high temperature oxidation rate and improved oxide scale adherence. This beneficial effect is well marked in the case of cyclic oxidation tests, since spallation which usually occurs on stainless steels is completely inhibited when a Nd_2O_3 film is applied on the alloy surface. It is suggested that changes in chromia grain morphology promote plastic deformation in the oxide scale allowing stress relaxation without any spallation. On Cr_2O_3-coated alloys, oxide scale fracture is retarded because of the duplex structure of the chromia scale; that is, with small equiaxed grains at the internal surface and large columnar grains in the outer part of the oxide scale. The small oxide particules making up the chromia coating are presumed to act as nucleation sites for chromia and promote growth of equiaxed grains. The outward cationie diffusion allows columnar chromia grains to grow above this initial structure. When this second layer reaches a critical thickness, spallation occurs as a result of the lack of plasticity of the outer part of the oxide scale.

References

1. H. E. Evans, *Int. Mater. Rev.*, 1995, **40**, (1), 1–40.
2. K.W. Chan, *Metal. Mater. Trans.*, 1997, **28A**, 411–422.
3. K. Przybylski, A. J. Garratt-Reed and G. J. Yurek, *J. Electrochem. Soc.: Solid-State Sci. Technol.*, 1988, **135**, 2, 509–517.
4. P. Papaiacovou, R. J. Hussey, D. F. Mitchell and M. J. Graham, *Corros. Sci.*, 1990, **30**, (4/5), 451–460.
5. C. M. Cotell, G. J. Yurek, R. J. Hussey, D. F. Mitchell and M. J. Graham, *Oxid. Met.*, 1990, **34**, 3/4, 173–199.
6. S. Chevalier, G. Bonnet, J. P. Larpin and J. C. Colson, *Oxid. Met.*, 1997, **47**, (1/2), 53–67.
7. D. Zhu, J. H. Stout, J. C. Nelson and D. A. Shores, *Mater. Sci. Forum*, 1997, **251–254**, 437–444.
8. G. M. Ecer, R. B. Singh and G. H. Meier, *Oxid. Met.*, 1982, **18**, (1/2), 55–81.
9. K. Przybylski and G. J. Yurek, *Mater. Sci. Forum*, 1989, **43**, 1–74.
10. A. Strawbridge and P. Y. Hou, *Mater. High Temp.*, 1994, **12**, (2–3), 177–181.
11. P. Y. Hou and J. Stringer, *Oxid. Met.*, 1988, **29**, (1/2), 45–73.
12. S. K. Mitra, S. K. Roy and S. K. Bose, *Oxid. Met.*, 1990, **34**, (1/2), 101–121.
13. S. Seal, S. K. Bose and S. K. Roy, *Oxid. Met.*, 1994, **41**, (1/2), 139–178.
14. S. Roure, F. Czerwinski and A. Petric, *Oxid. Met.*, 1994, **42**, (1/2), 75–102.
15. S. Chevalier, P. Dufour, G. Bonnet and J. C. Colson, *Oxid. Met.*, 1998, **50**, (1/2), 27–49.
16. M. Silva, S. Chevalier, G. Bonnet and J. C. Colson, *Mater. Sci. Forum*, 1997, **251–254**, 419–428.
17. S. Chevalier, G. Bonnet, J. P. Larpin and J. C. Colson, to be published.
18. B. Pieraggi, *Oxid. Met.*, 1987, **27**, (3/4), 177–185.
19. A. M. Huntz and M. Schütze, *Mater. High Temp.*, 1994, **12**, (2–3), 151–161.
20. A.M. Huntz, *Mater. Sci. Eng.*, 1995, **A201**, 211–228.
21. S. Chevalier, C. Valot, G. Bonnet, J. C. Colson and J. P. Larpin, to be published.
22. J. Jedlinski, M. Krasovec, G. Borchardt and J. W. Quadakkers, *Journal de Physique IV*, Colloque C9, supplément au Journal de Physique III, 1993, **3**, 591–597.
23. S. K. Mitra, S. K. Roy and S. K. Bose, *Oxid. Met.*, 1992, **37**, (1/2), 95–108.

28
Mullite Based Oxidation Protection for Silicon Carbide–Carbon/Carbon Composites in Air at Temperatures up to 1900K

H. FRITZE, A. SCHNITTKER, G. BORCHARDT, T. WITKE*
and B. SCHULTRICH*
Technische Universität Clausthal, Germany
*Fraunhofer-Institut für Werkstoff- und Strahltechnik, Dresden

ABSTRACT

For an industrial Si–SiC coated C/C material (reference material) the temperature dependence of the isothermal linear mass loss rate is interpreted in the temperature range 500°C < T < 1700°C. The Arrhenius plot of the thermogravimetrically determined oxidation rate shows four typical regimes, which can be individually modelled. Only in the temperature range 1050°C < T < 1550°C is the oxidation rate close to or even lower than the limit for long-term application.

In a second step a mullite coating was deposited by Pulsed Laser Deposition (PLD). As-ablated green layers do not yet show IR peaks typical for mullite. After a short oxidation treatment (15 min at 1400°C) the formation of mullite in the coating is completed as was confirmed by IR spectroscopy and XRD investigations.

PLD mullite coatings with a thickness of 2.5 μm on preoxidised samples improved the isothermal oxidation behaviour significantly. Because of SiO_2 formation at the mullite–SiC interface all samples exhibited a mass increase on oxidation. The inward diffusion of oxygen across the outer mullite-containing layer controlled the kinetics of the reaction as was deduced from ^{18}O diffusivity measurements in PLD mullite layers by secondary neutral mass spectrometry (SNMS). The calculated oxidation rates resulting from the diffusion parameters in SiO_2 and mullite are close to the thermogravimetric data. For oxidation durations of three days only amorphous SiO_2 is formed at the mullite–SiC interface.

In a further investigation, the material was tested under thermal cycling conditions between room temperature and 1100°C < T < 1600°C. An oxidation protection effect was observed which can be modelled on the basis of the isothermal oxidation rate and the cyclic testing conditions.

1. Introduction and Objectives

Because of their excellent mechanical properties at high temperatures carbon/carbon composites (C/C) are very attractive for structural applications. Without oxidation protection, however, carbon materials do not withstand oxidising atmospheres [1].

Depending on the duration of application the acceptable carbon loss determines the maximum oxidation rate which can be tolerated. As suggested by Luthra [2], we consider as acceptable a reduction of the wall thickness of a typical structural component by 0.3 mm. This criterion results in a maximum loss rate of 3×10^{-2} mg cm^{-2} h^{-1} for long-term (2000 h) applications (Fig. 1).

For these conditions, our thermogravimetric measurements yield a temperature limit of 450°C for long-term application of unprotected C/C materials in air. In order to improve the oxidation resistance at higher temperatures suitable oxygen diffusion barriers had to be developed.

Mullite ceramics ($3Al_2O_3 \cdot 2SiO_2$) are promising candidate materials for high temperature applications. Mullite coated SiC exhibits excellent oxidation resistance in dry air by forming a slowly growing SiO_2 scale at the mullite/SiC interface [3]. Depending on the SiO_2 content, mullite forms a liquid phase which can close cracks and pores [4].

Our work concentrated on the effect of protective outer mullite layers produced by Pulsed Laser Deposition (PLD). Based on our experimental data we discuss the oxidation behaviour in the high temperature range (1200°C ≤ T ≤ 1600°C) in air.

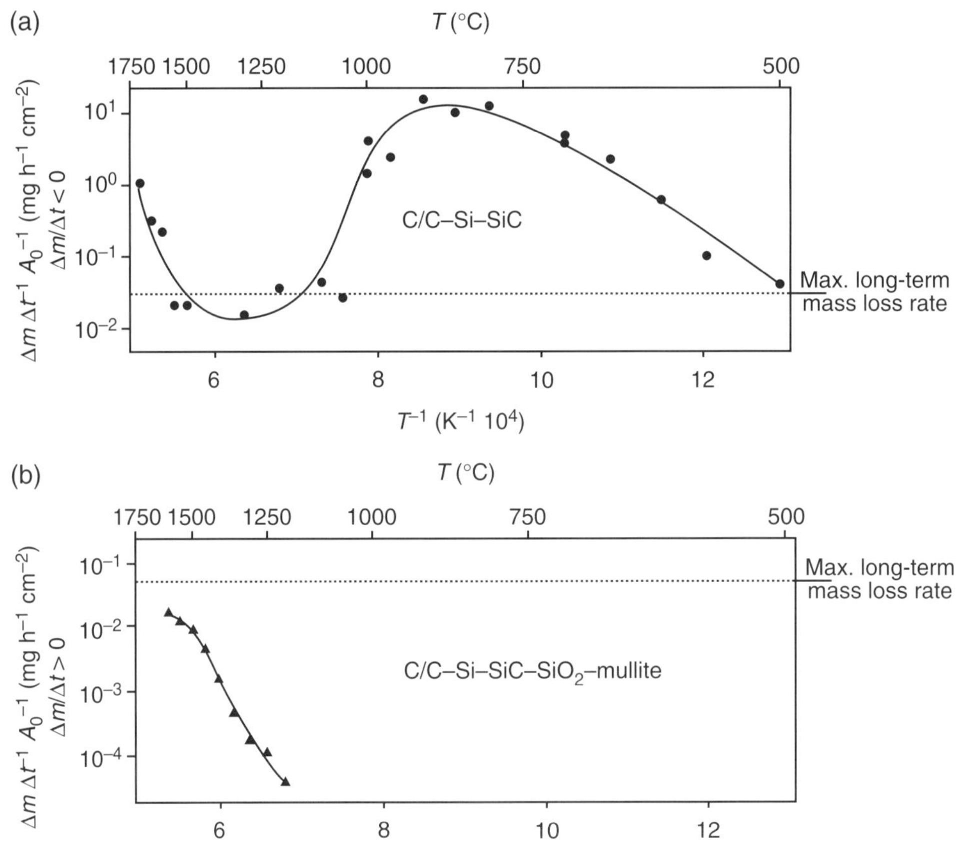

Fig. 1 *Arrhenius plot of the mass change rate* lg $|\Delta m\, \Delta t^{-1} A_0^{-1}| = f(T^{-1})$*: (a) reference material* ($\Delta m\, \Delta t^{-1} < 0$)*; (b) mullite coated material, high temperature range* ($\Delta m\, \Delta t^{-1} > 0$)*.*

2. Results and Discussion

2.1. Limited Oxidation Resistance of the Reference Material

The 2D-C/C samples were thin slabs with average dimensions $19 \times 19 \times 2$ mm^3. The material was infiltrated with liquid silicon under pressure. Subsequently, three CVD (chemical vapour deposition) silicon carbide layers were deposited with a total thickness of 50 µm. On the surfaces we observed a two-dimensional network of cracks with an average spacing of 0.4 mm. Because of the thermal coefficient of expansion mismatch between C/C and silicon carbide this network structure will always exist after cooling from the CVD processing temperature. The cracks extend to the C/C surface with an average width of 2 µm at room temperature. Heating up to the CVD processing temperature closes the cracks again.

The mass change of the reference specimens was measured as a function of time using a thermobalance. Only in the temperature range 1050°C $< T <$ 1550°C was the mass loss close to, or even lower than, the limit for long-term application. At all temperatures the Si–SiC coated samples showed an overall mass loss (Fig. 1a). Additional mullite coatings were expected to reduce the mass loss or even to yield an increase of mass. Therefore, we produced PLD mullite layers with a maximum thickness of 2.5 µm. Consequently, the cracks in the SiC layer were not covered in the low temperature range and it was therefore, only in the high temperature range that an improvement in the oxidation behaviour was to be expected — as reported below.

2.2. Reduced Oxidation Rates of the Improved Material

2.2.1. Deposition of mullite coatings

Mullite was deposited on the preoxidised C/C–Si–SiC substrates (thickness of the SiO_2 layer: 0.22 µm). High energy impact CO_2 laser pulses (λ = 10.6 µm, Δt = 170 ns, $j = 3 \times 10^7$ Wcm^{-2}) led to melting and evaporation of the target material in a single step. Therefore the flux of the metal components was stoichiometric. As we worked under reduced pressure ($P_{tot} = 10^{-5}$ mbar (10^{-3} Pa)) the oxygen content decreased. For our targets (sintered mullite powder) the typical deposition rate was 100 nm s^{-1}. Even with target rotation deposition rates up to 500 nm s^{-1} can be reached only by using larger targets. The substrate temperature did not rise to more than 100°C.

2.2.2. Phase analysis

Reflection spectra between 650 and 1300 cm^{-1} of the PLD coatings on SiC heated at 1400°C for 15 min show typical mullite line shapes. The formation of mullite is confirmed also by XRD investigations.

After heating runs at rather high temperatures (see thermogravimetric experiments) the reflection spectra indicate the formation of amorphous SiO_2 Thermogravimetric measurements yield a mass gain in the whole temperature range, which indicate that SiO_2 is formed. At 1400°C SiO_2 is not observable by means of surface sensitive IR spectroscopy in the reflection mode. Consequently, SiO_2 is formed at the inner interface SiC–mullite. At higher temperatures the interdiffusion at the interface SiO_2–mullite becomes important.

2.2.3. ^{18}O diffusion

Simultaneously, measurements were made of oxygen diffusion (^{18}O) in the mullite layers. Here, the coatings were deposited on SiC slabs so as to exclude effects caused by the reaction of ^{18}O with the C/C matrix. All samples were analysed using a secondary neutral mass spectrometer (SNMS). The oxidation rates calculated for the improved material from the diffusion data corresponded well with the thermogravimetric data (Fig. 2).

2.2.4. Oxidation of mullite coated C/C–Si–SiC material under isothermal conditions

The improvement of the sample preparation, i.e. the deposition of mullite coatings with a thickness of 2.5 μm and the pre-oxidation treatment, led to a mass gain (Fig. 1b). As can be seen from the marked region in Fig. 3 mullite can seal defects at temperatures close to the eutectic temperature (1587°C [4]).

2.2.5. Oxidation of mullite coated C/C–Si–SiC material under thermal cycling

The mullite coated-material was also tested under thermal cycling conditions (nominal cooling/heating rate of ± 20K s^{-1}) (Fig. 4). Because of the small difference in thermal expansion of SiC and mullite spallation of the mullite layer did not occur. During the high temperature steps the mass gain was comparable to the isothermal oxidation behaviour. Cooling down to room temperature led to opening of the cracks in the SiC layer which ran through the thin mullite layer which was not able to close cracks with an average width of 2 μm at room temperature. After prolonged thermal

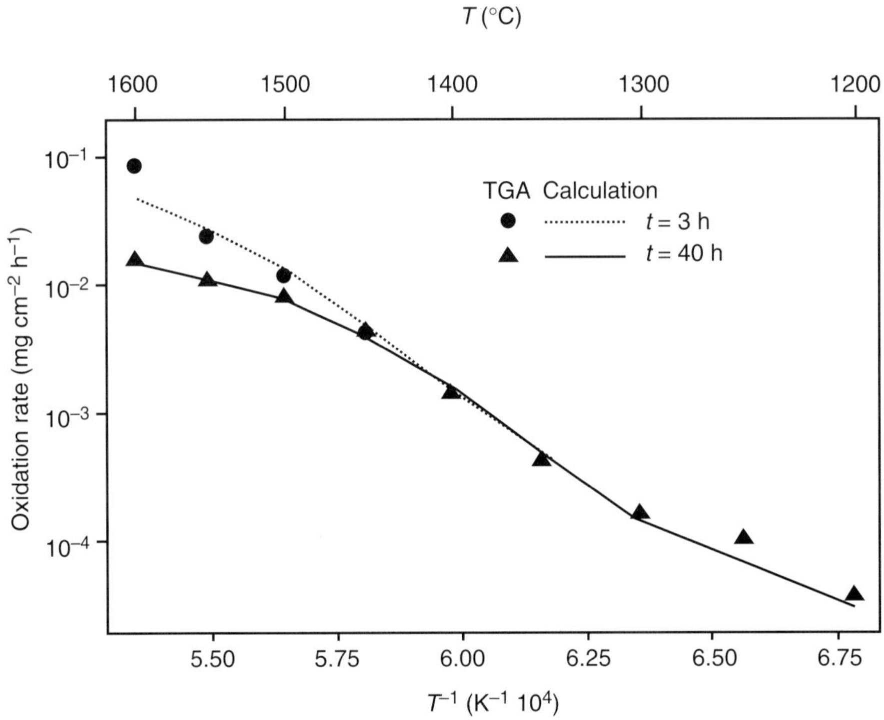

***Fig. 2** Calculated and thermogravimetrically (TGA) determined temperature dependence of the oxidation rate for pre-oxidised and mullite-coated C/C–Si–SiC material.*

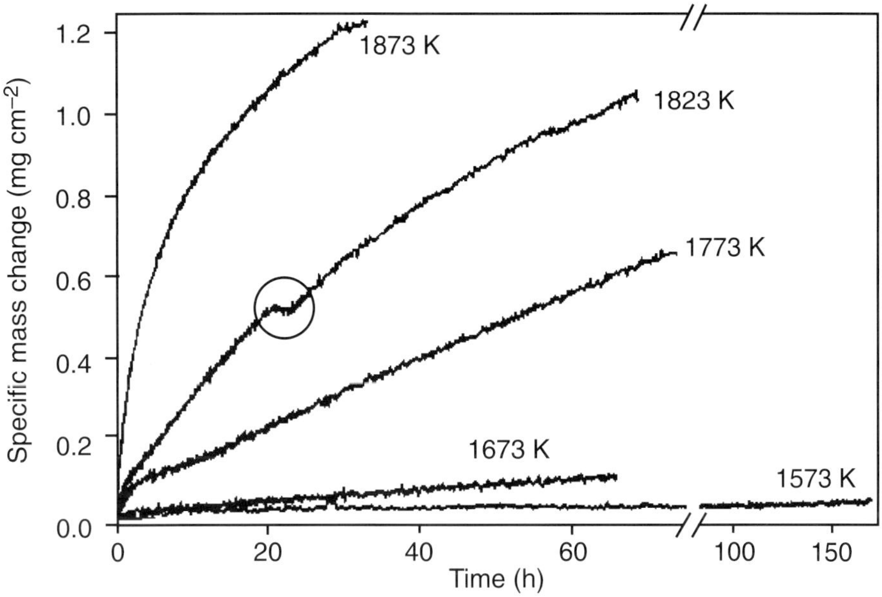

Fig. 3 *Specific mass change of pre-oxidised and mullite-coated C/C–Si–SiC material as a function of time.*

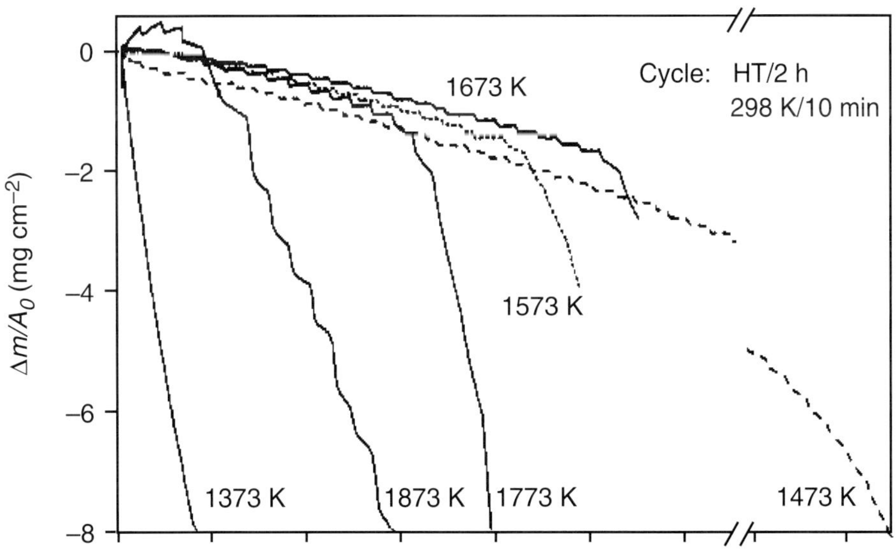

Fig. 4 *Specific mass change of pre-oxidised and mullite-coated C/C–Si–SiC matenal (mullite layer thickness 2.5 mm) under thermal cycling between room temperature (298K) and 1373K–1873K (HT).*

cycling we observed spallation of the SiC layer off the substrate material which was responsible for the ultimate breakdown of the oxidation resistance.

Both the isothermal and the cyclic oxidation behaviour could be quantitativly modelled using a phenomenological reaction model published elsewhere [5,6].

3. Conclusions

Mullite coatings with a thickness of 2.5 μm and a pre-oxidation treatment of the C/C–Si–SiC substrate material improve the oxidation behaviour of CVD–SiC coated C/C composites. As expected, all samples exhibit a mass increase. In the temperature range 1200°C < T < 1600°C the inward diffusion of oxygen across the outer mullite-containing layer controls the kinetics of the reaction, as was deduced from ^{18}O diffusivity measurements in PLD mullite layers. Under thermal cycling conditions the cracks in the SiC layer of the substrate material and the spallation of this layer (but not of the mullite layer!) are responsible for the ultimate breakdown of the oxidation resistance.

References

1. L. Luthra, R. N. Singh and M. K. Brun, Toughened silcomp composites — process and preliminary properties, *J. Am. Ceram. Soc.*, 1993, **72**, 79.
2. L. Luthra, Oxidation of carbon/carbon composites — A theoretical analysis, *Carbon,* 1988, **26**, 217.
3. S. Jacobson, Corrosion of silicon-based ceramics in combustion environments, *J. Am. Ceram. Soc.*, 1993, **76**, 3.
4. S. Aramaki and R. Roy, Revised phase diagram for the option Al_2O_3–SiO_2. *J. Am. Ceram. Soc.*, 1962, **45**, 229.
5. H. Fritze *et al.*, Mullite based oxidation protection for SiC-C/C composites in air at temperatures up to 1900K. *J. Eur. Ceram. Soc.*, 1998, **18**, 2351.
6. H. Fritze, Oxidation protection of C/C materials in the temperature range 1200°C to 1600°C based on laser-ablated mullite coatings (in German) Ph.D. thesis. TU Clausthal, 1996.

29
Cyclic Oxidation — Guidelines for Test Standardisation, Aimed at the Assessment of Service Behaviour

J. R. NICHOLLS and M. J. BENNETT*
Cranfield University, Cranfield, Bedford MK43 0AL, UK
*Materials Consultant, Three Chimneys, South Moreton, Oxford OX11 9AH, UK

ABSTRACT
This paper reviews current procedures used for cyclic oxidation testing and their ability to simulate service behaviour. Test complexity varies from a simple laboratory, slow cycle, furnace test to the use of high velocity dynamic combustion rigs. It is shown that the response of an alloy under cyclic oxidation testing varies, depending on the exact test conditions chosen and the length of test exposure. But for many alloy systems this response has common features. It is proposed that all alloys that are protected by a stable, slow growing oxide scale under isothermal conditions conform to a common, generic behaviour under cyclic oxidation conditions. This life cycle of such alloys involves first the formation of a stable, protective oxide. At a critical thickness this may start to spall increasing the rate of scale forming element consumption. When the activity of such elements falls below a critical level, in the near surface region, internal oxidation results together with the formation of less protective scales. Ultimately, it is no longer possible for the alloy to self repair following a thermal cycle and then breakaway corrosion ensues, marking the end of the life of the alloy.

Thus, the choice of cyclic test procedure must be tailored to the particular phase of the alloy life cycle under investigation.

1. Introduction and Background

Over the past two decades an increased emphasis has been placed on studying high-temperature alloy performance under cyclic oxidation conditions [1–15], in addition to the well accepted isothermal oxidation studies. This is particularly true for nickel- and cobalt-based alloys and coatings for use in gas turbine engines powering aircraft that operate on military or short haul civil duty cycles [1–3,5,10,12,13,16–19], and reflects that such duty cycles are likely to introduce many thermal cycles into the life of components. Table 1 summarises the typical duty cycles for various gas turbine engines (data taken from [19]).

The preferred testing methods are either based around a cyclic furnace test, of which the NASA thermal cycle test rig is an example [1–3,5], or based on a dynamic combustion rig, burning either aviation distillate, marine kerosene or natural gas as a fuel [20–25]. These burner rig tests may be of the 'high velocity' type, in the range Mach 0.3–1.0, or the 'low velocity' type with velocities in the range Mach 0.1–0.3 [24,25].

Table 1. *Typical duty cycle for various gas turbine engines [19]*

Gas turbine application		Duty cycle* (h)	
	Minimum	Mean	Maximum
Aero gas turbine/Civil	1	2	8
Aero gas tubine/Military	1	2	4
Industrial turbine/Base power generation	500	2000	8000
Industrial turbine/Peak power generation	2	4	12

* Duty cycle is defined as the typical mission time per engine start in hours.

The implications, advantages and disadvantages of these various test methods will be discussed as part of this paper, but broadly reflect the method of heating, the heating and cooling rates achieved, the maximum test temperature, the duty cycle at temperature and the test environment.

Before discussing the relative methods of the various test methods it is worth a brief review of what industry requires to aid materials selection and provide a basis for component life prediction. This need varies considerably with service application. For example, the majority of high temperature alloys used in industrial, high temperature, process or power plant are chromia formers, depending on the formation of a protective Cr_2O_3 scale for prolonged service life. However, the maximum service temperature for this class of material is limited, due to the propensity to form volatile CrO_3. Thus, chromia-formers are rarely used above 1000°C for prolonged high temperature service. At these more elevated temperatures alloys forming alumina scales are more attractive, due to the stability and slow rate of growth of the α-Al_2O_3 scale. For the arduous service conditions found within the turbine section of aero, industrial and marine turbines it is extremely difficult to develop structural alloys capable of providing the desired high temperature mechanical properties as well as achieving an acceptable resistance to environmental attack. For such arduous conditions these two requirements were decoupled, such that structural alloys were developed for optimal mechanical properties and coatings were developed to provide the environmental resistance. This in turn gave rise to a third degradation mode associated with the interaction and interdiffusion of the coating and substrate alloys.

From our industrial perspective, this need translates into the necessity to provide adequate protection against all the *active*, in service, degradation modes which may include mechanical effects such as creep, high cycle fatigue, low cycle thermo-mechanical fatigue, high temperature oxidation, cyclic oxidation, hot corrosion, erosion and possibly foreign object damage. This implies a knowledge of the durability of the chosen structural material, or coating, under normal and fault conditions to each of these degradation modes, with the shortest life determining the overall component life. For this current study discussion of materials degradation will be limited to the consideration of cyclic oxidation only.

Thus, the prediction of cyclic oxidation lifetimes requires:

- An understanding of the causes of cyclic oxidation failure.
- The determination of reliable, possibly repeatable, cyclic oxidation data, and
- The development of predictive models for cyclic oxidation, that can take into account the stochastic nature of oxide spallation, such that materials performance under normal and extreme (fault) plant operating conditions can be deduced.

If remnant life prediction is the driving aim, then it is not possible to define *one* standard cyclic test procedure. Such an approach would allow the ranking of materials performance, but would not result in data that were appreciable to many plant operating conditions. Instead, test methods should be developed that aim to simulate the critical parameters that determine the service behaviour. Such understanding comes from a knowledge of the plant operating conditions and the causes of cyclic oxidation failure.

It should, however, be possible to provide guidelines that define how a cyclic oxidation test should be undertaken, what measurements should be recorded and the precision to which each measurement/operating condition has to be held. Further, the inclusion of reference materials within any test would permit a direct comparison across different test rigs and rig operating conditions.

An implication of this test philosophy is that the mechanisms of cyclic oxidation failure may vary from material to material and with exposure, i.e. the time at temperature, throughout the component life.

2. Simulating Service Behaviour From a Cyclic Oxidation Perspective

Oxide spallation is the primary mode of failure under cyclic oxidation conditions. Whether, or more exactly when, spallation will occur depends on a complex interplay of plant operating conditions, including the maximum operating temperature, the duty cycle, the accumulated time at temperature, heating and cooling rates, and the test environment.

Loss of oxide scale by spallation is a major concern for all high temperature components, particularly when operating close to their maximum service temperatures; and is a particular problem for alumina- or chromia- formers, where loss of the scale-forming element from the near-surface region of the alloy, or from a coating, can lead to loss of protection when the concentration falls below a critical level (for example, the critical aluminium concentration to form an alumina scale is usually cited as 4–5mass% for Co or Ni based MCrAlY coatings [19], 1.3–2.9mass% for FeCrAl based structural alloys [6,26] depending on the alloy system, and has recently been evaluated as ~9mass% (19at.%) for conventional and platinum modified diffusion coatings [13].

As a result, the mechanisms of scale spallation have been extensively studied [10, 27–36]. Of particular note is the modelling work of Evans [31,32,34,36] who has shown the important role that stress relaxation processes, such as creep, play in reducing the driving force for oxide spallation. These studies have shown that the critical parameters controlling spallation are the oxide adhesive energy, the oxide thickness and the extent of stress relaxation possible in the early part of the cooling cycle. Evans has evaluated the spallation resistance of various alloy/oxide systems by measuring the critical temperature drop to spall [36]. Time at temperature determines the critical oxide thickness, while cooling rate influence the degree of stress relaxation possible and thus the driving force for spallation during each thermal transient.

Thus, in defining cyclic oxidation damage cooling rate is probably a more important parameter than the actual duty cycle (time at temperature per cycle increment) or the maximum service temperature. However, the quantification of cooling rates is often overlooked, or not quoted, in many laboratory cyclic oxidation studies. Table 2 gives the typical heating and cooling rates achievable over a range of power plant operating conditions, compared to those achievable within laboratory, furnace based, equipment [37].

From Table 2 it can be seen that it is difficult within a laboratory simulation to achieve the cooling rates expected for gas turbine service conditions. With the aim of more closely simulating severe cooling conditions a specialised thermal cycling rig has been built in the author's laboratory, capable of achieving cooling rates of $>10^6$ °C h^{-1} (300°C s^{-1}), close to that achieved in the gas turbine by using water aspirated air as a cooling medium. The capability of this rig is discussed elsewhere in this volume [38].

Thus, in defining a suitable cyclic oxidation test to suit a given service behaviour, one must not only specify the maximum test temperature, but also the heating rate, cooling rate and duty cycle (time at temperature) for the test. Table 3 summarises the range of heating rates, cooling rates and duty cycles observed in many high temperature power/process plants.

3. Guidelines for Cyclic Oxidation Testing

To our knowledge, two surveys of test methods that have direct relevance to cyclic oxidation testing have been undertaken over recent years. These are an evaluation

Table 2. *Typical heating and cooling rates [37]*

	Heating and cooling rates (°C s^{-1})*		
	0.05	(180)	Generating plant heat up rate
Furnace cool	0.1	(360)	
Air cool	0.5	(1800)	
	1.0	(3600)	
	2.0	(7200)	
Forced air cool	5.0	(18 000)	
	10.0	(36 000)	
	50.0	(180 000)	Gas turbine heating/cool down rate

* Heating rates/cooling rates in parentheses are in °C h^{-1}.

Table 3. *Heating rates, cooling rates and duty cycles to simulate service behaviour*

Service duty cycle	Heating rate	Time at temp. (h)	Cooling rate[a]
Gas turbine			
•Short range — military	H	0.3–1.0	H
•Short haul — civil	H	0.6–2.0	H
•Medium haul— civil	H	2.0–6.0	H
•Long haul — civil	H	6–18	H
•Industrial power plant — peak lopping	H	1.0–3.0	H
•Industrial power plant — base load	H	100–200	H
Coal fired plant			
•"Two shifting" duty cycle	L	12	L
•Continuous — base load	L	2000–8000	L
High temperature process equipment			
•Short duration heat treatment	M	0.3–6.0	M
•"in-line" plant	M/L	120–140	M/L
•Heating elements	M	0.1–0.3	M
Automotive			
•Catalytic converters	M	0.1–4.0	M
•Turbochargers	M/H	0.1–4.0	M/H
•Exhaust manifolds	M	0.1–4.0	M

[a]Definition values for heating and cooling rates are as follows:

	H	M	L
Heating rate	50°C s^{-1} (180 000°C h^{-1})	5.0°C s^{-1} (18 000°C h^{-1})	0.05°C s^{-1} (180°C h^{-1})
Cooling rate	50°C s^{-1} (180 000°C h^{-1})	5.0°C s^{-1} (18 000°C h^{-1})	0.05°C s^{-1} (180°C h^{-1})

of test methods used for the discontinuous measurement of high temperature corrosion by Nicholls in 1995 [39] and an evaluation of cyclic oxidation test procedures by Starr in 1997 [37]. Many of the papers* in the present volume [40] further add to this body of knowledge and it is therefore opportune to consider whether we should have a rigid standard test, or, in our view, the more flexible approach of recommended guidelines based on 'best practice'. Both of the above reviews [37,39] were based on surveys of many high temperature laboratories, proficient in high temperature oxidation and corrosion testing, and therefore the combination of information contained therein with the outcome of this cyclic oxidation workshop [40] can be deemed to represent 'best practice'.

*Presented at the EFC workshop on Cyclic Oxidation Testing as a Tool for High Temperature Materials Characteristics, held at DECHEMA e.v., Frankfurt, Germany, February 1999.

3.1. Best Practice in Cyclic Oxidation Testing

The first of the reviews, by Nicholls [39], was far more wide ranging than just cyclic oxidation testing, in that it looked at oxidation (both isothermal and cyclic), gas phase corrosion and corrosion under deposits and reviewed all relevant discontinuous methods of corrosion testing, but excluded thermogravimetry as this was covered by a separate paper in the same volume [41]. As a result only 10% of the total reported test conditions (one laboratory = one test condition) related to cyclic oxidation testing, even so 180 relevant data were recorded. The survey by Starr [37] was by nature more limited as it specifically focused on cyclic oxidation testing. Some 34 replies were received, with all focusing on the shorter duration, 'rapid thermal cycle' test typical of the NASA test or similar (see papers 1, 2 and 5 for example). This raises the first point, that is what is meant by 'cyclic oxidation testing'. In its general form any high temperature oxidation study that involves more than one shutdown must be cyclic, but this is not our classic understanding. A better definition would be that:

> *A cyclic oxidation test*
> requires regular, periodic shutdowns with the aim of investigating the role such shutdowns have on the on-going oxidation processes.

This implies that the shutdown procedure, i.e. the cooling step, must be well defined, controlled and hence reproducible. With this definition in mind the survey replies from both studies [37,39] were analysed to try to deduce any concensus as a basis for best practice. Table 4 presents a subset of the data from ref. [39], focused on isothermal and cyclic oxidation test procedures.

3.1.1. Test temperature and maximum duration

Table 4 shows that defining one reference temperature or a single exposure time for a cyclic oxidation test is not a viable option. The test temperature and exposure time are very much dependent on the purpose of the test and may often be specific to a particular application. It is interesting to note that many of the shorter duration tests (less than 1000 h) are undertaken within universities, whilst all of the very long duration exposures (greater than 10 000 h) are undertaken by industry or research

Table 4. *Summary of isothermal and cyclic test methods reported in the discontinuous corrosion test methods survey by Nicholls [39]*

Usage	Environment	Proportion of total number of test conditions (%)	Temp. range (°C)	Range of exposure times (h)
(a) Isothermal oxidation	Lab. air blended gas	15 19	400–1500 400–1300	100–50 000 100–5000
(b) Cyclic oxidation	Lab. air	10	250–1500	100–50 000

laboratories. From this survey the preferred test temperatures for cyclic oxidation testing lay in the range 900–1150°C; all tests were cooled to room temperature, often at different rates; and the preferred maximum cyclic exposure times were either 2000 or 5000 h, but this depended on the hot cycle time.

3.1.2. Hot cycle time and number of cycles

An interesting comparison can be drawn from the results of the two cyclic oxidation surveys [37,39] in the light of the definition of cyclic oxidation presented earlier, as reproduced in Fig. 1(a) and (b). Depending on whether the term 'discontinuous oxidation' or 'cyclic oxidation' was used in the survey different perceptions of 'hot cycle time' and 'number of cycles per test' come to mind, even though both test procedures meet the definition of a cyclic oxidation test presented earlier.

'Discontinuous oxidation', cyclic tests that involve periodically removing samples from a furnace, cooling to room temperature and then replacing have longer hot cycle times and fewer number of repeat cycles per tests [39] when compared to test termed as 'cyclic oxidation' [37]. The latter tend to be run on automatic rigs with short duration periodic sample removal and cooling to room temperature.

The 'discontinuous oxidation' cycle may have hot cycle times of 20–5000 h, with 5–50 cycles defining the total test duration. Such repeat periods make this types of test very suitable for manual operation, although automatic rigs are known to exist that complete tests to this frame of cycle durations. Many of the automatic, cyclic test rigs utilising long hot cycle times are based on combustion/burner rig test procedures.

The automatic, cyclic oxidation test is often based on the NASA cyclic oxidation test procedure (see [1], [2], [5] and [37]) or a test rig of similar design, offering comparable test capability. Hot cycle test durations are much shorter, 0.2–200 h, with a preferred one-hour hot cycle. As must be expected, the total number of test cycles significantly increase such that a reasonably long total exposure time is achieved. The preferred number of cycles lie between 20 and 2000, although cycles from 4–100 000+ were reported. The majority of replies quoted from 100 to 500 cycles per test.

3.1.3. Choice of test procedure

From the combined surveys, three cyclic oxidation test procedures dominate the replies. These are:

(a) *A manual procedure* suitable for cyclic oxidation tests with relatively long hot cycle durations. Numerous, discontinuous, cyclic oxidation test procedures have been reported in the literature as is clearly apparent from the procedures adopted in the various papers reported in this volume. Test hot gas cycles may be as short as 20 h, allowing inspection of test pieces on a daily basis [42,43]; or as long as 3 000 h, as used in some extended oxidation tests within the European COST 501 programme where total exposure times in excess of 20 000 h were accumulated on some samples [7,26,44], with parallel cyclic tests on tube bank components lasting in excess of 10 000 h [45]. A European funded research programme, Testcorr [46,47], is currently underway aimed at defining test guidelines for discontinuous oxidation/corrosion testing. This

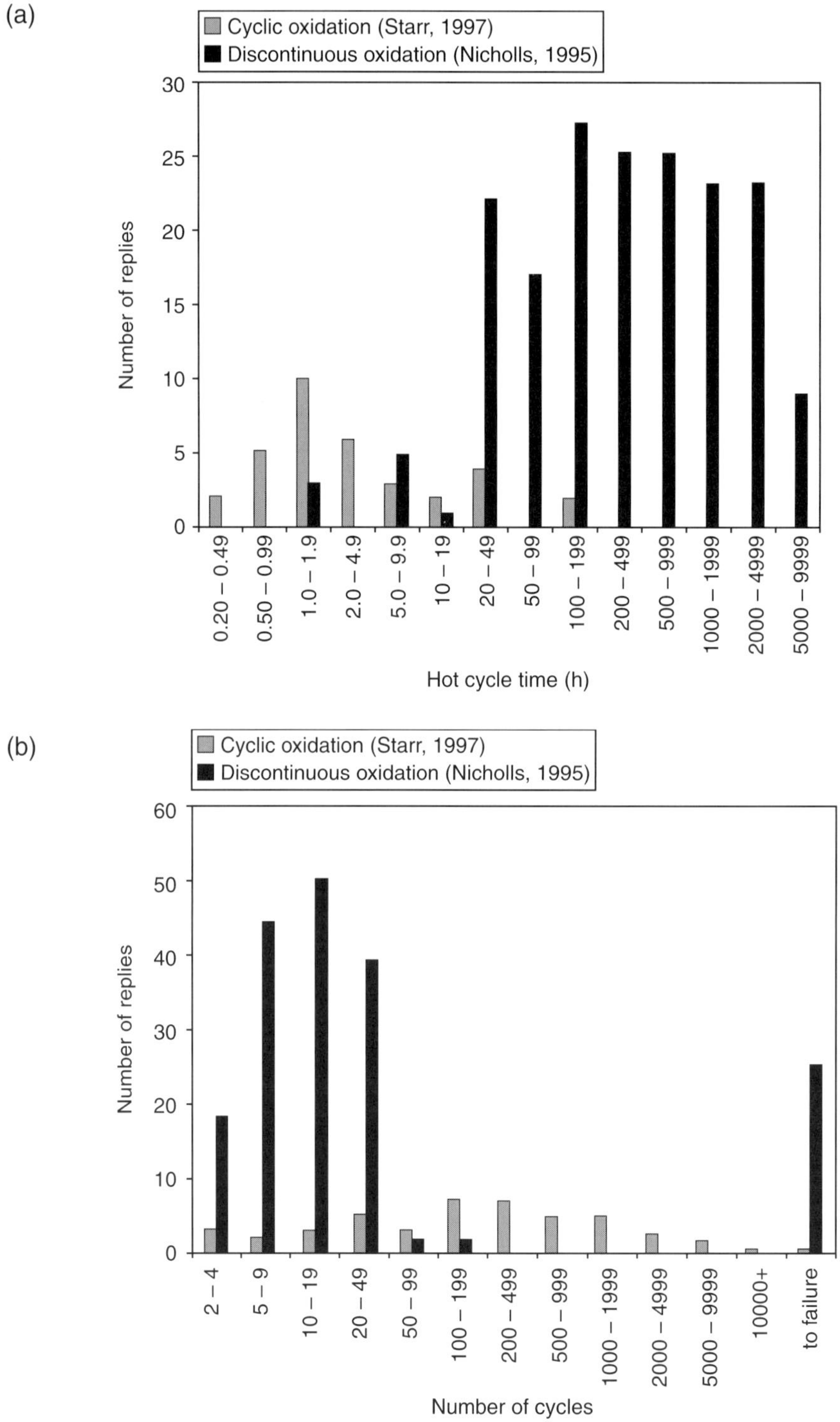

Fig. 1 *Results from two cyclic oxidation surveys [37,39]; (a) Distribution of hot cycle times; (b)Distribution in number of cycles of test.*

involves some 13 laboratories in various aspects of test procedure assessment and is due for completion later this year (1999). Recommendations from Testcorr will be reported in the next section on cyclic oxidation test methods (Section 4).

(b) *The NASA cyclic oxidation test* is an automatic cyclic test procedure based on a short duration, typically one-hour, test cycle. The test provides the capability to test multiple specimens and was the first test to automatically collect any spalled oxide during the cool down cycle (see [1], [2], [5] and [37]). The NASA cyclic oxidation test will be reviewed later in this paper.

(c) *Burner rig test methods* have been successfully used for cyclic oxidation (or more specifically cyclic corrosion) studies since the mid 1960s. Two international round robin tests have been completed. The first by ASTM in 1970 [48] and the second as part of the VAMAS initiative on measurements and standards [24,25,49]. Both of these comparisons address the hot corrosion test capability of such rigs, but in doing so also address the standardisation of test procedures for operating such rigs [25]. Burner rig test methods use hot combustion gases to heat the samples. The combustion environment may be relatively clean if fired on natural gas, but may contain SO_2/SO_3 as contaminants (often at levels between 123–325 vpm depending on the firing temperature) if fired on liquid fuels like NATO F76 1%S reference diesel fuel oil [24].

Cooled specimens can be used [22], or samples may be cooled by supplemental cold air blasts once removed from the hot gas stream [20,21,23]. This type of test allows the effect of temperature gradients to be studied at heat fluxes that approach those expected in service. A number of test pieces can be evaluated in a single burner rig run, varying between 6 and 32 depending on rig design.

Other test methods that have also been developed for cyclic oxidation assessment, but are not in wide usage, are also discussed in Section 4 on cyclic oxidation test methods.

4. Cyclic Oxidation Test Methods

4.1. Discontinuous, Cyclic Oxidation Test – Recommended Practice

(based on the guidelines proposed by Testcorr [46,47])

In this paper only a brief synopsis of the test methods recommended for discontinuous corrosion testing can be presented. Those of direct relevance to manual, cyclic, discontinuous oxidation testing are summarised below – for a draft of the full code of practice, please refer to ref. [47].

The code of practice covers the following items:

(i) the preparation of corrosion test specimens; including recommended specimen sizes, shape, surface condition and cleaning procedures.

(ii) the selection and characterisation of gaseous atmospheres.

(iii) the definition of suitable equipment, its calibration and control limits.

(iv) the procedures for carrying out discontinuous oxidation/corrosion tests, including start-up and shutdown procedures.

Additional points of relevance to cyclic oxidation testing are given in sections 4.1.1. and 4.1.2.

4.1.1. Preparation of specimens for cyclic oxidation testing

In the Testcorr draft code of practice [47] it is noted that the size and shape of test specimens may be governed by the type and form of material available, e.g. whether wrought products (bar, rod, plate or strip) or cast products (sticks of varying shape). However, within these limitations it is recommended that one of three basic shapes of specimen be used, either a rod, a disc, or a block. The recommended dimensions are illustrated in Fig. 2, and all three specimens have similar surface areas; 5 $cm^2 \pm 5\%$. The machining tolerance for the dimensions of all specimens should be ± 0.05 mm.

It is acknowledged that the specimen thickness is often dictated by the available material form and as such can be varied, but for severe test conditions 4 mm is recommended to be a minimum.

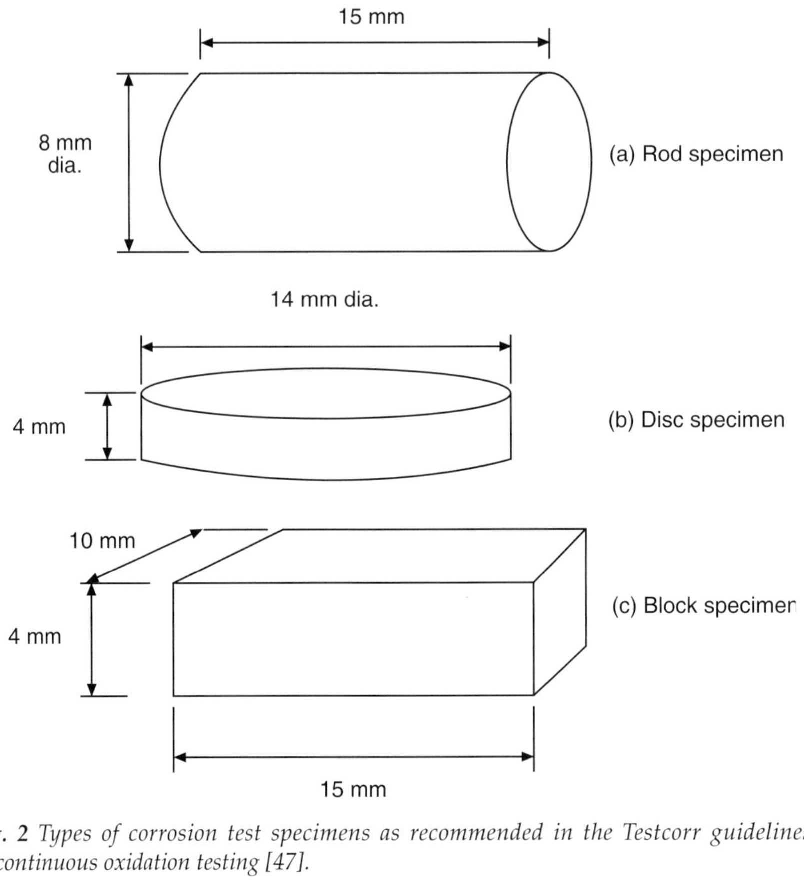

Fig. 2 *Types of corrosion test specimens as recommended in the Testcorr guidelines for discontinuous oxidation testing [47].*

For the NASA test, specimen dimensions used by Barrett and Lowell for sheet materials [1] were 12.7 mm × 19 mm × 1–3 mm thick, and for cast materials 13 mm × 23 mm × 2.3 mm thick [2]. These dimensions give surface areas between 5.46 and 7.636 cm^2, marginally larger than those recommended by Testcorr.

It is recommended that one of the three Testcorr specimen geometries be adopted for future cyclic oxidation studies.

The need for a reproducible surface finish and rigorous cleaning procedure is well recognised. The Testcorr guidelines [47] recommend a set sequence for grinding, from coarse to fine grades, using controllable grinding/polishing machines, with the final finish 0.8 μm Ra. Samples should be ultrasonically cleaned for a minimum of 10 min, using an organic solvent such as methanol or acetone and then hot air dried, or alternatively, a vapour degreasing column can be used to give a more reproducible cleaning procedure.

The Testcorr guidelines [47] require a reference mark at one end of each sample. For cyclic oxidation testing this could be the specimen suspension hole. The Testcorr guidelines [47] detail how to measure the samples and recommends a minimum of 8 repeat measurements per flat surface or per diameter (depending on specimen shape), with a measurement accuracy of ± 0.002 mm. The NASA procedure [1,2] recommends measurement to the nearest 0.001 mm.

Measurement of the initial mass, and subsequent mass changes, is very important in both studies, with a detailed weighing procedure identified in the Testcorr [47] guidelines. The Testcorr guidelines set the tight weighing limits requiring a traceable, calibrated balance with a minimum accuracy of weighing of 10 μg, together with a certified mass for checking calibration [47].

4.1.2. Suitability of furnace hardware, calibration and controls

The Testcorr guidelines [47] define the suitability of furnace equipment, furnace tubes and other hardware for discontinuous oxidation and corrosion testing, together with calibration of thermocouples, furnace temperatures and the control of humidity in the test environment. The recommendations as to hardware suitability were based on an earlier survey of laboratories undertaking discontinuous oxidation and cyclic oxidation studies [37,39].

4.2. The NASA Cyclic Oxidation Test Procedure [1,2]

Unlike the Testcorr procedure [47], which is aimed at manual, longer hot cycle time exposures, the NASA cyclic oxidation test is based on an automated procedure [1–3,5]. The test facility is illustrated schematically in Fig. 3 and consists of a vertically oriented alumina tube furnace, with a specimen carrier that can be automatically raised and lowered using a pneumatic actuator. The cyclic period is defined by re-settable timers. A particular innovation of this test was the inclusion of individual shields to prevent cross-spall contamination and the ability to collect the spall from individual samples, using six cups that automatically rotated underneath the samples, during the early part of the cool down cycle. The rotation and spall capture cycle took approximately 10 s. This mechanical spall capture sequence limited cross contamination to less than 1%. The complete enclosure of each sample in its own metal foil cup meant that spall loss from any sample was also very low, again of the order of 1%.

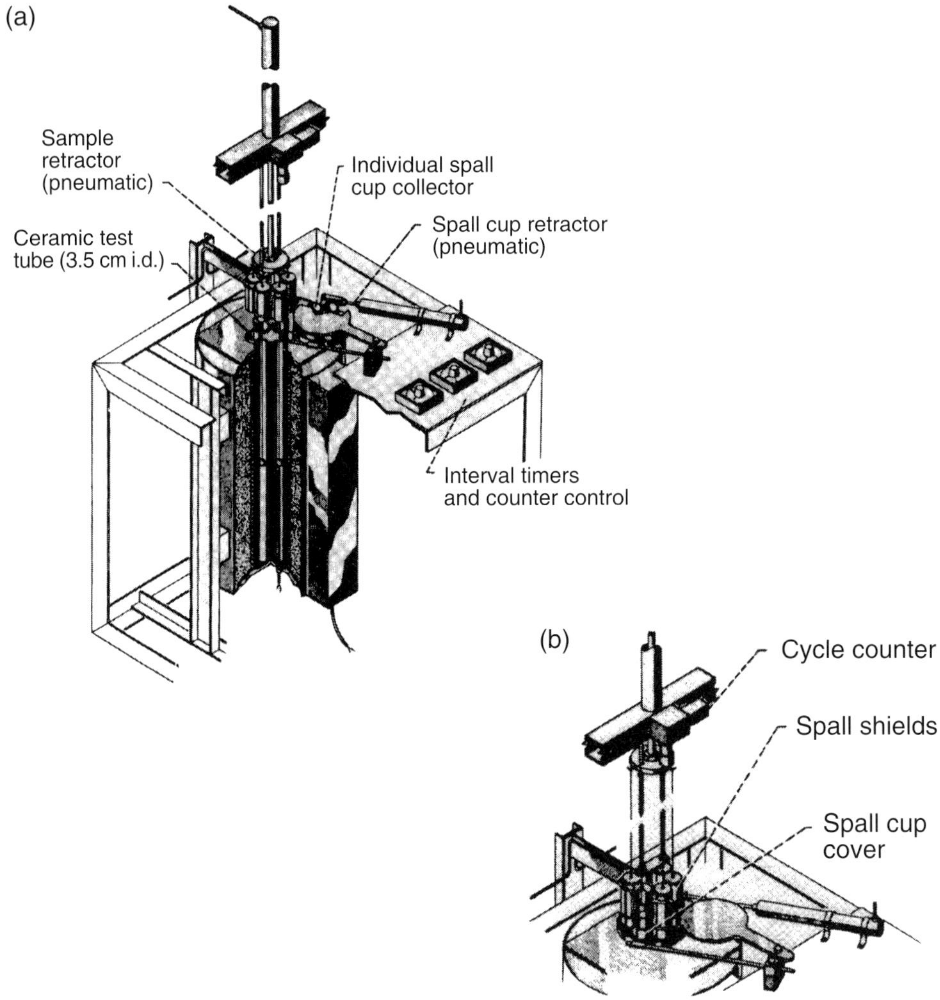

Fig. 3 Multi-specimen, high temperature, cyclic oxidation apparatus used for the NASA tests [1–3,5].(a) Samples at test temperature; (b) samples cooled to ambient.

The samples were removed and weighed once per day to generate mass change against time curves. The automatic cycle provided hourly heating cycles. Figure 4 illustrates the typical heating and cooling rates achieved for samples exposed at 1150°C. The time to heat to temperature was approximately 1 min, while the time to cool from 1150°C to below 100°C took 9 min. Thus, every one-hour hot cycle takes at least 70 min to complete including the cool down cycle; often, 40-min cool periods have been used [20], resulting in a total cycle time approaching 2 h [20]. A typical heating and cooling rate curve for the NASA thermal cycling rig is reproduced in Fig. 4 [1]. The maximum heating rate achieved is 40°C s^{-1} (1.44×10^5 °C h^{-1}) and the maximum cooling rate is 24°C s^{-1} (8.6×10^4 °C h^{-1}). Such cooling rates exceed that expected from a forced air cool, but are still a factor of two lower than the cooling rates achieved in gas turbine engines, or through the use of water aspirated air as a coolant [38].

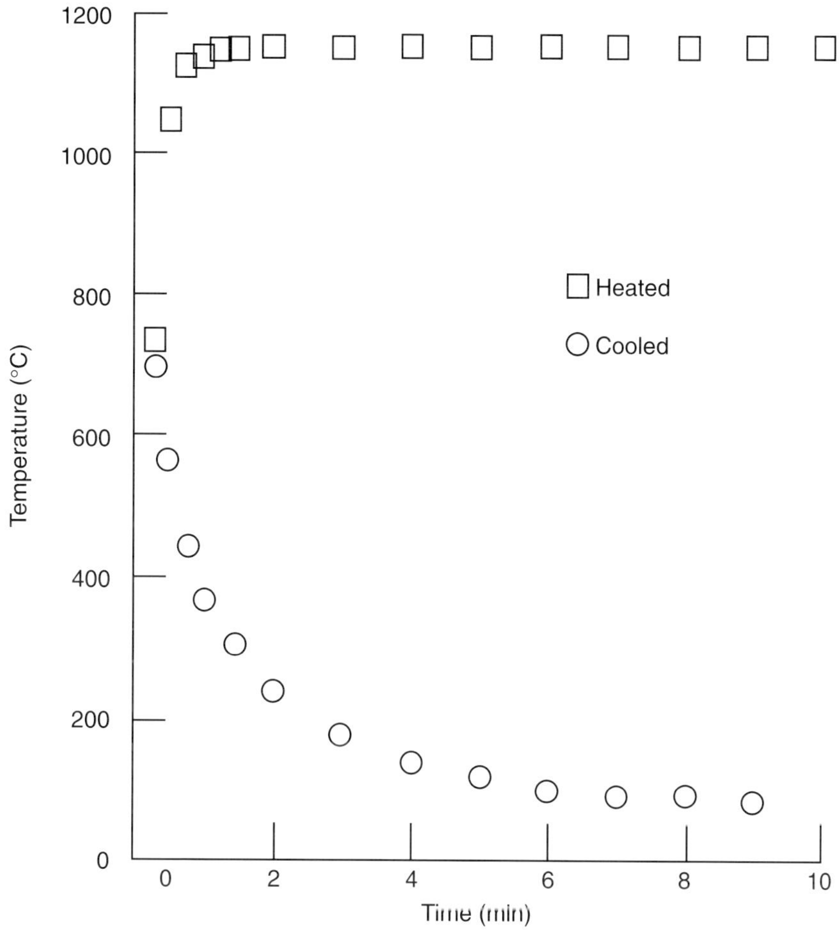

Fig. 4 *Heating and cooling rates, measured on the NASA cyclic oxidation test apparatus (sample IN-702, exposure temperature 1150 °C) [1].*

4.3. Burner Rig Test Method

Burner rigs have been used to provide a dynamic oxidation test facility [20–23] as well as the previously mentioned corrosion test capability. The most popular route has been to provide thermal cycles using a Mach 0.3 velocity or faster facility, as typified by the Becon rig, originally designed by Pratt and Whitney [21–23]. Between 6 and 32 samples can be tested in one rig test, using a rotating carousel, with the samples heated by the high velocity combustion flame, then cooled by high velocity (Mach 0.7 [20]) air jets. Specimen rotation speeds between 20 [24] and 1000 [20] rpm have been used. A typical test cycle could be similar to the NASA test [1,2] with one hour at temperature, but followed by a 3 min rapid air blast cooling [16]. Cooling rates of 4–6°C s^{-1} (1.4–2 × 10^4C h^{-1}) are achievable on pin samples using the standard Becon rig design [21], with rates up to 25× faster when using turbine simulation rigs that use aerofoil specimens [22].

Other test cycles have also been suggested, and one of the most rapid provides 0.05 h (3 min) heating to approximately 1100°C, followed by a 0.05 h fast cool [20].

Thus, burner rigs offer combustion environments, high heat fluxes, rapid thermal cycles and the ability to test multiple specimens. When using internally cooled specimens, very high rate cool down cycles can be achieved of up to 100°C s^{-1} (3.6×10^5 °C h^{-1}) comparable with those found in turbine service. Furthermore, internally cooled probes allow the effects of temperature gradients to be investigated.

4.4. Other Test Methods

Other test methods include rapid thermal cycling rigs (see ref. [37] (part of this volume)) capable of achieving cooling rates similar to the best of the combustion rigs, cyclic thermogravimetry, plus acoustic emission monitoring [8,50] and cyclic thermogravimetry plus thin layer activation [51,52]. The latter two permit continuous monitoring of mass change coupled with the early detection of cracking and spallation. Reference [8], which reviews 'measurement methods for the determination of spallation of oxide scales', provides an excellent overview of these alternative techniques.

Figure 5(a) illustrates a schematic diagram of an acoustic emission detector coupled to a controlled atmosphere microbalance [50]. On cooling, concurrent acoustic emission (AE) events and mass change can be recorded (Fig. 5b), thus the critical temperature drop to cause spalling can be identified. Figure 5(b) presents the data for a Ni–30%Cr–0.5Y sample [53].

5. Cyclic Oxidation Testing — What Should one Measure?

Having reviewed best practice (Section 3) and examined available test methods (Section 4), one can pose the question 'What parameters best define the test procedure?' This is a key question and its answer will determine whether we can define a *standard* or *guidelines*.

If a constant set of control parameters apply throughout the life of a component, then it *may* be possible to define procedures that test these in a *standard* way. If this is the case we can define a *test standard*. However, if the control parameters change throughout the component life, i.e. short morbidity differs from long life failures, then it is necessary to change the nature of the test throughout the component life and in this case it is only possible to define *guidelines* that define the preferred route for the test procedure.

Consider now cyclic oxidation testing in more detail. Table 5 defines the test parameters and materials characteristics which will determine how a particular alloy or coating will behave. (This table has been put together using data in the two review papers mentioned earlier [37,39] and information presented by Haanappel and Stroosnijder [54] in this volume.)

Clearly, the test rig operation conditions should be well defined, and can be controlled throughout any cyclic oxidation test programme. Repeatable results will depend on this control of *critical* parameters. Unfortunately, in many published papers, not all of the parameters are recorded and this makes reproducing the test difficult.

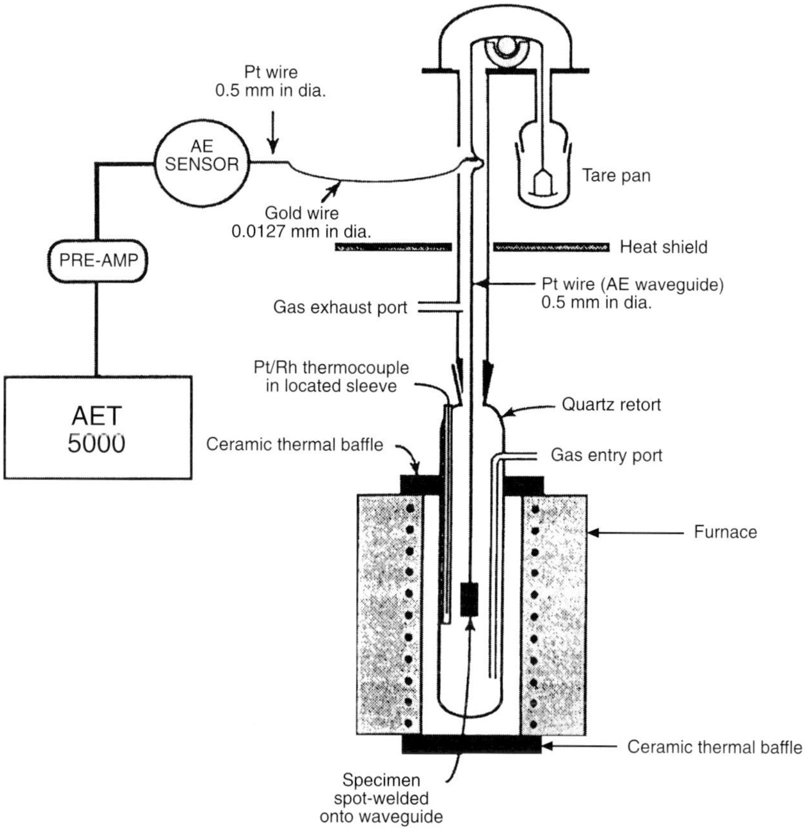

Fig. 5(a) *Schematic diagram of an acoustic emission (AE) detector coupled to a controlled atmosphere microbalance [50].*

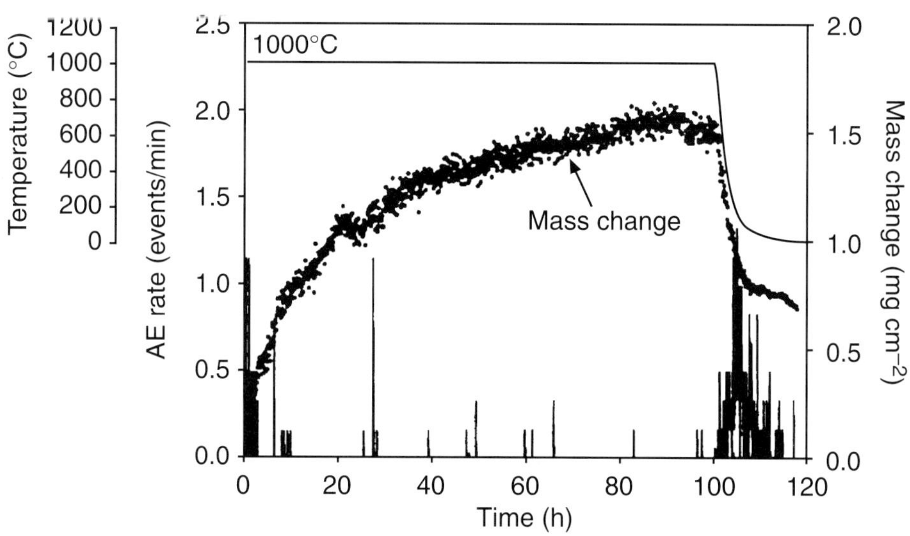

Fig. 5(b) *Mass change and AE counts during isothermal oxidation and cooling to room temperature of a Ni–30Cr–0.6Y alloy [50].*

Table 5. Necessary controlled parameters for cyclic oxidation testing [37,39,54]

Test rig operation	Characteristics of the Materials
• Oxidation temperature.	• Component geometry.
• Time increment at temperature.	• Alloy composition.
• Cooling rate.	• Nature of the protective oxide.
• Incremental temperature change (ΔT).	• Oxide growth rate.
• Heating rate.	• Oxide fracture toughness/bond strength.
• Number of cycles.	• Stress relaxation behaviour of the metal and oxide.
• Total exposure time.	
• Environment (atmosphere, stress, etc.).	

In addition, it is necessary to record various material characteristics such that understanding of the material behaviour can be well documented. Not all of the parameters in Table 5 may be known, but it is important to record all the parameters that are known relevant to the cyclic oxidation behaviour. Parameters that are relevant and known must include the alloy composition, the nature of the protective oxide and whether this changes through the component life, the rate of oxide growth (or metal consumption) and the extent of spallation, the component geometry, including stress concentrations and surface finish.

Interpretation of the cyclic oxidation behaviour would be aided from a knowledge of the oxide fracture toughness/bond strength and the degree of stress relaxation through creep for example. These latter factors are not critical to the operation of the test, but may play a major role in the response the material system gives to the cyclic test procedures adopted.

Using a holistic approach, suggested by Wilber *et al.* [38], the next section considers how the changing nature of the protective oxide alters how the material system responds to cyclic oxidation. This understanding is critical to the definition of operating procedures and guidelines for cyclic oxidation testing.

6. 'Birth', 'Life' and 'Death' of an Oxide

Wilber *et al.* [38] in their study of the cyclic oxidation behaviour of alumina-forming alloys identified three key stages governing the life of a protective alumina scale. However, it is the authors' belief that the observations made in ref. [38] are more generally applicable, and may be generic to all cyclic oxidation studies.

Using the terminology of Wilber *et al.* [38], then,

'Birth' — would be the formation of the initial protective oxide film, following a period of transient oxidation and the establishment of protective (parabolic, or sub-parabolic) oxidation kinetics.

'Life' — involves the continued growth of the scale. But as it matures (thickens) the scale becomes more prone to mechanical failure, via cracking and spallation. The onset and extent of mechanical failure depends primarily on an interplay between three factors.

(i) **defects**, such as voids, pores and microcracks, that develop in the scale as it grows [28–30,55–57].

(ii) **stresses** that act on the scale. These may be generated during oxide growth, but particularly result from the thermal expansion mismatch between the oxide scale and the alloy, during thermal transients. These stresses are locally concentrated by the defects in the scale.

(iii) **stress relief** within the oxide scale/substrate system, primarily by plastic deformation of the near-surface region of the alloy substrate while at temperature and during the early, high temperature part of the cooling cycle [31,32,34,36].

The interaction between these three factors will determine at what stage during the growth of the scale mechanical failure is likely to occur. Furthermore, predicting scale failure requires a holistic, systems approach involving knowledge of the substrate behaviour and mechanical properties, as well as those of the scale.

When the scale mechanically fails, the damage areas can be healed by the formation of new oxide. This re-healing process may involve the formation of protective scales, based on complex spinel phases, depending on the availability of protective scale-forming elements within the alloy.

'Death' — once the near-surface concentration of scale-forming elements depletes below some critical level, as a result of the continued oxidation, then mechanical damage to the protective oxide scale can no longer be re-healed. Non-protective oxidation ensuses, forming rapidly growing base metal oxides. These grow in a catastrophic manner at high temperatures, leading to what is termed as 'breakaway oxidation'. The formation of these non-protective oxides defines the end of the useful lifetime of the component.

Figure 6 presents a schematic diagram of this oxide scale life cycle. Critical stages in this life cycle that influence the behaviour under cyclic conditions are marked. Thus, from first exposure protective oxides will form, following an initial period of transient oxidation, until some critical oxide thickness is reached whereby the severity of the cool down cycle is sufficient to cause cracking and spallation of the oxide scale.

At this stage localised repair of damaged regions takes place, with each repair depleting the reservoir of protective scale forming elements more rapidly. Clearly, the more the mechanical damage to the scale during the 'life' of the component the quicker this reservoir of scale-forming elements will be depleted.

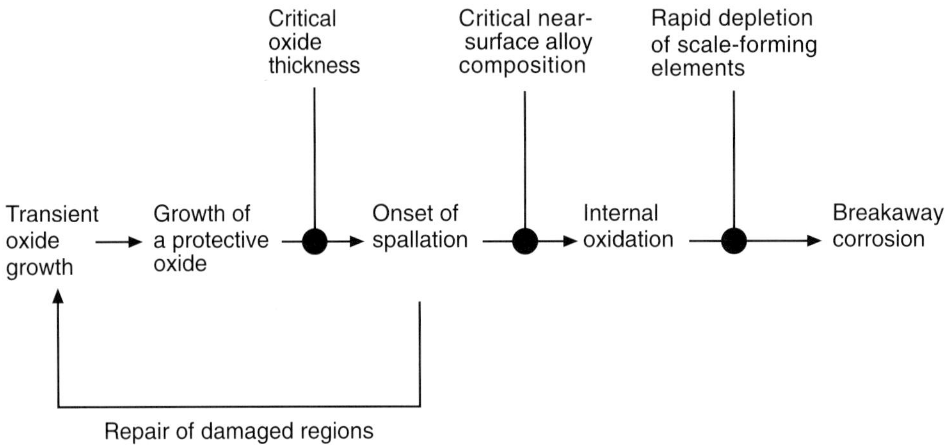

Fig. 6 *Schematic diagram of the generic behaviour for any alloy oxidation system.*

Once the near-surface alloy composition depletes to a critical level, the primary protective scale can no longer form. Depending on the alloy system, subscales may form (for example chromia and iron oxides underneath an alumina scale) or the alloy may oxidise internally rapidly consuming any remaining scale-forming elements. This rapid depletion of scale-forming elements leads to the catastrophic growth of base metal oxides in what is termed 'breakaway' oxidation.

Many papers in this volume (e.g. refs [38], [58–60], [67] and [76], show that the cyclic oxidation performance of alumina- and chromia-forming alloys and coatings follow this generic behaviour, some examples of which are cited in the remainder of this paper as one discusses the implication of such generic alloy oxidation behaviour on cyclic oxidation testing.

Figure 7 illustrates a representative example of the cyclic oxidation behaviour of an iron-based alumina-forming alloy, following 1-h cycles at 1200°C. The data were taken from the work of Tortorelli *et al.* [58]. Marked on this figure are the three transitions in behaviour as discussed earlier.

Position 1 identifies the onset of spallation, and thus defines the critical thickness of oxide (mass gain) at which spallation starts. A better way of identifying this transition would be to determine the mass of spall and net mass gain separately and then plot mass of spall against gross mass gain (mass spalled + specimen mass change). Examples of this type of plot are illustrated elsewhere in this volume [38].

Position 2 marks the onset of subscale formation of internal oxidation and associated specimen deformation. For alumina scale formation such a region would be associated with a change in surface colour of the oxide from grey to green/brown as chromia and iron oxides form under, or are incorporated within, the inner part of the alumina scale [59,60].

Position 3 illustrates the start of the rapid mass gain associated with 'breakaway' oxidation as base metal oxides form disrupting further the protective surface oxide. Figure 8 illustrates this behaviour, presenting a classic set of micrographs of iron oxide rods disrupting the alumina scale on an iron aluminide after 350 h at 1300°C [61].

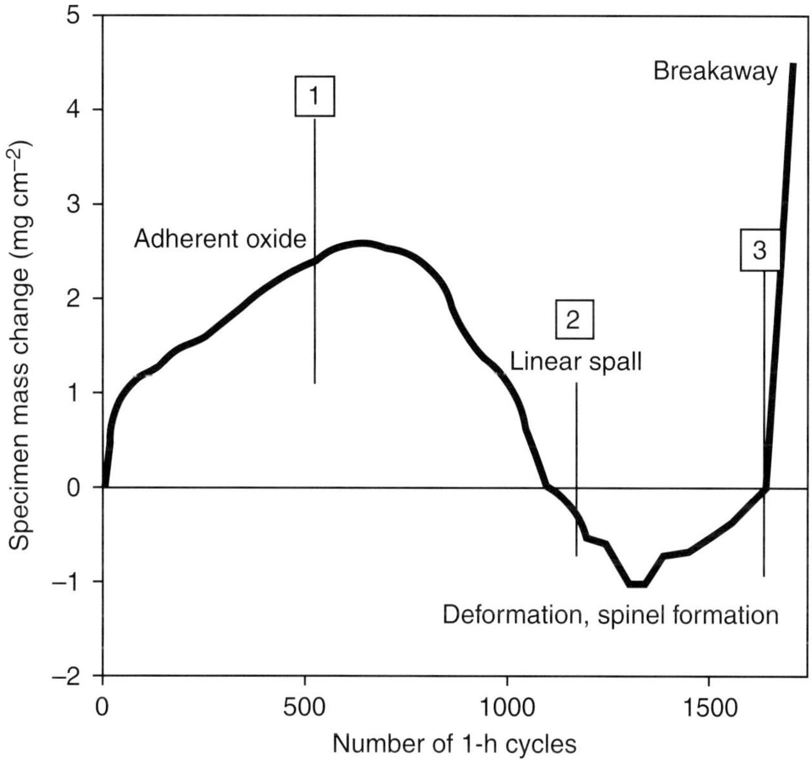

Fig. 7 Representative examples of specimen mass change for Fe-base alumina-forming alloys in 1-h cycles at 1200°C [58].

6.1. Implications of Such Generic Behaviour on Cyclic Oxidation Testing

If this generic behavioural pattern is true for all protective oxide systems when exposed under cyclic conditions then it should be possible to predict the characteristics to be expected as the cyclic oxidation test proceeds through a component's life cycle.

6.1.1. Prior to the onset of spallation

In this regime, following the initial transient oxide growth, a stable protective oxide growth regime should develop with parabolic or sub-parabolic oxidation kinetics. If this is the case, then

(i) cyclic oxidation rates should be independent of the cycle frequency.

(ii) the mass gain will depend on the accumulated time at temperature.

(iii) alloy depletion will depend on the accumulated time at temperature.

(iv) the onset of spallation will depend on cooling rate. The onset of spallation will occur earlier in the components life cycle, i.e. for thinner oxides, for the more rapidly cooled samples.

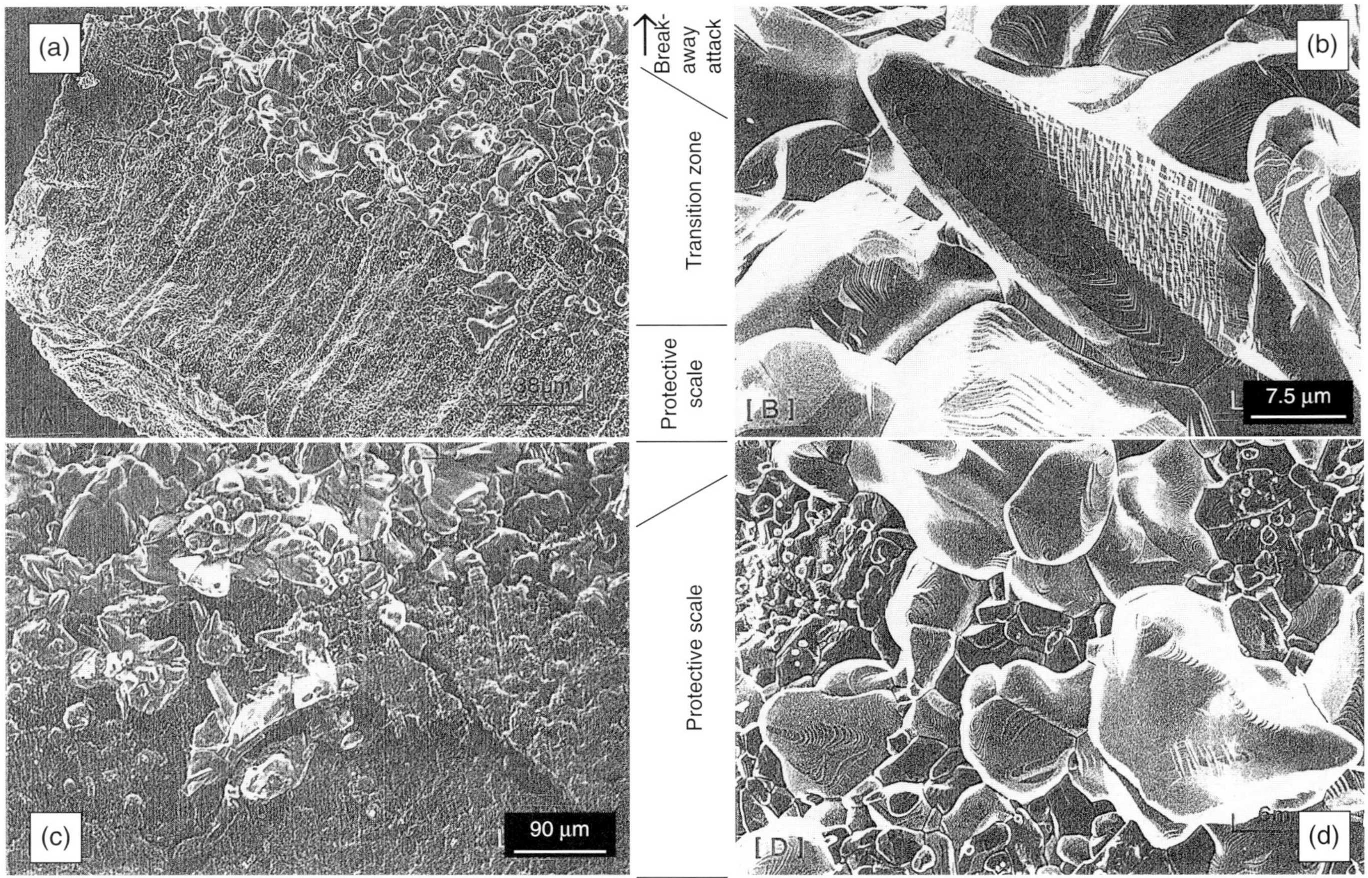

Fig. 8 *Scanning electron micrographs of FAL after 350 h oxidation in air at 1300 °C showing the topography of (a) the protective scale and the transition zone, (b) the breakaway attack, (c) and (d) the transition zone [61].*

(v) for a given protective oxide, spallation will occur more readily, and thus earlier in the life cycle, the stronger the alloy, i.e. the alloy is less able to relieve thermally induced stresses during cooling by creep and other relaxation processes the stronger the alloy.

This behaviour is well illustrated by the following examples taken from papers in this volume [40]. The compositions of the alloys in this paper are given in the appropriate chapters in this volume as cited in the references.

Work by Strehle *et al.* [59] has examined the cyclic oxidation of an alumina-forming FeCrAl based alloy (Aluchrom I SE) at 1100°C in air. These studies (Fig. 9) showed that the early oxidation kinetics were independent of cycle frequency when this was varied from 5 to 50 h. Furthermore, for this alloy, when tested in the form of thin foils, no spallation occurred before the onset of chemical failure (either the observation of chromia and iron oxide discolourations under the alumina scale, or scale disruption due to breakaway corrosion). Chemical failure is illustrated as filled symbols in Fig. 9 and was observed to occur at an oxygen uptake of 8.3 gm^{-2}, independent of cycle frequency.

Other works published in this volume show similar behaviour but also add to our body of knowledge. Quadakkers *et al.* [60] found that when oxidising 100 µm PM2000 foils (an ODS alumina forming alloy) at 1200°C under isothermal or cyclic oxidation conditions in synthetic air, the cyclic oxidation kinetics matched that for the isothermal samples. Again for these thin foil samples, no spallation was observed

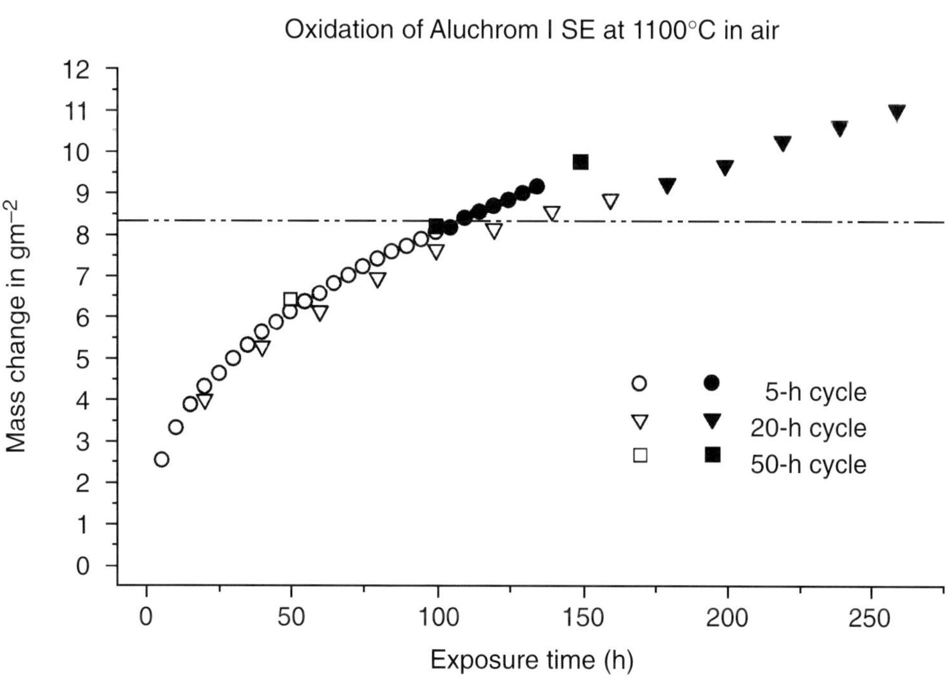

Fig. 9 *Cyclic oxidation of Aluchrom I SE foil at 1100°C in air [60].*

prior to breakaway, even though the PM2000 foil is expected to be significantly stronger than the Aluchrom I SE alloy reported by Strehle *et al.* [59].

By increasing the specimen thickness to 2 mm and testing at higher temperatures (1300°C), Wilber *et al.* [38] in their study of the oxidation of 'weak' and 'strong' alumina formers, clearly showed the onset of stage 2, the 'steady state spallation regime', as illustrated in Fig. 10. This work showed that duplicate PM2000 (an ODS FeCrAl alloy) samples had reached the critical oxide thickness to spall after 100 h at 1300°C. The weaker alloys, Kanthal APM (a powder metallurgy FeCrAl alloy) and Fecralloy JA13 (a wrought alloy) showed no propensity to spall at this stage in the cyclic oxidation test. Kanthal APM (stronger than Fecralloy JA13) started to spall after 700 h of cyclic testing, while Fecralloy JA13 (the weakest alloy) showed little evidence of spallation throughout the 2000 h of cyclic testing; when spallation did occur for this alloy it was on sample edges and at corners. These observations confirm the importance of substrate creep and other stress relief mechanisms, as proposed by Evans [31,32,36] in determining when a scale will spall on cooling. For PM2000 the critical scale thickness to spall is 8–10 μm (1.6–2.0 $mgcm^{-2}$). For Kanthal APM gross spalling occurred at 25–35 μm thickness (5.0–7.0 $mgcm^{-2}$) while spalling from sample edges and other discontinuities occurred somewhat earlier at ~20 μm thickness (~4.0 $mgcm^{-2}$). The Fecralloy JA13 alloy only spalled at discontinuities and this was observed for oxide thickness above ~30 mm (~6.0 $mgcm^{-2}$). Thus the creep resistant ODS alloy (PM2000) exhibited little stress relief during the cooling cycle and spallation occurred early in its life, while Fecralloy JA13, the least creep resistant alloy, was able to relieve stress dynamically during the early stages of cooling giving an extended life to first spall.

Finally, within this section, the role of cooling rate is well illustrated by the modelling work of Evans *et al.* [36]. In predicting the kinetics of wedge crack growth at the interface of an alumina-forming alloy (Haynes 214), with a 5 μm alumina scale, he showed that the critical temperature drop to spall increased from 360°C to 500°C, when cooling from 1100°C if the cooling rate changed from 10^4 to 10^2 °C h^{-1}. This behaviour is illustrated in Fig. 11 and shows that for much of the early cooling cycle creep relaxation can relieve the stress with minimal crack growth until a critical temperature drop is reached whereupon rapid crack growth ensues.

This modelling work [36] further shows that following a few thermal cycles this stress relief during cooling leads to in-plane tensile stresses within the oxide when reheated to temperature. This could lead to through thickness scale cracking and re-healing as proposed by Schütze [28,33,62], further consuming available scale-forming elements.

The work on Haynes 214 [36] found that during the first 100–200°C temperature drop the crack tip normal stress remains constant at about 200 MPa, it then increases with a further increase in temperature drop until the interfacial fracture stress is reached, whereupon spallation occurs. The temperature to achieve this is the critical temperature drop discussed earlier. Clearly, changes in the near-surface alloy composition (alloy depletion) will affect this behaviour as the change in alloy composition must affect the local creep properties of the alloy.

Interfacial fracture results when the thermally induced stress exceeds the fracture strength of the oxide in the vicinity of the interface (cohesive failure) or exceeds the metal-oxide bond strength (adhesive failure). Both of these processes will be

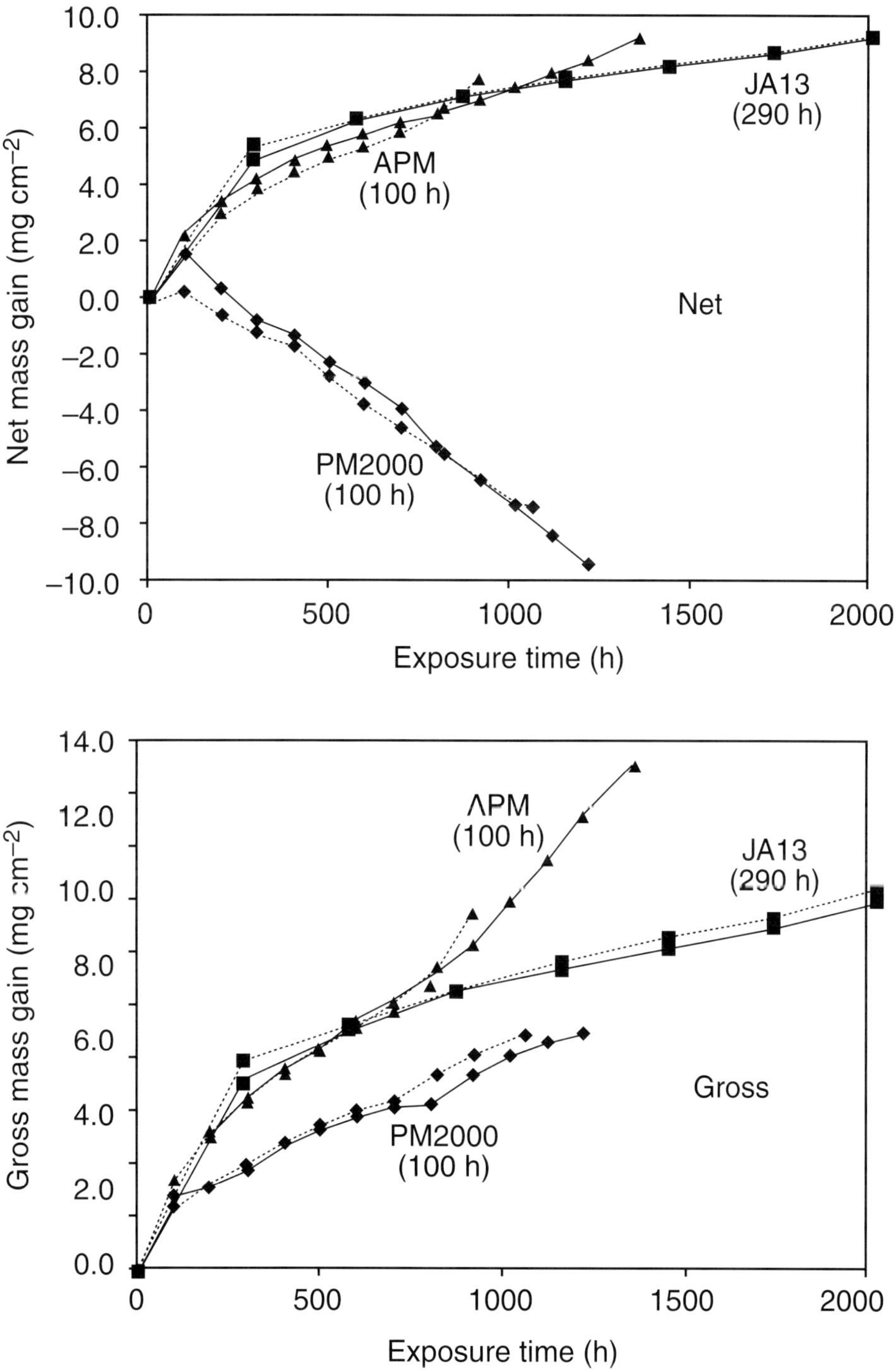

Fig. 10 *Cyclic oxidation of FeCrAl alloys at 1300°C [38].*

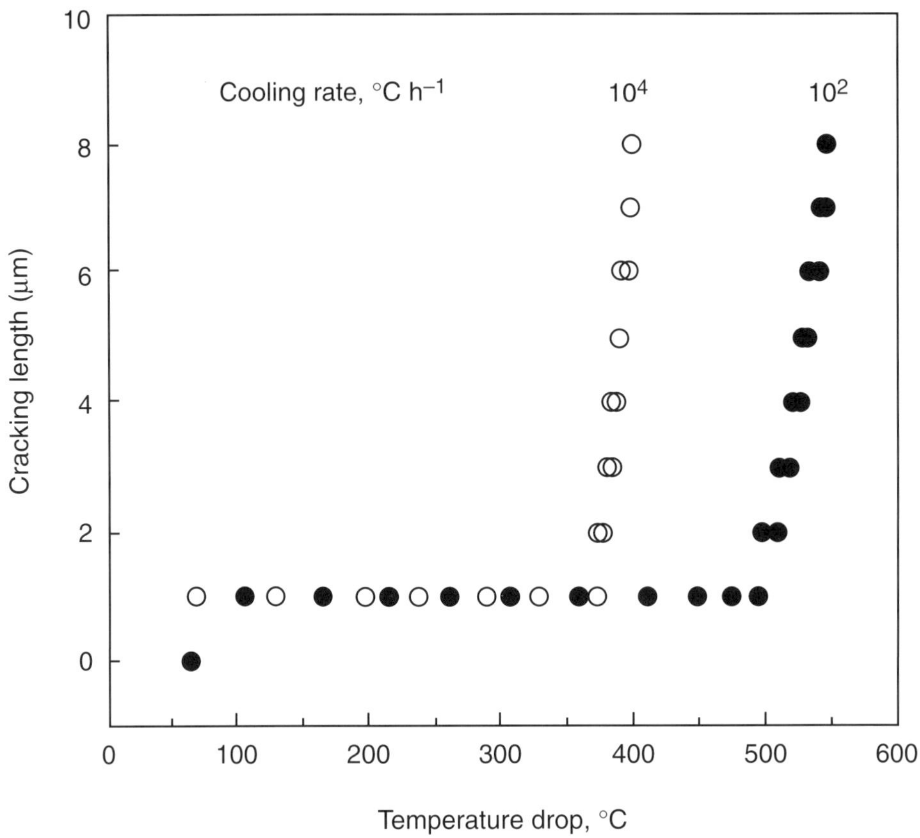

Fig. 11 Kinetics of wedge-crack growth for Haynes 214 with a 5 µm alumina layer cooled from 1100°C at 10^2 and 10^4°C h^{-1} [36].

influenced by local defects (stress concentrators) within the oxide or at the oxide/metal interface. The role of the oxide defect structure on scale fracture was first proposed by Hancock and Nicholls [54,55] who suggested that many small defects may interact to form a composite defect, leading to easier scale fracture. Figure 12 is a micrograph of a alumina scale on MA956 from the paper by Guttmann *et al.* [57], excellently illustrating these defect structures in a 25 µm alumina scale following 5200 h oxidation at 1100°C. Furthermore recent work by Wilber *et al.* [56,63] has shown that for alumina scales such defects are formed, and grow, as the scale ages. This process continues, with the defects acting collectively, until a critical defect size is reached whereupon the scale fails on the next thermal cycle.

It is believed that this mechanism, based on the growth and collective interaction of defects, equally applies to the spallation of chromia scales [54,55] and other oxide systems [62,64–66].

From the foregoing it can be seen that the onset of spallation during thermal cycling reflects a complex interplay of material parameters, involving both the oxide and substrate. Substrate behaviour, such as alloy depletion profiles and creep properties, are as important as scale growth, scale morphology and oxide adhesion/fracture strength.

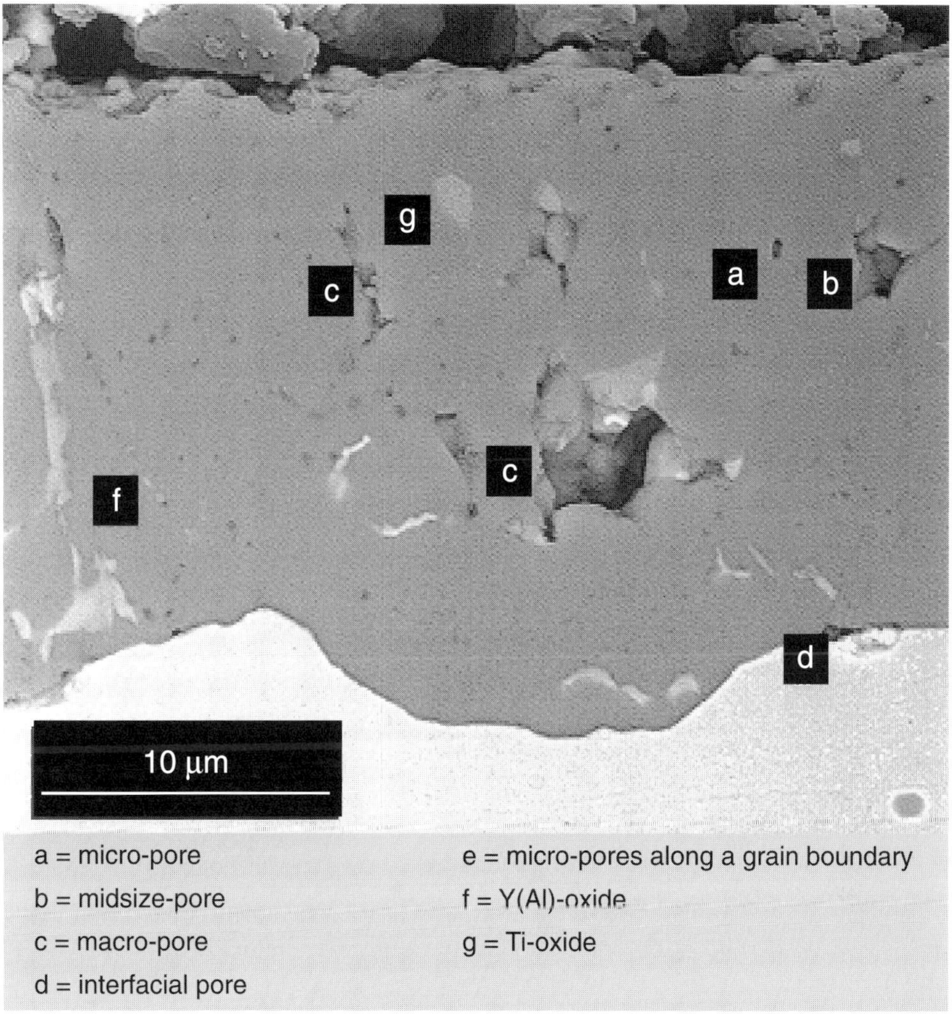

Fig. 12 *Defects in the Al_2O_3 scale on MA956 after 5200 h at 1100°C [57].*

6.1.2. The 'steady state' spall region

Once spalling has commenced, cyclic oxidation proceeds by continued spallation and re-healing. Within this region

(i) the mass of spall depends on the incremental time at temperature and is therefore a function of cycle frequency.

(ii) the occurrence of further spallation is determined by the local oxide thickness, the localised scale and substrate mechanical properties and the cooling rate.

(iii) alloy depletion of scale-forming elements is a function of the scale growth rate and the mass of spalled oxide.

Thus, within this regime one would expect the mass change to vary with cycle frequency and with alloy composition during a cyclic oxidation test.

The effect of alloy composition on oxide spallation was discussed in the previous section when considering the onset of first spall. The stronger the alloy the more likely it is to spall under a given cyclic condition. Thus, for similar scale growth rates the stronger alloys will exhibit a more rapid mass loss due to the spallation processes than the weaker alloys (see Fig. 10 (net mass change data) [38]).

This dependence on alloy composition has been observed for chromia formers [58,67–73] as well as alumina formers [38,57–60,74–82], whether as structural alloys [38,57–60,67–75] or coatings [76–84], as reported elsewhere in this volume.

At this point it is worth highlighting the role the test environment may play on the local scale growth rate, and possibly on a change in scale properties [69,72,76]. In particular, the work by Janakiraman *et al.* [76] has shown that both the onset of first spall and the 'steady state' spallation rate depends on the test environment. Figure 13 illustrates that the onset of spallation occurs earlier in the CoCrAlY coating's life when tested in moist air, compared to dry air [76]. Equally, the rate of mass change due to spallation in the 'steady state' region increases with moisture content and oxygen content in the test environment [76] (see Fig. 14). It is interesting to note that prior to the onset of spallation, in the early cyclic oxidation life, the rate of scale growth appears to be independent of the changes in the test environment under

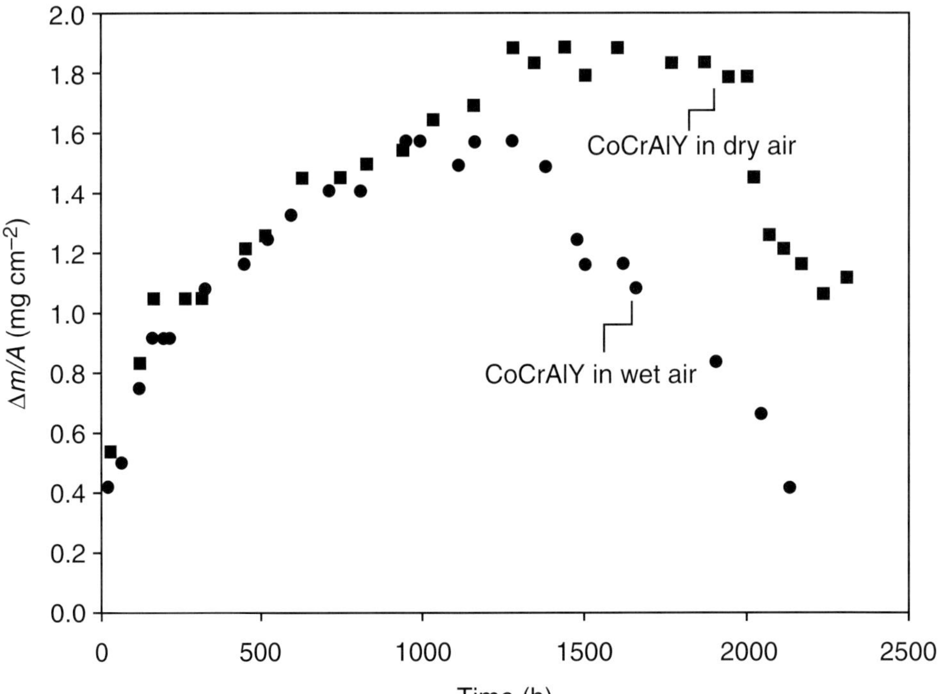

Fig. 13 *Cyclic oxidation of a CoCrAlY coating in wet and dry air at 1100°C [76].*

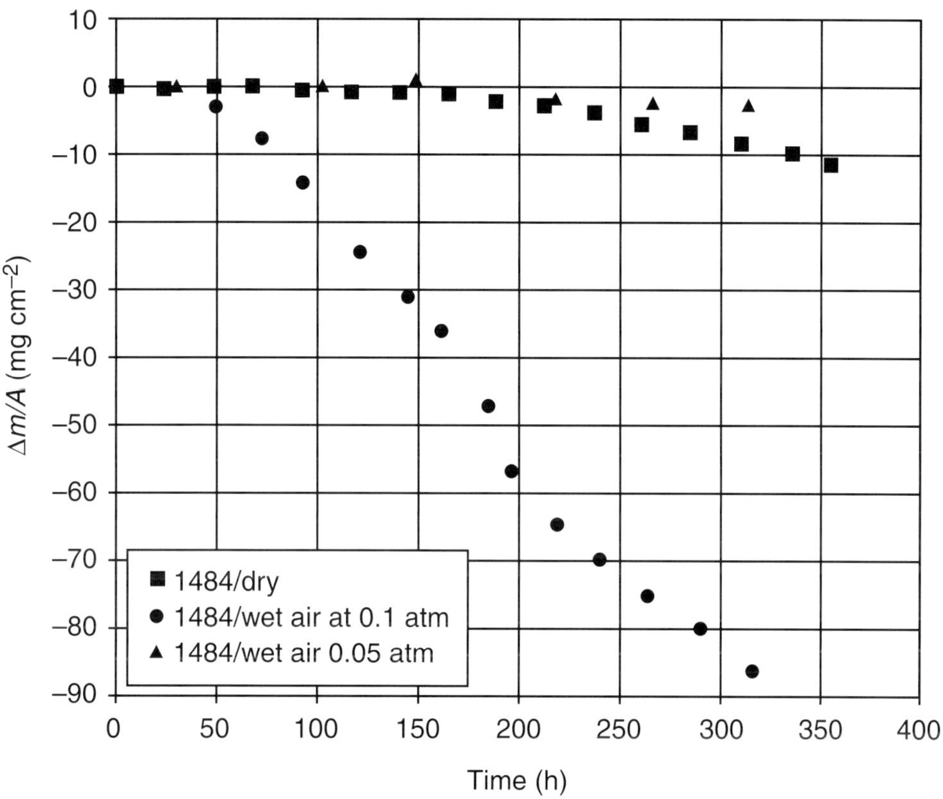

Fig. 14 *Mass change* Δm *per unit area* A *vs time for the cyclic oxidation of PWA 1484 at 1100°C in various environments [76].*

oxidising conditions [72,76]. This does not appear to be the case under reducing conditions [69].

Work by Pint *et al.* [58] and Wilber *et al.* [38] have studied the effect of cycle frequency, in this region where repeated spallation occurs. Both show that the mass change due to spallation is frequency dependent in this region. This is best illustrated in Fig. 15 which plots the mass change for a Fe–20Cr–10Al–0.8Y alloy that has been oxidised at 1200°C, as a function of cycle frequency and time [58]. The 100-h cycle data is close to that expected under isothermal conditions, indicating that very little spallation has occurred. In contrast, the 1-h cycles result in a rapid mass loss, associated with extensive spallation.

6.1.3. Onset of internal oxidation

Many of the papers in this volume noted that as spallation progressed extensive internal oxidation was observed with the alloy substrate (e.g. refs [58–60], [67] and [73]). This was particularly true for chromia-forming alloys as can be seen in Fig. 16 [67]. From these studies, and previous works, it would appear that

(i) internal oxidation occurs at, or below, a critical alloy addition activity level.

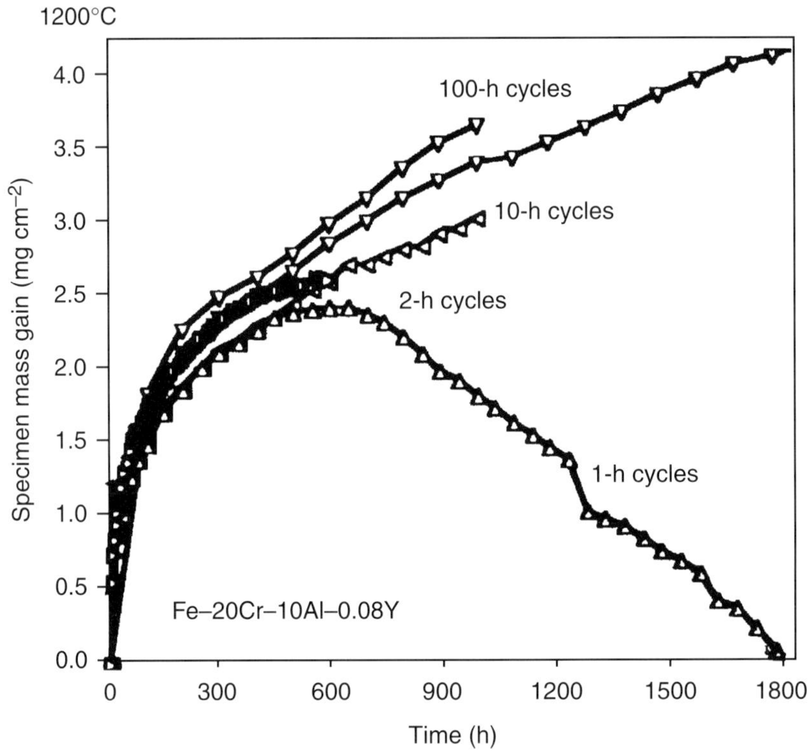

Fig. 15 *Influence of cycle frequency on the oxidation of Fe20Cr10Al0.8Y at 1200°C [58].*

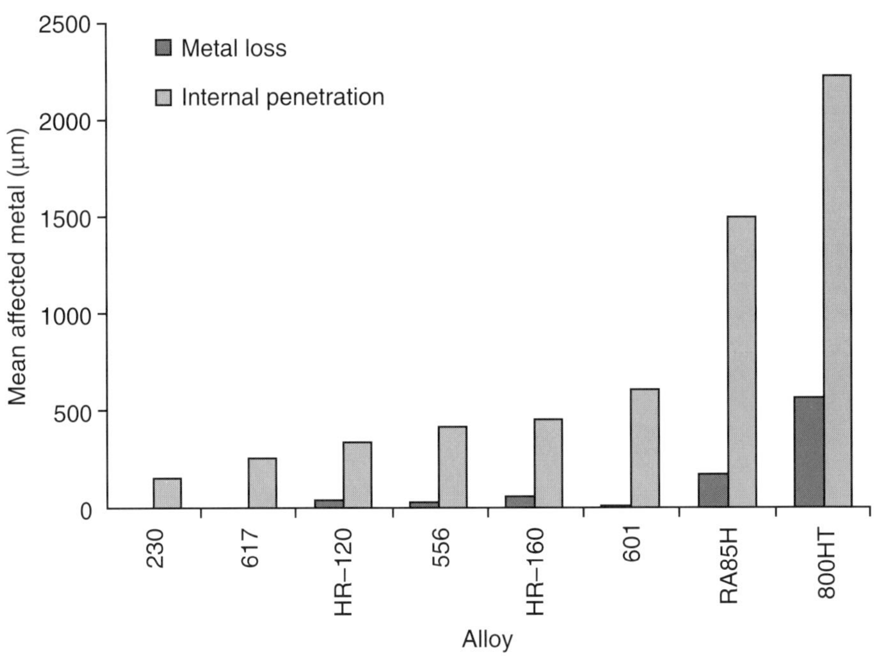

Fig. 16 *Long term cyclic oxidation of a range of chromia-forming alloys at 982°C for 17 280 h [67].*

(ii) the alloy may still by protected by a less protective oxide, for example chromia in the case of an alumina-forming alloy.

(iii) the critical activity level depends on alloy composition, scale composition, temperature and environment.

(iv) rapid depletion of scale forming elements accompanies this internal oxidation process, exacerbating the onset of breakaway corrosion.

6.1.4. Onset of breakaway corrosion

Once it is no longer possible to reform the most protective oxide, and the remaining scale-forming elements have been consumed by the internal oxidation processes mentioned above, then

(i) the next thermal cycle, triggers the onset of breakaway.

(ii) rapid alloy consumption ensues, through the formation and growth of base metal oxides.

It is possible for the depleted alloy to remain in an effectively dormant state; that is, the alloy is completely depleted of scale-forming elements and therefore has no repair capability, however, the oxide scale that has been formed prior to the loss of any remaining scale-forming elements still offers some protection (see [59] and [60]). This dormant state has been termed the No.S.H. region (no self-healing capability) and is often found for thin foil samples prior to the onset of breakaway, where deformation of the foils provides sufficient stress relief to accommodate any thermally induced strain. In this case the next severe thermal cycle that is capable of fracturing the protective oxide envelope triggers breakaway corrosion.

Figure 17 illustrates the morphology of breakaway oxidation on an Fe–15Cr–5Al alloy [85], following thermal cycle tests at 1000°C. Figure 8, presented earlier in this paper, illustrates similar breakaway attack of an iron aluminide (FAL) [61] following 350 h cyclic oxidation at 1300°C. Both sets of micrographs show the formation of base metal oxides (in this case iron oxides) in the breakaway region. Partially protective, chromia rich, scales are formed within a transition zone between the base metal oxides and the substrate. Internal oxidation is observed within the substrate.

7. Summary Remarks

This paper has reviewed many current works on cyclic oxidation and the procedures adopted for the cyclic oxidation tests.

It has been shown that the cyclic oxidation behaviour of any multi-component alloy follows *generic* trends, independent of the alloy system, provided the alloy is protected under isothermal conditions by a slow growing, protective oxide scale.

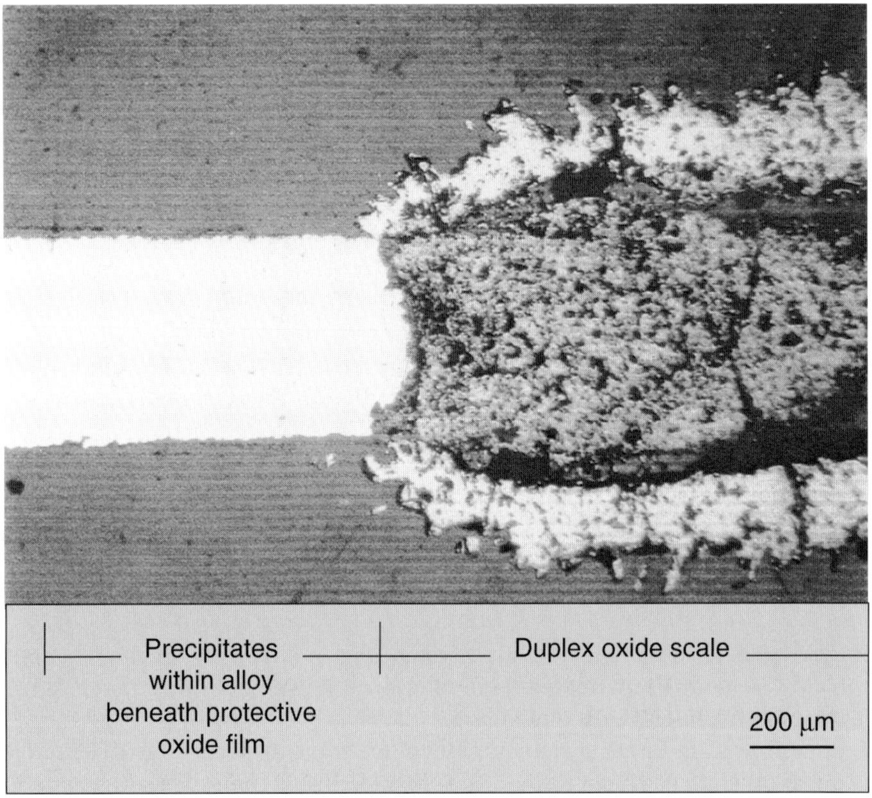

***Fig. 17** Breakaway oxidation of Fe15Cr5Al in air at 1000°C following thermal cycling.*

These trends in cyclic oxidation behaviour may be summarised as follows:

Prior to the onset of spallation

(i) a stable protective oxide growth regime is to be expected, with parabolic, or sub-parabolic, kinetics.

(ii) the cyclic oxidation rate will be independent of cycle frequency.

(iii) both the mass-gain and the alloy depletion will depend on the accumulated time at temperature.

(iv) the onset of spallation will depend on the cooling rate; occurring earlier in the life cycle (thinner oxides) for the more rapidly cooled samples.

(v) for a given protective oxide, spallation will occur more readily with the stronger alloy, that is, the alloy is less able to relieve thermally induced stresses during cooling by creep and other relaxation processes the stronger the alloy.

Within the 'Steady State' Spall Region

(i) once spalling has commenced, cyclic oxidation proceeds by continued spalling and rehealing.

(ii) the mass of spall produced depends on the incremental time at temperature and is therefore a function of cycle frequency.

iii) the occurrence of repeated spallation is determined by the local oxide thickness, the localised scale and substrate mechanical properties and the sample cooling rate.

(iv) alloy depletion of scale-forming elements, depends on the scale growth rate and the mass of spalled oxide.

The Onset of Internal Oxidation

(i) internal oxidation occurs at, or below, a critical alloy addition activity level.

(ii) the alloy may still be protected by a less protective oxide scale.

(iii) the critical activity level depends on alloy composition, scale composition, temperature and environment.

(iv) rapid depletion of scale-forming elements accompanies this internal oxidation process, exacerbating the onset of breakaway corrosion.

The Onset of Breakaway Corrosion

(i) once it is no longer possible to reform the most protective oxide, and the remaining scale-forming elements have been consumed by internal oxidation, then the next thermal cycle capable of rupturing the protective oxide envelope triggers breakaway corrosion.

(ii) breakaway corrosion results in the rapid consumption of any remaining metal through the formation of base metal oxides.

7.1. Choice of Test Procedure

The changes in material response to cyclic oxidation throughout the life cycle of an oxidation resistant alloy means it is not possible to define a single oxidation test as a standard. Recommended test procedures can be adopted, providing guidelines for 'best practice', and many of these have been reviewed in this paper.

It will, however, be necessary to alter the test procedures used to study the various aspects of cyclic oxidation most effectively. For example, studies of the early stages of cyclic oxidation before the first spall are not critically dependent on the cycle frequency, whereas in the 'steady state' spall regime this is a dominant parameter controlling material loss. Equally, cooling rate is important both in determining the onset of first spall, the rate of material loss under 'steady state' cyclic conditions and the onset of breakaway corrosion.

Careful consideration of the different stages in the cyclic life of a component should allow the best cyclic oxidation test procedure to be identified for a given study.

References

1. C. A. Barrett and C. E. Lowell, *Oxid. Met.*, 1975, **9,** 4, 307-355.
2. C. A. Barrett and C. E. Lowell, *Oxid. Met.*, 1978, **12,** 4, 293-311.
3. H. B. Probst and C E. Lowell, *J. Mater. Sci.*, 1988, **40,** 10, 18-21.
4. J. Nesbitt, *J. Electrochem. Soc.*, 1989, **136**, 1511.
5. C. E. Lowell, C. A. Barrett, R. W. Palmer, J. V. Auping and H. B. Probst, *Oxid. Met.*, 1991, **36,** 81.
6. M. J. Bennett, R. Perkins, J. B. Price and F. Starr, in *Materials for Advanced Power Engineering* (D. Coutsouradis *et al.*, eds). Kluwer, Dordrecht, 1994, p.1553.
7. W. J. Quadakkers and M. J. Bennett, *Mater. Sci. Tech.*, 1994, **10**, 126.
8. M. J. Bennett, H. E. Evans and D. A. Shores, *Mater. High Temp*, 1994, **12**, 127–133.
9. M. A. Harper, J. E. Barnes and G. Y. Lai, in *Corrosion, 97*, NACE, Houston, TX, 1997, paper 97-132.
10. K. S. Chan, *Met. and Mater. Trans.*, 1997, **28A**, 411–422.
11. B. A. Gleeson and M. A. Harper, *Oxid. Met.*, 1998, **49,** 3/4, 373–399.
12. N. S. Cheruvu, K. S. Chan and G. R. Leverant, ASME, 1998, paper 98-GT-468.
13. K. S. Chan, N. S Cheruvu and G. R. Leverant, *J. Eng. Gas Turbines and Power,* 1998, **120**,609–614.
14. N. S. Cheruvu, reprint from the IGTI Conference 1999.
15. K. S. Chan, N. S. Cheruvu and G. R. Leverant, reprint from the IGTI Conference 1999.
16. J. A. Nesbitt and R W. Heckel, *Thin Solid Films*,1984, **119**, 281–290.
17. R. Miller, *J. Am. Ceram. Soc.*, 1984, **67,** 8, 517–521.
18. P. Morrell and D. S. Rickerby, in *Thermal Barrier Coatings*, 1998, AGARD report 823, paper 20.
19. 'Coatings for High Temperature Structural Materials', report to the National Materials Advisory Board, NRC, Washington DC 1996.
20. C. A. Barrett, J. R. Johnston and W. A. Sanders, *Oxid. Met.*, 1978, **12,** 4, 343–377.
21. H. J. Ratzer-Schriebe, K. Fritscher and M. Seiersten, *High Temp. Technol.*, 1998, **7,** 4.
22. R. J. H. Wanhill, A. J. A. Mom, H. J. A. Hersbach, G. A. Kool and J. A. M. Rogers, *High Temp. Technol.,* 1989, **7**, 4, 202.
23. J. P. Immarigeon, V. R. Parameswaran, D. Chow, D. D. Morphy, P. Gougeon, M. Prystay and C. Moreau, in *Thermal Barrier Coatings, 1998,* AGARD Report No. 823, paper 15.
24. J. R. Nicholls and S. R. J. Saunders, *High. Temp. Technol.*, 1989, **7,** 4, 193–201.
25. S. R. J. Saunders and J. R. Nicholls, *High Temp. Technol.*, 1989, **7,** 4, 232–240.
26. W. J. Quadakkers, K. Bongartz, F. Schubert, and H. Schuster, in *Materials for Advanced Power Engineering* (D. Coursouradis *et al.* eds). Kluwer Academic Publishers, **2**, pp.1533–1542.
27. J. Robinson and M. I. Manning, *Mater. Sci. Tech.*, 1990, **6**, 81–91.
28. M. Schutze, *Corros. Sci., 1993,* **35**, 955–963.
29. M. M. Nagal and W. T. Evans, *J. Mater. Sci.*, 1993, **28**, 6247–6260.
30. P. Hancock and J. R. Nicholls, *Mater. High Temp.*, 1994, **12**, 209–218.

31. H. E. Evans, *Mater. High Temp.*, 1994, **12**, 219–227.
32. H. E. Evans, *Int. Mater. Rev.,* 1995, **40**, 1–40.
33. M. Schutze, *Mater. High Temp.*, 1994, **12**, 237–247.
34. H. E. Evans and R. C. Lobb, *Corros. Sci.*, 1984, **24**, 209–222.
35. K. S. Chan, *Symp. on Cleavage Fracture Minerals*, Metals and Materials Society, pp.207–220, 1997.
36. H. E. Evans, this volume, pp.1–16.
37. F. Starr and M. F. Stroosnijder, *Corrosion 98*, Paper No. 98–205, NACE International, Houston, Tx., 1998.
38. J. P. Wilber, J. R. Nicholls and M. J. Bennett, this volume, pp.133–147.
39. J. R. Nicholls, in *Guidelines for Methods of Testing and Research in High Temperature Corrosion* (H. J. Grabke and D. B. Meadowcroft, eds). EFC No. 14, The Institute of Materials, pp.11–36, 1995.
40. M. Schütze and W. J. Quadakkers (eds), *Cyclic Oxidation of High Temperature Materials,* EFC No. 27, The Institute of Materials, 1999, 496pp.
41. H. J. Grabke, in *Guidelines for Methods of Testing and Research in High Temperature Corrosion* (H. J. Grabke and D. B. Meadowcroft, Eds). EFC No14, Institute of Materials, pp. 52–61, 1995.
42. J. R. Nicholls, P. J. Smith and J. E. Oakey in *Materials for Advanced Power Engineering,* Part II (D. Coutsouradis *et al.* eds). Kluwer Academic Publishers, Dordrecht, Netherlands, p.1273–1289 (1984).
43. J. R. Nicholls in *Progress in the Understanding and Provision of Corrosion* (J. M. Costa and A. D. Mercer, Eds). The Institute of Materials, London, Vol. 1, pp.712–729, 1993.
44. M. J. Bennett, R. Perkins, J. B. Price and F. Starr, in *Materials for Advanced Power Engineering* (D. Coutsouradis *et al.* eds). Kluwer Academic Publishers, **2**, 1553–1562.
45. N. J. Wood and F. Starr, *In Microscopy of Oxidation–2* (S. B. Newcomb and M. J. Bennett, eds). The Institute of Materials, London, pp.298–309, 1994.
46. First Draft Code of Practice for Discontinuous Corrosion Testing in High Temperature Gaseous Atmospheres, Authors Testcorr partners, published by ERA Technology, Dec. 1998.
47. D. B. Meadowcroft *et al., Mater. Sci. Forum*, 1997, **251–254**, 973–978.
48. Hot Corrosion Task Force of the Gas Turbine Panel, ASTME Round Robin Test 1970; cited in S. R. J. Saunders and J. R. Nicholls, *Thin Solid Films, 1984,* **119**, 247–269.
49.S. R. J. Saunders and J. R. Nicholls, *Mater. High Temp.*, 1995, **13**, 3.
50. J. Asher, M. J. Bennett, R. W. M. Hawes, J. B. Price, A. T. Tuson, D. J. Savage and S. Sugden, *Mater. Sci. Eng.*, 1987, **88**, 143.
51. J. Asher, S. Sugden, M. J. Bennett, R. W. M. Hawes, D. J. Savage and J. B. Price, *Werkst. Korros.*, 1987, **38**, 506.
52. M. J. Bennett, D. J. Butler, P. D. Colledge, J. B. Price, C. B. Scruby and K. A. Stacey, *Mater. Sci. Eng*. A., 1987, **120**, 199.
53. Y. Zeng, D. Zou, J. H. Stout and D. A. Shores, unpublished results; cited in ref. 8.
54. V. A. C. Haanappel, and M. F. Stroosnijder, this volume, pp.225–239.
55. P. Hancock and J. R. Nicholls, *Mater. Sci. Tech.*, 1988, **4**, 398–406.
56. J. P. Wilber, J. R. Nicholls and M. J. Bennett *Microscopy of Oxidation—3* (S. B. Newcomb and J. A. Little eds). The Institute of Materials, London, 1997, pp.207–220.
57. V. Guttmann, F. Hukelmann and P. A. Beaven, this volume, pp.17–37.
58. B. A. Pint, P. F. Tortorelli and I. G. Wright, this volume, pp.111–132.
59. D. Naumenko, L. Singheiser and W. J. Quadakkers, this volume, pp.287–306.
60. G. Strehel, L. Rodriguez, H. Al-Badairy, G. Borchardt and G. Tatlock, this volume, pp.82–94.
61. M. J. Bennett, J. H. DeVan and P. F. Tortorelli, in *Microscopy of Oxidation — 3* (S. B. Newcomb and J. A. Little, eds). The Institute of Materials, London, pp.233–245, 1994.
62. M. Schütze, *Protective Oxide Scales and Their Breakdown,* (D. Holmes, ed). John Wiley, Chichester 1997.
63. J. P. Wilber, M. J. Bennett and J. R. Nicholls, in *Materials for Advanced Power Engineering—*

1998 (J. Lecomte-Beckers, F. Schubert and P. J. Ennis, eds). Forschungszentrum Julich, Germany p.835–946, 1998.
64. M. M. Nagl, W. T. Evans, D. J. Hall and S. R. J. Saunders, *Mater. Sci. Technol.*, 1998, **8**, 1043.
65. M. M. Nagl, S. R. J. Saunders, W. T. Evans and D. J. Hall, *Corros. Sci.*, 1993, **35**, 965.
66. M. M. Nagl and W. T. Evans, *Mater. Sci.*, 1993, **28**, 6247.
67. M. A. Harper and B. Gleeson, this volume, pp.273–286.
68. W. Steinkusch, this volume, pp.307–323.
69. M. Maier, and J. F. Norton, this volume, pp.240–258.
70. P. Vangeli, this volume, pp.198–208.
71. L. Antoni, and J. M. Herbelin, this volume, pp.187–197.
72. C. Leyens, I. G. Wright, B. A. Pint and P. F. Tortorelli, this volume, pp.169–186.
73. S. Y. Chang, M. Wollmann, U. Krupp and H. J. Christ, this volume, pp.63–81.
74. H. P. Martinz, F. Müller and K. Prandini, this volume, pp.95–108.
75. B. Jönsson, this volume, pp.324–335.
76. R. Janakiraman, G. H. Meier and F. S. Pettit, this volume, pp.38–62.
77. L. Singheiser, private communication, 1999.
78. W. Quadakkers, this volume, pp.339–356.
79. T. Schmackers and W. Reimers, this volume, pp.357–372.
80. K. Fritscher, this volume, pp.383–391.
81. L. Swadzba, this volume, pp.392–407.
82. J. A. Hearley, J. A. Little and A. J. Sturgeon, this volume, pp.408–420.
83. T. Weber and M. Schutze, this volume, pp.259–269.
84. G. Kerkhoff, R. Vaßen and D. Stöver, this volume, pp.373–382.
85. M. J. Bennett and M. R. Houlton, High Temperature Materials for Power Engineering (E. Bachelet *et al.*, eds). Kluwer Academic Publishers, 1990, **1**, pp.227–238.

List of Abbreviations

The following abbreviations occur in the text and in the Index of contents.

AE	Acoustic emission
APS	Atmospheric plasma spraying
AT	Ambient temperature
BC	Bondcoat
COSP	Cyclic oxidation spalling program
CSEC	Critical strain energy criterion
CVD	Chemical vapour deposition
CYCLOX	(a version of COSP for PC use)
DEC	Diffraction elastic constant
EB	Electron beam
EDS	Energy dispersive spectroscopy
EDX	Energy dispersive X-ray analysis
EFC	European Federation of Corrosion
EPMA	Electron probe microanalysis
FEM	Finite-element method
HVOF	High velocity oxy-fuel
ICP-OES	Inductively coupled plasma-optical emission spectroscopy
IMC	Intermetallic compounds
LEAFA	Life extension of alumina forming alloys
LEFM	Linear elastic fracture mechanics
LiMi	Light microscopy
LPPS	Low pressure plasma spray
MOCVD	Metal organic chemical vapour deposition
ODS	Oxide dispersion strengthened
PLD	Pulsed laser deposition
PVD	Physical vapour deposition
RE	Reactive element
SEC	Solid electrolyte-based coulometric technique
SEM	Scanning electron microscope
SNMS	Secondary neutral mass spectroscopy
SPS	Shrouded plasma spraying
TBC	Thermal barrier coatings
TGO	Thermally grown oxide
TMA	Thermo-mechanical analyser
TMF	Thermo-mechanical fatigue
TPO	Temperature programmed oxidation
TPR	Temperature programmed reduction
VPS	Vacuum plasma spraying
WDX	Wavelength dispersive spectroscopy
XRD	X-ray diffraction

Index